Carbon and Nitrogen in the Terrestrial Environment

Carbon and Nitrogen in the Terrestrial Environment

R. Nieder and D.K. Benbi

Springer

R. Nieder
Institut für Geoökologie
Technische Universität Braunschweig
Braunschweig
Germany

D.K. Benbi
Department of Soils
Punjab Agricultural University
Ludhiana
India

ISBN 978-1-4020-8432-4 e-ISBN 978-1-4020-8433-1

Library of Congress Control Number: 2008927744

Printed on acid-free paper

9 8 7 6 5 4 3 2 1

springer.com

Preface

One of the biggest reality before us today is the global climate change resulting from the emission of greenhouse gases (GHGs). There has been an unprecedented increase in the concentration of carbon and nitrogen containing GHGs in the atmosphere, resulting primarily due to intervention in terrestrial carbon (C) and nitrogen (N) cycles by human beings. Two anthropogenic activities viz. food production and energy production are mainly responsible for perturbation of C and N cycles. If drastic remedial measures are not taken, the concentration of GHGs is projected to increase further. According to Kyoto Protocol, industrial countries are to reduce their emissions of GHGs by an average of 5% below their 1990 emissions by the first commitment period, 2008–2012. Therefore, there is an increased focus to look for options for mitigating the emission of GHGs. Terrestrial C sequestration through biotic processes is being viewed as a plausible option of reducing the rates of CO_2 emissions while abiotic processes of carbon storage and alternatives to fossil fuel take effect.

The importance of the C and N transfer from soils to the atmosphere lies not only in global warming, but also on soil quality and the potential of soils to perform ecosystems functions some of which are related to the three major international conventions on Biodiversity, Desertification, and Climate Change. Soil organic matter (SOM) being the main reservoir of C of the continental biosphere, can either be a source of CO_2 during mineralization or a sink if C sequestration is favored. During the last two centuries, soils have lost a considerable amount of C due to land use changes and expansion of agriculture. These losses from soils are clearly of concern in relation to future productivity and environment. To ensure sustainable management of land, it is imperative that organic matter in the soil is maintained and sustained at satisfactory levels through improved management practices.

As pool changes of C and N are often very slow, and the full impact of a change in land management practice may take decades to become apparent, long-term perspectives are required. The cycling of C and N is intimately linked and the two cannot be studied effectively separately. This necessitates a thorough understanding of the interdependent and dynamic pools and processes of C and N in the terrestrial ecosystem. Models could help in formulating or assessing land use strategies, generating scenarios for optimizing SOM conditions and minimizing emissions and upscaling research findings at different levels of spatial and temporal aggregation.

Development and use of models require a comprehensive knowledge about several interdisciplinary processes.

Most of the currently available books on C and N cycling either deal with a single element of an ecosystem, or are limited to one or a few selected aspects. This book fills the gap by presenting a comprehensive, interdisciplinary description of C and N fluxes between the atmosphere and terrestrial biosphere, issues related to C and N management in different ecosystems and their implications for the environment and global climate change, and the approaches to mitigate emission of GHGs. This unique volume presents comprehensive literature drawn from books, journals, reports, symposia proceeding and internet sources to document interrelationships between different aspects of C and N cycling in terrestrial ecosystems. Following an introductory chapter, Chapter 1 presents distribution of C and N in the various terrestrial pools, with special emphasis on storage in plants and soils. Chapter 2 presents the basics of C and N cycling processes and a generalized overview of fluxes in terrestrial ecosystems so as to develop an understanding of the complex interrelationships among different processes and the emission pathways, which are discussed in subsequent chapters. Soils, particularly soil organic matter, play an important role in the bidirectional flow of C and N in terrestrial ecosystems. Therefore, knowledge about the composition and characteristics of soil organic matter, and its role in influencing soil functions is essential to exploit synergies between management practices, GHG mitigation and sustainable productivity. While Chapter 3 presents physical, chemical and morphological characterization of soil organic matter, Chapter 4 enunciates the influence of SOM on soil quality and its ability to perform ecosystem functions. To complement the information provided in Chapter 1 on C and N forms, Chapter 5 presents the transformations of organic and inorganic forms of carbon and nitrogen in soils and their role in influencing C and N fluxes between soils and atmosphere. The impact of anthropogenic activities, particularly land use and land use changes and agricultural management on C and N dynamics is presented in Chapter 6. Chapter 7 discusses leaching of reactive C and N forms from soils and contamination of groundwater. Chapter 8 provides a detailed description of bidirectional biosphere-atmosphere interactions with current estimates of GHG emissions, their sources, governing variables and the mitigation options. Finally, Chapter 9 presents modeling approaches adopted to simulate various components of C and N cycling processes. The use of models to upscale measurements and generate scenarios on a regional and global scale vis-à-vis management options are discussed.

We are thankful to the German Research Foundation (Deutsche Forschungsgemeinschaft) for funding the stay of D.K. Benbi at Braunschweig Technical University. We appreciate our families: Alexandra, Raphaela and Petra (R. Nieder), and Adwitheya and Meenu (D.K. Benbi) for their patience and understanding during the preparation of this book. We are grateful to Hans P. Dauck for help in the preparation of illustrations.

R. Nieder
D.K. Benbi

Contents

Introduction

Carbon (C) and nitrogen (N) are the building blocks of life on earth. Carbon delivers the framework for carbohydrates, fats and proteins and N as component of proteins is present in amino acids, enzymes and nucleic acids. These organic forms occur in living and dead organic materials of plants, animals and humans and are also important constituents of soil organic matter (SOM). Both C and N also exist in inorganic forms and are present in all ecosystems. In the atmosphere, carbon is present as carbon dioxide (CO_2). Minor amounts of gaseous C occur as methane (CH_4), carbon monoxide (CO) and other higher molecular C-containing gases. In the lithosphere C is a major constituent of limestone, occurring as carbonates of calcium and magnesium ($CaCO_3$ and $CaMg(CO_3)_2$). In ocean and fresh water, it is present as dissolved carbonates. Flow of carbon occurs between different spheres, leading to what is generally termed as carbon cycle. The dominant fluxes of the global C cycle are those that link atmospheric CO_2 to land biosphere and oceans.

About 98% of the world's nitrogen is found in the solid earth within rock, soil and sediment. The remainder moves in a dynamic cycle involving the atmosphere, ocean, lakes, streams, plants and animals. Nitrogen in the atmosphere mainly exists as molecular nitrogen (N_2), which comprises 78% of the atmospheric gases. Trace amounts of nitrogen oxides, gaseous ammonia, ammonium compounds, nitric acid vapor, particulate nitrate and organic nitrogen circulate through the atmosphere. Atmospheric nitrogen compounds cycle to the land and water through wet and dry deposition. Nitrogen is capable of being transformed biochemically or chemically through a number of processes termed as the nitrogen cycle. Most N transformations involve the oxidation or reduction by biological and chemical means. In the hydrosphere, N exists as soluble organic or inorganic nitrogen.

The global C cycle is one of the most important, complex and challenging cycles on earth as it influences several physical and biological systems directly and through its effect on global temperatures. The interest in the global C cycle has increased tremendously in the last 2 decades because of its role in global climate change and the recognition that human activities are altering the carbon cycle significantly. As early as 1896, Arrhenius indicated the importance of CO_2 in the air on the global temperature and calculated the alteration of temperature that would follow with the increase in CO_2 concentration. But the topic did not feature prominently in research agenda until 1958 when continuous measurements of CO_2

concentrations were initiated at Mauna Loa in Hawaii. However, the real impetus to C cycling research was provided in 1980s by the revelations of ocean core sediments and ice-core measurements, that atmospheric CO_2 concentrations were much lower in cold stages as compared to contemporary ones. These results brought to focus potential climatic consequences of human induced elevated CO_2 levels. New ice core records show that the present atmospheric concentrations of CO_2, or indeed of CH_4, are unprecedented for at least 650,000 years, i.e. six glacial-interglacial cycles (Denman et al., 2007). The increasing trend in the atmospheric CO_2 concentration still continues and over the last 250 years its concentration has increased globally by 100 ppm (36%) from about 275 ppm in the preindustrial era (AD 1000–1750) to 379 ppm in 2005 (Denman et al., 2007). The increase in global atmospheric CO_2 is mainly due to human activities; primarily combustion of fossil fuel and cement production though there is substantial contribution from land use changes and management such as deforestation, biomass burning, crop production and conversion of grassland to croplands. This has serious implications for all forms of life in terrestrial ecosystems. It has been predicted that there will be an increase in the Earth's average surface temperature, shifts in weather patterns, and more frequent extremes in weather events. Because of these concerns there is a tremendous effort underway to better understand the global C cycle, reduce anthropogenic emissions and to mitigate the atmospheric CO_2 concentration.

In addition to CO_2, methane (CH_4) and nitrous and nitric oxides (N_2O and NO) are also considered to cause global warming. In 2005, the global average abundance of CH_4 was 1,774 ± 1.8 ppb (Forster et al., 2007), which is more than three times the concentration during glacial periods. In recent years atmospheric growth rate of CH_4 seems to stagnate, or even decline but the implications for future changes in its atmospheric burden are not clear. While emissions from natural sources dominated the preindustrial global budget of atmospheric CH_4, anthropogenic emissions dominate the current CH_4 budget. Wetlands account for about 80% of the total natural emissions with small contributions from oceans, forests, wildfires, termites, and geological sources. The anthropogenic sources include rice agriculture, livestock, landfills and waste treatment, ruminants, biomass burning, and fossil fuel combustion. Since irrigated rice contributes about 70–80% of the CH_4 emission from global rice fields it provides the most promising target for mitigation strategies.

Nitrous oxide, N_2O, constitutes 6% of the anthropogenic greenhouse effect and its concentration in the atmosphere has been increasing by about 0.25% per year, from about 270 ppb in preindustrial times to 319 ppb in 2005. Nitrous oxide is emitted into the atmosphere both from natural (soil, ocean and atmospheric NH_3 oxidation) and anthropogenic sources. Anthropogenic emissions of N_2O originate from biological nitrification and denitrification in soils and biomass burning. Nitric oxide ($NO_x = NO + NO_2$) emissions, which are also environmentally important originate from surface and troposheric sources. The surface sources include fossil fuel and biomass burning and biogenic emissions from soils. For alleviating biogenic emissions of nitrogen oxides from soils, it is important to adopt practices leading to improved N use efficiency. The higher the N recovery efficiency in plants, the lesser is the amount of mineral N available for emission to the atmosphere.

Burning of fossil fuel and activities related to land use, primarily tropical deforestation and biomass burning cause major perturbation to terrestrial C and N cycles. During the 1990s deforestation occurred at a rate of about 13 million hectares year^{-1} and over the 15 year period from 1990 to 2005, the world lost 3% of its total forest area (FAO, 2007). Most of the C stored in the earth's biota and soils is associated with forests, when cleared and burned, much of this C ends up in the atmosphere as CO_2. During the period 1990–2005, C stocks in forest biomass decreased by about 5.5% at the global level (FAO, 2007). Obviously, through their destruction, forests can be serious sources of greenhouse gases but through their sustainable management they can be important sinks of the same gases. Conversion of forest cover to agriculture also leads to loss of C and N stocks from the land biosphere. During 1961–2002, agricultural land gained almost 500 million hectares from other land uses; on average annually 6 million hectares of forest land and 7 million hectares of other natural land were converted to agricultural land, particularly in the developing countries. The net effect of these land use changes is the reduction in C and N stocks in the landscapes. Agriculture also contributes to the emission of methane and nitrous oxide from livestock wastes, burning pastures and crop residues, rice paddies and the application of nitrogen-based fertilizers, besides contributing to other environmental issues such as groundwater pollution by nitrates and eutrophication of surface waters. Adoption of more sustainable production methods could minimize the negative impacts of agriculture and could also help in mitigating climate change through C sequestration in soils and vegetation. Currently, improved agriculture is being viewed as a potential route to the mitigation of climate change.

The importance of the C and N transfer between soils and the atmosphere lies not only in global warming, but also on soil quality and the potential of soils to produce food, fibre, and fuel. Soil organic matter, which is the main reservoir of C and N, influences soil functional ability and its response to environmental and anthropogenic influences. To ensure sustainable management of land and advancing food-security for resource-poor farmers, it is imperative that organic matter in the soil is maintained and sustained at satisfactory levels. At the beginning of permanent agriculture, fields were cropped for 2 years, followed by a fallow year that served to revamp soil fertility. As population pressure on land increased and the fallow was eliminated, soil organic carbon (SOC) and nitrogen (SON) declined on cultivated land. As a consequence, new management practices were introduced to augment soil fertility, and legume crops like clover and alfalfa became common rotation crops. In many agricultural systems, important means to maintain or increase soil organic matter (SOM) have been incorporation of crop residues, animal wastes and green manures and conservation tillage. In the 20th century, their significance has altered dramatically due to increased use of mineral N fertilizers.

Globally, soils contain about double the amount of C present in the atmosphere and most of it is in organic form. It has turnover times ranging from months to millennia, with much of it around several years and decades. Depending on the input-output balance, SOM can be both a source and sink of atmospheric CO_2. A soil source results when net decomposition exceeds C inputs to the soil, either as a result of human activities such as clearing of forests for agriculture or because of increased decomposition rates due to global warming. Net sinks of C in soils are postulated from increased C input to the soil through enhanced biomass production and exogenous supply of organic materials, and decreased output/losses through adoption of improved management practices for reducing soil respiration. Turnover of SOC and SON has been measured on both, short (within year) and long (years, decades) term scales, but it is the long-term trends that determine whether SOM will act as a net source or sink for C in ecosystems with respect to global environmental change. Changes in climate are likely to influence the rates of accumulation and decomposition of SOM, both directly through changes in temperature and moisture, and indirectly through changes in plant growth and rhizodepositions. Changes in agricultural management practices, land use and soil degradation may have even greater effects on terrestrial C and N pools, especially on SOM. As pool changes of C and N are often very slow, and the full impact of a change in land management practice may take decades to become apparent, long-term perspectives are required. In order to assess the impact of land management practices on organic matter turnover in soils several physical, chemical, biological, and functional pools have been postulated. Efforts have been made to relate some of the functional or conceptual pools to measurable soil organic matter fractions. This necessitates a thorough understanding of the interdependent and dynamic pools and processes of C and N in the terrestrial ecosystem. Much effort has gone into modeling potential soil-atmosphere-climate interactions. Models have been used in formulating/assessing land use strategies and generating scenarios for optimizing SOM conditions. Though a number of models have been developed, but their role in C and N optimization on a regional scale needs further elaboration.

During the last 2 decades, our knowledge on C and N pools and cycling has increased tremendously, particularly in relation to soil and environmental quality. Availability of improved measurement techniques have provided new and relatively precise estimates of global C and N fluxes. New computing tools and the development of several Atmospheric General Circulation Models have led to scenarios of unprecedented magnitude in the area of C and N cycling in terrestrial ecosystems. In efforts to develop strategies for mitigating the emission of greenhouse gases from soils, several process based models have been used to study the influence of management practices on emission of greenhouse gases and fertilizer use efficiency in different ecosystems. Meeting the challenge of sustainable management of C and N requires the widening of knowledge through basic and applied research.

This book provides a holistic and up to date view of all the aspects related to C and N cycling in terrestrial ecosystems. We hope that the book will be of immense value to ecologists, environmentalists, soil scientists, agronomists, action agencies, consultants, extension workers, and students.

Chapter 1
Carbon and Nitrogen Pools in Terrestrial Ecosystems

Carbon and nitrogen account for 95% of the biosphere and are two of six elements (C, H, O, N, P, S) being the major constituents of plant tissue. Carbon is constantly being absorbed, released, and recycled by a range of natural and human-induced biological and chemical processes. Of fundamental importance is the process of photosynthesis in which plants absorb atmospheric carbon as they grow and convert it to biomass. When plant residues and roots decompose, the carbon they contain is transformed primarily into soil organic matter (SOM) and carbon based gases. Soil organic matter is particularly critical in conditioning soil quality. Nitrogen is the limiting factor in plant growth in most ecosystems. Recent interest in the global C and N cycles has focused attention on the high proportion of terrestrial carbon and nitrogen stored in different pools. The carbon and nitrogen cycles include all life forms, inorganic C and N reservoirs and the links between them. This chapter deals with carbon and nitrogen forms and pools, with special focus on the C and N reservoirs in plant biomass and in soils. Carbon and nitrogen fluxes between the different reservoirs are discussed in Chapter 2.

1.1 Forms and Quantities of Carbon and Nitrogen on Earth

1.1.1 Carbon

1.1.1.1 Compounds of Carbon

There are over a million C compounds of which several thousands are necessary for life. Carbon in elemental form is known as amorphous C, graphite and diamond. Carbon atoms can change their oxidation status from +4 to −6, occurring mostly in the +4 state as carbon dioxide (CO_2) and in carbonate form. Carbonate is present in solid form in the lithosphere as $CaCO_3$, $CaMg(CO_3)_2$ and $FeCO_3$. In waters, carbonate exists as H_2CO_3, HCO_3^- and CO_3^{2-}. CO is present in the atmosphere as oxidation state +2. The most reduced form of carbon (−4) is Methane (CH_4).

Among the seven isotopes of carbon (^{10}C, ^{11}C, ^{12}C, ^{13}C, ^{14}C, ^{15}C, ^{16}C), two (^{12}C and ^{13}C) are stable and five (^{10}C, ^{11}C, ^{14}C, ^{15}C, ^{16}C) are radioactive with half-live times varying between 0.74 s (^{16}C) and 5,726 years (^{14}C) (Holmen, 2000). ^{12}C is the most abundant isotope constituting 99% of the C in ecosystems. Isotopic variations are an important tool for calculating C fluxes between carbon reservoirs. Differences in the isotopic C composition are caused either by isotopic fractionation (e.g. preferential uptake of ^{12}C by plants) or by radioactive decay in the upper atmosphere (formation of ^{14}C). As a consequence, the radiocarbon content of plant or soil material depends on the exchange rate with the atmosphere. Carbon reservoirs with a high geological age (lignite: 10^3–10^5 year, hard coal and carbonate rocks: 10^6–10^9 year) are free of radiocarbon because their residence times are significantly longer than the half life time of ^{14}C.

1.1.1.2 Forms of Carbon in Soil

Carbon in soil exists in organic (soil organic carbon: SOC) and inorganic (soil inorganic carbon: SIC) forms. Total carbon is defined as being the sum of both. Soil organic C is part of humic and non-humic substances. Their composition and properties (chemical, physical and morphological) are described in Chapter 3. Organic carbon to some extent may be occluded within charcoal and phytoliths. The latter is also referred to as plant opal (Parr & Sullivan, 2005). Phytoliths are silicified materials that are formed as a result of biomineralization within plants. They constitute up to 3% of the total soil mass (Drees et al., 1989). Rates of phytolith production and the long-term sequestration of C occluded in phytoliths vary according to the overlying plant community. In some cases such as in sugarcane monocultures phytolith organic C accumulation rate may reach up to 180 kg C ha^{-1} $year^{-1}$. Phytolith organic C is very persistent in soils (Mulholland & Prior, 1993).

Inorganic carbon in the soil occurs largely in carbonate minerals, such as calcium carbonate ($CaCO_3$) and dolomite ($CaMg\ (CO_3)_2$). Large concentrations of carbonates are typical for soils, which have developed on calcareous parent materials and under arid or semiarid climate (e.g. Calcisols, Rendzic Leptosols, some Regosols, Chernozems; according to World Reference Base for Soil Resources; FAO, 1998a). The carbonate-C content has been a criterion to distinguish the FAO-UNESCO soil unit Calcisol from other soil units (FAO, 1971–1981, 1990, 1998a). It has also been used for differentiation between various subunits of a particular soil subunit (e.g. calcic vs. dystric Cambisol). Some types of soils, especially coarse-textured, acid and strongly weathered ones, do not contain appreciable amounts of carbonates, because the carbonates originally present in the parent material have been dissolved and leached.

1.1.1.3 Quantities of Carbon on a Global Scale

The earth contains about 10^8 Pg (1 Pg = petagram = 10^{15} g = 1 billion tons) of carbon (Schlesinger, 1997). Only a small portion is part of terrestrial sediments where it is found in organic compounds and carbonates (Table 1.1). Soil organic carbon is

Table 1.1 Global pools of carbon

Reservoir	Pg C	Source
Atmosphere	8×10^2	Schimel et al. (1995)
Terrestrial sediments (including soils)		
Organic compounds	1.56×10^7	Des Marais et al. (1992)
Carbonates	6.5×10^7	Li (1972)
Soils		
Soil organic matter	1.5×10^3	Table 1.18
Soil carbonates	7.2×10^2	Sombroek et al. (1993)
	6.95×10^2	Batjes (1997)
Vegetation	$5–7 \times 10^2$	Houghton & Skole (1990)
		Melillo et al. (1990)
		Sombroek (1990)
		Schimel et al. (1995)
Fossil organic C		
Coal	4×10^3	Lal (2000)
Gas	5×10^2	Lal (2000)
Oil	5×10^2	Lal (2000)
Oceans	3.8×10^4	Schimel et al. (1995)
Sum	1×10^8	Schlesinger (1997)

the largest carbon reservoir at the earth surface. The sum of the active pools near the Earth's surface (C in soils and vegetation) is about 3×10^3 Pg C. The sea contains about 50 times more carbon than the atmosphere. Soils, vegetation and oceans link the carbon dioxide exchange with the atmosphere and are therefore most important for the global carbon cycle (see Chapter 2).

1.1.2 Nitrogen

1.1.2.1 Compounds of Nitrogen

Nitrogen exists in many different forms with an oxidation state between +5 and –3 (+5: HNO_3; +4: NO_2; +3: HNO_2; +2: NO; +1: N_2O; 0: N_2; –3: NH_3, NH_4^+) (Jaffe, 2000). Numerous N compounds have nitrogen bonded to carbon, hydrogen (H) or oxygen (O). When N is bonded to C or H, the oxidation state of the nitrogen is negative because N is more electronegative than C or H. In contrast, nitrogen bonded to O, has a positive oxidation state.

1.1.2.2 Forms of Nitrogen in Soil

Nitrogen in soil is mainly stored in organic form (soil organic nitrogen: SON). Only small amounts are stored as inorganic nitrogen in soils capable of fixing NH_4^+

Table 1.2 Global pools of nitrogen

Reservoir	Tg N	Source
Atmosphere	3.9×10^{12}	Schlesinger (1997)
Terrestrial biomass	3.5×10^{3}	Schlesinger (1997)
Plant biomass	1.0×10^{3}	Davidson (1994)
Microbial biomass	2.0×10^{3}	Davidson (1994)
Soil organic matter	1.33×10^{5}	Batjes (1997)
Lithosphere	1.64×10^{11}	Pierzynski et al. (2000)
Igneous rocks (crust and mantle)	1.63×10^{11}	Pierzynski et al. (2000)
Sediments (fossil N)	4.50×10^{8}	Pierzynski et al. (2000)
Core of the earth	1.30×10^{8}	Pierzynski et al. (2000)
Coal	1.00×10^{5}	Pierzynski et al. (2000)
Hydrosphere (oceans, estuaries, lakes, rivers, streams	2.30×10^{7}	Pierzynski et al. (2000)
Ocean sediments	5.40×10^{5}	Pierzynski et al. (2000)

(fine-textured soils containing illite and vermiculite). Around 90% of the nitrogen stored in soils is part of the organic matrix, 6–12% exists as mineral fixed NH_4^+, and 1–3% can be found as plant-available mineral nitrogen (NO_3^- and NH_4^+) (Benbi & Richter, 2003). In coarse-textured soils having little capacity to fix NH_4^+ in clay minerals, the proportion of organic N is >97% and the inorganic fraction is 1–3% (Baldock & Nelson, 2000). On a global scale, the organic N fraction may account for 95% of the total soil N pool (Söderlund & Svensson, 1976). The C/N ratio of SOM depends on the chemical composition of the inputs by vegetation, their C/N ratios and the degree to which they are decomposed. The nitrogen compounds of SOM are described in Chapter 3.

1.1.2.3 Quantities of Nitrogen on a Global Scale

Containing 3.9×10^{12} Tg (1 Tg = teragram = 10^{12} g = 1 million tons), the atmosphere contains the largest pool of nitrogen (Table 1.2). The amount of soil organic nitrogen is smaller than atmospheric N_2, but larger than the amounts of N in biomass and the surface oceans.

1.2 Carbon and Nitrogen in Terrestrial Phytomass

Estimates of global terrestrial plant biomass vary considerable, ranging from ~500–700 Pg C (Table 1.1). Terrestrial biomass is divided into a number of subreservoirs with different turnover times. Forests contain ~90% of all carbon in living matter on land, but their net primary production (NPP, expressed as Mg dry matter ha^{-1} year^{-1}) is only 60% of the total (Holmen, 2000). About half of the primary production in forests yields twigs, leaves, shrubs, and herbs that make up 10% of the biomass. Carbon in wood has a turnover time of about 50 years. Turnover times of carbon in leaves, flowers, fruits, and fine root biomass are less than a few years. Average turnover time for carbon in litter is about 1.5 years. In tropical ecosystems, the litter decomposition rate is equal to or greater than the supply rate, indicating that significant storage of organic matter is not possible. In contrast, in colder climates, NPP exceeds the rate of decomposition in the soil. In the tropics, NPP is extremely high but soil organic carbon and nitrogen stocks are relatively low. All higher latitude areas show the opposite relationship.

1.2.1 Estimates of Phytomass C and N Stocks for Natural Ecosystem Types

1.2.1.1 Arctic

Most part of the polar zone (about 75%) is covered with ice. The ice-free areas of the Arctic can be classified according to different ecosystem types. In the ice-free high Arctic (polar desert and semidesert) only the semidesert is vegetated, whereas the Tundra of the low Arctic has an almost closed vegetation cover (Fig. 1.1). Arctic and alpine vegetation have relatively low stocks of plant and soil carbon and nitrogen

Fig. 1.1 Discontinuously vegetated tundra site with stone rings. (Jotunheim, southern Norway). Cryoturbation accounts for the orientation of stones and patterned features at the soil surface (Photo: F. Bailly)

Table 1.3 Estimates of average net primary production (NPP) of vegetation, and carbon in plant biomass and soil of different arctic ecosystem types (Compiled from Oechel & Billings, 1992; Schultz, 2000)

Arctic ecosystem type	NPP (Mg ha^{-1} $year^{-1}$)	Carbon in biomass (Mg C ha^{-1})	Carbon in soils[a] (Mg C ha^{-1})	Present increase in soil organic matter (Mg C ha^{-1} $year^{-1}$)
High Arctic				
Polar desert	0.022	0.05	20.5	0
Semidesert	0.31	6.5	158	0.04
Low Arctic				
Wet sedge/mire	2.38	21	295	0.27
Tussock/sedge dwarf shrub	1.98	73	638	0.23
Low shrub	2.97	17	84	0.10
Tall shrub	6.93	57	9	0.29

[a] Only organic layers plus A horizon. Organic matter of subsoils (including permafrost) was not considered

because of the lack of a tree stratum and the often spotty vegetation (Table 1.3). The stock of carbon fixed in biomass per unit area corresponds to about 8% of the global average (McGuire et al., 1997). In the arctic tundra, green phytomass and aboveground shoots contribute to about 35% of the total biomass. Another 65% is found in roots and belowground shoots (Larcher, 2003). On average 8 Mg C ha^{-1} may be fixed annually in the phytomass of arctic ecosystems (Schlesinger, 1997). The SOC content in arctic soils is estimated to be about 14% of the total global soil carbon (Post et al., 1982). There is a general trend from the northern to the southern arctic of increasing C amounts in plant biomass and soils. It also appears that the soil organic matter pool decreases from oceanic toward continental regions (Christensen et al., 1995). The ratio of C to N is typically about 20 for soils and 60 for arctic vegetation (McGuire et al., 1997).

1.2.1.2 Boreal Forests

Boreal forest (also called taiga) cover most parts of Alaska (USA), Canada, Scandinavia and northern Russia (Treter, 1993). The dominating needleleaf species of the Boreal zone are *Picea* spp., *Abies* spp., *Pinus* spp. and *Larix* spp. (Hare & Ritchie, 1972), the most important broadleaf species are *Betula* ssp. and *Populus* ssp. The aboveground dry matter of biomass in mature needleleaf forests in the Boreal zone is about 150 Mg ha^{-1} in the north (Fig. 1.2) and up to 300 Mg ha^{-1} in the southern part (Schultz, 2000). Assuming a carbon content of 42% in phytomass (IPCC, 1997), 63–126 Mg C ha^{-1} (on average 95 Mg C ha^{-1}; Schlesinger, 1997) may be fixed in the standing biomass of the Boreal zone. The nitrogen contents in boreal needleleaf forests on average amount to 0.23% in total biomass and 0.48% in needles (Cole & Rapp, 1981). On the basis of 150 Mg dry matter, the aboveground biomass of boreal forests contains at least 350 kg N ha^{-1}.

Fig. 1.2 Taiga scenery dominated by coniferous forest (*Abies sibirica*). Salair mountains east of Novosibirsk, Siberia (Photo: F. Bailly)

The belowground biomass may contribute 20–40% of the total biomass (Vogt et al., 1996). Besides the climatic conditions (increase of growth period from north to south), soil quality and geomorphology exert a great influence on the development of the biomass in forest stands. Van Cleve et al. (1983) reported aboveground phytomasses in Boreal forests of Alaska (stands with similar stand age) ranging from 26 to 250 Mg ha^{-1}. Vogt et al. (1996) reported biomasses ranging from 30 to 144 Mg ha^{-1} in different regions of the same climatic forest type. The NPP is limited by the harsh climatic conditions. Moreover, most of the soils under coniferous forests are acidic. Between 2 and 15 Mg ha^{-1} phytomass (0.8–6.3 Mg C) are produced annually, with the higher values found on nutrient-rich and/or warmer sites of this region (Larcher, 2003). In Boreal forests of Finland, the soil carbon density has been found to increase with the effective temperature sum (Liski & Westman, 1997).

The litter is decomposed slowly in the Boreal zone because of the (i) high acidity of the litter derived from the conifers and the ground vegetation (*Calluna*, *Vaccinium*, *Erica* and *Andromeda*), (ii) low temperatures, and (iii) high soil water contents during most part of the year. High amounts (40–300%) of the soil organic matter are, therefore, located in organic layers on the surface of the mineral soil (Vogt et al., 1996). Dystrophic humus forms like Mor or peat are typical. In some old stands of the Boreal zone, the organic layers may amount to 1,000 Mg dry matter ha^{-1} (Schultz, 2000), corresponding to about 500 Mg C ha^{-1} and 17 Mg N ha^{-1} (C/N: 30). On average, the organic matter in forest floor and Ah horizon may sum up to 400 Mg ha^{-1} (~200 Mg C ha^{-1} and 7 Mg N ha^{-1}) (Schultz, 2000).

1.2.1.3 Broadleaf Deciduous Forests of the Humid Temperate Climate Zone

Broadleaf deciduous forests of the humid temperate climate zone (Fig. 1.3) are distributed over Central and Eastern Europe, Northeast China, Korea, North Japan, Northeastern USA, West Canada and Northwest USA, West Patagonia (South Chile), Southeast Australia and the southern island of New Zealand. The tree species are numerous. Amongst others, broadleaf deciduous forests include *Acer* spp., *Alnus* spp., *Betula* spp., *Fagus* spp., *Fraximus* spp., *Juglans* spp., *Magnolia* spp., *Populus* spp., *Quercus* spp., *Salix* spp., *Tilia* spp. and *Ulmus* spp. The aboveground biomass of broadleaf deciduous forests in the humid temperate climate zone on average amounts to 190 Mg ha^{-1}, corresponding to 80 Mg C ha^{-1} (Schlesinger, 1997). The NPP on average amounts to 12 Mg dry matter ha^{-1} $year^{-1}$ (Larcher, 2003). The nitrogen contents in broadleaf forests on average amount to 0.29% in total biomass and 2.06% in leaves (Cole & Rapp, 1981). On the basis of 190 Mg dry matter, the aboveground biomass may contain about 500 kg N ha^{-1}. The belowground biomass may contribute to 30–50% of the total biomass (Proctor, 1983).

In broadleaf deciduous forests, a great proportion of the soil organic matter is stored in the uppermost mineral horizon (Ah horizon). The dominating humus forms are Mull and Moder. As the litter is decomposed more rapidly than in needle-leaf forests, only 5–25% of the soil organic matter is located in organic layers on the surface of the mineral soil (Vogt et al., 1996).

Fig. 1.3 Near-natural broadleaf deciduous forest (*Fagus sylvatica*), Elm mountains near Braunschweig, North Germany (Photo: R. Nieder)

1.2.1.4 Temperate Grasslands

Temperate grasslands or steppe ecosystems cover major parts of Central Eurasia (Ukraine, Turan, Kazakhstan, Sinkiang, Tibet, Mongolia), the Midwest (Saskatchewan and Alberta in Canada, Great Plains of the USA), and West (Columbia Basin, Great Basin) of North America, and East Patagonia (Argentina). In temperate grasslands, the phytomass is increased rapidly in the course of the production period (commonly spring time), but at the same time parts of the shoots and roots die off or are removed by consumers. In steppes, this loss accounts for more than half of the phytomass formed during the year, and in desert plant communities consisting primarily of ephemeral species, it may be as much as 60–100% (Larcher, 2003). Depending on the degree of aridity, different types of steppes have developed. In Eurasia, the forest steppe is a transition zone between the Temperate or the Boreal zone in the north and the tall grass steppe in the south. The phytomass amounts to about 100–300 Mg ha^{-1} (~40–120 Mg C ha^{-1}). The nitrogen content of the biomass depends on the growth stage of the grassland and may range from <1% (dried grass) to 4% (young biomass). The NPP amounts to 17–30 Mg ha^{-1} $year^{-1}$ (Schultz, 2000).

In the Eurasian tall grass steppe (Fig. 1.4) with annual rainfall of 400–600 mm, grasses (>90%) form the main part of the biomass. Compared to the total (aboveground plus belowground) phytomass (12–25 Mg ha^{-1}, corresponding to roughly 5–10.5 Mg C ha^{-1}), the NPP (11–16 Mg ha^{-1} $year^{-1}$) is high (Woledge & Parsons, 1986). About two thirds of the total phytomass is below ground. In the short grass steppe with annual rainfall of 200–400 mm, the phytomass amounts to 5–12 Mg ha^{-1} (~2–5 Mg C ha^{-1}), the NPP ranges from 6 to 11 Mg ha^{-1} $year^{-1}$. In the desert steppe with annual rainfall < 200 mm, plant biomass is between 2.5 and 5 Mg ha^{-1} (~1–2 Mg C ha^{-1}), the NPP varies from 4 to 6 Mg ha^{-1} $year^{-1}$.

For all of the above grass steppe types, similar amounts of phytomass were found in East Patagonia (Argentina) (Schulze et al., 1996), whereas in North America, total phytomass was estimated to be 10 Mg ha^{-1} in tall grass steppe and 5 Mg ha^{-1} in short grass steppe (Sims & Coupland, 1979). As the aboveground biomass dries after the end of the vegetation period, the annual production of litter is similar to the NPP of the same year. An active soil fauna incorporates high amounts of organic matter into the mineral soil (bioturbation). The soil microflora and soil fauna decomposes the easily decomposable organic material within 1 year (Andrén et al., 1994), and organic layers are commonly not developed. The typical humus form is Mull (see Chapter 3).

1.2.1.5 Mediterranean Ecosystems

Mediterranean ecosystems are characterized by cool and wet winters and hot and dry summers, i.e., temperature and humidity optima occur at different seasons of the year. They are found in regions with Mediterranean climate such as California (USA), Central Chile, South Africa and southwestern Australia. Mediterranean ecosystems are, therefore, relatively low in NPP (Table 1.4).

Fig. 1.4 Natural tall grass steppe in a Chernozem landscape near Kurzk, Russia (Photo: G. Fritsche)

Most natural plants are evergreen and/or hard-leaved species. The highest growth rates are obtained during spring months, when temperatures are relatively high and soil moisture contents are sufficient for plant growth. Besides the length of the period for optimum growth, the structure of the plant community determines NPP and total phytomass. The highest values can be observed in natural evergreen oak forests (Fig. 1.5).

In secondary systems such as evergreen shrub and semi-shrub vegetation (Fig. 1.6), only about one tenth of the biomass produced in evergreen oak forests is observed. The latter systems are developed particularly on degraded land and/or under extreme aridity (Schultz, 2000). In the Phrygana system, most of the plant biomass is developed as root mass.

Table 1.4 Examples for production characteristics of typical Mediterranean plant communities (Adapted from Mooney, 1981)

Plant community	Region	NPP (Mg ha^{-1} $year^{-1}$)	Stand age (year)	Total biomass (Mg ha^{-1})	Root biomass (% of total)	Carbon in total biomass (Mg C ha^{-1})
Evergreen oak forest Quercus ilex	South France	6.5	150	320	16	135
Evergreen shrub vegetation						
Chaparral	California, USA	4.1	18	33	36	14
Garrigue	South France	3.4	17	–	–	–
Semi-shrub vegetation						
Phrygana	Greece	4.1	–	27	60	12

Fig. 1.5 Primary evergreen hard-leaved oak forest (*Quercus ilex*), island of Rab, Croatia (Photo: R. Nieder)

Fig. 1.6 Secondary evergreen shrub vegetation in a carst landscape, island of Crete, Greece, with grazing goats (Photo: R. Nieder)

Quercus ilex stands may produce more than 10 Mg litter annually. In spite of the high litter production, the organic layers on the mineral soil surface under *Quercus ilex* are shallower as compared to the broadleaf deciduous forests of the humid temperate climate zone and the Boreal needleleaf forests. This is explained by the more rapid decomposition (usually within 3 years) of the organic material (Hernández et al., 1992). The dominating humus form is Mull. In contrast, the decomposition of the litter produced under shrub vegetation (Garrigue or Phrygana) is slower. Besides the drought during the summer months some physiological characteristics of the vegetation (e.g. hard and thick leaves, high specific leaf weight, low permeability of the leaves, low nutrient contents) aggravate decomposition. The typical humus form under these plant communities is Moder.

1.2.1.6 Subtropical Broad-Leaved Evergreen Forests

Subtropical broad-leaved evergreen forests are part of the permanently wet subtropics most of which are found in the southeastern USA, Central China, southern Korea, South Japan, southern Brazil, southeastern South Africa, eastern Australia and the northern island of New Zealand. The tree and shrub species are numerous and include *Lauraceae*, *Fagaceae*, *Oleaceae*, *Cupressaceae*, *Myrsinaceae*, *Rosaceae*, *Araliaceae*, *Palmae*, *Rutaceae*, *Magnoliaceae*, *Proteaceae*, *Myrtaceae*, etc. The composition of different species has a significant influence on the standing biomass. The trees extend in maximum height from 20 to 25 m. The stock of a subtropical broad-leaved evergreen forest may represent a range of aboveground phytomass (including liana) of 150–450 Mg ha^{-1} (e.g. Monk & Day, 1988; Hegarty, 1991), corresponding to 60–190 Mg C ha^{-1}. The belowground biomass may amount to 50 Mg ha^{-1}, corresponding to 20 Mg C ha^{-1}. A mature evergreen forest stores more than 500 kg N ha^{-1} in aboveground biomass (Monk & Day, 1988). Estimates for NPP of subtropical broad-leaved evergreen forests are within a range of 15 Mg ha^{-1} $year^{-1}$ (Kira et al., 1978; Satoo, 1983; Monk & Day, 1988). The representative humus forms are Mull and Moder. Most of the SOM is found in the uppermost mineral soil horizon. Only a small part is located in organic layers.

1.2.1.7 Permanently Humid Tropical Forests

Most of the permanently humid tropics occur between 10° N and 10° S latitude. Eighty percent of the humid tropical forest area is concentrated on only nine countries (Bolivia, Brazil, Colombia, Peru, Venezuela, Indonesia, Malaysia, Congo and Gabun). In some regions, they are extended to 20° N (Central America, East India, Bangla Desh, Nepal) and 20° S (Southeast Brazil, Madagaskar) latitude. The canopy (upper story) of a tropical lowland forest is built up of several layers and extends in height from 35 to 40 m (Fig. 1.7). Some individuals reach >50 m (Vareschi, 1980). The stock of a mature lowland rainforest represents a total (belowground plus aboveground) biomass of 300–650 Mg ha^{-1} (Klinge et al., 1975; Walter, 1979), corresponding to 125–210 Mg C ha^{-1} and roughly 1–2 Mg N ha^{-1}. The NPP is about 20–30 Mg dry matter ha^{-1} $year^{-1}$ (Kira, 1978), 30% of which is litter. In tropical mountain forests, the total phytomass amounts to 200–350 Mg ha^{-1} (Proctor, 1983).

Fig. 1.7 Tropical lowland forest, eastern Ecuador (Photo: R. Nieder)

Fig. 1.8 Dry savanna ("short-grass savanna") in southern Burkina Faso. The region is semiarid today, but has formerly been exposed to more humid climatic conditions causing intensive soil development in the past. The topsoil has been eroded due to overgrazing, so that locally plinthite has become exposed to the surface (Photo: F. Bailly)

Grubb & Edwards (1982), for a tropical mountain forest, estimated an aboveground N stock of 750 kg ha^{-1}. The belowground phytomass of the total biomass may contribute to 15–20% in tropical lowland forests and 15–40% in tropical mountain forests (Proctor, 1983).

About 20–50% of the entire root system of tropical lowland forests lies in the upper 10 cm of the mineral soil, 80–90% lies within the upper 30 cm of the soil, and only 10–20% below 30 cm (Walter, 1964). The litter layer is commonly only 1–5 cm deep because the turnover of forest litter is rapid. About 80% of the leaves are cycled yearly, which represents about 4 Mg of leaf litter ha^{-1} $year^{-1}$ or ~90 kg N ha^{-1} $year^{-1}$ (Klinge et al., 1975). The typical humus forms are Mull and Moder.

1.2.1.8 Tropical Savannas and Dry Forests

Tropical savannas and dry forests cover areas between the permanently humid tropics and the tropical semideserts and deserts. They are characterized by a rainy season in summer and a distinct dry season in winter. Tropical savannas and dry forests exist as a continuum with increasing tree density at increasing annual rainfall (Holdridge, 1947; Lauer, 1975; Walter, 1979). Biomass levels are strongly related to mean annual rainfall, soil factors and the successional stage of the forest. At 200–500 mm annual rainfall thorn savanna and dry savanna (arid eutrophic savanna) with shrub and grassland and small multi-trunk trees (Fig. 1.8) occur which intergrades to closed canopy thorn forests with small multi-trunk trees at 500–900 mm annual rainfall.

Closed canopy forests occur at 900–1,600 mm mean annual rainfall (moist savanna) (Fig. 1.9). Global data show a wide range of total aboveground biomass.

Fig. 1.9 Savanna forest with closed canopy ("wet savanna") in a mountainous landscape of southern Zambia (Photo: F. Bailly)

Martínez-Yrízar (1995) gives ranges of 30–270 Mg ha^{-1} for aboveground and 10–45 Mg ha^{-1} for belowground biomass in tropical deciduous forests. The associated NPP is 8–21 Mg ha^{-1}, roughly 2–5 Mg ha^{-1} of which are belowground. Global estimates for NPP of tropical dry forests range from 2 to 25 Mg ha^{-1} year^{-1} (Larcher, 2003). Tropical dry forests on average store 65 Mg C ha^{-1} (Schlesinger, 1997) and 300–1,100 kg N ha^{-1} (Nye & Greenland, 1960) in vegetation.

The aboveground biomass stocks under grassy vegetation vary between 0 during the dry season and 5–15 Mg ha^{-1} (corresponding to 2–6 Mg C ha^{-1}) when the growth peak is reached during the rainy season (Tiessen et al., 1998). The total NPP, including underground vegetation organs such as rhizomes and tubers may amount to 2.5–10 Mg ha^{-1} year^{-1}. The proportions of grassy and woody vegetation components are highly variable between savanna types and patchy within savannas.

1.2.1.9 Tropical Semideserts and Deserts

Most of the tropical semideserts and deserts (Fig. 1.10) occur in Asia (Arabian peninsula and parts of Iran and Pakistan), Africa (Sahara and Namib), South America (Atacama), south of the USA, Mexico and Australia (Central, South and West Australia). Because of the extreme drought, only a few plant species (e.g. *Acacia*, *Prosopis* and *Cactus* varieties) survive, chiefly in basins with run-in. Plants with very short life cycles (ephemeral species) are also present, springing up in response to the irregular rainfall. In extreme deserts (0–100 mm rainfall), the NPP ranges from 0 to 0.2 Mg ha^{-1}. The standing biomass of semiarid vegetation (100–400 mm annual rainfall) is about 6 Mg ha^{-1} (~2.5 Mg C ha^{-1}), the NPP amounts to approximately 0.6–1.2 Mg ha^{-1} annually (Whyte, 1976). Besides the

Fig. 1.10 Desert landscape with V-shaped valley exhibiting strong thermal rock destruction and black weatherings crusts, filled by wind-blown sand (Wadi Maftuh, South Egypt). The sand originates from far-distance aerial transport (Photo: M. Facklam)

drought, soil formation and organic matter accumulation in the tropical semideserts and deserts are strongly impeded by wind erosion. Therefore, soil organic matter contents rarely exceed 1% (w/w).

1.2.2 Estimates for Agroecosystems

The total C storage in vegetation of major agroclimatic zones lies within a range of 62–173 Pg (Table 1.5).

Carbon storage in agroecosystem vegetation is limited because agricultural biomass densities are lower than those of forests and natural grasslands. Biomass produced in agriculture is commonly exported from the fields and used in ways that release the stored C and N. Only if the share of deep-rooted, woody or tree crops was significantly increased, agriculture would notably contribute to long-term global vegetation carbon and nitrogen storage. Vegetation densities are generally higher in humid and warmer environments. The highest vegetation carbon stocks are observed in the humid and subhumid, warm subtropics and tropics, and agroecosystems of the temperate climate zone. Maximum yields of dry matter, harvest indices, and C and N contents in harvest products of crop plants are given in Table 1.6. Besides swamp grasses, C_4 grasses such as sugarcane have the highest yield potential. However, cereals (C_3 grasses) like wheat, rice and maize are the most important food crops. The annual world production of raw sugar (roughly 118 million Megagrams) is largely exceeded by that of wheat (552 million Megagrams), rice (553 million Megagrams) and maize (516 million Megagrams) on a dry matter

Table 1.5 Storage of carbon in agricultural vegetation and soils of major climatic zones compared to total global C stocks in vegetation and soils (Compiled from Olson et al., 1983; Batjes, 1996; GLCCD, 1998)

Major agroclimatic zone	Pg C Vegetation (range)	Pg C soils (0–100 cm) (mean)	Pg C total (range)	Area (million square kilometres)
Boreal	1–3	5	6–8	0.3
Temperate				
Humid/subhumid	12–27	81	93–109	6.4
Semiarid/arid	11–28	78	90–106	7.0
Moderate cool/cool subtropics				
Humid/subhumid	9–21	48	57–69	4.5
Semiarid/arid	3–8	18	21–26	2.5
Moderate cool/cool tropics				
Humid/subhumid	2–6	9	11–15	0.9
Semiarid/arid	1–3	4	5–7	0.5
Warm subtropics and tropics				
Humid/subhumid	17–59	83	100–142	8.6
Semiarid/arid	7–19	42	49–60	5.6
Agriculture (crops and grassland)	62–173	368	431–542	36.2
Global total (including agricultural and (near-) natural ecosystems)		1,555	1,823–2,456	

basis (FAO, 1998b). On a global scale, average crop yields remain far below maximum values because of inadequate management measures.

1.2.3 Net Primary Production and Phytomass Stocks in Different Climatic Zones

High NPP is concentrated to regions that offer plants a favorable combination of temperature, water, nutrients and light (Table 1.7). These regions are found in the humid tropics (between 20° N and S latitude) and in the temperate climatic zone in transition to the boreal zone between 40° and 60° N and S latitude. Plant communities with the highest NPP are found in areas where land and water meet (i.e., semiterrestrial ecosystems), e.g. in shallow waters near the coasts, in swamps and swamp forests of the warm subtropics and tropics. In terrestrial ecosystems of the humid tropics, production is limited by a deficiency in mineral substances and by deficiency of light in dense stands. On a large proportion of the Earth's surface, only moderate production is possible. On 41% of the terrestrial surface, water is the most growth-limiting factor. Unfavorable temperatures limit plant growth on 8% of the terrestrial surface (short growing seasons and low temperature during the summer months). After swamps and marshes, the widest ranges for NPP can be observed for agricultural crops. The wide ranges given for maximum yields suggest that annual production of biomass in agriculture could be increased significantly with special treatments, such as

Table 1.6 Maximum dry matter yields, harvest index and C and N contents in harvest products of crop plants (Compiled from IPCC, 1997; Bruinsma, 2003; Larcher, 2003; Templer et al., 2005)

Crop plant	Maximum yield (Mg ha^{-1} year^{-1})	Harvest index (economic yield in % of total yield)	C content in harvest products (% dry matter)	N content in harvest products (% dry matter)
C_3 grasses				
Wheat	10–30	25–45	48.5	1.9
Barley	10–20	32–52	45.7	1.7
Rice	20–50	40–55	41.4	1.3
Meadow grasses	20–30	70–80	39.5	
Swamp grasses	50–100	70–80		
C_4 grasses				
Sugar cane	60–80	85		0.2
Maize	20–40	40–50	47.0	1.4
Millet	40–50	40		1.5
Tropical fodder grasses	30–80	70–80	39.5	
Root crops				
Cassava	30–40	70		0.2
Sugar beet	20–30	45–67	40.7	0.2
Potatoes	20	82–86	42.3	0.3
Sweet potatoes	20	80		0.3
Topinambur	20	75		0.3
Legumes				
Lucerne	30	70–80		3.0
Soybeans	10–30	30–35		3.5
Oil palm	20–40		45	1.5

fertilizer applications at levels adjusted to the requirements of each growth stage, irrigation, plant breeding and pest control.

1.3 Carbon and Nitrogen in Soils

1.3.1 Global Soil Organic Carbon and Nitrogen Pools

Soil organic carbon and nitrogen budgets are only moderately accurate because calculations of the global pools are complicated by factors like spatial variability in the SOC and SON content of soils, limited knowledge of the extent of different kind of soils, unavailability of data on bulk density and coarse fragments, and the confounding effect of vegetation and land use changes (Nieder et al., 2003a). The information for soil organic matter (SOM) is especially incomplete for organic soils (Histosols) of northern latitudes. Estimates of the soil C pool are also constrained by the lack of information for charcoal C in soils. Relative amounts of charcoal C may be substantial in fire-dependent ecosystems (e.g., tropical savannas). Estimates of the size of the global SOC pool in the 1970s for 0–100 cm soil depth have varied between 700 and 3,000 Pg (Table 1.8).

Table 1.7 Net primary production of vegetation in different ecosystems (Compiled from Houghton & Skole, 1990; Schlesinger, 1997; Larcher, 2003)

Ecosystems	NPP ($Mg\ ha^{-1}\ year^{-1}$) (range)	NPP ($Mg\ ha^{-1}\ year^{-1}$) (mean)	Phytomass ($Mg\ C\ ha^{-1}$) (mean)	Carbon in vegetation (Pg C)	Area (million square kilometres)
Rock and ice	0–0.1	0.03	–	–	15.2
Tundra (mean of different types)	0.1–4	1.4	8	9.0	11.0
Boreal forests	2.0–15	8	95	143.0	15.0
Temperate forests	4–25	12	80	73.3	9.2
Temperate grassland (mean of different types)	2–15	6	30	43.8	15.1
Tropical rain forests	10–35	22	150	156.0	10.4
Tropical dry forests	16–25	18	65	49.7	7.7
(Sub)tropical wood-land and savanna	2–25	9	20	48.8	24.6
Deserts and semideserts	0.1–3	0.9	3	5.9	18.2
Cropland	1–40	6.5	14	21.5	15.9
Wetlands	10–60	30	27	7.8	2.9
Inland waters	1–15	4			2.0
Total				558.8	147.2

Table 1.8 Estimates of the size of the global SOC and SON pools (0–100 cm)

Pg SOC	Source
700	Bolin (1970)
1,392	Bazilevich (1974)
1,080	Baes et al. (1977)
2,946	Bohn (1978)
2,070	Ajtay et al. (1979)
1,395	Post et al. (1982)
1,515	Schlesinger (1984)
1,500	Woodwell (1984), Eswaran et al. (1993); IPCC (1996); Watson et al. (1995); Batjes (1996, 1997)
3,200 (0–300 cm)	Jobaggy & Jackson (2000)
Pg SON	
92–117	Zinke et al. (1984)
95	Post et al. (1982)
100	Davidson (1994)
96	Eswaran et al. (1995)
133	Batjes (1997)

At present, a value of ~1,500 Pg (0–100 cm) is commonly accepted. At least one third of the SOC is stored in Histosols. Jobaggy & Jackson (2000) extended the size of the global SOC pool to 3 m depth, adding about 55% to the known stock. This depth exceeds the average depth of the rooting zone which is less than 2 m for most plants. Jobaggy and Jackson (2000) furthermore included carbon in wetland and permanently frozen soils which were ignored in previous studies. However, the exact magnitudes of these stocks are still very uncertain. Zinke et al. (1984) and Post et al. (1982) estimated the global SON pool using an ecosystem approach. Their data are similar to those obtained by Davidson (1994) and Eswaran et al. (1995). The higher value obtained by Batjes (1997) is due to the fact that his database contains values for a large number of agricultural soils, where N levels have been increased by long-term mineral fertilizer N application.

1.3.1.1 Distribution of Soil Organic Carbon and Nitrogen According to FAO-UNESCO Soil Classification

In mineral soils, the major part of soil organic carbon and nitrogen is stored in the uppermost horizon (A horizon). The A horizon is a mineral horizon formed or forming adjacent to the surface that has an accumulation of completely humified organic matter intimately associated with the mineral fraction. Its morphology is acquired mainly by earthworm activity and it lacks the properties of E and B horizons. Incorporated organic matter in A horizons is in form of fine particles or coatings on the soil minerals. Distribution of the organic matter throughout the A horizon is by biological activity and not through translocation. This mixing gives A horizons a darker color than underlying mineral horizons. The FAO soil classification (FAO, 1988, 1990, 1998a) and the US Soil Taxonomy (Soil Survey Staff, 1995) classified diagnostic properties of different A horizons. The simplified characteristics of mollic, umbric and ochric A horizons can be drawn from Table 1.9.

Table 1.9 Characteristics of diagnostic A horizons (Compiled from Soil Survey Staff, 1995)

Horizon	Properties (simplified) of A horizons		Thickness (cm)
Mollic	OC	0.6–12.0%	>25 cm; >10 cm[a]
	Munsell value and chroma	<3.5 (wet) <5.5 (dry)	>10 cm[a]
	BS	>50%	
	Well-aggregated		
	Not compacted		
Umbric	like "mollic", except for BS	(<50%)	>25 cm
Ochric	OC	<0.6%	<25 cm
	Munsell value and chroma	>3.5 (wet) <5.5 (dry)	
Anthropic	Like "mollic", except for citrate-soluble P_2O_5	(>0.25 mg g^{-1})	>18 cm
Melanic	OC	>6.0%	>30 cm
	Munsell value and chroma	<2 (wet)	
	Melanic index	<1.7	
	Rest like "andic"		

[a]Only with underlying rock material
OC: organic carbon; BS: base saturation

SOC and SON pools in the upper 30 and 100cm of the FAO-UNESCO soil units (FAO, 1971–1981) can be drawn from Table 1.10. The table includes recent changes of FAO-UNESCO soil classification to the World Reference Base for Soil Resources (FAO, 1998a). Mean soil organic carbon content in the upper 100cm of the various soils ranges from 3.1kg C and 0.52kg N m^{-2} for sandy Arenosols to 77.6kg C and 4.01kg N m^{-2} for Histosols (see also section 1.3.1.3. The large values for the latter are due to the slow decomposition of organic material under water saturated conditions, particularly when mean soil temperatures are low.

Table 1.10 Estimates of organic carbon (C) and nitrogen (N) pools for the depth intervals 0–30cm and 0–100cm (C/N for 0–30cm, 30–50cm and 50–100cm) by FAO-UNESCO Soil Units, with adaptation to FAO, 1998a (Adapted from Batjes, 1997)

	Depth interval (cm)						
	0–30		0–100		0–30	30–50	50–100
	(kg m^{-2})						
Soil unit	C	N	C	N	C/N	C/N	C/N
Acrisols	5.1	0.48	9.4	1.10	13.2	10.1	8.9
Cambisols	5.0	0.58	9.6	1.12	11.5	9.7	9.0
Chernozems	6.0	0.88	12.5	1.70	10.8	10.7	9.4
Phaeozems	7.7	0.71	14.6	1.51	11.4	10.0	8.9
Podzoluvisols[a]	5.6	0.54	7.3	0.76	13.6	7.4	7.5
Rendzinas[b]	13.3	1.05	–	–	11.2	–	–
Ferralsols	5.7	0.46	10.7	0.97	14.3	12.6	11.8
Gleysols	7.7	0.75	13.1	1.34	12.6	11.2	10.4
Lithosols[b]	3.6	0.42	–	–	11.1	–	–
Fluvisols	3.8	0.50	9.3	1.23	11.2	11.3	10.4
Kastanozems	5.4	0.68	9.6	1.78	10.6	8.8	8.6
Luvisols	3.1	0.45	6.5	1.03	11.6	9.9	9.4
Greyzems[c]	10.8	0.96	19.7	1.92	8.9	11.0	8.6
Nitosols[d]	4.1	0.49	8.4	1.00	12.6	9.8	8.6
Histosols	28.3	1.61	77.6	4.01	25.8	29.8	22.3
Podzols	13.6	0.81	24.2	1.39	23.8	21.5	24.5
Arenosols	1.3	0.22	3.1	0.52	14.2	12.6	9.5
Regosols	3.1	0.45	5.0	0.70	13.5	9.6	10.2
Solonetz	3.2	0.45	6.2	1.11	12.2	10.5	8.8
Andosols	11.4	0.91	25.4	1.99	13.3	13.8	14.3
Rankers[b]	15.9	2.18	–	–	17.1	–	–
Vertisols	4.5	0.50	11.1	1.23	13.3	12.5	12.5
Planosols	3.9	0.41	7.7	1.00	11.5	10.3	7.9
Xerosols[e]	2.0	0.33	4.8	0.58	9.9	9.2	7.0
Yermosols[e]	1.3	0.15	3.0	0.37	11.1	10.5	10.9
Solonchaks	1.8	0.27	4.2	0.75	11.7	9.2	8.5

[a] Albeluvisols according to FAO (1998a)
[b] Leptosols according to FAO (1998a)
[c] Now FAO (1998a) merged to Phaeozems
[d] Now FAO (1998a) merged to Nitisols
[e] Soils of deserts and half-deserts (FAO, 1971–1981) according to FAO (1998a) were merged to other soil units (e.g. Calcisols, Gypsisols, Leptosols, Arenosols, Regosols)

According to earlier literature (Campbell & Claridge, 1987) SOM and its accumulation was of minor importance with respect to soil formation in permafrost soils (Cryosols, according to FAO, 1998a). This may be the reason why e.g. Antarctica is not considered in global estimates of SOC and SON (Nieder et al., 2003a). Blume et al. (1997) and Beyer et al. (1998) documented that the SOM contents in mineral topsoils of the ice-free coastal antarctic region was greater as expected from the earlier literature. In the formation of soils of the periglacial arctic regions, freeze-thaw cycles and accompanied cryoturbation lead to sorted and non-sorted patterned ground features on the soil surface. The thickness of the "active" layer is controlled by soil texture and moisture, thickness of surface organic layer, vegetation, and latitude. Cryoturbated soil profiles are characterized by irregular or disrupted soil horizons (Fig. 1.11) and oriented stones in the soil. Some Cryosols contain large amounts of organic matter. According to Deckers et al. (1998) carbon sequestration in Cryosols in some cases can be significant. Under the vegetation cover of the low Arctic, organic layers can develop as Mor (for description of humus forms see Chapter 3) or as peat (for description see section 1.3.3). This is because the decomposition of organic matter is slow due to low mean temperatures and frequently stagnic conditions in the temporarily thawed part of the soil. Soils of the Arctic have a perennial frozen subsoil (permafrost).

Very small amounts of organic carbon and nitrogen are encountered in soils of half-deserts (Xerosols: 4.8 kg C and 0.58 kg N m^{-2}) and deserts (Yermosols: 3.0 kg

Fig. 1.11 Cryoturbation (relictic) in a Podzol, near Ülzen, North Germany. Freeze-thaw cycles in (ant)arctic regions are responsible for cryoturbation. At the beginning of the frost phase, freezing fronts in the thawed layer move both from the soil surface downwards and from the permafrost table upwards. As a result, unfrozen materials are displaced and soil horizons are contoured and broken (Photo: R. Nieder)

C and 0.37 kg N m^{-2}) where plant growth is limited. The soil units Xerosols and Yermosols go back to FAO (1971–1981). In the World Reference Base for Soil Resources (FAO, 1998a), these were merged to soil units such as Leptosols, Arenosols and Regosols. Soils with mollic and umbric (dark) A horizons (for definition see Chapter 3) have a moderate to high organic matter content. The mollic A is base-rich, typically occurring in soils of steppe ecosystems. It is typical for Chernozems (Fig. 1.12), Phaeozems, and Kastanozems, as well as for soils overlying calcareous material like Rendzic Leptosols (Fig. 1.13) or base-rich material like Mollic Leptosols. In these soils, the presence of a highly active soil biomass, including megafauna, tends to intensively mix organic materials into the upper mineral soil horizon. This process is called bioturbation.

The umbric A horizon is base-poor and occurs in some Leptosols, Fluvisols, Gleysols, Andosols, and, most typically in Umbrisols (Fig. 1.14). Soils that are repeatedly wetted and dried and that contain clays with a large capacity for expansion tend to crack widely and deeply, allowing topsoil particles and organic materials to fall into lower soil layers, so that over time the whole soil is turned over (Driessen & Dudal, 1991). This process is called peloturbation (Fig. 1.15). Soils which are formed in this way are called Vertisols (Fig. 1.16) most of which are characterized by uniform dark colors which is due to very pronounced organomineral bonds. However, their SOC and SON contents are only moderate (Table 1.10).

Vertisols occur especially in warm climates with an alteration of distinct wet and dry seasons such as semiarid to subhumid tropical and Mediterranean climates (FAO, 1990). Humic soils (not given in Table 1.10) like well-drained Humic Nitisols (18.2 kg C m^{-2}) or poorly drained Mollic and Umbric Gleysols (29.3 kg C m^{-2}) contain much larger amounts of organic C than the means for Nitisols (8.4 kg C m^{-2}) and Gleysols (13.1 kg C m^{-2}). The large values (25.4 kg C and 1.99 kg N m^{-2}) for Andosols (Fig. 1.17) can be explained by the protection of SOM by allophane (Mizota & Van Reeuwijk, 1989). Generally, the stabilizing effect of inorganic

Fig. 1.12 Haplic Chernozem in loess with a mollic Ah horizon (earthworm mull), near Halle, eastern Germany (Photo: R. Nieder)

Fig. 1.13 Rendzic Leptosol; a mollic Ah horizon (earthworm mull) with a crumby structure overlies mesozoic limestone, Elm mountains near Braunschweig, North Germany (Photo: R. Nieder)

particles on SOM decreases in the sequence allophane > amorphous and poorly crystalline Al-silicates > smectite > illite > kaolinite (Van Breemen & Feijtel, 1990).

The mean C/N ratios across the soil units range from 8.9 for Greyzems (now merged to Phaeozems, according to FAO, 1998a) to 29.8 for Histosols. As a result of most input of C and N to a soil profile being introduced from the overlying standing biomass, SOC and SON generally decrease down the soil profile. The degree to which SOM is concentrated in different compartments of the soil is a function of climate, soil type, and rooting depth. In mineral soils, as much as 50% of the total SOC and SON inventory to 1 m may be present in the upper 30 cm of the mineral soil (Table 1.10; Bird et al., 2001). The general trend of values in Table 1.10 shows a decrease in C/N ratio with depth, which reflects a greater degree of breakdown and older age of the humus stored in the lower parts of the profile.

1.3.1.2 Carbon and Nitrogen in Soil Microbial Biomass

Next to living plants, microorganisms constitute the largest biomass on our planet. They carry out the greatest range physiological processes ranging from decomposition to the numerous reactions in the C, N, S and P cycles. Soil microbial biomass

Fig. 1.14 Humic Umbrisol with an umbric Ah horizon overlying palaeozoic sandstone, near Gatumba, Central Rwanda (Photo: R. Nieder)

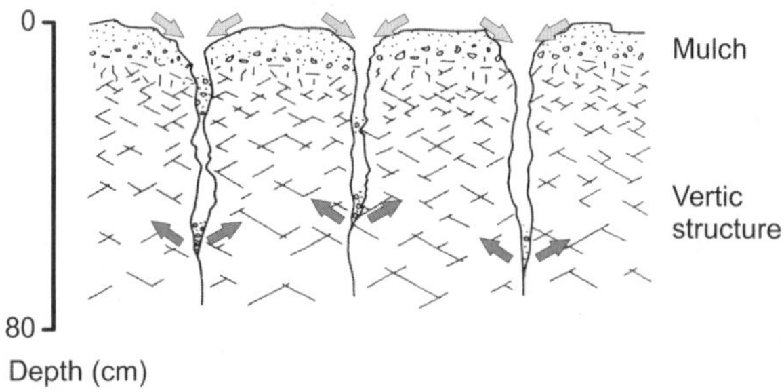

Fig. 1.15 Peloturbation occurring in Vertisols. The drawing shows the shrinking state during the dry season, with soil material (granules or crumbs) falling from the surface mulch layer into the cracks. Subsequent re-wetting generates pressure which results in sliding of soil masses along each other (Adapted from Driessen & Dudal, 1991)

accounts for 1–3% of the organic C and 2–6% of the organic N in soil (Jenkinson, 1987). Major differences in terms of mean microbial C and N content exist when different ecosystems are compared (Table 1.11).

Fig. 1.16 Haplic Vertisol showing dark clay (organic complexes with smectites) and a crumby to polyedric structure in the Ah horizon and a polyedric to prismatic structure with slickensides in the vertic B horizon, Morocco, North Africa (Photo: R. Nieder)

Fig. 1.17 Landscape with Mollic Andosol developed from volcanic pumice. The organic matter (up to 8%) in the dark colored Ah horizon is protected by allophane, Chimborazo area, Ecuador (Photo: R. Nieder)

It appears that microbial biomass stores a higher portion of total organic C and N in tropical ecosystems as compared to temperate and (sub)arctic systems. Within the tropics, the availability of water plays a major role for microbial life which is demonstrated by the decrease in microbial C and N from the permanently wet tropics

Table 1.11 Estimates of microbial biomass C and N in different ecosystems (Adapted from Wardle, 1992; Groffman et al., 2001; Templer et al., 2005)

Ecosystem	Microbial C ($\mu g\ g^{-1}$ soil)	Number of studies	Total number of sites	Microbial N ($\mu g\ g^{-1}$ soil)	Number of studies	Total number of sites
Antarctic sandstone	126	1	1	11	1	1
Subarctic arable	800	1	1	66	14	1
Boreal coniferous forest	736 ± 661	10	18	93 ± 65	7	18
Temperate forest	877 ± 757	17	34	93	7	18
Natural heathland (Europe)	1,373 ± 220	3	3			
Temperate managed grassland	1,011 ± 559	37	137	170 ± 102	17	61
Cool temperate arable	463 ± 328	44	155	66 ± 41	14	26
Warm temperate arable	331 ± 245	16	54	47 ± 29	6	12
Steppe (NorthAmerica)	846 ± 56	3	3			
Subtropical pasture	611	1	1			
Tropical rainforest	986 ± 834	3	3	100 ± 75	3	5
Tropical pasture (Central America)	2,000 ± 700	1	6	240 ± 75	1	3
Tropical abandoned pasture (Central America)	2,800 ± 320	1	3	420 ± 30	1	3
Tropical young mixed gardens (Central America)	2,560 ± 320	1	2	420 ± 30	1	2
Tropical old mixed gardens (Central America)	3,420 ± 410	1	2	530 ± 40	1	2
Tropical agriculture: Cacao (Central America)	2,560 ± 460	1	3	250 ± 40	1	3
Tropical agriculture: Oil palm (Central America)	780 ± 100	1	3	120 ± 10	1	3
Dry tropical forest	653 ± 133	6	6	65 ± 14	6	6
Tropical savanna	342 ± 173	3	3	35 ± 12	3	3
Desert shrubland (Israel)	340	1	1			

over the dry tropical forest to the tropical savanna. Forest conversion to agricultural land can decrease soil microbial biomass. However, the level of microbial C and N is higher under pasture as compared to arable land. In some cases land converted to pasture may have amounts of microbial biomass C and N that equal or even exceed pre-cultivation levels (Templer et al., 2005). There is some evidence that in the tropics an introduction of mixed cultures and abandonment of agriculture with subsequent regeneration of forest may increase microbial biomass.

Global storage of microbial carbon and nitrogen were estimated to be 0.18 Pg C and 0.03 Pg N for tundra, 1.82 and 0.25 Pg C for boreal forests, 1.48 Pg C and 0.26 Pg N for temperate forests, 3.03 Pg C and 0.48 Pg N for temperate grassland, 3.68 Pg C and 0.43 Pg N for tropical forests, and 3.73 Pg C and 0.38 Pg N for savanna ecosystems (Wardle, 1992). The total soil microbial C and N pools were estimated to be 13.9 and 1.83 Pg, respectively.

1.3.1.3 Histosols as Sink for Carbon and Nitrogen

Accumulation of organic matter in Histosols represents a geological sink for carbon and nitrogen (Table 1.12).

In the FAO guidelines for soil profile description (FAO, 1990), organic horizons of Histosols are referred to as H horizons. The H horizon is formed by an organic accumulation that is saturated for prolonged periods or is permanently saturated unless artificially drained. The H horizon should have a thickness of more than 20 cm but less than 40 cm and contain 18% or more organic carbon if the mineral fraction contains more than 60% of clay; lesser amounts of organic carbon are permitted at lower clay contents. The H horizon may be between 40 and 60 cm thick if it consists mainly of sphagnum, or has a bulk density when moist of 0.1 Mg m^{-3}. Histosols store the highest carbon quantities among all soil units. While mineral soils may contain between 3 and 25 kg C m^{-2} (30–250 Mg ha^{-1}) in 0–100 cm depth (see Table 1.10), reported values for Histosols typically are an order of magnitude greater, with some values as high as nearly 2,000 Mg C ha^{-1} (Table 1.12). The quantities of C stored in some very deep Histosols is undoubtably even higher. The organic nitrogen content of these soils ranges from 0.5% to 2.5%.

Histosols are formed of peat that consists of lignin, cellulose, hemicellulose, and small quantities of proteins, waxes, tannins, resin, suberin, etc. (Driessen et al., 2001). In northern regions, peats are predominantly ombrogenous. Many occur in a 30–50 cm thawed (active) layer on top of permafrost subsoil. In temperate regions, topogenous low moor peat is mainly woody forest peat and peat derived from grassy marsh vegetation. The high moor peat of ombrogenous raised bogs is mostly *Sphagnum* moss peat (Fig. 1.18).

Blanked peat in upland areas is rain-dependent peat formed under heather and other low shrubs. In the tropics, lowland peat is almost exclusively ombrogenous and

Table 1.12 Carbon storage in organic soils

Location	Site characteristics	Quantity of stored C (Mg ha^{-1}) Mean	Range	Reference
Different organic wetlands of the temperate zone		2,000		Armentado & Menges (1986)
Maryland, USA	Coastal Marsh	590	180–1,660	Griffin & Rabenhorst (1989)
USA	Coastal Marsh, Atlantic and Gulf Coasts	640	90–1,910	Rabenhorst (1995)

Fig. 1.18 Ombrogenous sphagnum moss peat Histosol (Fibric Histosol) of a raised bog with buried Podzol, Teufelsmoor near Bremen, North Germany. The peat consists mainly of *Sphagnum* moss which is weakly decomposed in the upper part (H1) and strongly decomposed in the lower part (H2) of the profile (Photo: R. Nieder)

made up of woody rain forest debris. Topogenous peat in the tropics and subtropics is confined to comparatively small occurrences in coastal plains and lagoon areas and to filled-in lakes at high elevation. This peat is less woody than ombrogenous forest peat (e.g. *Papyrus* swamps, sawgrass peat, etc.) but richer in mineral constituents.

Northern peatlands (of cool and temperate climate) contain about one third (450 Pg; Gorham, 1991) of the global store of soil organic C, with carbon accumulation rates of 20–40 g C m^{-2} $year^{-1}$ over the last 5,000–10,000 years (Tolonen & Turunen, 1996). A collation of studies of carbon dioxide exchange in northern peatlands by Frolking et al. (1998) reveals that the summer uptake rates through photosynthesis are small (15–30 g CO_2 m^{-2} day^{-1}) compared with uptake in forests, grassland and crops (50–200 g CO_2 m^{-2} day^{-1}; Ruimy et al., 1995). Some Histosol areas are also found in warmer (subtropical and tropical) climate. Most northern latitude Histosols occupy regions which were covered by glaciers during the last ice age and have formed following the glacial retreat. Reported average rates of peat accumulation in northern bogs and fens have been as high as >1 mm $year^{-1}$, but more typically fall in the range of 0.5–1.0 mm $year^{-1}$ (Table 1.13). This means an annual increase of 5–10 m^3 of peat per hectare.

Annual carbon accumulation rates on average may range from 0.2 Mg ha^{-1} $year^{-1}$ in subarctic regions to 2.0 Mg ha^{-1} $year^{-1}$ in temperate (e.g. Maryland, USA) and 2.5 Mg ha^{-1} $year^{-1}$ subtropical (e.g. Louisiana, USA) regions (Table 1.14). Histosols formed in organic soil material under the permanent influence of groundwater ("low moor peat") occupy the lower parts of fluvial, lacustrine and marine landscapes, mainly in temperate regions (Fig. 1.19). During the glacial maximum (about 20,000 year BP) of the last ice age, when large quantities of water were tied up in the glacial ice, sea level worldwide was more than 100 m below the present level. Melting of the ice and concurrent ocean warming caused sea level to rise at such a rapid rate (10–20 mm $year^{-1}$) that initially vegetation could not colonize the tidal regions. Approximately 3,000–5,000 years ago, sea level rise slowed to a more modest pace so that marsh vegetation could become established and organic parent

Table 1.13 Values for peat accumulation in organic soils

Location	Site characteristics	Peat accumulation (mm $year^{-1}$)	Reference
North Russia	Raised bogs	0.6–0.8	Botch & Masing (1986)
Sweden	Raised bogs	0.3–1.0	Almquist-Jacobson & Foster (1995)
North Canada	Raised bogs	0.3–0.6	Kuhry & Vitt (1996)
Central Finland	Raised bogs	0.75	Tolonen (1979)
North Germany	Raised bogs	0.70	Tolonen (1979)
Minnesota, USA	Low moor	0.85–1.15	Gorham (1987)
Louisiana, USA	Coastal marsh	6.5–9.5	Nyman & DeLaune (1991)
Louisiana, USA	Coastal marsh	7.0–13.0	Hatton et al. (1983)
Massachusetts, USA	Coastal marsh	1.1–2.6	Redfield & Rubin (1962)

Table 1.14 Values for carbon accumulation in organic soils

Location	Site characteristics	C accumulation (Mg ha^{-1} $year^{-1}$)	Reference
Alaska, USA	Subarctic region	0.11–0.61	Billings (1987)
Canada	Subarctic region	0.14–0.35	Kuhry & Vitt (1996)
Canada	Boreal region	0.23–0.29	Gorham (1991)
Finland	Boreal region	0.20–0.28	Francez & Vasander (1995)
Russia	Boreal region	0.12–0.80	Botch et al. (1995)
Maryland, USA	Coastal marsh	1.2–4.2	Kearney & Stevenson (1991)
Louisiana, USA	Coastal marsh	1.7–2.7	Nyman & DeLaune (1991)
Louisiana, USA	Coastal marsh	1.8–3.0	Smith et al. (1983)
Florida, USA	Coastal marsh	0.7–1.05	Craft & Richardson (1993)

Fig. 1.19 Topogenous low moor peat Histosol (Eutric Histosol) showing a terric H1 and a sapric H2 horizon, Ochsenmoor near Osnabrück, North Germany (Photo: R. Nieder)

materials began to accumulate (Redfield, 1972). In addition to the eustatic sea level rise, sediments in transgressing coastal regions are subsiding.

As sea level has continued to rise, organic materials have accumulated in Histosols, and coastal marshes and mangroves generally have been thought to have accreted approximately the rate of sea. The combination of rising sea level (presently estimated at 1 mm year^{-1} worldwide) and coastal subsidence can be joined to yield an apparent sea level rise, which is substantially greater. Current estimates of peat accretion in coastal areas generally range from 3 to 8 mm year^{-1}, which are much higher than in noncoastal regions, with even higher rates reported in rapidly subsiding areas (Table 1.13). Current evidence suggests that the highest rates of sea level rise may be too great for marsh systems to maintain, and that some of these areas are suffering marsh loss.

1.3.1.4 Associations of Histosols with Other Soil Groups

Organic soil materials in northern regions could accumulate there because the decay of organic debris is retarded by frost in the cold season and by prolonged water-saturation of the thawed surface soil during summer. Permafrost-affected Histosols are associated with Cryosols and with soils that have gleyic or stagnic properties, e.g. Gleysols in Alaska and in the northern part of the former USSR. Where (sub)arctic region grades into the cool Temperate zone, associations with Podzols can be expected. Other soils in the same environment are Fluvisols, Gleysols, and, in coastal regions, Solonchaks (e.g., adjacent to coastal mangrove peat). Histosols in lacustrine landforms are commonly associated with Vertisols. Rain-dependent Histosols are found in environments with sufficiently high and evenly spread rainfall, e.g., in raised "dome" peat formations (high moor peat) in lowland areas and in upland areas with blanked peat, where paucity of nutrient elements, acidity and near-permanent wetness retard decay of organic debris. Lateral linkages exist with a variety of soil groups, including Andosols, Podzols, Fluvisols, Gleysols, Cambisols and Regosols.

1.3.2 Global Soil Inorganic Carbon and Nitrogen Pools

1.3.2.1 Soil Inorganic Carbon

The current global estimate of soil inorganic C (SIC: 695–720 Pg, see Table 1.1) was derived from average carbonate-C contents for soil types published by Schlesinger (1982) and soil area estimates derived from the Soil Map of the World (FAO, 1991). Table 1.15 shows mean SIC contents according to the Holdridge (1947) classification scheme. Processes governing the dynamics of SOC and SIC pools differ among Holdridge life-zones. They further interact with land use, farming system, soil tillage and crop management practice (see Chapter 6). Compared to the SOC pool, the SIC pool is the smaller in soils of humid regions, whereas SIC is the predominant pool in many soils of arid and semiarid regions (desert, chaparral, tropical semiarid regions) where annual precipitation is <500 mm. The total area of arid and semiarid regions is estimated at 48.5×10^6 km^2 (37.3% of the total land area) with a relative distribution of 43.5% in semiarid climates 46.5% in arid regions (UNEP, 1992). Dominant soil units conditioned by an accumulation of carbonates include Xerosols and Yermosols (Table 1.16).

Soils of arid and semiarid regions are characterized by accumulation of soluble salts and carbonates in the profile with ascending water because evapotranspiration exceeds precipitation (Driessen & Dudal, 1991). Plant growth and accumulation of SOC are reduced by low rainfall (Batjes, 1997). Carbonate-C does not participate in the C flux to other C systems as rapidly as SOC. Exceptions are irrigated soils or systems which become acidified by increased S and N inputs (Lal et al., 1995). Changes in pedogenic carbonate may have an influence on phosphorus (P) availability

Table 1.15 Estimates of soil inorganic carbon (SIC) pool for the depth interval 0–100 cm aggregated as per Holdridge life-zone (Compiled from Batjes, 1997)

Holdridge life zone	Pg SIC	SIC (% of total C)	Area[a]
Tundra	16.4	9.4	10.16
Cold parklands	15.3	30.0	2.78
Forest tundra	14.4	7.4	8.72
Boreal forest	34.9	9.2	14.94
Cool desert	49.6	71.8	4.00
Steppe	73.1	51.4	7.35
Temperate forest	31.9	20.9	9.90
Hot desert	231.6	77.8	20.80
Chaparral	53.7	55.8	5.62
Warm temperate forest	6.3	17.2	3.20
Tropical semiarid	82.2	60.8	9.53
Tropical dry forest	55.9	32.1	14.85
Tropical seasonal forest	21.1	13.7	15.08
Tropical rain forest	8.6	8.8	8.46
All ecosystems	695.0	32.2	135.39

[a] Area in 10^6 km^2, excluding land glaciers

Table 1.16 Estimates of soil inorganic carbon (SIC) pools of main dryland soils of the world for the depth intervals 0–30 cm and 0–100 cm by FAO-UNESCO Soil Units, with adaptation to FAO, 1998a (Adapted from Batjes, 1997)

	0–30 cm depth interval		0–100 cm depth interval	
Soil unit	Pg SIC	SIC (% of total C)	Pg SIC	SIC (% of total C)
Arenosols	6.2	43.3	24.9	57.9
Cambisols	9.7	35.1	30.7	45.9
Fluvisols	8.6	44.3	28.1	49.9
Lithosols[a]	41.8	51.9	41.8	51.9
Regosols	8.6	43.7	18.5	46.2
Rendzinas[a]	2.0	33.9	3.5	47.3
Solonchaks	12.3	72.3	45.0	80.5
Solonetz	4.5	39.5	17.8	56.7
Vertisols	7.6	36.9	26.5	45.4
Xerosols[b]	26.7	71.6	124.5	85.2
Yermosols[b]	30.7	74.3	147.7	84.8
All dryland soils	158.6	53.8	508.9	66.9

[a] Leptosols according to FAO (1998a)
[b] Soils of deserts and half-deserts (FAO, 1971–1981) according to FAO (1998a) were merged to other soil units (e.g. Calcisols, Gypsisols, Leptosols, Arenosols, Regosols)

and P cycling in dryland systems (West et al., 1994) because in these soils most inorganic P is present Ca-phosphates (Lathja & Bloomer, 1988). Calcisols are high or very high in $CaCO_3$ content. They have a calcic (>15% $CaCO_3$) or a hypercalcic (>50% $CaCO_3$) horizon in the uppermost 100 cm of the soil which is >15 cm thick (Fig. 1.20). A hard, cemented petrocalcic horizon is called "calcrete". Important associated soils include Regosols, Vertisols, Arenosols, Cambisols and a range of shallow soils limited in depth such as Lithosols, Rendzinas and Rankers (the latter three soil units were merged to Leptosols according to FAO, 1998a). Calcaric Chernozems, Kastanozems and Luvisols occurring in more humid climate types (e.g. steppe) also contain considerable amounts of SIC.

1.3.2.2 Soil Inorganic Nitrogen

Mineralization of SOM and release of nitrogen from organic and mineral fertilizers and subsequent nitrification provide reactive, inorganic N forms such as ammonium (NH_4^+), nitrite (NO_2^-) and nitrate (NO_3^-). Nitrate and soluble and exchangeable NH_4^+-N are readily available for plant and microbial use. Processes related to reactive N forms in soil are discussed in Chapter 5. A high portion of ammonium-N in soil is bound in a non-exchangeable form as so-called "fixed" ammonium (Nieder & Benbi, 1996). After soil organic nitrogen, it represents the second largest soil N pool (see section 1.1.2.2). Up to now, data on global estimates of fixed NH_4^+ in soils are not available. Mineral soils vary in their capacity to fix NH_4^+ ranging from a few kilograms to several thousand kilograms per hectare in the plow layer (Table 1.17). For example, the non-exchangeable NH_4^+-N content in soils has been reported to range from 25 to 850 mg kg^{-1} soil in Germany (Scherer, 1993, and references cited therein), 45–190 mg kg^{-1} in Austria, 180–490 mg kg^{-1} in clay soils of Spain (Moyano & Gallardo, 1988), 35–210 mg kg^{-1} in the US, 6–107 mg kg^{-1} in Queensland,

Fig. 1.20 Alluvial landscape with Calcisols derived from base-rich sediments near the Murray River, Australia. The carbonates have originally accumulated at some depth below the former soil surface due to capillary rise from groundwater. Next to the river, the continuously cemented calcareous accumulation zone is seen to crop out at the recent surface due to erosion of the uppermost part of the former soil profile (Photo: U. Schwertmann)

Table 1.17 Contents of fixed NH_4^+-N in soils as to different parent materials

Authors	Country	Soil type (FAO)	Parent material	Clay content (%)	NH_4^+-N concentration (mg kg^{-1})	(kg ha^{-1} 30 cm^{-1})
Petersburgsky & Smirnov (1966)	Russia	Podzols	Diluvial sand	?	26–100	115–450
Nommik (1967)	Sweden	Podzols	Diluvial sand	<5	10–17	45–77
Fleige & Meyer (1975)	Germany	Podzol	Diluvial sand	<2	80	360
Scherer & Mengel (1979)	Germany	Cambisol	Red sandstone	17	12.5	50
Scherer & Mengel (1979)	Germany	Diff. soils	Basalt	17–38	60–80	270–360
Mengel et al. (1990)	Germany	Vertisol	Basalt	24	100	445
Hinman (1964)	Canada	Diff. soils	Loess	12–35	130–300	580–1,350
Petersburgsky & Smirnov (1966)	Russia	Chernozems	Loess	?	90–140	410–630
Scheffer & Meyer (1970)	Germany	Luvisol	Loess	10	170	760
Mba-Chibogu et al. (1975)	Germany	Luvisol	Loess	15	100	450
Fleige & Meyer (1975)	Germany	Chernozem	Loess	?	140	650
Scherer & Mengel (1979)	Germany	Luvisols	Loess	12–18	120–160	540–720
Beese (1986)	Germany	Luvisol	Loess	13	100	450
Nieder et al. (1996)	Germany	Luvisols	Loess	19	115–170	520–760
Fleige & Meyer (1975)	Germany	Leptosol	Limestone	?	340	1,530
Fleige & Meyer (1975)	Germany	Cambisol	Limestone	?	460	2,070
Fleige & Meyer (1975)	Germany	Cambisol	Boulder clay	30	170–220	765–1,000
Mohammed (1979)	Libya	Diff. soils	Clay stone	38–46	90–100	415–450
Schachtschabel (1961)	Germany	Fluvisols	Marsh	20–70	150–850	675–3,825
Atanasiu et al. (1967)	India	Fluvisol	River sediment	14	200	900
Atanasiu et al. (1967)	Egypt	Fluvisol	River sediment	34	210	945
Fleige & Meyer (1975)	Germany	Fluvisol	River sediment	20	260	1,170
Scherer & Mengel (1979)	Germany	Fluvisols	River sediment	10–40	70–270	315–1,215
Gorlach & Grywnowicz (1988)	Poland	Fluvisol	River	17	122	549
Mengel et al. (1990)	Germany	Fluvisol	Marsh	41	195	880

Australia, 40–490 mg kg^{-1} in the former USSR, 35–573 mg kg^{-1} in China (Qi-Xiao et al., 1995), 30–60 mg kg^{-1} in Sudan, 8–98 mg kg^{-1} in Nigeria, and 57–367 mg kg^{-1} in six soil profiles from southern Ontario, Canada (Doram & Evans, 1983). Investigations on top soils from Antarctic also showed that fixed NH_4^+ occurs in amounts (0–322 mg kg^{-1}) similar to elsewhere in the world (Greenfield, 1991).

As percent of the total N, the non-exchangeable NH_4^+-N constituted a minimum of 2% in soils from Israel (Feigin & Yaalon, 1974) to as high as 85% in subsurface soils from the US (Smith et al., 1994). It ranged from 14–78% in British Caribbean soils (Rodrigues, 1954), 2–79% in main soil groups of Israel (Feigin & Yaalon, 1974), 3–44% in southern Ontario soils (Doram & Evans, 1983), 16–59% in Vertisols and 13–31% in Cambisols from India (Sahrawat, 1995), and 21% in Italian Fluvisols (Benedetti et al., 1996).

Generally, non-exchangeable NH_4^+-N content as percent of the total N increases with soil depth probably due to blocking effect of native K, decreasing total N (Black & Waring, 1972) and organic matter content down the profile. The native fixed NH_4^+-N in 24 soils from Queensland, Australia averaged 4% of total N in surface soil and 6.4% for deeper samples (Black & Waring, 1972). The magnitude of increase with depth was much higher in soils from southwestern Saskatchewan where it ranged from 7% of the total N in the surface soil to 58% at 120 cm depth (Hinman, 1964). It ranged from 21–33% in the surface layers and 30–83% in subsurface layers of soils from Spain with illite as the dominant clay mineral (Moyano & Gallardo, 1988).

The content of native fixed NH_4^+ and the differential NH_4^+ -N fixation capacity of the soils from different regions may be linked to parent material (Table 1.17), texture, clay mineral composition, potassium status of the soil and K-saturation of the interlayers of 2:1 clay minerals, and moisture conditions. In soils formed on diluvial sand and red sandstone, the non-exchangeable NH_4^+-N content ranged 10–100 mg kg^{-1}, whereas it varied between 60–100 mg kg^{-1} in basalt, 90–300 mg kg^{-1} in loess, 90–460 mg kg^{-1} in limestone and clay stone, and 150–850 mg kg^{-1} in marsh soils (Table 1.17). In soils with clay fraction derived from embedded aeolian dust, the non-exchangeable NH_4^+-N exceeded 140 mg kg^{-1} clay. Whereas in soils derived from calcareous rocks-limestone, dolomite, or marls, or from basalt and basaltic tuff containing little mica, its content was generally less than 140 mg kg^{-1} clay (Yaalon & Feigin, 1970).

Clay fraction, containing 2:1 clay minerals, is the dominant factor that influences the content of non-exchangeable ammonium in soils. In central European loess soils, about 65% of the fixed NH_4^+ is interlayer ammonium (Niederbudde & Friedrich, 1984), but some is also fixed in the silt fraction (Jensen et al., 1989). The native fixed NH_4^+-N content in different primary minerals has been reported to vary between nil in quartz to 266 mg kg^{-1} in biotite (Wlotzka, 1961). Yaalon & Feigin (1970) found that illite contains approximately 600 mg N kg^{-1} illite, whereas montmorillonite clays contain small and kaolinite clays negligible amounts only. In mixed soil clays, the non-exchangeable- NH_4^+ level is determined by the quantity of illite (Feigin & Yaalon, 1974) and illite plus vermiculite (Sparks et al., 1979; Doram & Evans, 1983) present. Bajwa (1985) found that in five soil-clays montmorillonitic clay fixed the maximum (98%) followed by vermiculite (88%) and least (34%) in clay containing hydrous mica, chlorite and halloysite. This shows that it is not only the amount of clay but also its mineralogical composition that governs the NH_4^+ fixation in soils.

1.4 Global Vegetation-Soil Organic Matter Interrelationships

Climate and vegetation are major factors controlling global SOC and SON pools. Temperature and precipitation control the levels of input from biomass into the soil, and the rate at which carbon and nitrogen added to the soil are cycled through the SOM pool and finally mineralized. Climate also rules the rates at which C as dissolved organic carbon (DOC) and N as NH_4^+, NO_2^-, NO_3^- and dissolved organic nitrogen (DON) are lost via leaching or cycled back to the atmosphere via denitrification (N_2, N_2O), ammonia volatilization (NH_3) or respiration (CO_2). Ladd et al. (1985) compared the loss of ^{14}C labeled plant residues from four soils in South Australia with those obtained by Jenkinson & Ayanaba (1977) for soils in UK and Nigeria. The rate of substrate C mineralized doubled for an 8–9°C increase in mean annual temperature. The trend of decreasing SOM content with increasing temperature implies that the temperature sensitivity of decomposition is greater than that of primary productivity. Kirschbaum (1995) in controlled incubation studies showed that the Q_{10} value of C mineralization from soil was greater than that for net primary production developed by Lieth (1973), particularly at temperatures <15°C. Climate also affects the chemical structure of SOM. Bracewell et al. (1976) using pyrolysis-gas chromatography in a climosequence of nine soils in New Zealand observed significant correlations between changes in the intensity of peaks in the chromatograms and mean annual temperature and precipitation. Amelung et al. (1997) in a study covering different climatic zones of North America observed an increased content of polysaccharides under more humid conditions. The authors suggested that the increase in polysaccharide C might have resulted from increased plant production and enhanced microbial and earthworm activity.

In combination with other factors, climate controls litter quality (nitrogen and lignin content, etc.; Melillo et al., 1982) and processes that influence the quality of SOM. Climate controls SOC and SON distribution in the soil profile by influencing bioturbation and illuvation of soluble C and N compounds in the soil profile (Holt & Coventry, 1990). At the long term, climate also affects the texture and mineralogy of the inorganic fraction of the soil (Goh et al., 1976). However, it should be noted that soil-forming processes operate on time scales from several decades (e.g. podzolization) over several hundred (e.g. eluviation) to millions of years (e.g. deep weathering in the humid tropics).

The effect of vegetation on SOC and SON depends on amount, biodegradability and placement of plant residues. In a soil profile, the upper soil horizon (A) is mostly influenced by inputs of organic materials. Except for sites with deep-rooting grasses, accumulations of organic matter in lower soil horizons are mostly due to pedogenetic processes which occur over longer time periods (Bridges &

Mukhopadhyay, 2003). For Brazilian soils, Volkhoff & Cerri (1988) showed that vegetation only influenced the organic matter in A horizons, whereas subsoil organic matter was not influenced by the current vegetation. Organic matter accumulated in subsoils is less accessible to decomposers. As a consequence, radiocarbon ages increase with soil depth (Pressanda et al., 1996).

Other factors that affect SOC and SOM levels are geomorphology, and anthropogenic disturbances (Chapter 6). Mean soil organic carbon and nitrogen contents estimated using the Holdridge (1947) classification scheme are presented in Table 1.18. Global SOC and SON inventories depend on the interactions between climate, vegetation and soil texture. Across the tropical, temperate and boreal forests, plant biomass and litter decrease continuously. Lowland tropical forests are not greatly different from temperate forest soils in terms of soil carbon and nitrogen content. The highest amounts of carbon and nitrogen are found in boreal forests. Most of this variation can be related to the effects of temperature on litter decomposition (Scharpenseel et al., 1992). Changes in litterfall quality and morphology among different climatic zones are also evident. High rates of SOM production in the tropics are accompanied by high rates of decomposition. Although plant production is lower at high elevations, larger SOM accumulations occur in montane tropical forests because decomposition is inhibited (Grubb, 1971). A similar pattern is true for mountains of the temperate zone (Hanawald & Whittaker, 1976). Low temperatures retard decomposition in Tundra and Boreal areas. Soils of these regions (including Histosols) worldwide contain the largest SOM accumulations. Because turnover in the surface layers is markedly slower at high latitudes, it is not surprising that litter on the soil surface increases from 1% of the total detritus in tropical forests to 13% in boreal forests (Schlesinger, 1984).

Table 1.18 Estimates of soil organic carbon (C) and nitrogen (N) pools for the depth interval 0–100 cm (corrected for coarse fragments), aggregated as per Holdridge life-zone (Compiled from Batjes, 1997)

Life zone	Pg C	Pg N	Area[a]
Tundra	158.8	8.3	10.16
Cold parklands	35.7	3.4	2.78
Forest tundra	180.2	11.7	8.72
Boreal forest	345.9	23.4	14.94
Cool desert	19.5	3.0	4.00
Steppe	69.1	9.2	7.35
Temperate forest	120.3	11.7	9.90
Hot desert	66.1	11.4	20.80
Chaparral	42.5	5.4	5.62
Warm temperate forest	30.4	3.1	3.20
Tropical semiarid	53.1	7.1	9.53
Tropical dry forest	118.3	13.7	14.85
Tropical seasonal forest	133.0	13.5	15.08
Tropical rain forest	89.0	8.0	8.46
All ecosystems	1462.0	133.0	135.39

[a] Area in 10^6 km^2, excluding land glaciers

Temperate steppe (prairie) soils such as those in Canada, the USA and the former USSR, contain very large amounts of SOM. After extensive plant and root growth in spring, decomposition of the produced organic material is inhibited due to drying of the soil in summer and autumn months, and subsequent freezing in the winter period. The SOM contents of tropical grasslands and dry forests are much lower, partly because the frequent fires may limit the amount of plant debris in these ecosystems. Grassland soils tend to have a higher proportion of SOC and SON in deeper soil layers than in comparable forested regions, probably related to deeper rooting in grassland ecosystems.

Hot, dry regions have low SOC and SON inventories because plant production is low. The information on the polar regions in Table 1.18 includes only soils with vegetation cover (Tundra), whereas non-vegetated Cryosols are not considered. The exact magnitude of SOC (and SON) for the whole area of Cryosols is still uncertain.

Besides the climatic control in any region, fine-textured soils tend to have higher SOC and SON inventories than coarse-textured soils, due to organomineral interactions between organic matter and clay particles which lead to chemical stabilization or physical protection of SOM (Chapter 3). Texture is therefore of primary importance for the distribution of SOM across landscapes (Parton et al., 1987). Local topographic effects tend to lead to higher SOM concentrations at locations that are lower in the landscape because of downslope movement of the organic material. Soil organic carbon and nitrogen inventories also tend to be higher along watercourses. In regions with frequent fires, SOM inventories are higher in local topographic depressions which are protected against fire.

The rate of incorporation of surface deposited residues depends on the activity of soil microflora and fauna and their ability to mix these materials into the A horizon. In well-drained soils containing available Ca^{++}, the activity of microorganisms, earthworms and other soil fauna is high, leading to diminution of residue particle size, ingestion, casting, bioturbation, and microbial transformation. Under such conditions, the Mull type of humus develops. In contrast, soils low in or free of Ca^{++} are low in microbial and faunal activity, and plant residues tend to accumulate on the soil surface, forming either Moder or Mor (Chapter 3).

Where climatic and soil factors are constant, the kind of placement of plant residues becomes important. Scharpenseel et al. (1992) compared the amounts of SOC contained in plant biomass and soils of temperate grassland and forest ecosystems. Despite a much smaller amount of plant biomass in the grassland, litter C inputs and SOC contents were roughly twice that of the forests. The effects of agricultural management practices, including residue return, on SOC and SON levels are discussed in Chapter 6.

Chapter 2
Carbon and Nitrogen Cycles in Terrestrial Ecosystems

Cycling of carbon (C) and nitrogen (N) are of paramount interest to biogeochemistry and to life on earth. The global C and N cycles have been extensively studied despite being complicated elemental cycles. In the atmosphere, C is mainly present as CO_2. Minor amounts of gaseous C occur as CH_4, CO and other higher molecular C-containing gases. The most abundant N form in the atmosphere is N_2. Other gaseous N forms comprise nitrous (N_2O) and nitric (NO) oxide, and NO_x. The least reactive N species is N_2.

Photosynthetic organisms capture sun energy, convert atmospheric CO_2 to organic compounds and account for the presence of atmospheric O_2. The presence of O_2 sets the redox potential for organic metabolism in terrestrial habitats. Oxidation and reduction processes enable organisms to transform other elements essential for life such as sulfur, phosphorus, and particularly N. Nitrogen is an integral part of enzymes in living tissues, that control the biochemical processes in which carbon is involved, such as photosynthesis and respiration. A large variety of microbes are able to transform N in compounds with different oxidation status and to use the released energy by the changes in redox potential. Nitrogen often limits the rate of net primary production. This realization has led to the massive use of synthetic N fertilizers during the 20th century, resulting in significant increase in crop yields.

Global climate change brought about by the enhanced greenhouse effect of fossil fuel CO_2 in the atmosphere during the last few decades has prompted much carbon-related research. The widespread use of fossil fuels and the increased use of N fertilizers also enhanced the release of reactive N compounds, which lead to important implications in various regions including photochemical smog, acid precipitation, stratospheric ozone, water pollution, eutrophication and change in biodiversity.

2.1 The Global Carbon Cycle

2.1.1 Biosphere-Atmosphere Exchange of Carbon Dioxide

Large amounts of carbon are found in the terrestrial ecosystems and there is a rapid exchange of C between the atmosphere, terrestrial biota and soils. The dominant fluxes of the global C cycle are those that link atmospheric CO_2 to primary biomass and to oceans (Fig. 2.1). Global net primary production (NPP) on land is estimated at 60 Pg CO_2-C year^{-1}, and corresponds with the annual respiration rate. On the basis of these fluxes and a CO_2-C pool in the atmosphere of about 765 Pg, the mean residence time of CO_2 in the atmosphere is about 5.3 year. The current atmospheric CO_2 concentration is 379 ppmv (Denman et al., 2007) with an annual increase of about 0.4% caused mainly by the release of CO_2 in fossil fuels. The annual uptake by the oceans (92.2 Pg C year^{-1}) is slightly greater than the return to the atmosphere (90.6 Pg C year^{-1}). This ocean sink of about 1.6 Pg C year^{-1} is relatively small compared with the overall fluxes.

Concentrations of atmospheric CO_2 vary according to a seasonal pattern. Continuous records of seasonal CO_2 oscillations have started in 1958 on Mauna Loa (elevation: 3,400 m), a volcanic mountain on the island of Hawaii (Pales & Keeling, 1965) and in the same year at the South Pole (Keeling et al., 1976). All continuous records show a CO_2 peak in late winter and a minimum in late summer. During the summer in both hemispheres there is a net fixation of C because photosynthesis exceeds respiration. During the rest of the year total respiration exceeds gross photosynthesis, i.e. there is a net CO_2 release. This seasonal pattern of CO_2 is mirrored by oscillations in atmospheric O_2 which has a larger pool and a longer residence time in atmosphere (Keeling et al., 1995, 1996). Oscillations in the CO_2 content of the atmosphere vary in amplitude with latitude and elevation (Bolin & Keeling, 1963). About two thirds of the vegetation occurs in regions with seasonal periods of growth, and one third in the moist tropics, where photosynthesis occurs throughout the whole year (Box, 1988). The oscillations in the CO_2 content are

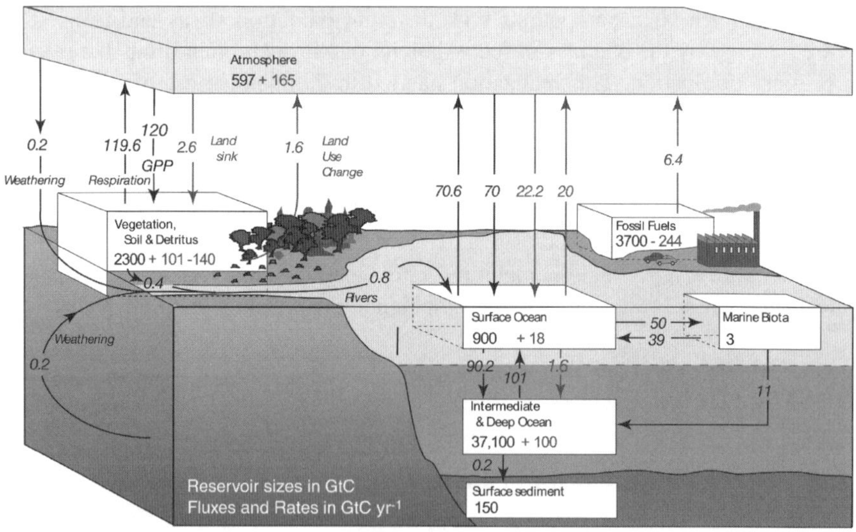

Fig. 2.1 The global carbon cycle for the 1990s showing the main annual fluxes in Pg C year^{-1}. The preindustrial fluxes are given in black and anthropogenic fluxes in grey (Denman et al., 2007, p. 515. Reproduced with kind permission from Cambridge University Press)

therefore most pronounced in the northern hemisphere where the major part of the continental area occurs. Smaller fluctuations of CO_2 in the atmosphere of the southern hemisphere are also due to exchange with oceans (Keeling et al., 1984).

Most of the C stored in the earth's biota and soils is associated with forests (Chapter 1). Until about 1960, the CO_2 release from land clearing may have exceeded the release from fossil fuel combustion (Houghton et al., 1983). In 1990, the estimated CO_2 release from deforestation in the tropics (1.6 Pg C $year^{-1}$) was partially compensated by CO_2 accumulated through the regrowth of forests in temperate regions (0.7 Pg C $year^{-1}$) (Dixon et al., 1994).

The possible effect of increased atmospheric CO_2 on photosynthesis were reviewed by Rosenberg (1981). Increasing CO_2 enhanced the assimilation rate of some plants in greenhouse experiments. The fertilization effect of the atmosphere with CO_2 may be limited since most plants are limited by other environmental factors such as light, water, nutrients and water. However, recent estimates show net CO_2 uptake in an undisturbed tropical rain forest, which may be an indication of the CO_2 fertilization effect (Grace et al., 1996). Elevated N deposition from the atmosphere may also stimulate plant growth and CO_2 uptake in some regions (Townsend et al., 1996). Despite widespread forest destruction in tropical regions, enhanced uptake of CO_2 in regions of undisturbed vegetation could add to the pool of carbon on land.

2.1.2 Biosphere-Atmosphere Exchange of Methane, Carbon Monoxide and Other C-Containing Gases

About 1% of the atmospheric C budget is maintained by methane (Ehhalt, 1974). Global background levels of CH_4 are estimated at 1.7 ppmv, corresponding to 3 Pg C (Blake & Rowland 1988). Sources of methane are natural and anthropogenic. Natural sources include fluxes from wetlands and enteric fermentation in wild animals. Sources from anthropogenic activities include rice paddies, livestock production, coal mines, leakage from gas fields, and biomass burning. The main sink for methane in the atmosphere is the hydroxyl radical (OH). Oxidation of CH_4 is an important source of atmospheric carbon monoxide (CO). The CO concentration in the atmosphere ranges from 0.05 to 0.2 ppmv, with considerable differences between the northern and southern hemispheres. The global storage of CO is estimated at 0.2 Pg (Holmen, 2000). Besides CO_2, CH_4 and CO other C-containing gases like terpenes, isoprenes and compounds of petrochemical origin are present in the atmosphere. The latter are estimated at 0.05 Pg in total (Holmen, 2000).

2.1.3 Ocean-Atmosphere Exchange of Carbon Dioxide

Carbon dioxide moves between the atmosphere and the ocean by molecular diffusion when there is a CO_2 gas pressure (pCO_2) gradient between the atmosphere and the sea. Based on several hundred thousand measurements of the surface water

pCO_2 in the global surface waters since the 1960s and using a spatial resolution of 4° latitude × 5° longitude, Takahashi et al. (2002) for 1995 calculated a net ocean uptake of ~1.5 Pg C $year^{-1}$. The main uncertainties in such calculations are the estimate of the surface-near gradients in wind speed and the coastal zone CO_2 fluxes (Sabine et al., 2004). Ocean models suggest that the interannual variability in the global ocean flux is roughly ±0.5 Pg C $year^{-1}$ (Greenblatt & Sarmiento, 2004). The gross annual exchanges of CO_2 with the oceans are much larger than the net CO_2 flux (Fig. 2.1). The total net flux from the atmosphere to the sea is 91.9 Pg C $year^{-1}$, that from the sea to the atmosphere is 90.6 Pg C $year^{-1}$, including an anthropogenic contribution of 21.9 and 20 Pg C $year^{-1}$, respectively (Sabine et al., 2004). The resulting ocean CO_2 sink of 1.3 Pg C $year^{-1}$ is consistent with the estimate of Takahashi et al. (2002).

During the last few years, significant advances have been made in separating the anthropogenic component from the background of ocean dissolved CO_2. A global inventory of anthropogenic CO_2 in the oceans yielded an amount of 112 ± 17 Pg CO_2-C with the highest CO_2 concentrations in the mid-latitudes and the lowest near the equator and in the high latitude of the southern ocean (Sabine et al., 2004). About 25% of the total inventory of anthropogenic CO_2 is in the North Atlantic, an important region for deepwater formation. Another ~56% of the total anthropogenic CO_2 is stored in the southern hemisphere. Deep ocean waters have not yet been exposed to CO_2 originating from human activities because of the slow ventilation of the deep waters. The ocean may, therefore, have a long-term potential to take up the major part of anthropogenic CO_2. Marine systems may take up CO_2 as long as the atmospheric CO_2 concentration will continue to increase. As this reaction is not irreversible, the surface-near waters would start to release part of the anthropogenic CO_2 to the atmosphere if the atmospheric concentration was to decrease in the future.

2.1.4 Transport of Carbon to Oceans via Fluvial Systems

The transport of carbon by rivers and groundwater from continents to oceans is an important flux in the global C cycle (Fig. 2.1). Rivers also move inorganic C products to oceans, such as carbonates which result from rock weathering. Estimates of carbon transfer to oceans are complicated by the diverse dynamics of carbon compounds and by the fact that data are scarce. Humans have increased significantly carbon and nutrient (particularly nitrogen) concentrations in rivers. Through intensified land use, up to 100 times more sediment and its associated carbon is transported to oceans compared to pre-cultivation conditions (Sabine et al., 2004). Moreover, the use of sewage sludge, slurry and other organic fertilizers drastically increased the concentrations in soluble organic C and nutrients in rivers. Not all of the organic C moves passively through rivers. Part of the organic C is mineralized during transport which leads to elevated concentrations of CO_2 in rivers, lakes and estuaries. Another proportion of the organic C may be transported as recalcitrant carbon or retained as particulate organic C. In total, 1.5 Pg C $year^{-1}$ may be exported to ocean through soil erosion. About 1 Pg C $year^{-1}$ is lost via river outgassing. The resulting global annual net transport rate of C by rivers to oceans accounts to ~1.1 Pg C $year^{-1}$, of which are ~0.4 Pg C $year^{-1}$ dissolved inorganic C, 0.2 Pg C $year^{-1}$ particulate inorganic C, 0.3 Pg C $year^{-1}$ dissolved organic C and 0.2 Pg C $year^{-1}$ particulate organic C (Chen, 2004). Groundwater discharge makes up a significant part (about 10%) of the surface flow to oceans. However, its contribution to C import to oceans is poorly known.

2.2 The Global Nitrogen Cycle

Nitrogen is capable of being transformed biochemically or chemically through a number of processes conceptually summarized as the nitrogen cycle. Nitrogen can be found in many different forms, including molecular N, organic molecules, minerals, gases and mobile ions. Most N transformations involve the oxidation or reduction by the N atom by both biological and chemical means. In the atmosphere, N exists mainly as N_2 which comprises 78% of the atmospheric gases. The transformation of N_2 to fixed nitrogen requires energy to break the N≡N triple bound. Physical and biological processes provide this energy. Nitrogen that is available to biota was originally derived from N_2 fixation by lightning or by symbiotic and asymbiotic microorganisms (Chapter 8). In the hydrosphere, N exists as soluble organic or inorganic nitrogen. The global pools of nitrogen have been shown in Table 1.2. The fluxes in the global nitrogen cycle during the 1990s are given in Table 2.1.

2.2.1 N_2 Fixation by Lightning

The global rate of N_2 fixation by lightning, which allows N_2 and O_2 to combine to NO under high temperature and pressure is estimated at 3–5.4 Tg N $year^{-1}$ (Table 2.1). Nitric oxide is oxidized to NO_2 and subsequently to HNO_3. In the latter form it is introduced into ecosystems by dry and wet deposition, particularly in the wet tropics. The estimated rate of N_2 fixation by lightning is small relative to biological N_2 fixation but may be important for ecosystems that lack other important N sources. Galloway et al. (2004) have given a generalized overview of different N fluxes which differ from those presented in other studies. The estimates of fluxes from other studies are discussed in Chapter 8.

2.2.2 Biological N_2 Fixation

Methodological differences and uncertainties in spatial coverage of important N_2-fixing species complicate global estimates of biological N_2 fixation (Cleveland et al., 1999). Moreover, for tropical regions of Africa, Asia and South America, data on

Table 2.1 Estimates of the fluxes in the global nitrogen cycle in the 1990s. The units are Tg N year^{-1}

	Schlesinger (1997)	Jaffe (2000)	Galloway et al. (2004)
1. *N_2 fixation in soils*			
(a) Lightning	3	3	5.4
(b) Natural biological		90–130	107
(c) Cultivation-induced	40	40	31.5
Total	140	133–173	140
2. *Industrial N_2 fixation*			100
3. *N fertilizer consumption*	80		86
4. *N deposition*			
(a) To continents		54	63.5
(b) To oceans		24	39.0
Total		78	103.5
5. *NO_x-N emission*			
(a) Energy production	20	(a) Plus (b)	27.2
(b) Food production		31	9
(c) Natural		5	9.7
Total		36	45
6. *N_2O-N emission*			
(a) Natural		6	6.6
(b) Anthropogenic		3	3.2
Total		9	9.8
7. *Denitrification* (NO_x, N_2O plus N_2)			
(a) Natural	70	80–180	
(b) Cultivation-induced	90	40–100	
Total	160	120–280	115
8. *NH_3-N emission*			
(a) Natural		8	11
(b) Anthropogenic		47	47.3
Total		55	58.3
9. *N transport to groundwater*	11		
10. *River runoff to oceans*			
(a) Inorganic N	36	34	
(b) Organic N		17	
Total		51	48
11. *Oceanic N_2 fixation*	15	10–200	86.5–156
12. *N precipitation over oceans*	30	31	33.4
13. *Oceanic denitrification*	110	25–180	147–454

biological N_2 fixation are rare. However, the high abundance of N_2-fixing legumes in tropical forests suggests that symbiontic N_2 fixation is high relative to temperate regions (Crews, 1999). According to Cleveland et al. (1999) tropical rain forest regions represent about 24% of the global annual biological N_2 fixation. Galloway et al. (2004) estimated the biological N_2 fixation rate in the natural world at ~120 Tg N year^{-1} for 1860 and ~107 Tg N year^{-1} for the presence.

Before the Haber-Bosch process was invented, the only N_2 fixed due to human activities was by the cultivation of legumes. Galloway & Cowling (2002) estimated that in 1900, ~15 Tg N year^{-1} was fixed by cultivation-induced N_2 fixation. The present global total biological N_2 fixation from cultivation may be ~33 Tg N year^{-1} (Smil, 1999). Total global biological N_2 fixation on land is thus about ~140 Tg N year^{-1} globally (Burns & Hardy, 1975; Schlesinger, 1997; Galloway et al., 2004) which is equivalent to a mean fixation of ~10 kg N ha^{-1} year^{-1} on the land surface.

2.2.3 Ammonia Production with the Haber-Bosch Process

In the 1990s, 100 Tg N year^{-1} was produced with the Haber-Bosch process (Table 2.1) for food production and other industrial activities (Kramer, 1999; Galloway et al., 2004), 14% of which was either emitted to the environment during processing or used in manufacture of synthetic fibers, refrigerants, explosives, plastics, nitroparaffins, etc. (Smil, 1999).

Most NH_3 (86%) was used to produce N fertilizers. The world N consumption is increasing rapidly. The global N fertilizer demand is presently projected to expand at an annual rate of 1.7% year^{-1} (Heffer, 2004). In 2003, Asia and America were the main producing regions (Table 2.2). The bulk of their production is for domestic consumption. The other main producing regions are dedicated to export. Commonly used N fertilizers produced from NH_3 gas synthesized by the Haber-Bosch process are listed in Table 2.3.

Ammonia can be used directly as pressurized gas, or in water solution ($NH_3 \times H_2O$), combined with chloride to form NH_4Cl, combined with NO_3 to form NH_4NO_3, reacted with sulfuric acid to form $(NH_4)_2SO_4$, combined with various types of phosphoric acid to form ammonium phosphates, reacted with CO_2 to form

Table 2.2 Estimates of industrial N fixation in different regions for 2003 (Prud'homme, 2005)

Region	Tg N	Global share (%)
East Asia	30.4	28
South Asia	19.9	18
Eastern Europe and Central Asia	15.4	14
North America	12.4	12
West Europe	9.7	9
Rest of Asia	8.0	7
Latin America	6.5	6
Central Europe	4.4	4
Africa	1.1	1
Oceania	0.9	1
World	108.8	100

Table 2.3 Chemical composition and N content of major synthetic N fertilizers (Adapted from Pierzynski et al., 2000)

N fertilizer	Chemical composition	N content (%)
Anhydrous ammonia	NH_3	82
Hydrous ammonia	NH_3 x H_2O	20–25
Ammonium chloride	NH_4Cl	25
Ammonium nitrate	NH_4NO_3	33
Ammonium sulfate	$(NH_4)_2SO_4$	21
Ammonium phosphate	$NH_4H_2PO_4$	11
Diammonium phosphate	$(NH_4)_2HPO_4$	18–21
Ammonium polyphosphates	$(NH_4)_3HP_2O_7$	10–11
Urea	$CO(NH_2)_2$	45
Urea-ammonium-nitrate solutions	30–35% urea; 40–43% NH_4NO_3	28–32
Ureaform	Urea-formaldehyde	38
Calcium nitrate	$Ca(NO_3)_2$	15
Potassium nitrate	KNO_3	13
Sodium nitrate	$NaNO_3$	16

Table 2.4 Nitrogen contents of organic fertilizers (Windt & Wollenweber, 2005)

Organic fertilizer	Unit	Dry matter content (%)	N (kg)
Poultry manure	10 t fresh matter	55.0	280
Turkey manure	10 t fresh matter	55.0	230
Cattle slurry	$10\,m^3$	10.0	47
Pig slurry	$10\,m^3$	6.0	39
Poultry slurry	$10\,m^3$	12.0	118
Cattle dung water	$10\,m^3$	2.0	30
Pig dung water	$10\,m^3$	2.0	50
Sewage sludge	1 t	<10.0	33
Bio compost	1 t	60.0	8.5

urea. Various nitrate-N fertilizers are also available. A variety of N solutions and slow-release solid fertilizers that are coated with resins are also produced. Nitrogen contents of some important organic fertilizers are given in Table 2.4.

2.2.4 Atmospheric N Depositions

Since the 19th century, N deposition rates increased over large world regions. For 1860, total NO_x deposition was estimated at ~12.8 Tg N $year^{-1}$ with ~6.6 Tg N $year^{-1}$ to continents and ~6.2 Tg N $year^{-1}$ to oceans (Galloway et al., 2004). Deposition of total NH_y (including non-local sources plus rapid deposition of local NH_3 emission) in 1860 may have amounted to ~10.8 Tg N $year^{-1}$ to continents and ~7.9 Tg N $year^{-1}$ to oceans. Deposition of NH_y from non-local sources alone in 1860 was ~4.8 Tg N $year^{-1}$ to continents and ~2.3 Tg N $year^{-1}$ to oceans. In the 1990s, total NO_x deposition was ~45 Tg N $year^{-1}$ with about half of it to continents

(~25 Tg N $year^{-1}$) and the other half (~20 Tg N $year^{-1}$) to oceans (Table 2.1). Total NH_y deposition in the 1990s was estimated at ~56 with about ~38 Tg N $year^{-1}$ to continents and ~18 Tg N $year^{-1}$ to oceans, demonstrating that NH_y-N is deposited after short-distance transport and NO_x-N is deposited after longer-distance transport. Galloway et al. (2004) generated a global map of atmospheric N depositions for 1860 and the early 1990s, including a projection for 2050, using a 5° by 3.75° grid, subdivided into 50 × 50 km sub-grids. N deposition in 1860 was particularly pronounced near populated areas with rates up to or slightly more than 1 kg N ha^{-1} $year^{-1}$ (e.g. in China), whereas in remote regions, N deposition was <0.1 kg N ha^{-1} $year^{-1}$.

In the 1990s, large regions of Europe, North America and Asia received N deposition rates >1 kg N ha^{-1} $year^{-1}$ (Galloway et al., 2004). The estimates by Galloway and coworkers are most important for large-region observations and for comparison of long-term developments. However, they hardly reflect local situations. Many areas in Europe, North America and Asia partly receive extremely high N depositions (at least one order of magnitude higher than the above large-scale estimates) with an increasing proportion of nitrogen originating from agriculture (especially NH_y-N from livestock production, storage and application of manures, increasing mineral N fertilizer (particularly urea) application). The NH_y-N emitted from these areas is deposited in the neighborhood (commonly some 100 m to several kilometers) of its source and thus remains in terrestrial ecosystems (Table 2.1). For example, in intensively cultivated areas of the Guanzhong Plain of the Chinese Loess Plateau (winter wheat-summer maize double cropping system with manure application and mineral N fertilization (urea) of 260–335 kg N ha^{-1} $year^{-1}$; Richter & Roelcke, 2000), atmospheric N deposition (only wet deposition) was estimated at 22 kg N ha^{-1} $year^{-1}$ (Li et al., 1993). The total N deposition, including dry deposition, may be much higher. Similar rates of wet N deposition (up to 20 kg N ha^{-1} $year^{-1}$) were observed at several locations in Jiangsu Province and in the Taihu Region (winter wheat-summer rice double-cropping system with total mineral N fertilization (urea) of 465–635 kg N ha^{-1} $year^{-1}$; Richter & Roelcke, 2000) of Eastern China (Zhu & Wen, 1992). In Germany, the mean N deposition (wet and dry) may amount to 10–20 kg N ha^{-1} $year^{-1}$ of NO_x-N and >20 kg of NH_y-N N ha^{-1} $year^{-1}$ (Meesenburg et al., 1994). In parts of North-West Germany where the livestock density is one of the highest in the world (>3.0 life weight units of 500 kg ha^{-1}), the N surplus may exceed 300 kg N ha^{-1} $year^{-1}$ in numerous farms (Barunke, 2002). Similar conditions can be found in parts of the Netherlands and Denmark (Erisman et al., 2001; Mosier et al., 2004). In and around areas dominated by intensive livestock production, total N deposition is mainly determined by NH_y-N and in some cases may exceed 100 kg N ha^{-1} $year^{-1}$ especially in forest areas (Heinsdorf & Krauss, 1991). It can be expected that such "hot spots" in many parts of the world will increase in number and size due to further intensification of food production, including meat products.

Atmospheric organic N compounds like organic nitrates, aerosol amines and urea, and particulate atmospheric organic nitrogen (e.g. bacteria and dust) may be a significant component of the terrestrial N cycle. However, estimates of these fluxes still reflect substantial unresolved uncertainties. For the present time Neff et al. (2002) estimate a range of 10–50 (on average 30) Tg $year^{-1}$ of organic N deposited and emitted on a global basis.

2.2.5 Emissions of NO_x, N_2O, N_2, NH_3 and Organic N

Emissions of NO_x and NH_3 result from natural processes, food and energy production. While a detailed description of nitrogen emissions in different ecosystems and regions of the world is presented in Chapter 8, a generalized overview of the fluxes is presented here. Natural emissions of NO_x occur from processes in the soil, biomass burning and stratospheric injection. Total NO_x emissions in 1860 were ~13.1 of which ~2.0 Tg N $year^{-1}$ resulted from food production, ~0.6 Tg N $year^{-1}$ resulted from energy production, and ~10.5 Tg N $year^{-1}$ were natural (Galloway et al., 2004). In the 1990s, NO_x emissions totaled 35–45 Tg N $year^{-1}$, with 5–10 Tg N $year^{-1}$ from food production, ~25 Tg N $year^{-1}$ from energy production and 5–10 Tg N $year^{-1}$ from natural (microbial) sources (Table 2.1). Food production NO_x emissions were composed of combustion of agricultural waste, forests and savanna grass, and emissions from fertilized soils.

Terrestrial ecosystems in 1860 emitted ~6.6 Tg N_2O-N $year^{-1}$. Anthropogenic N_2O emissions from agricultural soils, animal waste management systems, biomass burning, biofuel combustion, energy/transport sources and industrial processes summed up to ~1.4 Tg N $year^{-1}$ (Kroeze et al., 1999). N_2O-N emissions in the 1990s totaled ~10 Tg N $year^{-1}$ (Table 2.1). While the anthropogenic N_2O-N emissions increased from <1 to ~3 Tg N $year^{-1}$ from 1860 until the 1990s (Bouwman et al., 1995; Kroeze et al., 1999; Galloway et al., 2004), the N_2O-N emissions from natural terrestrial ecosystems (~7 Tg N $year^{-1}$) remained almost constant in the same time period.

N losses by denitrification in managed and unmanaged ecosystems are the biggest unknown in the terrestrial N cycle. Total N losses by denitrification (NO_x, N_2O plus N_2) are highly variable in space and time and depend on a number of factors (Chapter 8). Estimates of global denitrification range from 13 to 280 Tg N $year^{-1}$ (Bowden, 1986; Table 2.1). Most of the loss occurs as N_2, but the small fraction that is lost as N_2O during nitrification and denitrification contributes significantly to the global budget of this gas. Production of N_2 via denitrification can be an important loss of nitrogen from agroecosystems (Mosier et al., 2002). If N_2 fixation and denitrification prior to large-scale anthropogenic disturbance were once in a balance, then a terrestrial denitrification rate of ~130 Tg N $year^{-1}$ was most likely in natural terrestrial ecosystems prior to large-scale anthropogenic disturbance. The current rise in atmospheric N_2O can be used to estimate the overall increase in global denitrification. Under the assumption that the mean N_2 to N_2O ratio is 22 and that the recent increase of N_2O concentration in the atmosphere of ~4 Tg N $year^{-1}$ derives from increased denitrification, the cultivation-induced denitrification may amount to as much as 90 Tg N $year^{-1}$ (Schlesinger, 1997). On the basis of these calculations, the present overall denitrification rate may sum up to 220 (130 + 90) Tg N $year^{-1}$ (Table 2.1).

Ammonia emissions to the atmosphere from energy production in 1860 may have amounted to ~0.7 Tg N $year^{-1}$ from the combustion of biofuels and to ~6.6 Tg N $year^{-1}$ from food production and human nutrition. Of the latter, ~0.6 Tg N $year^{-1}$ resulted from combustion of agricultural wastes, ~0.2 Tg N $year^{-1}$ from forests, ~0.2 Tg N $year^{-1}$ from savannas, ~5.3 Tg N $year^{-1}$ from domestic animal waste, ~0.1 from human sewage, and ~0.2 from crops (Galloway et al., 2004). About 7.6 Tg NH_3-N $year^{-1}$ were emitted from natural sources (~1.6 Tg N $year^{-1}$ from natural fires and ~6.0 Tg N $year^{-1}$ from natural soils and vegetation). Thus, globally emitted NH_3 emissions from terrestrial ecosystems in 1860 may have summed up to ~15 Tg N $year^{-1}$. The corresponding data for the 1990s were ~55 Tg N $year^{-1}$ for global total NH_3-N emissions, of which ~47 originated from anthropogenic sources (~44 Tg N $year^{-1}$ from food production and human nutrition; ~3 Tg N $year^{-1}$ from the energy/transport sectors), and >8 Tg N $year^{-1}$ from natural sources (Table 2.1).

Global anthropogenic emissions of NO_x and NH_3 in the 1990s were ~40 Tg N $year^{-1}$ and ~47 Tg N $year^{-1}$, respectively. More than two thirds of the NO_x emissions were a consequence of energy production and about 95% of the NH_3 emissions were from food production and human nutrition, with about half being from animal waste. For emissions of total N, about 70% are a consequence of food production and human nutrition (Galloway et al., 2004).

2.2.6 *Leaching of Nitrogen to Groundwater*

During the last decades, loss of nitrate to groundwater has become a significant fate of nitrogen in agroecosystems. Leaching of nitrate can lead to significant, potentially harmful concentrations of NO_3^- in groundwater. Eutrophication can occur when NO_3^- accumulates in lakes, ponds or estuaries. Global estimates of nitrate leaching are complicated by numerous factors. Schlesinger (1997), on the basis of the global annual flux of groundwater of 11,000 km^3 $year^{-1}$, estimated the global sink in ground water at ~11 Tg N $year^{-1}$ (Table 2.1).

2.2.7 *Transport of Nitrogen to Oceans by Rivers*

In recent decades, N concentrations and loads have increased drastically in numerous rivers of the world (Seitzinger & Kroeze, 1998), particularly of Europe (Tsirkunov et al., 1992; Isermann & Isermann, 1997), North America (Gilliom et al., 1995) and Asia (Duan et al., 2000; Galloway, 2000). Turner & Rabalais (1991) reported that the NO_3^- concentrations in the Mississippi River since 1965 have more than doubled due to various human activities. The N transport by rivers is an important factor of N loss from continents and eutrophication of oceans (Smil,

1997). In 1860, N transport to coastal systems by rivers amounted to ~27 Tg N $year^{-1}$ (Galloway et al., 2004).

For the 1990s, it was estimated at ~50 Tg N $year^{-1}$ (Table 2.1), including all important N fractions such as dissolved N, particulate N and organic N forms. The transport of inorganic nitrogen was estimated at ~35 Tg N $year^{-1}$, that of organic N at ~15 Tg N $year^{-1}$. While the flux in rivers (~50 Tg N $year^{-1}$) is a rather small component of the terrestrial N cycle, it contributes a major part (60%) of the total nitrogen delivered annually to oceans (Table 2.1). On a global basis, the river-ocean continuum is a permanent nitrogen sink. However, the role of denitrification in stream systems is not completely understood.

2.2.8 *Ocean N Budgets*

2.2.8.1 Nitrogen Sinks

The transport of nitrogen by rivers to oceans is small compared to the main N fluxes on continents but it contributes a significant proportion of the nitrogen imported into marine ecosystems. The riverflux of nitrogen (~50 Tg N $year^{-1}$) is most important in coastal seas and estuaries, while N inputs from the atmosphere occur mostly in the open oceans. The atmospheric N deposition (NO_x-N plus NH_y-N) currently may be around 30 Tg N $year^{-1}$. For 1860, Galloway et al. (2004) estimated the atmospheric N deposition at 8.5 Tg N $year^{-1}$. The total atmospheric N deposition over ocean regions may be somewhat greater than 30 Tg N $year^{-1}$ because most workers ignored inputs of organic N (Cornell et al., 1995) which means that the relative importance of atmospheric N deposition has increased from ~24% in 1860 to ~40% in the 1990s.

Similar to terrestrial ecosystems, biological N_2 fixation is the most important N flux from the atmosphere into oceans. Because of the lack of coherent N_2 fixation data, extrapolation of local measurements plays an important role for calculations of marine nitrogen budgets. A compilation of numerous measurements conducted in different ocean habitats of pelagic and shelf zones yielded a range of total oceanic biological N_2 fixation from 86.5 to 156 (mean: 122) Tg N $year^{-1}$ (Galloway et al., 2004).

2.2.8.2 Nitrogen Sources

The major part of the N input to oceans is returned to the atmosphere by denitrification. The largest pool of inorganic N (mostly derived from the decomposition of organic matter) is concentrated in the deep ocean. Estimates of denitrification in deep sediments vary greatly, ranging from 8.4 to 17 (Galloway et al., 2004) to 130 Tg N $year^{-1}$ (Middleburg et al., 1996).

Shelf sediments are globally one of the most important sites for denitrification (Christensen et al., 1987; Devol, 1991). Most estimates of denitrification in shelf areas are still uncertain because data used for extrapolations in many cases originate from sediment incubation experiments or geochemical modeling of sediment nitrate distributions which are at relatively small scales. In recent estimates by Galloway et al. (2004) and Codispoti et al. (2001), the shelf contribution was up to 287 and 300 Tg N year^{-1}, respectively.

Pelagic (open ocean) denitrification is thought to play an important role in anoxic plumes of the tropical Pacific ocean and the Arabian Sea which develop as a consequence of vertical flux of organic matter and poor ventilation. Circulation of deep water having low O_2 and high NO_3^- contents also contributes to pelagic denitrification. Galloway et al. (2004) estimated total global pelagic denitrification at 81–150 Tg N year^{-1}.

2.2.9 Summary of the Major Global N Fluxes

Human activity has drastically changed cycling of nitrogen during the last centuries. Of the ~240 Tg N fixed annually in the 1990s (N_2 fixation in soils: ~140 Tg N year^{-1}; industrial N_2 fixation: ~100 Tg N year^{-1}), 55% can be drawn back to anthropogenic processes. More than 80 Tg N was emitted annually to the atmosphere. Cultivation-induced losses of N_2 via denitrification are not considered in this calculation because there is a large uncertainty about the total N_2 production rate. Of the total N emissions, ~59 Tg N year^{-1} was deposited back to continents and ~33 Tg N year^{-1} to oceans. An additional ~50 Tg N year^{-1} was transported from continents to oceans by river runoff. The total amount of nitrogen that was lost from the terrestrial to the marine environment may be ~83 Tg N year^{-1}.

For marine ecosystems, biological N_2 fixation flux is still unknown and may be significantly underestimated for big ocean areas by current methods (Brandes & Devol, 2002). Assuming the mean rates for marine biological N_2 fixation (~122 Tg N year^{-1}), precipitation N flux to oceans (~35 Tg N year^{-1}) and ocean denitrification (~319 Tg N year^{-1}) given in Table 2.1, there is an apparent excess of sources over sinks of the marine nitrogen cycle of ~160 Tg N year^{-1}. If this imbalance is real, it may have important implications to the global N cycle. There still exist large uncertainties concerning estimates of global N fluxes in both the terrestrial and the marine environment. A major part of the uncertainties may be drawn back to the limitations by the methods currently used for regional and global N flux estimates. Another uncertainty may base on the non-inclusion of management and land-use changes into global N-flux estimates which may induce important sinks or sources for reactive nitrogen (Chapter 6).

The world population is projected to increase from 5.8 billion in 2000 to ~9 billion in 2050. There will likely be increases in fossil fuel combustion and in fertilizer N application in many parts of the world. Depending on the scenario used, IPCC (2001a) give a range of 38.8–94.9 Tg N year^{-1}. The increase in world population by ~50% will cause an increase in food production, and may therefore require the conversion of natural ecosystems to agricultural land. As a consequence, biological

N_2 fixation will decrease. Galloway et al. (2004) scaled the reduction in natural biological N_2 fixation inversely to the increase in population between 1990 and 2050, resulting in an estimate of ~98 Tg N $year^{-1}$ for 2050. The present global total biological N_2 fixation from cultivation may be ~30 Tg N $year^{-1}$ (Smil, 1999). Estimates for the future are complicated by several factors. For example, much of the additional land used for cultivation-induced N_2 fixation will be in the tropics where larger per area losses of natural biological N_2 fixation makes uncertain the fate of the replaced land. Another fact is that many areas in the tropics are poor in soil quality which sometimes is unfavorable for the process N_2 fixation. Considering these and other uncertainties, Galloway et al. (2004) estimate that N_2 fixation from cultivation may be within a range of 45–55 Tg N $year^{-1}$ in 2050.

Increases in NO_x (particularly from energy and transport sectors and the application of mineral fertilizers) and NH_3 (particularly from livestock production, urea application on alkaline soils and the storage and application of manures) emissions will drastically change the magnitude of N deposition. Compared to the 1990s, the total area receiving >1 kg N ha^{-1} $year^{-1}$ is expected to increase significantly, and to expand from Asia to other regions (e.g. Central America and Africa) of the world. For South and East Asia it is expected that the large-scale N deposition will exceed 5 kg N ha^{-1} $year^{-1}$. For West Europe and North America, increases in large-scale annual N deposition are not expected, while for eastern Europe (especially the new members of the EU), significant increases in N deposition are most probable.

2.3 Carbon and Nitrogen Cycling in Soils

Carbon and nitrogen dynamics differ greatly between upland and wetland soils. Upland soils constitute the vast majority of agricultural and forest areas of the world. Soils are conditioned by climate and vegetation. The location of a soil relative to elevation and latitude have a strong influence on a soil's carbon and nitrogen storage and fluxes. Compared to soils under (near-)natural vegetation, C:N ratios of the soil organic matter (SOM) is lower in cultivated upland soils. Losses of soil C due to land use result from accelerated decomposition and soil erosion. Due to erosion, upland soils often exhibit pronounced erosive C losses in upper positions and C accumulation in lower parts of landscapes.

Wetland ecosystems, also called semiterrestrial ecosystems, are located between terrestrial and aquatic ecosystems and often possess the properties of both systems (Schlesinger, 1997). They occupy 2.8×10^6 km^2 or 2.2% of the earth's surface (Reddy et al., 2000). Natural wetlands (swamps, marshes, peat lands and bogs) are important habitats. Under permanently saturated soil conditions, vertical layering of different metabolic activities can be present. By their low-lying position in the landscape, wetland ecosystems receive water and nutrient inputs from adjacent uplands. Wetlands are mediated by soils with low redox potential. Nutrients received from adjacent landscapes are often transformed during their passage through wetlands. In many swamps, net primary production (NPP) is directly related to nutrient inputs and NPP is highest in wetlands that are seasonally exposed to oxygen. Bogs that receive little nutrient input usually have very low productivity. Because of their low decomposition rate, wetlands are commonly seen as net sinks for organic C and N. Thick and dark surface layers consisting of more or less decomposed plant remains are therefore common for wetland soils. If the organic layers are thick enough such as they comprise the greatest portion of the soil they are referred to as Histosols. In the following sections, important aspects of C and N cycling in uplands and wetlands are discussed.

2.3.1 Carbon and Nitrogen Cycling in Upland Soils

The presence of oxygen plays a key role in C and N dynamics of upland ecosystems. Compared to wetland soils, organic materials undergo higher decomposition rates. Drying and wetting cycles temporarily cause increases and decreases in O_2 availability which influence directions of processes and turnover rates. The exposure influences the temperature and water regime of an upland soil. The composition of the parent material and the position in a landscape affect infiltration and drainage of water, and the movement of nutrients in an upland soil. The upper position in the landscape and a positive water balance promote drainage of upland soils with the consequence that nutrients, dissolved organic matter, and particles such as clay minerals are moved downward. The exposure can affect vegetation, soil temperature, water retention regimes and thus may influence soil carbon and nitrogen storage and mineralization.

2.3.1.1 Soil Microbial Biomass in Upland Soils

In upland soils, microbial biomass may account for 0.3–5% of the organic C and 0.5–15% of the total N (Anderson & Domsch, 1980). Soil microbial biomass and its activity depends highly on the presence of available carbon substrates. Along vertical soil profiles, it shows a steep decline (Lavahun et al., 1996). In upper soil horizons, it increases with soil pH (Weigand et al., 1995). In sandy soils higher metabolic coefficients are found as compared to clay soils (Winter & Beese, 1995). Soil microbial biomass attached to the clay fraction is most susceptible to drying and wetting processes (van Gestel et al., 1996).

In upland systems, fungal biomass tends to predominate over bacterial biomass, with ratios of 3:1 for C, 2:3 for N, and 6:7 for P, resulting in an average C:N:P ratio of 100:15:12. This ratio can change with season and cropping system (Patra et al., 1995). Temporary increases in biomass induced by single organic inputs are considered as relatively short-lived (Nannipieri et al., 1983). The mean turnover time for microbial biomass C is 0.13–0.24 year in soils of the humid tropics and 1.4–2.5 year in soils of temperate regions (Houot et al., 1991), which demonstrates the important role of temperature in controlling soil carbon dynamics. Growing plant roots, a source of readily decomposable photosynthates for microbial biomass, have a priming effect on the breakdown of organic residues added to soil (Lynch, 1990). Carbon to N ratios of soil microbes of 4–12 (with lower ratios for bacteria and higher ones for fungi) exceed by far the mean C:N ratios of crop plants of 60–80. Microbial biomass increases with increasing available C content, e.g. due to crop residue incorporation into the soil. Due to the gradient in C:N ratios, the microbial demand for N during growth at the expense of decomposing crop residues is proportionally higher than the N demand of crops. Nitrogen assimilation by microbes is likely to increase the microbial N pool, which is released during subsequent mineralization of soil microbes (Kelly & Stevenson, 1987). Long-term management influences the SOC pool and the microbial biomass. The latter increases with increasing pool size of SOC.

For regions with cold seasons (arctic, alpine and cold-temperate climates), until recently it was assumed that microbes were effectively frozen into dormancy for the winter. Most of the studies on microbial biomass were, therefore, concentrated on the vegetation period. Recent studies on microbial processes in cold soils with a snow cover show that there are distinct shifts in microbial populations and processes that occur between winter and summer. As long as there is unfrozen water, microorganisms can generally remain physiologically active (Mikan et al., 2002). During the winter, the soil microbial community is dominated by fungi (Schadt et al., 2003), while bacteria appear to dominate during the growing season (Lipson et al., 2002). As most of the winter community dies off in the spring, it releases N-rich compounds that are subsequently mineralized and may provide most of the annual plant N needs (Jaeger et al., 1999; Lipson et al., 2000). Soil respiration under cold-season conditions is extremely sensitive to temperature, with Q_{10} values (relative change in rate for a 10°C temperature change) for organic soils of between 60 and 200 below 0°C compared to a maximum of 9 above 0°C (Mikan et al., 2002).

2.3.1.2 Soil Organic Carbon and Nitrogen in Upland Soils

The C and N cycles are closely linked and cannot be effectively studied separately (McGill & Cole, 1981). Most of the soil C and N occurs as SOM. Organic C and N in soils are a continuous product of microbial processes. Soil organic matter contributes most actively to gaseous exchanges of C and N with the atmosphere and to nutrient cycling in the soil-plant system. Land use by humans led to dramatically increased exchanges of C and N between the land and atmosphere (see Chapter 6). Until the 1940s, changes in C and N emissions from terrestrial ecosystems were dominated by expanding agriculture in the middle and high latitudes. Since the 1950s, the changes in land use are of particular importance for tropical regions where the human population is increasing most rapidly. The mean residence time of SOM is about 22 year (Post et al., 1992). The turnover time of SOM increases with soil depth, ranging from several years for litter to 15–40 year in the upper 10 cm and over 100 year below a depth of 25 cm (Harrison et al., 1990).

Cycling of Carbon in Upland Soils

Mean residence times for SOC of 2,000–5,000 years have been reported for soils rich in allophane (Wada & Aomine, 1975), reflecting the importance of mineralogy, active aluminum and iron hydrous oxides in preserving organic materials in the soil. Similarly, Ferralsols derived from basic or ultrabasic parent materials have consistently higher organic matter contents than those derived from acid materials. Important processes of the upland soil C cycle are given in Fig. 2.2.

Soil organic matter can be divided into various physical, chemical or microbial pools (Chapter 3). Numerous studies have shown that the fractionation of soil separates gives significantly different SOM values for the sand, silt and clay fractions, with the fine-textured fractions containing the oldest and highest amounts of SOM (Batjes & Sombroek, 1997). Carbon dating based on naturally occurring ^{14}C and use of the ^{13}C signal supplied by photosynthetic discrimination by C_3 and C_4 plants can be effectively used for determination of root versus soil respiration, plant residue decomposition rates, the source of mineralized CO_2, and the turnover rate of different SOM fractions (e.g. Martin et al., 1990; Paul et al., 1995). The distribution of SOM in the soil can be considered as a continuum of numerous pools in which the quality is equated with substrate availability to microorganisms. The labile pool averages about a quarter to a third of the total SOM in temperate region soils, but is probably smaller in tropical soils (Duxbury et al., 1989). Labile constituents of SOM decompose within a few weeks or months and are suggested to include comminuted

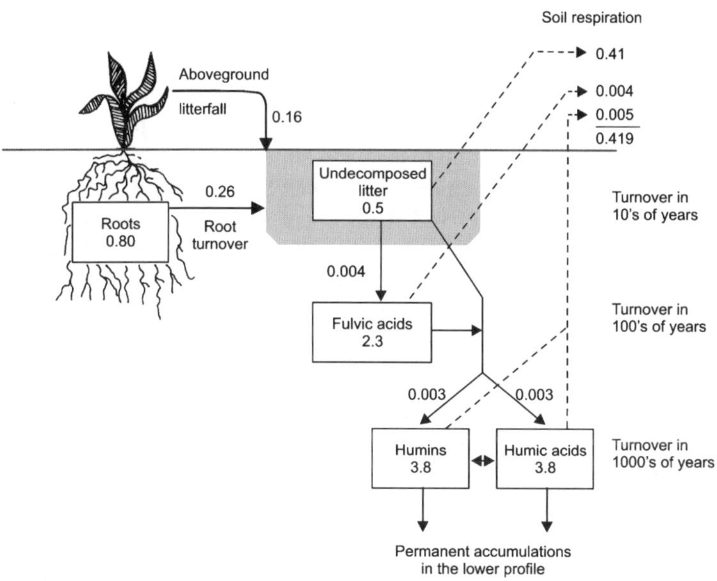

Fig. 2.2 The carbon cycle in upland soils showing the C pools (kg C m^{-2}) and the annual transfers (kg C m^{-2} $year^{-1}$) (Holmen, 2000, p. 296. Reproduced with kind permission from Elsevier)

plant litter, macroorganic matter, non-humic substances not bound to mineral constituents, water-soluble forms, macroorganisms (fauna) and microbial biomass. The stable components of SOM may persist for hundreds to thousands of years and are largely represented by humic substances and other organic macromolecules which are intrinsically resistant to microbial attack or physically protected by association with mineral surfaces or trapped within mineral aggregates or clay layers.

The relative pool sizes, i.e. the amount and composition of the various SOC and SON constituents, are mainly controlled by residue inputs and climate gradients, edaphic factors, soil mineral types and contents, and other physical and chemical properties of the soil that affect the activity of decomposer organisms (Theng et al., 1989). The chemical composition and physical state of the residue, the site of placement in soil, and tillage operations are additional factors which affect SOM stability. Animal manures are much more effective than plant residues in contributing to the labile SOM pools, because they have already been subjected to the rapid initial decomposition. The size of the labile SOM pool derived from plant residue additions will be smaller and will respond quicker to changes in residue inputs in tropical than in temperate soils (Sauerbeck & Gonzales, 1977). With constant, long-term annual residue inputs of 1.0 Mg C ha^{-1}, the labile SOM pool is estimated to contain a maximum of 2.5 Mg C ha^{-1}, 250 kg N ha^{-1}, and 25 kg P ha^{-1}, if the C/N and C/P ratios are 10 and 100, respectively.

Each year, enormous quantities of organic residues are deposited into terrestrial environments. The relative contributions of aboveground and belowground parts of plants to SOM formation depend on the respective ecosystem. Except for grassland and tundra systems, where SOM is built up particularly from root decay, most other systems receive most of their organic matter from plant residues produced above and below ground. Plant litter accounts for between 1.0 and 15.3 Mg of organic material ha^{-1} $year^{-1}$ (Williams & Gray, 1974). The root biomass in the top 30 cm of soil is estimated in the order of 4.4–15.8 Mg ha^{-1} dry matter. Animals and their excreta and dead microbial cells provide a significant residue as well. In order to maintain a steady state condition between assimilation and mineralization a continuous degradation of SOM is needed. Upland agricultural systems were usually derived from grassland or forested ecosystems. For SOC comparisons, their origin and location (latitude and altitude) are important. Soil carbon also cycles between organic and inorganic pools (carbonates, bicarbonates, carbonic acid and CO_2). Cycling times range between a few minutes to hours for CO_2 and plants or microorganisms, to a few weeks for plant material and light SOC, to several months for more resistant SOC, to thousands of years for stable SOC and to millions of years for carbonate rocks. The chemical composition of plant residues varies widely between plant species, and some typical ranges are shown in Table 2.5.

The degradability and residence time of common organic substrates are given in Table 2.6. The mean residence time of organic substrates varies from <1 day (glucose) to ~1,000 years or more (stable SOM). The organic matter in roots underestimates the addition of organic matter by a crop to the soil as the figures exclude the loss of parts of the roots during the growing season and root exudates.

Table 2.5 Gross chemical composition of plant residues (Adapted from Reddy et al., 1986)

Constituent	Typical range (%)
Water-soluble simple sugars, amino acids and aliphatic acids	5–30
Cellulose	10–50
Hemicellulose	10–30
Lignin	5–30
Fats, waxes, oils and resins	1–8
Proteins	1–20

Table 2.6 Degradability and residence time of natural organic substances (Adapted from Nieder et al., 2003a)

Organic substance	Degradability	Residence time (year)
Soil organic matter	High to very low	<5–10^3
Farmyard manure	High	<5
Hay, grass, litter, etc.	High	<5
Deciduous leaf litter	High	<5
Coniferous forest floor	Moderate	1–10
Straw	Moderate	<1–10
Pine needle litter	Moderate	<1–10
Bark	Low	10–10^2
Wood	Low	10–10^2
Chemical compounds		
Glucose	Very high	<1–10[a]
Cellulose	Moderate	<1–10
Lignin	Low	10–10^2

[a]Days

Recent research has suggested that root turnover is relatively short for many temperate species, for example ~30% of grass and clover roots survive for <3 weeks under UK field conditions (Watson et al., 2000). Although there are now reliable estimates of root turnover for many tree and agricultural species (Black et al., 1998; Watson et al., 2000), there is still a lack of quantitative data on organic matter inputs from this source. Rhizosphere processes play an important role in C sequestration in terrestrial ecosystems (Helal & Sauerbeck, 1989). Rhizodeposition is almost always measured using ^{14}C to distinguish recently deposited C from the large unlabelled SOM background. Both continuous (Meharg, 1994) and pulse-labelling (Jensen, 1993, 1994) have been widely used. Rates of rhizodeposition during a whole growth period vary between crops ranging from 1.0 to 2.9 Mg C ha^{-1} under cereals and from 0.8 to 4.4 Mg C ha^{-1} under permanent grassland (Table 2.7). Typically, root respiration was not distinguished from rhizodeposition and estimates (of rhizodepositon + root respiration) for a variety of plant species and experimental conditions averaged 58% of the total C imported into the soil (Whipps, 1990). Recovery of C in the soil at harvest was considerably lower averaging 20% of the total C imports.

Most below-ground studies have been conducted under ambient CO_2 levels. Elevated atmospheric CO_2 concentration may lead to an increase of C input to soils

Table 2.7 Total below-ground translocated C (roots, exudates and CO_2 produced by root respiration) during the whole vegetation period of some cereals and grasses quantified with tracer techniques (Adapted from Nieder et al., 2003a)

Crop	Input (kg C ha^{-1})	Country	References
Spring wheat	1,500	Australia	Martin & Puchridge (1982)
Winter barley	2,370	Denmark	Jensen (1994)
Winter wheat	1,000–1,600	Germany	Knof (1985)
Winter wheat and spring barley	1,460–2,250	Netherlands	Swinnen et al. (1995)
Lolium perenne	840–1,660	Netherlands	van Ginkel et al. (1997)
Winter wheat	1,200–2,900	UK	Whipps (1990)
Different grasses	2,450–4,430	New Zealand	Saggar et al. (1997)

through increased plant productivity and enhanced allocation of C to below-ground components (Darrah, 1996). Most experiments show that significant changes in pool sizes could not be detected due to elevated CO_2 because SOM pools are so large in comparison to annual inputs from vegetation (Cardon, 1996).

Animal wastes are an important source of SOM in livestock production systems. The faeces of farm animals consist mostly of undigested food that has escaped bacterial action during passage through the body. This undigested food is mostly cellulose or lignin fibers (Table 2.8), although some modification of the lignin to humic substances has occurred. The faeces also contain the cells of microorganisms. Nitrogen in manure solids occurs largely in organic forms (undigested proteins and the bodies of microorganisms. The C/N ratio of farmyard manure is usually 15–30. Liquid manure may also contain significant amounts of NH_4^+ which has been formed from urea through hydrolysis. Animal wastes are more concentrated than the original feed in lignins and minerals. Lipids are present along with humic like substances. Manures also contain a variety of trace organics, such as antibiotics and hormones. The manure applied to cropland varies greatly in nutrient content, depending on animal type, ration fed, amount and type of bedding material, and storage condition. Both N content and availability of the N to plants decreases with losses of NH_3 through volatilization and NO_3^- through leaching. The two other major plant nutrients, P and K, are as available as fertilizer sources of these nutrients. Manures aged by cycles of wetting and drying and subjected to leaching with rainwater may have lost so much N that very little will be available to the crop in the year of application.

Sewage sludge, obtained from biological treatment of domestic sewage, is a stabilized product with an earthy odor, which does not contain raw, undigested solids. It is the material available for land disposal from most municipal treatment plants. Liquid sewage sludge is blackish and contains colloidal and suspended solids. Most sludges, as produced in a sewage treatment plant, contain 2–5% solids (Stevenson, 1986). The solid portion of sewage sludge consists of roughly equal parts of organic and inorganic material. The latter includes N, P, K, S, Cl, Zn, Cu, Pb, Cd, Hg, Cr,

Table 2.8 Gross chemical composition of animal manures (Adapted from McCalla et al., 1977; Pierzynski et al., 2000)

Constituent	Typical range (%)
Ether-soluble compounds	1.8–2.8
Cold water-soluble compounds	3.2–19.2
Hot water-soluble compounds	2.4–5.7
Hemicellulose	18.5–23.5
Cellulose	18.7–27.5
Lignin	14.2–20.7
Ash	9.1–17.2
Total N	1.3–6.0
Beef	1.3–1.8
Dairy	2.5–3.0
Poultry	4.0–6.0
Swine	3.5–4.5
C/N	15–30

Table 2.9 Approximate chemical composition of sewage sludge (Adapted from Varanka et al., 1976)

Constituent	% of organic material
Fats, waxes, and oils	19.1–19.8
Resins	3.8–8.2
Water-soluble polysaccharides	3.2–14.4
Hemicellulose	4.0–6.0
Cellulose	3.2–3.5
Lignin	14.5–16.8
Total N	3.9–6.3

Ni, Mn, B and others. The heavy metals like Zn, Cu, Pb, Cd, Hg, Cr, Ni can be toxic at some concentration. They may occur in quantities sufficient to adversely affect plants and soils. The availability of any given metal in soil will be influenced by pH, SOM content, type and amount of clay, content of other metals, cation exchange capacity, variety of crops grown, and others.

The organic component is a complex mixture consisting of (i) digested constituents that are resistant to anaerobic decomposition, (ii) dead and live microbial cells, and (iii) compounds synthesized by microbes during the digestion process. The gross chemical composition of sewage sludge is approximately as shown as in Table 2.9. The composition of individual sludges can vary appreciably from the values shown. The organic material is rather rich in N, P, and S. The C/N ratio of digested sludge ranges from 7 to 12, but is usually about 10. N availability in sludges decreases as the content of NH_4^+ and NO_3^- decreases and as the organic N becomes more stable as a result of digestion during biological waste treatment. Conservation of the N that often volatilizes as NH_3 could greatly increase the value of sewage sludge as an N source.

Decomposition of SOM is influenced by numerous physical, chemical and biological factors controlling the activity of microorganisms and soil fauna (Andrén et al., 1993). Soil temperature and moisture are the most important environmental factors. The increase in decomposition rate by temperature is usually described by a power function (Q_{10}, Arrhenius equation). Optimal moisture conditions in most soils are around 55–60% water-filled pore space (Doran et al., 1988), with decomposition decreasing as the soil dries. Water contents near or at saturation inhibit decomposition due to reduced diffusion and availability of oxygen. Both soil temperature and moisture are affected by management including cropping intensity, crop type, residue management, irrigation and tillage.

The rate of organic matter breakdown also depends on the relative proportion of single organic substrates. The order of rapidity of decomposition is given as follows: water soluble organics > hemicelluloses > celluloses > lignin (Tenny & Waksman, 1929). Cellulose, hemicellulose and lignin were found to have decomposed more slowly under anaerobic than aerobic conditions. Decomposition rates were found to be significantly correlated with the initial lignin content or the initial C/N ratio. The lower the N content of a substrate or the wider its C/N ratio, the slower the rate of decomposition. This is expected because of the dependency of the microflora on N. Hunt (1977) obtained the following equation to calculate the easily decomposable fraction of plant residues:

$$S_0 = 0.070 + 1.11\ \sqrt[3]{(N/C)};\quad R^2 = 0.98 \tag{2.1}$$

where S_0 is the initial proportion of easily decomposable C fraction, and N/C is the nitrogen to carbon ratio. Decomposition of detritus plant tissue in aquatic systems and subsequent N release were also found to be dependent on the C/N ratio of the plants.

Because of the complex nature of organic remains, numerous species of microorganisms are involved in the decomposition process. Some of the organic material is completely mineralized, some is incorporated into microbial tissue, and some is converted into humic substances. Native humus is mineralized simultaneously. Thus, although prodigious quantities of organic residues may be returned to the soil, decomposition does not necessarily lead to an increase in SOM content. The latter depends on the difference between the amount of C that enters the soil and the amount of C that leaves the soil through leaching, erosion or decomposition. Most of the C is lost from the system through decomposition.

Several stages can be delineated in the decomposition of organic residues (Stevenson, 1986). Earthworms and other soil animals play an important role in reducing the size of fresh plant material. Soil animals are reported to contribute 1–5% to total heterotrophic respiration in coniferous forest soils, 3–13% in deciduous forest soils and 5–25% in mesic grasslands (Persson, 1989). They generally contribute less to energy turnover in dry grasslands than in mesic ones. The low figures in coniferous forest soils reflect a situation where fungivores, bacterivores and carnivores are dominant components of the soil fauna, while litter and root feeders are scarce. In deciduous forests, litter feeders such as earthworms are abundant, and in grasslands the abundant occurrence of both litter feeders and root-feeding arthropods may explain the very high figures of respiration (Persson, 1989).

Further transformations are carried out by enzymes produced by microorganisms. The initial phase of microbial attack is characterized by rapid loss of readily decomposable organic substances. Simple sugars, amino acids, most proteins, and certain polysaccharides decompose very quickly. Depending on the nature of the soil microflora and quantity of synthesized microbial cells, the amount of substrate C utilized for cell synthesis will vary from 10% to 70%. Molds and spore-forming bacteria are especially active in consuming proteins, starches, and cellulose. In subsequent phases, organic intermediates and newly formed biomass tissues are attacked by a wide variety of microorganisms, with production of new biomass. The final stage of decomposition is characterized by gradual decomposition of the more resistant plant parts, such as lignin, for which the actinomycetes and fungi play a major role.

Major factors influencing decomposition include positioning of the plant material beneath or on the soil surface, soil water regime, temperature, and length of the growing season as affected by climate (Stevenson, 1986). Studies on the decomposition of plant residues in the field have shown that the residues are attacked rapidly at first but after a few months the rate slows down to a very low value. The same pattern has been observed in laboratory incubations where the environment has been kept constant through careful control of temperature and humidity. The slowdown in decomposition with time is mostly due to differences in the rate at which the various plant components decompose. Also, part of the C of the more easily decomposable constituents is resynthesized into microbial components more resistant to decomposition than the original plant material. Results from field experiments obtained for C retention after the first year or growing season are summarized in Table 2.10. From 20% to 45% retention of applied C has been observed, depending on location (climate and soil) and plant material involved. Considerable differences in C retention are particularly found for soils of colder and warmer regions.

The mass change of C in the soil (from residues plus native humus) can be expressed mathematically in many different equations, usually involving one or more kinetic rate constants of decomposition, represented as k values. One equation is

$$C_{ti} = C_0 e^{-k_1 t} + C_{added} e^{-k_2 t}, \tag{2.2}$$

with C_{ti} as the amount of soil C at time i, C_0 is the amount of soil C at time 0, k_1 is the decomposition rate constant of the total soil C pool before amendment of C added, C_{added} is the amount of C added, and k_2 is the decomposition rate constant of the added C. The decomposition process is often seen as a series of first-order reactions for the

Table 2.10 Carbon retained in soil after 1 year from plant material buried in field soils (Adapted from Nieder et al., 2003a)

Plant material	% C retained	Country	Reference
Winter wheat Straw	23–34	Germany	Nieder & Richter (1989)
Winter wheat straw	31	Germany	Sauerbeck & Gonzales (1977)
Maize straw	33	Austria	IAEA (1968)
Winter wheat Straw	31	UK	Jenkinson (1971)
Wheat straw	35–45	Canada	Shields & Paul (1973)
Ryegrass	20	Nigeria	Jenkinson & Ayanaba (1977)

various C fractions, each with its own size and decomposition rate. For modeling purposes, SOM is commonly divided into various pools, involving at least the microbial fraction (alive and dead), the labile SOM fraction, and the recalcitrant SOM pool. Adequate methods to experimentally establish the partitioning of SOM over the different pools conceptualized in many of the modeling studies are still lacking. Although the decomposition of organic materials in soil is a complex process, such simple models have successfully described the long-term dynamics of SOM (see Chapter 9).

Cycling of Nitrogen in Upland Soils

Together with C, nitrogen in soils occurs in crop residues and in different SOM fractions. Nitrogen in soils is also present as urea, NH_4^+, NO_2^- and NO_3^-, and in gaseous forms (N_2, N_2O and NO_x). The relative proportions in upland soils commonly decrease in the order SOM > plant residue N > NO_3^-- N > NH_4^+-N > gaseous N forms. Sources for soil nitrogen are plant residues, organic fertilizers (including manures, sewage sludge, compost), biological N_2 fixation, synthetic fertilizers and N in precipitation (wet and dry) (Fig. 2.3). N loss processes include erosion and runoff, nitrate leaching, ammonia volatilization and denitrification (Benbi & Richter, 2003). On agricultural land, N is also removed in crops. Under aerobic conditions, residues and SOM are decomposed to CO_2 and NH_4^+ via the process of mineralization. Local or temporary anaerobic conditions can also result in C mineralization with the by-product CH_4.

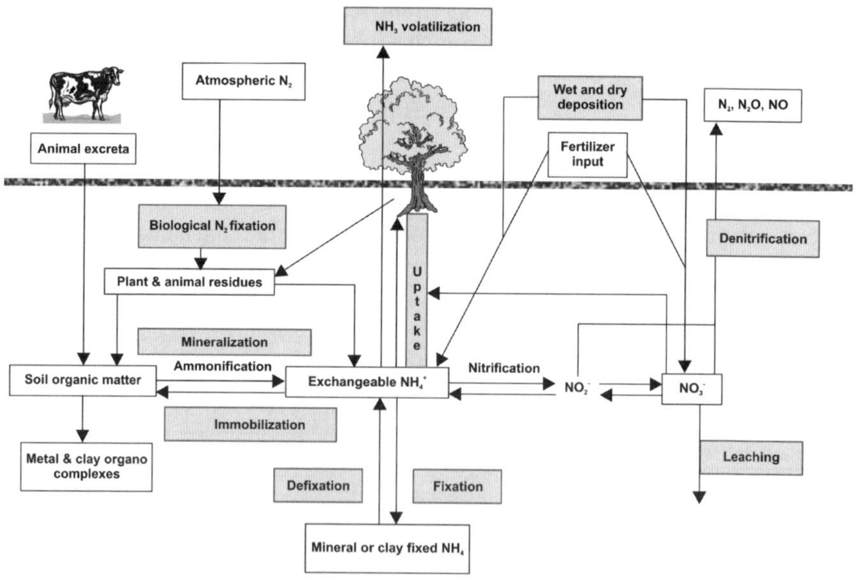

Fig. 2.3 The nitrogen cycle in upland soils (Benbi & Richter, 2003, p. 410. Reproduced with kind permission from Haworth Press)

The heterotrophic microflora utilizes C and N from the decomposing residues and NO_3^- and NH_4^+ from the mineral nitrogen pool to produce microbial biomass. In agricultural soils microbial biomass varies from 250 to 900 kg C ha^{-1} for different soil and management types (Doran, 1987). Incorporation of residues with a wide C:N ration such as barley or wheat straw (C:N ~80–120) causes a temporary uptake of mineral N by microbes (C:N ~6) to supply their N demand for growth. This microbially immobilized nitrogen is remobilized when the microbes die after most of the decomposable portions of the straw are mineralized. In general, microbial N immobilization occurs when residues with C:N ratios greater than ~25 are decomposed. At lower C:N ratios, N mineralization dominates over immobilization. This cycling between inorganic and organic N forms occurs rapidly and can be completed within a few hours. In contrast, some fractions of SOM may have an age of several thousands of years (Paul et al., 1997b).

Nitrification, the conversion of NH_4^+ to NO_3^-, is mediated by the aerobic and chemolithoautotrophic microorganisms *Nitrosomonas* and *Nitrobacter*. They use the energy derived in the oxidation of NH_4^+ to NO_2^- and NO_3^- to assimilate the CO_2. *Nitrosomonas* converts NH_4^+ to NO_2^-, and *Nitrobacter* oxidizes NO_2^- to NO_3^-. Nitrification rate depends strongly on NH_4^+ concentration, temperature, soil water content and pH (Benbi & Richter, 2003). Authotrophic nitrifiers can be considered "keystone species" and their disappearance, because of negative impacts by pollutants such as acids or improper agricultural management, can comprise the metabolic capacity of soil (Nannipieri et al., 2001). However, biological production of NO_3^-, for a long time believed to be carried out only by the small group of authotrophic nitrifiers, can also occur through the activity of heterotrophic bacteria and fungi, capable of oxidizing NH_4^+ and certain organic nitrogen compounds to substituted hydroxylamines, to nitroso compounds and, eventually, to NO_3^- (Knowles, 1986). The reactions occurring in heterotrophic nitrifiers are not ATP-coupled and thus do not provide energy. The finding that methane monooxygenase (the enzyme catalysing the oxidation of CH_4 to CH_3OH) of methanotrophic bacteria can also oxidize NH_4^+ to NH_2OH, the first intermediate of the authotrophic nitrification process, further testimonies the complex metabolic activity of the soil system and its enormous capacity to metabolize any compound introduced to it. During nitrification, some conversion of NO_2^- to NO_x can occur which thus may contribute to greenhouse gas production (Xu et al., 1998).

The mobility of reactive nitrogen depends mainly on the mineral N form (NH_4^+ or NO_3^-) and the electrostatic adsorption of the ions. In permanent (negative) charge soils, nitrate anions move freely with the soil solution. Nitrogen leaching can be of serious environmental concern in areas with positive water balance where N can enter the groundwater. The rate and extent of NO_3^- loss through leaching depend on climatic, soil, plant, and management factors. Among the climatic factors, rainfall, evaporation and temperature are the most important, affecting the downward flux of water and the NO_3^- concentration in the leaching water. Among the soil factors, soil texture and soil structure interact to influence leaching of NO_3. Generally NO_3^- is leached more rapidly from sandy than silt and clay soils.

There is a linear relationship between leaching and N fertilization (Benbi, 1990). Kolenbrander (1981) developed diagrams based on a number of fertilization experiments to show the relationship between the amount of leaching and fertilization, on the basis of groundwater recharge of 300 mm year^{-1}. In contrast to permanent charge soils, nitrate movement is retarded on the surfaces of highly weathered variable charge subsoils of subtropical and tropical regions (e.g. Bellini et al., 1996; Katou et al., 1996; Qafoku et al., 2000). In the A horizons of variable charge soils, high concentrations of soil organic matter (SOM) create negatively charged surfaces that make nitrate highly mobile. Loss of NO_3^- from surface horizons strongly depends on the amount and quality of SOM. Nitrogen lost from A horizons can be retained in subsoils as exchangeable NO_3^-. As a consequence, increases in NO_3^- leaching from surface horizons due to e.g. greater N deposition must not lead to similar increases in the NO_3^- loading of aquatic ecosystems. Prediction of the role of nitrate mobilization is further complicated by the biogeochemistry of other elements. Wang et al. (1987) found that NO_3^- adsorption in both a Brasilian and a Chinese Ferralsol was 15–20% greater when the accompanying cation was calcium as opposed to potassium. Given that soil acidification should result in base cation losses, NO_3^- retention may vary with time as the balance of cations in soil solution shifts (Matson et al., 1999). Besides water balance and accompanying (cat)ions the leaching rate of a respective ion is mainly determined by the net surface charge of the whole soil profile and the gradient in surface charge between the A horizon and the underlying subsoil horizons.

Volatilization of ammonia, i.e. the gaseous loss of NH_4^+, in upland soils is an abiotic process that occurs mainly at the soil surface. The process is highly pH dependent and only indirectly mediated by microbes, because they play an important role in NH_4^+ production. The NH_4^+ ion in soil solution is in a chemical equilibrium with NH_3 gas. Application of high amounts of NH_4^+-based fertilizers (including urea and manures), high pH values (>7.0) and temperatures are preconditions for high losses of NH_3. Incorporation of NH_4^+ fertilizers into the soil after application can significantly reduce NH_3 volatilization rates (Roelcke et al., 2002; Pacholski, 2003).

The process of denitrification is mediated primarily by a wide range of facultative anaerobic and heterotrophic bacteria, such as *Paracoccus, Pseudomonas, Alcaligenes, Flavobacterium* and *Bacillus*. They oxidize organic carbon and reduce nitrogeneous oxides (NO_3^- and NO_2^-) in absence of O_2. Partial or transient anaerobiosis results in the production of N_2O and NO_x, while complete reduction of nitrogeneous oxides to N_2 is an anaerobic process. Conditions for denitrification are favorable at high concentrations of NO_3^- in soil solution, the occurrence of an available carbon source, warm temperatures, and at high soil water contents, e.g. due to precipitation or irrigation (Tiedje, 1988; Nieder et al., 1989).

Mass flow and diffusion are the two major processes by which NO_3^- and NH_4^+ are transported to the root. Mass flow has been assumed to be the main mechanism for moving NO_3^- towards the roots (Renger & Strebel, 1976; Benbi et al., 1991b). This is based on the assumption that the amount of NO_3^- extracted from the soil depends only on the amount of water taken up by the crop. However, Liao and

Bartholomew (1974) showed that N uptake by corn could be less than, equal to, or greater than that predicted by mass flow alone, depending on a number of soil and plant factors. In situations, where plant N content is high, plant processes can discriminate against N uptake so that the actual uptake is less than the amount transported to the root surface by the transpiration stream. In sandy soils mass flow could account for only 40–60% of the N uptake by corn (Watts & Hanks, 1978). Strebel et al. (1980) observed that during early development of spring wheat 50% of the ^{15}N-labeled fertilizer NO_3^- was transported by mass flow, but the percentage dropped with crop age so that in later stages more NO_3^- was transported by diffusion than by mass flow.

Plant uptake of mineral N is an important sink for agricultural systems. Typical uptake rates of NO_3^- and NH_4^+ by crops may range from about 50 to more than 400 kg N ha^{-1} year^{-1}. Many farmers are tempted to pre-apply large doses of N fertilizers to ensure that mineral N is available when the plant needs nitrogen and the crop will not become N deficient. However, this practice increases the risk of N leaching and gaseous N emissions.

2.3.1.3 Dissolved Organic Matter in Upland Soils

Dissolved organic matter (DOM) originates from plant litter, soil humus, microbial biomass and root exudates. It represents the most reactive and mobile form of organic matter in soils and plays an important role in the biogeochemistry of C, N, and P, the transport of soil pollutants and in the pedogenesis. Dissolved organic matter is a continuum of organic molecules of different sizes and structures that pass through a filter of 0.45 μm pore size. It is mainly composed of high molecular weight complex humic substances. Only small proportions of DOM, mostly low molecular weight substances such as organic acids, sugars, and amino acids can be identified chemically (Herbert & Bertsch, 1995). The origin and fate of DOM in soil profiles is related to soil biological activity and sorption/desorption processes with the soil matrix. The knowledge about the formation and fate of DOM in soils and its response to changing environmental conditions is often inconsistent. Most of the information available is on soils of temperate and cool mixed deciduous forests, and temperate, cool, montane and boreal coniferous forests.

Although the release of DOM has been researched extensively, it is still not clear in which part DOM originates from recent litter or from stable organic matter of A horizons. Zsolnay (1996) suggested that humified organic matter is the major source of DOM because of the relatively high proportion of humus in relation to litter in soils. McDowell & Likens (1988) also hypothesized that leaching and microbial decay of humus rather than of recent litter are largely responsible for DOM production in the forest floor. In contrast, Qualls et al. (1991) observed that the greatest net increases in the fluxes of DOM occurred in the upper part of the forest floor (L horizon; for definition of forest humus forms see Chapter 3) in a deciduous forest. Recent litterfall contributes significantly to the temporal variation in the DOM production rate. Highest levels of hydrophilic neutral fraction of DOM (mainly simple sugars and nonhumic-bound polysaccharides) occurred after litter-fall. According to Michalzik & Matzner (1999), the greatest amounts of DOM occurred in the L horizon. The lower organic layers (Of and Oh) functioned rather as a sink for the organic solutes, leading to a substantial decrease of DOM in the downward flux. Substrate quality influences significantly DOM production. The rhizosphere is often associated with a large carbon flux attributable to root exudation and turnover.

In the form of DOM, organic matter can also penetrate to lower soil depths, and to some extent, mobilization of DOM may also contribute to C accumulation in groundwater. In forest ecosystems, the flux of DOC from the forest floor into the mineral subsoil has been estimated at 115–500 kg C ha^{-1} $year^{-1}$ (Guggenberger & Zech, 1993; Michalzik & Matzner, 1999; Kaiser et al., 2001). Concentrations of DOC in deep soil horizons and its export from the rooting zone are commonly small. Typically, 40–370 kg DOC ha^{-1} $year^{-1}$ are retained in the mineral subsoil (Guggenberger & Kaiser, 2003). Soil mineral components are important in adsorption of DOM. These include Al and Fe oxides and hydroxides, allophane and 2:1 layer clays (Greenland, 1971). The most important mechanisms involved in the adsorption of DOM include ligand exchange, electrostatic attraction to anion exchange sites and hydrogen bonding (Qualls, 2000). Soluble organic carbon adsorbed by soils may contribute to the stock of organic C accumulating during soil development. Guggenberger & Kaiser (2003) hypothesized that sorptive stabilization in the mineral soil is restricted to juvenile mineral surfaces which implies that the contribution of DOM to the formation of stable SOM is related to the availability of mineral sorption sites. Experiments by Qualls & Bridgham (2005) conducted on 75, 255 and 616 year old mudflows of andesitic material have shown that the proportion of DOM adsorbed by the soils increased with age. The DOM appeared to be comprised of two fractions, one that comprised 32% of the total that mineralized with a half time of decay of 7 days and a second fraction comprising 68% with a half time of decay of about 1.6 years. Thus, the major part of the DOM added to mineral surfaces contributed to the protection of SOM. Studies by Kalbitz et al. (2005) on a Haplic Podzol under Norway spruce (*Picea abies*) in Bavaria (Germany) yielded that the fraction of mineral-adsorbed organic C mineralized during incubation was only one third to one sixth of that mineralized in solution. The main stabilization processes were supposed to be sorption of intrinsically stable compounds and strong chemical bonds to the mineral soil and a physical inaccessibility of organic matter to soil microbes. Aromatic and complex compounds, probably derived from lignin, were preferentially stabilized by adsorption of DOM. The adsorption of DOM also stabilized indigenous organic matter.

Dissolved organic nitrogen (DON) has been hypothesized to constitute a major component of the terrestrial N cycle (Nashölm et al., 2000; Neff et al., 2002). High levels of DON have been reported to occur in some natural ecosystems (e.g. Jones & Kielland, 2002; Perakis & Hedin, 2002; Hood et al., 2003) which is of particular interest because DON can be readily leached from soil (Siemens & Kaupenjohann, 2002). The size of the DON pool in most studies was correlated with litter quality. Dissolved organic N is comprised mainly of complex, humic-rich organic matter which is poorly bioavailable (Marschner & Kalbitz, 2003; Jones et al., 2004). Amino acids generally represent only a small fraction of DON which is due to a rapid removal from soil solution by either plant roots or microorganisms (Jones et al., 2004). Thus, DON may contribute significantly to plant and microbial productivity in soil.

The influence of different agricultural land use on the concentration of DOC and DON in soil up to now remains understudied although there are some studies to suggest that differences can be expected to occur (Murphy et al., 2000; Chantigny, 2003). A recent study by Christou et al. (2005) considering seven contrasting agricultural land use types in Greece and UK (citrus, vegetable, arable, forest, grassland, heathland and wetland) yielded that DON on average constituted 57 ± 8% of the total dissolved nitrogen pool (mineral plus DON) across all land use types. Land use had a significant impact on the concentration of DON in soil solution and followed the series citrus > vegetable > forest = arable > grassland = wetland > heathland. For all sites, the concentration of DON was positively and linearly correlated to the amount of DOC. The average DOC: DON ratio of the soil solutions from all land uses was 16 ± 4.

2.3.2 Carbon and Nitrogen Cycling in Wetland Soils

In wetland systems, O_2 is introduced into the soil by fluctuations in water table depth, by diffusion through the floodwater, and by diffusion and mass flow from the atmosphere through plants into the rooting zone. Oxygen diffusion through water is about 10^4-fold slower than in air. After flooding, O_2 present in the soil pore water is rapidly consumed due to aerobic respiration. As a consequence, aerobic microbial processes are replaced by predominantly anaerobic processes. During this switch, bacteria start to obtain energy by oxidizing organic and inorganic compounds through several intermediate steps. Due to a high biological O_2 demand compared to the supply, two distinctly different soil layers are developed. The upper soil layer is an oxidized horizon, whereas the underlying horizon is O_2–free. Oxygen diffusion through floodwater maintains aerobic conditions at the floodwater/soil interface.

The reduction of electron acceptors as a function of depth follows the order of O_2 reduction (Eh > 300 mV), NO_3^- and Mn IV reduction to N_2 and Mn II (Eh 100–300 mV), Fe III reduction to Fe II (Eh 100 to −100 mV), SO_4^{2-} reduction to S^{2-} (Eh −100 to −200 mV), and methanogenesis (Eh < −200 mV). Redox gradients can have daily fluctuations as a result of plant growth. In wetlands with plant cover at the floodwater surface, production of O_2 during photosynthesis can result in an aerobic zone. Respiration during nighttime may convert this layer to an anaerobic horizon. Vascular plants (e.g. rice plants) are specially adapted to wetland systems. They have a well-developed system of intracellular air spaces (aerenchyma) in stems, leaves and roots, which allows the transport of O_2 from the atmosphere to the root meristems and also serves as a pathway of CH_4 from the soil into the atmosphere

(Lloyd et al., 1998). Details on the properties of a rice soil and the CH_4 and O_2 pathways in rice fields are discussed in Chapter 8 (Figure 8.8).

2.3.2.1 Microbial Biomass in Wetland Soils

Microbial biomass has been ascribed important roles in wetland soils as a nutrient pool, a driving force of nutrient turnover, and an early indicator of crop management (Shibahara & Inubishi, 1997). While biomass in aerated upland soils depends mainly on heterotrophic processes, chemoautotrophic biomass production dominates in wetlands. The presence of aerobic and anaerobic zones in wetlands supports a wide range of microbial populations with different metabolic functions, including oxygen reduction in the aerobic interface, and reduction of alternative electron acceptors in the anaerobic zone (Reddy & D'Angelo, 1994). Under water-saturated conditions, vertical layering of different metabolic activities can be present. Most of the decomposition of the plant residues occurs in the aerobic interface overlying the soil. However, a reduction in O_2 supply may drive certain groups of microbes to use alternate electron acceptors (e.g. nitrate, sulfate, bicarbonate, Fe and Mn oxides). The catabolic energy yields are lower for bacteria utilizing alternative electron acceptors are lower than for O_2. As a consequence, microbial growth rates are significantly lower in anaerobic environments (Westermann, 1993).

Oxygen is strongly reduced at and just below the floodwater/soil interface. The O_2 supply to a wetland soil is limited by the rate of O_2 diffusion through the water layer. In some wetlands, the influence of NO_3^-, Mn^{4+} and Fe^{3+} on organic matter decomposition is minimal because of the high demand for electron acceptors with great reduction potential. Microbial activity is then mostly supported by electron acceptors of lower reduction potential such as SO_4^{2-} and HCO_3^-. The reduction of sulfate is viewed as the dominant reduction process in coastal wetlands, whereas methane production can be viewed as the terminal step in anaerobic decomposition in freshwater wetlands. Both processes can also occur simultaneously in the same ecosystem.

Phototrophic primary production in flooded rice systems may account for 0.2–1.9 g C m^{-2} day^{-1} (Roger, 1996). For various Japanese paddy field soils, Shibahara & Inubishi (1995) identified 1.24–5.56% of total C as microbial biomass C and 1.49–4.55% of total N as microbial N. The size and activity of the microbial biomass in wetland soils correlate with net N mineralization rates (White & Reddy, 2000). Microbial activity therefore affects wetland surface water quality. Carbon to N ratios of soil microbial biomass in wetlands change according to drainage degree during dry (natural wetlands) or fallow (paddies) seasons. In submerged rice systems, microbial biomass plays a particular role as sink and source of nutrients (Inubishi et al., 1997). In rice paddies, microbial biomass turnover times are high, varying between 20 and 95 h (Reichardt et al., 1998). An input of more than 4.5 Mg C ha^{-1} $year^{-1}$ is required to sustain the pool size at a maintenance coefficient of 0.012 μg glucose C μg^{-1} biomass C ha^{-1} (Anderson & Domsch, 1985b). The dynamics of microbial biomass are strongly affected by organic matter supply or by changes in the redox regime. Losses of microbial biomass due to drying can amount to more than a third of the total (van Gestel et al., 1996). Draining of a wetland soil accelerates organic matter decomposition because O_2 diffuses deeper into the soil.

2.3.2.2 Soil Organic Carbon and Nitrogen in Wetlands

Soil organic carbon and nitrogen in wetlands undergo complex cycling. Organic materials originating from different sources (algal and microbial biomass, plant material) are deposited in both the water layer as well as the soil. In wetland ecosystems, the stores of organic C and N in detrital tissue and soil organic matter comprise the vast majority of C and N. Organic N consists of complex proteins and humic compounds containing amino acids and amines.

The decomposition process in wetlands differs from that in uplands in many ways. Due to frequent anaerobic conditions resulting from flooding, the decomposition rates are significantly lower compared to uplands. As a consequence, organic matter accumulates. Net C accumulation in different peatland ecosystems may be in a range of 0.1–4.2 Mg C ha^{-1} $year^{-1}$ (Chapter 1, Table 1.14). The rate of organic matter turnover depends on several factors like the quantity and quality of organic substrates (DeBusk & Reddy, 1998), the length of the hydroperiod (Happell & Chanton, 1993), the supply of electron acceptors (D'Angelo & Reddy, 1994), and nutrient availability (Amador & Jones, 1995) which is important for the growth of decomposers. Although nutrient (particularly N and P) concentrations (temporarily) may be greater in many wetlands as compared to their surrounding uplands, their availability is commonly low relative to the pool of available C which may limit microbial growth (Westermann, 1993). Breakdown of detrital tissue results in release of dissolved organic N to the water layer, most of which is resistant to decomposition. By this way, water leaving wetlands may contain higher levels of organic N. The limitation in decomposition rates is an efficient mechanism of protecting adjacent ecosystems against eutrophication. The organic materials in wetlands resulting from plant biomass, algal and microbial biomass consist of complex nonhumic substances which are deposited in the water column and surface soil due to natural die off (Fig. 2.4).

Low-molecular-weight organic compounds are preferentially used by microbes, while slowly degradable compounds accumulate with time (Melillo et al., 1989). The depletion of O_2 causes a shift in microbial metabolism of monomeric C compounds (e.g. glucose) from aerobic (oxidation) to anaerobic pathways (fermentation). Methanogenic bacteria depend on the activity of fermenting bacteria that produce short-chain C compounds from the breakdown of mono- and polysaccharides (Howarth, 1993). The cell wall construction in vascular plants includes cellulose, hemicellulose and lignin, which occurs in cells of supportive and conductive tissue.

Although at a reduced rate, cellulose decomposition readily occurs under anoxic conditions, primarily mediated by bacteria (*Clostridium*) (Swift et al., 1979). The presence of lignin is a limiting factor in the decomposition of vascular plant tissue

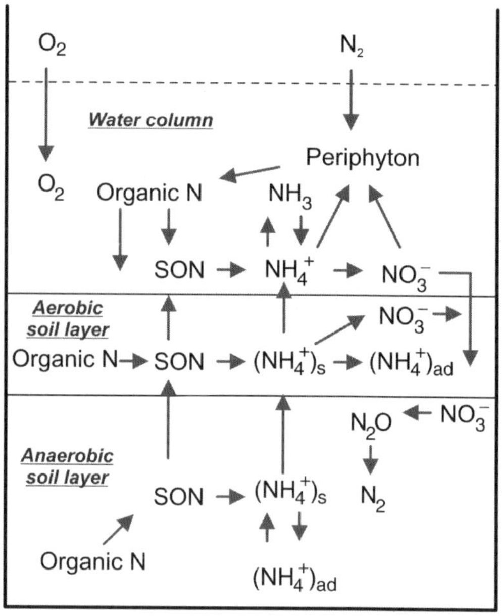

Fig. 2.4 The nitrogen cycle in wetland soils showing the N transformations in aerobic and anaerobic layers (DeBusk et al., 2001, p. 37. Reproduced with kind permission from Taylor & Francis, Copyright Clearance Center)

(Zeikus, 1981). White-rot fungi are the most active decomposers of lignin (Eriksson & Johnsrud, 1982) and lignin degradation by these organisms requires O_2, because oxygen radicals are responsible for the chemical oxidation of aromatic ring structures. Different species of fungi capable of lignin degradation have been found in oxidized layers of wetlands (Westermann, 1993). Major fungal genera involved in decomposition in these systems include *Alternaria*, *Cladosporium*, *Penicillium*, *Fusarium*, *Trichoderma*, *Alatospora* and *Tetacladium* (Reddy et al., 2000). Bacteria are dominant decomposers in anaerobic ecosystems. Important genera in wetlands include *Cytophaga*, *Vibrio*, *Achromobacter*, *Bacillus*, *Micrococcus*, *Chromobacterium*, *Streptomyces*, *Arthrobacter*, *Actinomyces* and others (Reddy et al., 2000). Some species of bacteria (e.g. *Bacillus*, *Nocardia*, *Azotobacter*, *Pseudomonas*) are also known to decompose lignin. This may explain why lignin decay is observed even in anoxic marsh and mangrove sediments (Benner et al., 1984). In the latter study, bacterial lignin degradation dominated over fungal decomposition. The ratio of lignin to cellulose degradation has been shown to be similar under anaerobic and aerobic conditions (Benner et al., 1984). Production and activity of enzymes are influenced by a number of factors, including pH, O_2, and nutrient availability. Production of phosphatases and proteases was enhanced due to N limitation in wetlands (Sinsabaugh et al., 1993). Lignin degradation was inhibited and cellulose degradation was enhanced by N amendments (Fog, 1988). Biodegradation rates of low molecular weight organic acids and sugars decreased due to interactions with mineral surfaces (Gordon & Millero, 1985).

The terminal step of decomposition in wetlands is the uptake and use of small molecular weight compounds by the heterotrophic microflora. A number of microorganisms use organic C compounds as electron donors and subsequently reduce electron acceptors (e.g. O_2, CO_2, NO_3^-, Mn IV, Fe III and SO_4^{2-}) for energy production. Most decomposition in wetlands may be by aerobic bacteria and fungi that use O_2 as final electron acceptor. This is because plant and other residues are typically deposited in the aerobic water column and on the soil surface. In contrast, aboveground organic material undergoes anaerobic decomposition where microbes use electron acceptors alternative to O_2 (CO_2, NO_3^-, Mn IV, Fe III and SO_4^{2-}). Anaerobic decomposition runs slower than aerobic decay. Energy differences determine the order ($O_2 > NO_3^- >$ Mn IV > Fe III > $SO_4^{2-} > CO_2$) in which electron acceptors are used. The above facts explain the stratification of specific microbial activities and chemicals in wetland soils. Recalcitrant organic compounds tend to accumulate in wetlands as humic substances or as undecomposed plant tissue (peat). Under anaerobic conditions, these materials are resistant to decomposition and tend to accumulate.

As with C, the more refractory organic N compounds become buried into the soil and accrete over time. Besides organic N forms, there exist stores of inorganic N such as NH_4^+, NO_3^- and NO_2^-. These forms are also called reactive nitrogen (Galloway et al., 2004). Different N transformations process inorganic N through nitrification, denitrification and ammonia volatilization. This means that the inorganic N forms are not very stable with time. They comprise only up to 1% of the total nitrogen in a wetland soil (Howard-Williams & Downes, 1994). The extent of these N transformation processes commonly increases with N loading. The production of NH_4^+ in wetland soils is determined by the balance between ammonification and immobilization. Anaerobic microbes demand less nitrogen as compared to aerobic microorganisms. Ammonification rates in wetlands were observed to be in a range from 0.004 to 0.357 g N m^{-2} day^{-2} (Martin & Reddy, 1997). Ammonium can be lost through ammonia volatilization. The latter process is controlled by the pH of the soil-water system. In wetlands, significant portions of nitrogen can be lost by this process after application of ammonium-based fertilizers (e.g. to paddies), at high ammonification rates, if the influent water contains high concentrations of NH_4^+, and if algal activity shifts pH above ~7.5.

Ammonium can also be oxidized to NO_3^- (nitrification). However, except for the aerobic portion of the wetland, the vast majority of inorganic N in flooded soils is present as NH_4^+, whereas NO_3^- and NO_2^- are typically found only in trace amounts. The relative increase in NH_4^+ concentration is also due to the absence of O_2, which prevents nitrification. Wetland soils therefore accumulate NH_4^+. Nitrification is an aerobic process and therefore occurs in the water layer, the aerobic soil layer and in aerated parts of the rooting zone. Ammonium supply to aerobic zones of wetland soils occurs through import from anaerobic soil layers. Nitrification rates in wetlands have been observed to range from 0.01 to 0.161 g N m^{-2} day^{-1} (Martin & Reddy, 1997), which are lower than that observed for ammonification. This suggests that O_2 availability limits nitrification rates.

Nitrate in wetlands diffuses into anaerobic soil layers where it is used as an alternative electron acceptor. The relatively high organic C content of wetlands promotes denitrification. Reported denitrification rates in wetlands were in a range of 0.03–1.02 g N m^{-2} day^{-1} (Martin & Reddy, 1997). Significant relationships have been found between denitrification rates and soluble C (Gale et al., 1993). At high soluble C contents and under anaerobic conditions in wetland soils, the NO_3^- concentration becomes the limiting factor in denitrification (White & Reddy, 1999). Rates of denitrification are commonly higher in wetland soils receiving continuous quantities of NO_3^- compared to soils receiving low levels (Cooper, 1990).

Biological N_2 fixation in wetlands is driven by several genera of bacteria and cyanobacteria. They occur in the water layer and in the soil symbiotic associations with vegetation, as plankton and as filamentous mats. One of the best-known associations is the symbiosis between the fern *Azolla* with the cyanobacterium *Anabaena azollae*. Rates of N_2 fixations in wetlands vary widely, depending on the environmental conditions. Nitrogen fixation rates ranged from 1.8 to 18 mg N m^{-2} $year^{-1}$ in the Everglades marsh of Florida (Inglett, 2000). In wetland systems without macrophyte coverage or cyanobacterial mats, N_2 fixation rates ranged from 0.002 to 1.6 g N mg N m^{-2} $year^{-1}$, while dense cyanobacterial mats exhibited rates varying between 1.2 and 76 g N m^{-2} $year^{-1}$ (Howarth et al., 1988).

Plants are of major importance for the C and N cycles of wetlands by photosynthetic C assimilation and N uptake, release of C and N through mineralization of plant residues, and providing an environment in the rhizosphere for N transformations such as nitrification (e.g. transport of O_2 into the soil via aerenchyma) and denitrification (e.g. production of C-rich rhizodeposits). Nitrogen use efficiency by aquatic vegetation depends on N availability, temperature and the type of vegetation. Nitrogen demand in vegetation and microbial biomass of several wetlands was demonstrated not to be met by external N inputs alone (White & Howes, 1994), which points to the role of soil internal N cycling. Several studies reflect a high proportion (54–95%) of plant N uptake from soil-born sources (DeLaune et al., 1989; White & Howes, 1994).

In wetlands of temperate climates, as with uplands, most C assimilation and N uptake by plants occurs during the vegetation period. During the winter period, plants die up and their residues accumulate on the soil surface. A significant portion of the residues are translocated into the soil where they are mineralized particularly during the summer period. In contrast to herbaceous vegetation (e.g. *Typha* ssp.) wetland forests store C and N at the long term in woody biomass. However, C and N stored in leaves is returned to the forest floor and partly mixed into the mineral soil. Turnover rates in wetland soils are significantly lower compared to uplands. Drainage of wetlands results in a decrease in surface elevation (subsidence). Drained organic soils are subsiding several cm (range: 1–10) $year^{-1}$ (Nieder et al., 2003a). Microbial oxidation is the predominant cause of soil subsidence. After drainage, wetlands generally act as a source of C and, except for Dystric Histosols, additionally as a source of N.

2.3.2.3 Dissolved Organic Matter in Wetlands

Dissolved organic matter in wetlands is known as a relatively stable component both in size and quality (Wetzel, 1984), but its ecological significance has not been clearly defined. Like in uplands, DOM decomposition involves both labile and recalcitrant organic matter. Turnover of labile DOM may be high, which means that the actual concentration of active DOM is generally low. During transport in soil, selective removal of DOM occurs due to microbial decay or interactions with the mineral component of the soil. As a consequence, the proportion of recalcitrant compounds of DOM increases with depth. The decay of detrital tissue results in a release of dissolved organic N to the water layer. As most of the DON is resistant to decomposition, water leaving the wetlands may contain elevated levels of N in organic forms. Wetlands, therefore, function as a sink for inorganic N, and as a source for soluble organic N.

2.4 Global Climate Change and C and N Cycling

The increase in atmospheric CO_2 and N_2O due to fossil fuel emissions, land clearing and biomass burning has been identified as a major driving force for global climate change. Conversely, the terrestrial biosphere is thought to be sequestering up to 2 Pg C $year^{-1}$ as a result of enhanced photosynthetic C fixation (Dixon et al., 1994). Standing biomass is thought to be responsible for the enhanced uptake required to balance part of the anthropogenic CO_2. In Chapter 8, the role of forests in CO_2 mitigation, the potential for C sequestration by agriculture, and the influence of global climate change on crop yields have been discussed.

Soil organic matter is thought to provide a long term transient sink for both, atmospheric carbon and nitrogen (Ciais et al., 1995; Schimel, 1995) which is due to the comparatively long time required for the SOM pool to establish a new equilibrium with the enhanced rates of delivery of C from standing biomass and N from atmospheric deposition. At elevated temperatures, however, the soil may act as an additional source for CO_2 if it is accessible to microbial decomposition. During decay of plant biomass, less than 1% of photosynthetically assimilated CO_2 enters the more stable SOM pool. Despite this low rate, the SOM pool has accumulated roughly 1.500 Pg C (0–100 cm) over centuries and millennia.

The very close coupling of carbon and nitrogen cycles in ecosystems indicates that there may be many avenues for interactions and feedback as one or the other cycle is altered via elevated CO_2 and climate change. The cycles of C and N may be altered through changing litter decomposition rates, plant N uptake or internal cycling of nutrients within plants (Graham et al., 1990). The potential of increased C acquisition by plants under elevated CO_2 can be limited by the availability of soil nutrients, which in turn is controlled by decomposition. While the process of decomposition is relatively well-known in view of factors like soil moisture, temperature, and nutrient quality of the litter, the knowledge of the effects of changing CO_2 levels on decomposition and C and N cycling is still limited. Elevated CO_2 could have an impact on decomposition rates in ecosystems through changes in the species composition, through direct effects on decomposer communities, or through changes in the chemical composition of litter. Increased amounts of cellulose and lignin are hypothesized to be a consequence of elevated CO_2 and could reduce decomposition rates. Litter characteristics like lignin and nutrient contents strongly influence decay patterns. The overall litter quality of the ecosystem may be altered by elevated CO_2 either by changes induced directly in the litter produced or by changes in species composition of plant communities and associated litter characteristics. For many species, C/N ratios in plant tissues increase with CO_2 enrichment (Couteaux et al., 1991). However, the C/N ratios of senescent tissues may not reflect those of living tissue. In many CO_2 enrichment studies, green leaf N concentration decreased with elevated CO_2 (e.g. Norby et al., 1992; Koch & Mooney, 1996). In perennial plants, N concentrations usually differ between green and senescent foliage. In woody deciduous species, approximately half of the N content of green foliage is withdrawn from senescing tissue prior to leaf fall and retranslocated to rapidly growing tissues or stored in stem or roots until new growth is initiated (Chapin et al., 1990). The proportion of N translocated varies by species and may be correlated with the N status of the soil. During senescence, recalcitrant C compounds frequently increase in concentration.

The above features were particularly found in pot studies (e.g. Melillo, 1983; Cotrufo et al., 1994). In contrast, such clear relationships could not always be identified in field studies (see review by O'Neill & Norby, 1996). In summary, contradictory results appear to be related to experimental approach (e.g. single vs. mixed-species decomposition experiments and scale of observation or pot experiments vs. field studies). In order to determine the potential of CO_2-induced changes in decomposition rates to affect ecosystems, research must be conducted at the ecosystem level. This has been practiced in some ecosystems but is still lacking for forests where the long life times of trees make long-term research necessary.

Chapter 3
Soil Organic Matter Characterization

At the outset it is important to clarify the terms soil organic matter (SOM) and humus. Sometimes it is a matter of confusion as chemists and biologists look into soil organic matter with different perspective. In the glossary of soil science terms (SSSA, 1997) soil organic matter is defined as the organic fraction of the soil exclusive of undecayed plant and animal residues and is considered synonymous with humus. However, other definitions of SOM have been used by numerous authors. Schnitzer (2000) referred to soil organic matter as the sum total of all organic carbon-containing substances in the soil, which comprises of a mixture of plant and animal residues in various stages of decomposition, substances synthesized microbiologically and/or chemically from the breakdown products, and the bodies of living and dead microoragnisms and their decomposing remains.

Conceptually organic component of soil can be defined as consisting of both living and dead organic matter (Fig. 3.1). The living organic matter is represented by plant roots, soil animals and microbial biomass and the dead organic matter is formed by chemical and biological decomposition of organic residues. The dead organic matter may be differentiated into unaltered material (in which morphology of the original material still exists) and the altered or the transformed products (also called humus). Generally, soil humus is defined as a mixture of dark, colloidal polydispersed organic compounds with high molecular weights and relatively resistant to decomposition.

For characterization and functional purposes, SOM is generally subdivided into different fractions or compartments. The approaches for fractionation may broadly be categorized as chemical, physical and biological or biochemical. Additionally, some morphological characteristics are also used to distinguish the development of different humus forms in terrestrial ecosystems. Since SOM is a continuum of complex heterogeneous material, no single fractionation approach may be expected to adequately characterize the turnover rates of the whole soil. In this chapter, we discuss different chemical and physical organic matter fractions and morphological humus forms. The biological or functional pools, that are mostly model-defined and may or may not be related to some chemically or physically defined fractions are discussed in Chapter 9.

R. Nieder, D.K. Benbi, *Carbon and Nitrogen in the Terrestrial Environment,*

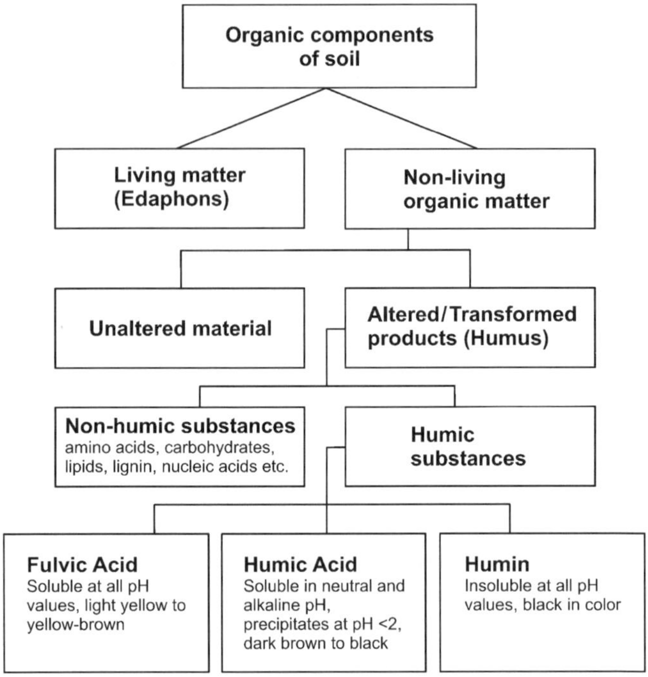

Fig. 3.1 Schematic representation of organic components of soil

3.1 Chemical Characterization of Soil Organic Matter

Chemical separation methods are mostly based on the solubility and affinity of certain organic carbon compounds in different solvents or extracting solutions. The solutions range from water and polar and nonpolar solvents (such as alcohols, acetonitrile, acetone, and hexane) to inorganic salt solutions (such as KCl and K_2SO_4) acids and bases of varying strengths, and chelating agents (Cheng & Kimble, 2001). The most effective and commonly used extracting solution is 0.5 M NaOH. The extracted solution is further separated by selective precipitation, solvent affinity, chromatographic, electrophoretic and size exclusion techniques. Alternatively, specific structural components and functional groups of organic carbon may be identified and measured by applying techniques such as infrared (IR) and ultraviolet spectroscopy or nuclear magnetic resonance (NMR), etc.

Generally, humus is distinguished between non-humic and humic substances. Non-humic substances comprise compounds belonging to the well-known classes of biochemistry such as amino acids, proteins, carbohydrates, lipids, lignin, nucleic acids, pigments, hormones and a variety of organic acids. As discussed later in this chapter humic substances are further subdivided into fulvic acid, humic acid and humin (Fig. 3.1).

3.1.1 Non-Humic Substances

Non-humic substances consist mainly of carbohydrates (monosaccharides, oligosaccharides and polysaccharides), amino acids, amino sugars, alkyl compounds and lignin (Kögel-Knabner et al., 1992a). The turnover time of non-humic substances varies from low (lignin) through moderate (e.g. oligosaccharides and polysaccharides) to very high (e.g. monosaccharides) (Nieder et al., 2003a). Most abundant in nature are polysaccharides, such as cellulose hemicelluloses and chitin.

3.1.1.1 Carbohydrates

Carbohydrates store most of the carbon (100–250 g kg^{-1} soil organic C) found in non-humic substances (Cheshire, 1979). They cover a broad range of molecules consisting of mainly five (pentose) or six (hexose) carbon atoms, which form oxygen-containing ring structures. Plants and soil organisms are the main sources for soil carbohydrate formation. The degree of polymerization of carbohydrates is linked to different cellular and biological functions. Plants deposit directly sugars, hemicellulose and cellulose in particulate residues and transfer soluble carbohydrates by root exudation and deposition of mucilaginous materials into the soil. Carbohydrates originating from soil microorganisms are part of extracellular mucilages, cellular tissue (e.g. chitin) and the cytoplast. If decomposition processes are not limited by specific environmental conditions (e.g. high soil water content, water stress, low pH) and processes (e.g. physical stabilization by adsorption onto mineral particles; spatial separation from decomposer communities), most carbohydrates are degraded rapidly. The decomposition of polysaccharides results in the formation of neutral sugars, amino sugars, acidic sugars and sugar alcohols. The neutral sugars consist of hexoses (glucose, galactose and mannose), pentoses (arabinose and xylose) and deoxyhexoses (rhamnose and fucose).

3.1.1.2 Amino Acids

Amino acids are essential molecules of organisms because they are substrates for protein synthesis and enzymes. Most nitrogen in organisms and in soil organic matter is found as amino groups. As nitrogen is generally a limiting factor for terrestrial ecosystems, organisms store this restricted element in the form of amino acids. Amino acids are the major constituents of microbial cell walls. Microorganisms also liberate amino acids as exoenzymes to degrade complex organic matter outside their cells to smaller monomers. Proteins and enzymes are readily decomposed by proteolytic enzymes that hydrolyze the peptide links.

3.1.1.3 Amino Sugars

Amino sugars, which are assumed to be mainly of microbial origin, account for 20–60 g kg^{-1} of soil organic C (Cheshire, 1979). Important amino sugars found in soils are D-glucosamine (component of chitin), N-acetylglocosamine (found in the tissue of fungal mycelia), muramic acid (found in microbes) and D-mannosamine. The ratio D-glucosamine/N-acetylglocosamine may indicate the composition of the microbial decomposer community in soils (Sowden, 1959). Glucuronic acids and galacturonic acids are the most important acidic sugars. The latter may account for at least 1–5% of the soil organic C (Greenland & Oades, 1975).

3.1.1.4 Alkyl Compounds

Alkyl compounds in soil consist of macromolecules synthesized by microorganisms, solvent and bound lipids (fatty acids and waxes originating from plants and soil microorganisms), and insoluble polyesters (cutin and suberin) and nonpolyesters (cutan and suberan) derived from plant cuticles and cork cells in roots and bark (Kögel-Knabner et al., 1992a). Alkyl C accounts for 15–20% of the organic C contained in litter (L) horizons of organic layers and 30–40% for humified organic layer horizons (Oh) and mineral Ah horizons (Kögel-Knabner et al., 1992b). Lipids include a great variety of substances (waxes, steroids, terpenoids, carotenoids, porphyrins, glycerides, phospholipids and organic acids (Stevenson, 1994) that are all soluble in nonpolar solvents such as hexane or chloroform. In forest soils they account for 30% of the ^{13}C NMR signal intensity in the alkyl C region (Ziegler & Zech, 1989). Lipids are highly decomposable, e.g. in forest soils, and thus do not contribute significantly to the accumulation of alkyl C in humic substances (Ziegler, 1989). Insoluble polyesters can be readily decomposed by soil microorganisms that produce cutinase. In contrast, insoluble nonpolyesters are resistant to microbial degradation and may, therefore, contribute to the accumulation of alkyl C in soils (Kögel-Knabner et al., 1992a, b). In soils with highly reactive surface areas, other forms of alkyl C (cutin and suberin, free and bound lipids) may be protected against degradation by interaction with fine particle size fractions (e.g. Vertisols, Chernozems and Luvisols) or oxides of iron and aluminum (Ferralsols) (Baldock et al., 1992; Oades et al., 1987).

3.1.1.5 Lignin

Lignin is more resistant to microbial degradation than other biopolymers found in plant material. The role of different organisms and processes for lignin degradation have been discussed by Shevchenko and Bailey (1996). White rot fungi, belonging to the group of filamentous basidiomycetes, are the most efficient lignin degrading organisms (Haider, 1992). Hatcher (1987) examined the chemical state of lignin using ^{13}C NMR spectra of isolated natural lignin. Of the 10–11 C atoms contained in the two major lignin monomers (guaiacyl and syringyl), on average 2–3 were phenolic, 3–4 were aromatic, 3 were O-alkyl, and 1–2 were methoxyl. Deviations from these distributions would indicate the alteration of the average lignin molecule during the decomposition process. The ratio of the quantity of acidic to aldehydic forms (Ac/Al) of guaiacyl and syringyl gives an indication of the state of structural alteration of each monomeric unit within the lignin polymer (Moran et al., 1991). The extent of lignin alteration in woody (dominated by undecomposed litter) and nonwoody (humified) forest horizons was studied by deMontigny et al. (1993). From the litter to the humified horizon, only little change in the total amount of phenolic C was detected (range: 31–33.5 g kg^{-1} organic C). Although the lignin content changed only little, the guaiacyl Ac/Al ratio increased significantly from 0.43 to 1.01, indicating that the lignin became more structurally modified with increasing soil depth. Lignin associated with mineral particle size fractions (sand, silt and clay) of mineral soils from different ecosystems showed a decrease in the extent of lignin alteration with decreasing particle size (Guggenberger et al., 1994).

3.1.2 Humic Substances

Humic substances (HS) are heterogeneous mixture of natural organic substances that are widely distributed in soil, water, sediments and fossil organic resources. These represent the largest pool of organic carbon on the surface of the earth. Stevenson (1994) defined HS as unspecified, transformed, dark colored heterogeneous, amorphous and high-molecular weight material formed by secondary synthesis reactions. Based on their solubility in acidic and alkaline solutions, HS are classified into fulvic acids (FAs) humic acids (HAs), and humins (Fig. 3.1). The FAs comprise the fraction of humic substances that remain soluble under all pH conditions or the fraction that stays in solution when alkaline soil extracts are adjusted to pH < 2. Fulvic acids are light yellow to yellow-brown in color. The HAs are the fraction of HS that are soluble in neutral or alkaline solution and precipitate when solution pH is reduced to <2 by acid addition. Humic acids are the major extractable component of soil humic substances. They are dark brown to black in color. Humic acids can be further fractionated into hymatomelanic acid through extraction in alcohol. Humin is the fraction of HS that is not soluble in water at any pH value and because of the difficulty in extraction it has not been studied as extensively as HAs and FAs. Humins are black in color. The percentage of the humus which occurs in the various humic fractions varies considerably from one soil type to another. The humus of forest soils is characterized by a high content of FAs while the humus of peat and grassland soils is high in HAs (Kononova, 1966; Stevenson, 1994). The humic acid/fulvic acid ratio usually decreases with depth. Though the terms fulvic acid, humic acid and humin have been in use for a long time, yet a number of scientists question the validity of their usage as these do not represent distinct chemical substances and are closely related materials (Aiken et al., 1985; Schnitzer, 2000).

3.1.2.1 Origin of Humic Substances

Humic substances are formed during the decomposition of plant and animal remains in soil through a process called humification. The litter or other organic materials added to soil generally follow a two-phase decomposition process: a rapid initial phase involving readily decomposable organic fraction followed by a much slower phase involving recalcitrant fraction (Benbi & Richter, 2002; Haer & Benbi, 2003). The decomposition of readily mineralizable components during the first phase leads to selective preservation of refractory components. The preserved organic substances in altered and unaltered forms, and resynthesized microbial products supposedly lead to the formation of humus by degradative and synthetic processes. Several pathways have been suggested for the formation of HS in soil. Maillard (1913) introduced the 'browning' reaction or the 'melanoidin' theory according to which monomeric reducing sugars, such as glucose, could condense with amino acids, such as glycine, to form brown macromolecular substances. A number of studies since then have shown that some browning reaction products have certain properties similar to those of soil HS. Waksman (1936) proposed the so-called classical lignin-protein theory according to which HS result from the condensation of modified lignin with microbially synthesized protein. Stevenson (1994) summarized the different humus formation pathways as (a) lignin breakdown with by-product polymerization, (b) lignin fragmentation, (c) cellulose dehydration, oxidation, and polymerization, and (d) condensation of sugars and amino acids. According to lignin breakdown theory, lignin is incompletely utilized by microorganism and during decomposition there is loss of methoxyl (OCH_3) groups with the generation of o-hydroxyphenols and oxidation of aliphatic side chains to form COOH groups. The modified material is subject to further unknown changes to yield first HAs and then FAs. According to lignin fragmentation theory, phenolic aldehydes and acids released from lignin during microbiological attack undergo enzymatic conversion to quinones, which polymerize in the presence or absence of amino compounds to form humic like macromolecules. The third pathway is characterized by the synthesis of polyphenols by microorganisms from nonlignin C sources such as cellulose. The polyphenols are then enzymatically oxidized to quinones and converted to humic substances. Both these pathways form the basis of polyphenol theory according to which quinones of lignin origin, together with those synthesized by microorganisms, are the major building blocks from which HS are formed. Flaig (1988) emphasized the role of quinones from di- and polyhydroxybenzene structures, with –OH groups in the 1,2- and 1,4-ring positions, on the synthesis of HS. Lignins can give rise to the appropriate phenols, and fungi are also known to synthesize phenols, many of which are components of melanins, the colored secondary metabolites formed during fungal degradation of saccharides.

According to sugar and amino acid condensation theory, reducing sugars and amino acids, formed as by-products of microbial metabolism, undergo nonenzymatic polymerization to form brown nitrogenous polymers. The initial reaction in sugar-amine condensation involves addition of the amine to the aldehyde group of the sugar to form the N-substituted glycosylamine. The glycosylamine subsequently undergoes to form the N-substituted-1-amino-deoxy-2-ketose. This is subject to: fragmentation and formation of 3-carbon chain aldehydes and ketones, such as acetol, diacetyl, etc.; dehydration and formation of reductones and hydroxymethyl furfurals. All of these compounds are highly reactive and readily polymerize in the presence of amino compounds to form brown-colored products.

Kögel-Knabner et al. (1991) suggested that predominating processes during humification in forest soils include selective preservation of refractory plant (and animal) components, the direct transformation of some of these to humic macromolecules, and synthesis by microorganisms proliferating on organic residues. Polysaccharides of plant litter are decomposed intensively and substituted by microbial polysaccharides. Lignin is partly mineralized and the remnant molecule is transformed directly by ring cleavage and side chain oxidation. As humification proceeds, the contribution of phenolic groups such as those in lignin, to the humic structures decreases with a concomitant increase in aromatic C. The model proposed by Ziechmann (1996) considers the formation of HS as a three- phase (metabolic, radical and conformation) synthetic and degradative process (Fig. 3.2). The metabolic phase comprises basic biochemical processes that are enzymatically controlled. During this phase partial microbial degradation and biosynthesis of aromatic and nonaromatic compounds occurs. Humic acid precursors (HAP) are formed from the aromatic compounds of these materials in the radical phase, via radical intermediates. Along with the HAP, the non HS are also channeled to the conformation phase leading to the establishment of HS.

Thus it is clear that soil HS can be formed from a variety of precursors with a broad range of biological and chemical processes. One, more or all the above pathways may operate in different soils, but not to the same extent or in the same order of importance. Different pathways may predominate in different ecosystems. Because of large heterogeneity of the soil environments in which the HS are formed, it is highly unlikely that two molecules in any batch are exactly the same (Hayes, 1991). Each humic component in each environment possesses an individuality that distinguishes it from other components in the same environment, and from the same humic components in different environments (Malcolm & McCarthy, 1991).

3.1.2.2 Analytical Characteristics and Structure of Humic Substances

At least for a century, research has focused on understanding the HS functional group chemistry and the macromolecular structure of HS that control their behavior in the terrestrial environment. However, a universally accepted molecular structure of soil organic matter is still elusive. The difficulties encountered in chemically defining the structure and reactivity of humus derives from its extremely large chemical heterogeneity, geographical variability and its dependence on general properties (such as vegetation, climate, topography, etc.) of the ecosystem in which it is formed (Piccolo, 2002). Hayes (1991) contended that since HS (as compared to proteins) do not meet different criteria for structure, they are unlikely to possess

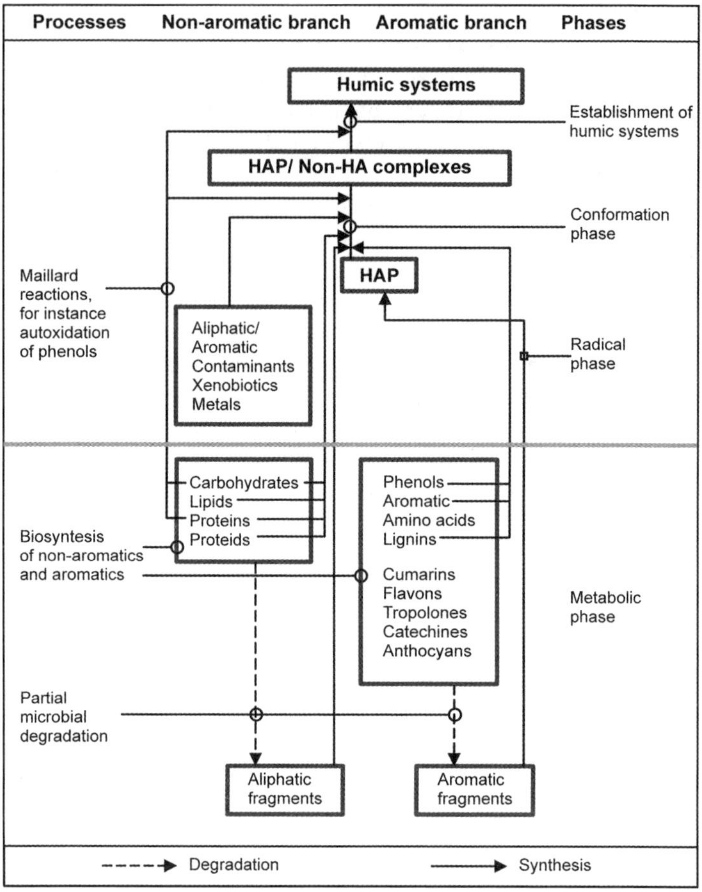

Fig. 3.2 Genesis of humic substances, HAP: humic acid precursors, HA: humic acids (Ziechmann & Müller-Wegener, 1990, p. 39. Reproduced with kind permission from Springer)

the degree of order needed to maintain rigid conformations and may have random coil conformation in solution.

While the early concepts depicted humic substances as consisting mainly of modified plant macromolecules, further research has shown their distinctive chemical features. A number of analytical techniques such as physical/chemical (acid-base titration, hydrolysis, oxidative degradation, reduction, pyrolysis), spectroscopic (ultra violet/visible (UV/VIS), nuclear magnetic resonance (NMR), fluorescence, mass spectrometry), and chromatographic/fractionation (gel chromatography, flow-field-flow-fractionation) methods have been employed for the qualitative and quantitative characterization of HS. Numerous reviews and detailed studies have been published on different aspects of methodology (Wilson, 1987; Arshad et al., 1988; Kinchesh et al., 1995; Preston, 1996; Schmidt et al., 1997; Simpson et al., 2002) and it has been suggested to apply as many independent methods as possible to gather information on identical HS samples from different points of view (Abbt-Braun et al., 2004).

The chemical composition of HS shows that as we proceed from FAs through HAs to humin there is an increase in degree of polymerization, molecular weight, and carbon (C) content and there is a decrease in oxygen (O) content, exchange acidity and degree of solubility (Stevenson, 1994). Flaig (1958) using ultra centrifuge showed that molecular weights for FAs were about 10,000 Da (Dalton; 1 Da is equal to weight of hydrogen atom) whereas that for HAs were in the range of 30,000–50,000 Da. Generally the molecular weights of FAs are considered to range between 400 and 1,500 (Aiken & Gillam, 1989; Wershaw, 1989; Stevenson, 1994). The two materials do not vary much with respect to hydrogen (H), nitrogen (N) and sulfur (S) contents (Table 3.1).While the oxygen in FAs can be accounted for largely in known functional groups (COOH, OH, C=O), about 74% of the total O in the HAs is accounted for in the functional groups (Schnitzer, 2000). Both the materials contain significant concentrations of phenolic OH, total C=O, and OCH_3 groups, but the FAs contain more functional groups of an acidic nature, particularly COOH. The total acidities of FAs (11.8–14.2 me g^{-1}) are considerably higher than for HAs (5.7–10.2 me g^{-1}) (Table 3.2). Total acidity is attributed to the sum of carboxyl and phenolic-OH group contents, and indicate the cation exchange and complexing capacities of humic matter. A high total acidity value is indicative of a high cation exchange capacity (CEC) and complexing power.

The C:H ratio is an index of aromaticity, the minimum value being 1 for benzene. Values >1 reflect the degree of condensation of rings and the substitution of other elements for H in the structure (White, 1997). Humic acids have more aromaticity

Table 3.1 Elemental composition of humic substances in soil (Adapted from Schnitzer & Khan, 1972)

Fraction	C (%)	H (%)	O (%)	N (%)
Fulvic acids	43–51	3.3–5.9	45–47	0.7–2.8
Humic acids	54–60	3.7–5.8	32–37	1.6–4.1
Humin	55–56	5.5–6.0	32–34	4.6–5.1

Table 3.2 Aliphatic C, aromatic C, COOH-C (%), total acidity and functional group contents (me g^{-1}) of three humus fractions (Adapted from Schnitzer & Khan, 1972; Tan 1994)

Functional group	Fulvic acid	Humic acid	Humin
Aliphatic C (%)	61.0	48.7	–
Aromatic C (%)	25.3	36.4	–
COOH C (%)	13.7	14.9	–
Total acidity	11.8–14.2	5.7–10.2	5.0–5.9
Carboxyl group	8.5–9.1	1.5–4.7	2.6–3.8
Phenolic OH	2.7–5.7	2.1–5.7	2.1–2.4
C=O group	1.1–3.1	0.9–5.2	4.8–5.7
OCH_3 group	0–0.5	0.3–0.4	0.3–0.4

that FAs. Fulvic acids contain more aliphatic compounds than HAs (Table 3.2.). Beyer (1996a) from a literature review summarized that FAs mainly consist of polysaccharides (carbohydrates, parts of the O-alkyl fraction) and variable amounts of alkyl carbon compounds with minor amounts of aromatic moieties. The polysaccharides may be modified and/or oxidized into carboxylic, ketonic and/or uronic acids. These (oxidized) polysaccharides and the fatty acids are the source of carboxyl groups whereas the noncyclic saccharides contain the aldehyde group. The HA fraction consists of little modified polysaccharides, aromatic lignin derivatives and long-chain alkylic moieties. The humin fraction contains long chain aliphatic, recalcitrant polymethylenes and lignin fragments are enriched in this fraction (Hatcher et al., 1980). It contains a high percentage of little modified litter compounds (Hempfling & Schulten, 1989) and the organomineral complexes (Kögel-Knabner, 1993). The non-extractable humins are thought to be HA type compounds that are strongly adsorbed, or precipitated on the mineral surfaces as metal salts or chelates. Spectra from nuclear magnetic resonance (NMR) and pyrolysis-field ionization mass spectrometry (Py-FIMS) of humin fractions are very similar to those of bulk soil samples (Preston & Schnitzer, 1984).

More than 100 compounds have been identified in the digests of oxidative degradation of HS (Hayes, 1991). Major compounds produced by the oxidation of methylated and unmethylated HS are aliphatic carboxylic, phenolic, and benzenecarboxylic acids (Schnitzer, 1978; Griffith & Schnitzer, 1989). Aliphatic dicarboxylic acids are the most abundant structures in the oxidative digests of HS and these include mono- to tetracarboxylic acids. Major aromatic oxidation products are benzenedi- and benzenepolycarboxylic (tri to hexa forms) acids, whereas phenolic acids include compounds containing between one and three OH groups and between one and five CO_2H groups per aromatic ring. Some examples of type of compounds identified in the alkaline permanganate and the alkaline cupric oxide media are illustrated in Fig. 3.3 (Hayes, 1991). In the reductive degradation digests,

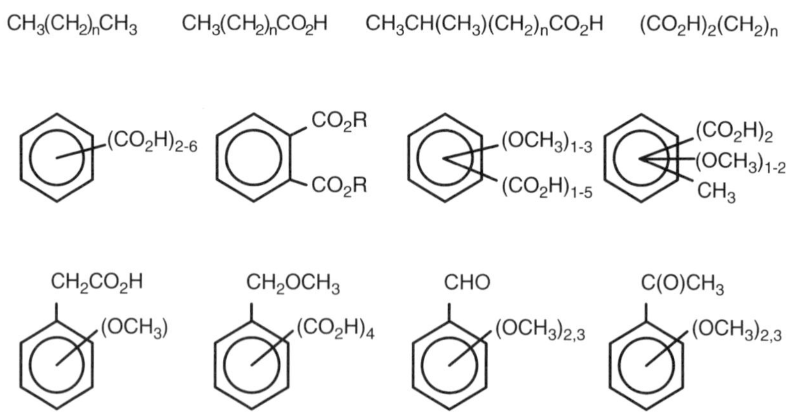

Fig. 3.3 Oxidation degradative products (Hayes, 1991, p. 12. Reproduced with kind permission from Woodhead Publishing Limited)

the type of compounds identified include phenols or their derivatives. Although alipahtic constituents were indicated in the digest by infra red spectroscopy but these could not be identified because of methodological limitations. Pyrolysis of the soil HS show that HAs are rich in compounds of polypeptide, and of lignin or polyphenol origin whereas that of FAs might have origins in substances with polysaccharides and pseudopolysaccharides, and to a lesser extent to the lignin-derived substances. By Py-FIMS the most abundant compounds identified in the humic fractions are carbohydrates, phenols, lignin monomers, lignin dimers, n-fatty acids, n alkylesters, and n-alkylbenzenes. Minor components include n-alkyl mono- and diesters, n-alkylbenzenes, methylnapthalenes, methylpenanthrenes, and N containing compounds. Humic acids tend to be enriched in n-fatty acids and the humin in n-alkylbenzenes (Schnitzer, 2000).

Mahieu et al. (1999) collected solid-state ^{13}C NMR data from a number of studies on 311 whole soils (varying in organic C content from 0.42% to 53.9%), physical fractions and chemical extracts to study the chemical composition of SOM under different systems of management and climatic conditions. They observed a remarkable similarity between all soils with respect to the distribution of different forms of C despite the wide range of land use (arable, grassland, uncultivated, forest), climate (from tropical rainforest to tundra), cropping and fertilizer practices. Functional groups in whole soils were always in the same order of abundance with a mean composition of: O-alkyls- 45%, followed by alkyls- 25%, aromatics- 20%, and finally carbonyls- 10%. Humic and fulvic acids contained relatively smaller proportions of O-alkyls and a larger proportion of carbonyls than whole soils (Table 3.3.). Humic acids contained more aromatics than the FAs and the whole soils. In FAs all the four functional groups were approximately in equal proportions. Clay-size fractions were the most different from whole soils, being more aliphatic (+8%). Sand size-fractions were generally similar to whole soils.

Based on the information on composition and functional group chemistry of HS a number of chemical structures have been proposed for HAs. Fuchs (1931) suggested that HAs consist of condensed aromatic and saturated rings substituted on the periphery by carboxyl and hydroxyl groups. The aromatic rings are linked by $-CH_2O$ and –C-N groups. Carbohydrates and peptides are bonded to the carbon linking the rings, and to CH_2 groups bonded to the rings. The model proposed by Flaig (1964) contains aromatic and quinone rings substituted by hydroxyl, carboxyl

Table 3.3 Distribution of alkyl, O-alkyl, aromatic, and carbonyl functional groups in whole soils, HAs and FAs (as % of total C in sample) visible to ^{13}C NMR (Compiled from Mahieu et al., 1999)

Functional group	Whole soil (n = 311)	Humic acids (n = 208)	Fulvic acids (n = 66)
Alkyls	24.8 ± 7.1	26.6 ± 11.0	26.5 ± 10.0
O-alkyls	44.8 ± 8.5	26.5 ± 7.9	25.9 ± 11.9
Aromatics	20.2 ± 6.0	30.5 ± 9.0	23.0 + 6.7
Carbonyls	10.1 ± 3.7	16.5 ± 4.5	24.7 ± 4.4

± indicates standard deviation

and methoxyl groups. Buffle's (1977) model consists of naphthalene rings substituted by hydroxyl, carboxyl, and short aliphatic chains containing alcohol, methyl, carboxyl, and carbonyl groups. Steelink (1985) proposed a tetramer HA model containing aromatic rings, phenols and quinones linked by aliphatic units with many OH groups. The COOH groups in this model are linked exclusively to aliphatic groups. The model was modified by Jansen et al. (1996) who proposed that building-block for HA has seven chiral centers and thus 128 stereoisomers. Instead of quinones, the model exhibits ketones or aldehydes.

Schulten et al. (1991) based on Py-FIMS and Curie-point pyrolysis gas chromatography/mass spectrometry (Py-GC/MS) data, proposed that HAs consist of isolated aromatic rings linked covalently by aliphatic chains. Schulten & Schnitzer (1993) developed a two-dimensional (2D) model structure of HA in which, n-alkyl aromatics play a significant role. Oxygen is present in the form of carboxyls, phenolic and alcoholic hydroxyls, esters, ethers, and ketones, whereas nitrogen occurs in nitriles and heterocyclic structures. The resulting carbon skeleton shows high microporosity with voids of various dimensions, which can trap and bind other organic and inorganic soil constituents as well as water. Schulten and Schnitzer (1997, 1998) converted the 2D HA structure to a three dimensional (3D) model by using HyperChem software (Fig. 3.4a). The model consists of 755 atoms (Table 3.4) with a molecular mass of 6,365 and contains 5 aliphatic and 21 aromatic carboxyl groups, 17 phenolic hydroxyls, 17 alcoholic hydroxyls, 7 quinonoid and ketonic carbonyls, 3 methoxyls, and 1 sulfur function. The authors (Schulten & Schnitzer, 1997) also proposed an SOM (containing 3% water) model structure consisting of 950 atoms and having a molecular mass of 7,760 g mol^{-1} (Table 3.4). The SOM model was improved subsequently (Schulten & Leinweber, 2000) to include one trapped trisaccharide, one hexapeptide and 12 water molecules one of which is protonated. The model structure contains 24-H bonds emphasizing the role of hydrogen bonding and dipole/dipole interactions in organic matter chemistry in soils (Fig. 3.4b).

Piccolo (2002) argued that the polymeric model of HS as proposed by many authors cannot explain some of their analytical results. Piccolo and associates (Piccolo et al., 1996a, b), therefore described HS as micellar associations that are stabilized by predominantly hydrophobic forces at pH 7. They suggested that the organic acids penetrate into the inner (hydrophobic) core of the micellar structure while neutralizing the HS acidic functions from pH 7 to 2. The association between the organic acids and HS occurs because of the amphiphilic properties of the acids which are able to interact with both the hydrophilic and the hydrophobic domains of humic aggregates. The authors proposed that the structures presented in Fig. 3.5 should be present in the original humic superstructure. By this concept HS may be considered as relatively small and heterogeneous molecules of various origin, which self-organize in supramolecular conformations. Humic superstructures of relatively-small molecules are not associated by covalent bonds but stabilized only by weak forces such as dispersive hydrophobic interactions (van der Waals, π–π, and CH–π bondings) and hydrogen bonds, the latter being more important at low pHs. In humic supramolecular organizations, the intermolecular forces determine

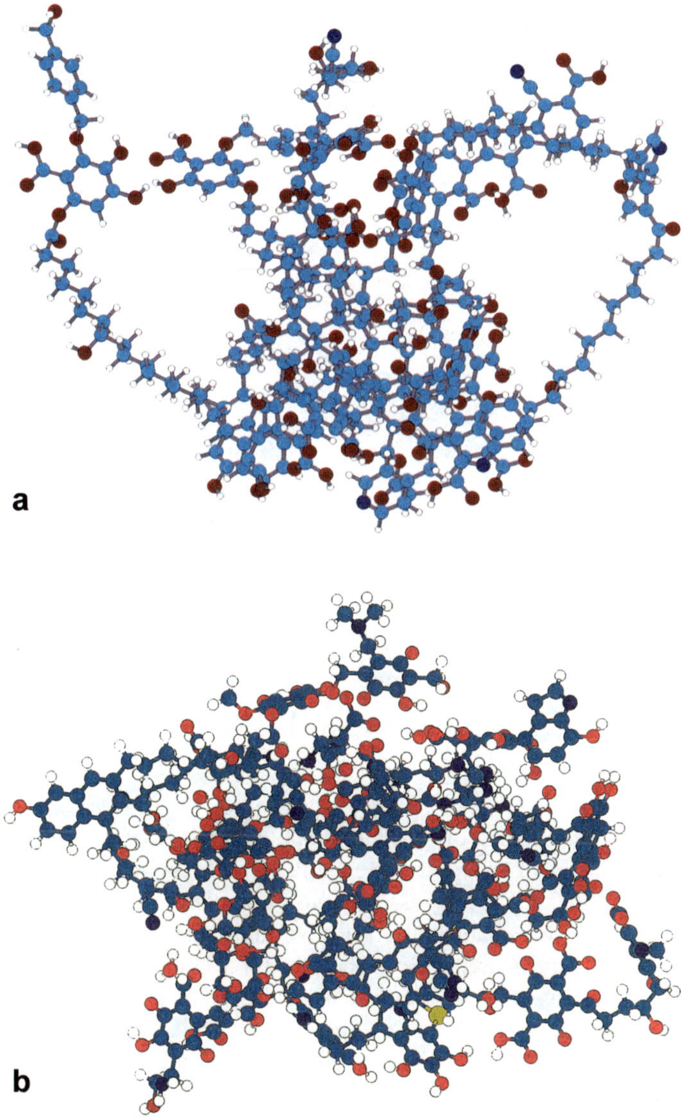

Fig. 3.4 Geometrically optimized three-dimensional structure of (a) soil humic acid and (b) soil organic matter. The element colors are: white H, cyan C, red O, blue N, yellow S (Schulten & Schnitzer, 1997, p. 120; Schulten & Leinweber, 2000, p. 414. Reproduced with kind permission from Wolters Kluwer Health; Lippincott Williams & Wilkins)

the conformational structure of HS and the complexity of the multiple non-covalent interactions control their environmental reactivity (Piccolo, 2002). Piccolo et al. (2003) proposed that based on the concept of supramolecular association, the classical definitions of humic and fulvic acids should be reconsidered. Fulvic acids may

Table 3.4 Chemical characteristics of humic substances (HS) and soil organic matter (SOM) with 3% water (Adapted from Schulten & Schnitzer, 1997)

Characteristic	HS	SOM + 3% water
Elemental composition	$C_{305}H_{299}N_{16}O_{134}S_1$	$C_{349}H_{401}N_{26}O_{173}S_1$
Elemental analysis (%)		
C	57.6	54.0
H	4.7	5.2
N	3.5	4.7
O	33.7	35.7
S	0.5	0.4
Molecular weight (Da)	6,364.8	7,760.2

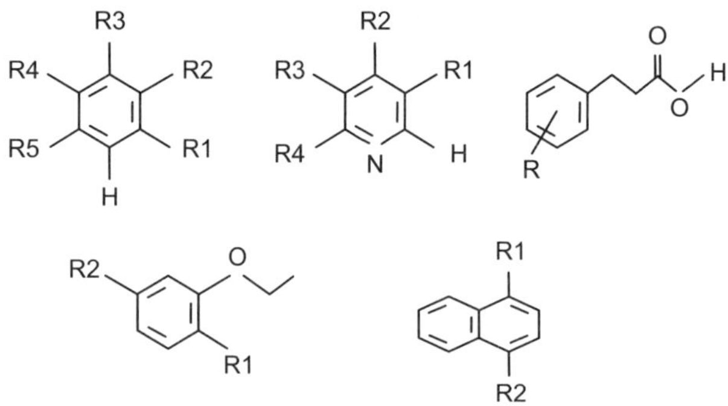

Fig. 3.5 Structural components of humic substances (Piccolo, 2002, p. 99. Reproduced with kind permission from Elsevier)

be regarded as associations of small hydrophilic molecules in which there are enough acidic functional groups to keep the fulvic clusters dispersed in solution at any pH. Humic acids are made by associations of predominantly hydrophobic compounds (polymethylenic chains, fatty acids, steroid compounds), which are stabilized at neutral pH by hydrophobic dispersive forces. Their conformations grow progressively in size when intermolecular hydrogen bondings are increasingly formed at lower pHs, until they flocculate.

Mao et al. (2000) used solid-state ^{13}C NMR to compare the chemical composition of HAs (from various Histosols), plant-extracted materials, and whole peat soil with different structural models of HAs. None of the eight models evaluated matched the composition of soil HAs completely, though a few models showed partial agreement. Therefore, search for a structural model of HS that can match the composition of soil HS still continues.

3.1.2.3 Nitrogen Compounds in Soil Organic Matter

Nitrogen exists in many different forms in soils, plants and animals. Soils form a major repository of N within terrestrial ecosystems. In soils, more than 90% of the nitrogen is organically combined with soil organic matter and it accounts for about

83% of the total N in the terrestrial biosphere (Anderson et al., 1991). The organic N in soil occurs in a variety of organic compounds, the main identifiable ones being amino acids and amino sugars. With the advancement of instrumental analytical techniques some nucleic acid bases and heterocyclics have been identified. Nitrogen in soil is usually characterized by acid or alkali hydrolysis. Generally, 20–30% of soil N cannot be solubilized by acid hydrolysis (termed acid insoluble N). But the acid hydrolysable fraction can be increased by pretreatment of the soil by hydrofluoric acid. Cultivation, manuring and other agricultural practices can alter the proportions of hydrolysable and nonhydrolysable N. Sowden et al. (1977) observed that 11–16% of soil N could not be hydrolyzed by hot 6 M HCl. Sharpley & Smith (1995) and Sulce et al. (1996) observed relatively high proportions of nonhydrolyzable N, to a maximum of 47% of total N.

The distribution of the major N compounds in soils formed under widely different climatic and geological conditions shows that amino acid N constitutes 33–42%, amino sugar N from 4.5% to 7.4% and ammonia from 18.0% to 32% (Sowden et al., 1977). Some of the ammonia probably originated from amino acid amides, amino sugars, and the release of fixed NH_4^+ from clays. The unidentified hydrolysable N constituted 16.5–17.8%. Estimates of nonprotein N ranged from 55% for the tropical soils to 64% for the arctic soils, averaging 61% for all soils meaning thereby that 40% of the total soil N was protein N (Sowden et al., 1977). Senwo & Tabatabai (1998) reported that total amino acids ranged from 10.9% to 32.4% of soil organic carbon and 12.0–27.4% of soil N. Though the hydrolysis of soils integrates the products from various components such as organic and mineral, living and dead yet hydrolysis of isolated humic and non-humic fractions gives broadly similar results (Table 3.5). This suggests that the easily-characterized products arise from co-extracted materials which are evenly distributed, possibly as a consequence of the strong reagents used in the extraction process, and also of the absorptive capacity, shape and surface activity of the humic macromolecules (Anderson et al., 1991).

The amino acid composition of soils has been found to be similar to that of bacteria (Sowden et al., 1977) indicating a major role of soil microbes in the synthesis of proteins, peptides, and amino acids from plant and animal residues. A number of protein and nonprotein amino acids have been identified in soils (Table 3.6) and there are possibly other amino acids present in soils that are yet to be identified

Table 3.5 Percentage distribution of forms of N in acid hydrolysates of whole soil, humic acids and fulvic acids (Adapted from Anderson et al., 1991)

Nitrogen form	Arable soils	Humic acids	Fulvic acids
Unhydrolysed	15 ± 6	12 ± 3	6 ± 2
Hydrolysed, unidentified	19 ± 6	30 ± 10	38 ± 10
Ammonium	21 ± 5	20 ± 5	18 ± 6
Amino acid	40 ± 7	40 ± 10	35 ± 5
Amino sugar	7 ± 2	5 ± 3	3 ± 2
Nucleic acid	0.7 ± 0.3	0.7 ± 03	1 ± 0.8

± indicates standard deviation

(Stevenson, 1994). Climatic conditions under which soils are formed, long-term cropping systems or agriculture management may alter the qualitative and quantitative composition of the amino acid fraction in soils. For example, tropical soils have been found to contain relatively higher amounts of acidic amino acids as compared to arctic soils (Sowden et al., 1977). Cropping systems involving legumes have been shown to increase the N content of SOM (Campbell, 1978; Praveen-Kumar et al., 2002). Similarly, treatments that increase the return of organic N residues to soils result in increased amount of hydrolysable amino compounds in soils. The most prominent amino sugars detected in soils are D-glucoasmine and D-galactosamine with the former occurring in greater amounts. Other amino sugars detected in relatively small amounts, are muramic acid, D-mannosamine, N-acetylglucosamine and D-fucosamine (Table 3.6)

Nucleic acid bases are generally considered to account for less than 1% of total soil N. However, Cortez & Schnitzer (1979) reported that nucleic acid bases constitute 3.1% of the total N in agricultural soils and 0.3% of the total N in organic soils. Nucleic acids identified in acid hydrolysates mainly include purines and pyrimidines (Table 3.6). Cortez & Schnitzer (1979) determined the distribution of purines (guanin and adenine) and pyrimidines (uracil, thymine, and cytocine) in 13 soils and humic materials. Quantitatively the distribution in soils was: guanin > cytosine > adenine > thymine > uracil. Humic acids were richer in guanine and adenine but poorer in cytosine, thymine and uracil than fulvic acids.

There remains a large amount of unidentified soil N that possibly results from interactions between amino acids and phenols or sugars. The effect of

Table 3.6 Amino acids, amino sugars, nucleic acid bases and other organic N compounds identified in soils and humic acids (Compiled from Stevenson, 1994; Schulten & Schnitzer, 1998)

Amino acids	Glycine, alanine, leucine, isoleucine, valine, serine, threonine, proline and hydroxyproline, phenylalanine, tyrosine and tryptophan, aspartic acid and glutamic acid, arginine, lysine and histidine, α-amino-n-butyric acid, α,ε-diaminopimelic acid, β-alanine, and γ-amino-butyric acid, ornithine, 3,4-dihydroxyphenylalanine and taurine, cysteine, methionine sulfone, and methionine sulfoxide
Amino sugars	D-glucoasmine and D-galactosamine, muramic acid, D-mannosamine, N-acetylglucosamine and D-fucosamine
Nucleic acid bases	Guanine, adenine, cytosine, thymine, and traces of uracil
Other N compounds	Pyrroles, imidazoles, pyrazoles, pyridines, pyrimidines, pyrazines, indoles and quinolines, N-containing derivatives of benzene (benzeneamines, benzonitriles, isocyanomethylbenzen benzothiazol, indole), aliphatic amines, and alkyl nitriles

strong alkali on these reactions is largely unknown. Schulten et al. (1997) applied Curie-point pyrolysis-gas chromatography/mass spectrometry (Py-GC/MS) and in-source pyrolysis-field ionization mass spectrometry (Py-FIMS) to characterize unidentified organic N in acid hyrolysates and hydrolysis residues of a Gleysol and a Podzol. They detected the presence of heterocyclic N-containing compounds (pyrroles, pyridines) and N-derivatives of benzene. Schulten & Schnitzer (1998) presented a detailed description of organic N compounds in soils and HAs, their molecular weights and chemical structures as determined by Py-FIMS and Py-GC/MS. A summary of the compounds is given in Table 3.6. The following distribution of total N in HS and soils has been proposed: proteinaceous materials (proteins, peptides, and amino acids)- ca. 40%; amino sugars- 5–6%; heterocyclic N compounds (including purines and pyrimidines)- ca. 35%; NH_3 – 19% out of which about 25% is fixed NH_4^+ (Schulten & Schnitzer, 1998). However, it has been suggested that the formation of heterocyclic N is associated with gradual humification occurring over years and is related to soil management. For example, in lowland rice soils, the proportion of heterocyclic N declines with increasing duration of submergence (Mahieu et al., 2000). Knicker et al. (2000) using ^{15}N NMR technique showed that most of the N compounds in HS are in the form of proteins that are trapped in the HS macromolecule. The origin of these N containing compounds is still unknown and needs further investigation.

3.2 Physical Characterization of Soil Organic Matter

While characterization of SOM by chemical procedures is useful for pedogenic studies, its division into different physical fractions or functional compartments in terms of persistence, availability or decomposability is important in situations related to soil fertility and plant productivity. Evidence accumulated in the last 3 decades have shown that fractionation of SOM according to particle size or density provides a useful tool for the study of its functions and dynamics in the terrestrial ecosystem. Separation of coarse or light fractions from fine fractions has been found to provide a relationship between density or the size of fraction and its turnover rate (Balesdent et al., 1987, 1988; Martin et al., 1990). As we will see later in Chapter 9, SOM fractions isolated by physical fractionation procedures, have been related to conceptual pools considered in some SOM turnover models (Cambardella & Elliott, 1992; Buyanovsky et al., 1994). Physical fractionation methods such as wet sieving, density flotation or chemical dispersal have been used to separate SOM into fractions of different sizes and stability classes. Numerous fractions varying in size or density or both have been defined by different authors. Depending on the severity of treatment, size separation can be achieved at aggregate or particle levels. Broadly SOM may be differentiated into two main fractions viz. particulate organic matter (POM) and organomineral complexes with further subdivisions based on size and/or density.

3.2.1 Particulate Organic Matter

Particulate organic matter also called uncomplexed organic matter mainly consists of partially decomposed plant and animal residues, root fragments, fungal hyphae, spores, fecal pellets, faunal skeletons, seeds and charcoal (Gregorich & Janzen, 1996; Christensen, 2001). Charcoal can constitute a significant proportion of POM in soils with a history of frequent vegetation burning (Skjemstad et al., 1990; Cadisch et al., 1996) and geomorphology (Di-Giovanni et al., 1999). Based on size or density or a combination of both, different fractions of POM such as coarse fraction (CF), light fraction (LF), free or inter-aggregate and occluded POM have been defined in the literature. Coarse fraction typically refers to SOM that is sand sized or larger (>53 µm) and common subdivisions include separation into 53–250 µm and >250 µm-sized material. Light fraction is isolated by density flotation in liquids ranging in density from 1.6 to 2.6 g cm^{-3} after a certain degree of dispersion of the soil. The yield of LF depends on the density used and the level of soil dispersion before the density flotation. The LF yield increases with increasing solution density and use of lower densities favors recovery of larger POM constituents (Ladd & Amato, 1980). The quantity and quality of LF depends on soil (e.g. pH, mineralogy, aeration and nutrient status), plant (e.g. litter quality) and climatic variables (e.g. temperature, moisture). In forest soils, the LF carbon is reported to constitute about 28% of the total soil C (Khanna et al., 2001). Free or inter-aggregate POM occurs in soil as loose organic particles and as adhering to the exterior of secondary organomineral complexes. The fractions: coarse, light and free POM represent the unprotected pool of SOM as these are not associated with soil minerals. The unprotected POM represents the labile fraction of SOM and it consists of plant residues in various stages of decomposition along with microbial biomass and microbial debris. It has high lignin content, high O-alkyl content, high C/N ratio, low N mineralization potential, and low mannose plus galactose/arabinose plus xylose ratio (Six et al., 2002).

Occluded organic matter is the intra-aggregate fraction of POM that is trapped and physically protected within micro- (<250 µm) and macro- (>250 µm) aggregates (Christensen, 2001). It differs considerably in composition as compared to free organic matter. While free organic matter consists mainly of partially decomposed litter residues, the occluded organic matter has undergone more decomposition during its physical protection within aggregates (Golchin et al., 1997) and has lower amounts of O-alkyl C (Kölbl & Kögel-Knabner, 2004). The extent of degradation and content of occluded POM is related to clay content. Clay content influences POM through its effect on soil aggregation. Kölbl & Kögel-Knabner (2004) found that in arable Cambisols from southern Germany, the amount of SOM stored in the occluded POM fraction increased with increasing clay content whereas it was not so for the free POM fraction (Fig. 3.6). The effect of clay content on the amount of occluded POM was most pronounced at clay contents between 5% and 30%. Higher soil clay contents promoted the conservation of POM with a low degree of alteration (Kölbl & Kögel-Knabner, 2004). This is probably because at high clay contents, protection of SOM against microbial decay occurs at an early stage of

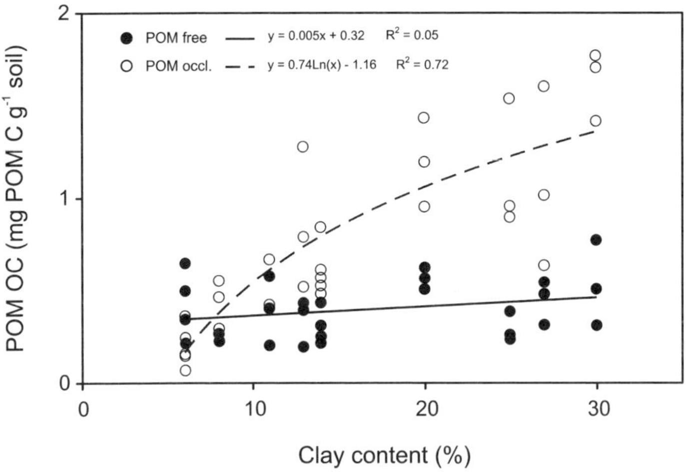

Fig. 3.6 Relationship between organic carbon content for free and occluded POM fractions and clay content (Kölbl & Kögel-Knabner, 2004, p. 49. Reproduced with kind permission from Wiley-VCH)

decomposition (Hassink & Whitmore, 1997), therefore, less degraded POM is occluded in the soil aggregates.

Aggregate occluded POM has a slower turnover rate than does unprotected POM (Beare et al., 1994a; Gregorich et al., 1995; Besnard et al., 1996; Jastrow et al., 1996; Wander & Yang, 2000). As a result there is greater C stabilization in the occluded POM as compared to free POM. Further, the shoot- and root-derived residues move between the two organic matter fractions at different rates and root derived materials are more rapidly occluded by aggregates (Besnard et al., 1996; Wander & Yang, 2000). It has been suggested that root-derived C in occluded POM may be more persistent in the long-term.

Because of variations in organic inputs and management practices, the proportion of SOM recovered as POM and its quality varies widely both in time and space. The POM content is affected by climate, land use, cultivation methods, soil and vegetation type, plant inputs, soil depth and a number of other factors that influence organic input and decomposition (Wander & Traina, 1996; Fließbach & Mäder, 2000). Its accumulation is favored in situations that slow down decomposition such as cold and dry climates, and where there is a large return of plant litter such as forests and grasslands. For example in native grassland soil, it can account for upto 48% of total soil organic carbon and 32% of the total soil N (Greenland & Ford, 1964). In soils with permanent vegetation POM can account for 15–40% of the SOM in surface horizons, whereas in long cultivated arable soil, the uncomplexed fraction usually makes up less than 10% of the organic matter (OM) in the tilled layer (Christensen, 2001). Typically POM has a C:N ratio of 20:1 with higher ratios in forest ecosystems. The C:N ratios of POM vary with fertilization practice and type of vegetation or crops grown. POM C:N ratios are reported to be higher in soils where crop production relies mainly on inorganic fertilizer N sources than in systems that include legumes or added organic manures (Kandeler et al., 1999a; Aoyama et al., 1999; Nissen & Wander, 2003). In soils where POM constitutes a large proportion of total SOM, there is a direct relationship between POM C:N and whole-soil C:N ratios. However in arable mineral soils, where POM-C usually accounts for a small proportion of the SOC, there is no clear relation between POM C:N ratios and whole-soil C:N ratios (Wander, 2004).

Different fractions of POM are strongly influenced by soil management (Christensen, 1992; Quiroga et al., 1996) and are considered to be good indicators of labile SOM or soil quality. Using ^{13}C natural abundance technique, Balesdent (1996) concluded that POM has a short mean residence time relative to C associated with clay- and silt- sized organomineral complexes, indicating the relatively high lability of POM. Garten & Wullschleger (2000) estimated the turnover times of coarse fraction POM-C in four switchgrass (*Panicum vigatum* L.) field trials in the southeastern US to be 2.4–4.3 years whereas those for mineral associated OM were 26–40 years. Many studies have shown that short-term soil C and N mineralization rates or the size of the microbial biomass are positively related to POM (Hassink, 1995; Monaghan & Barraclough, 1995; Fließbach & Mäder, 2000). The relationship between POM-C and biomass C has been used as an indicator of C availability (Alvarez et al., 1998). Because of ready availability of C in POM, it may be associated with immobilization of N in early stages of decomposition. Thus in some situations free POM could act as a sink rather than a source of plant-available mineral N (Whalen et al., 2000). POM seems to play an important role in the functioning of coarse-textured soils (Feller et al., 2001). It is especially important to N retention and availability in sandy soils, as the proportion of total N in POM is higher than in finer textured soils (Hook & Burk, 2000). Carbon content in POM appears to be more dynamic than the N content, therefore management effects on SOM are generally more apparent in the POM-C than in the POM-N fraction (Dalal & Mayer, 1986; Wander, 2004). While POM-N has been suggested to represent slow N pool (Delgado et al., 1996), POM-C is considered to be an effective measure of active SOM pool provided contaminants such as charcoal are not present or are accounted for (Gijsman, 1996; Gerzabek et al., 2001).

3.2.2 Organomineral Complexes

Most of the organic matter in soils is intimately associated with the mineral components, particularly with clay and silt-sized particles. The presence of pH dependent, or variable charge enables the humic molecules to form chemical complexes or chelates with metals, and interact with soil mineral particles to form organomineral complexes (Fig. 3.7). The mechanisms for the formation of organomineral complexes are postulated to be through van der Waal's forces, bonding by cation bridging, oxy or hydroxy bridges for hydroxyl and carboxyl functional groups in humic substances (Schnitzer, 1986), adsorption on interlamellar spaces of clay minerals, and through hydrogen bonding for neutral and negatively charged polysaccharides

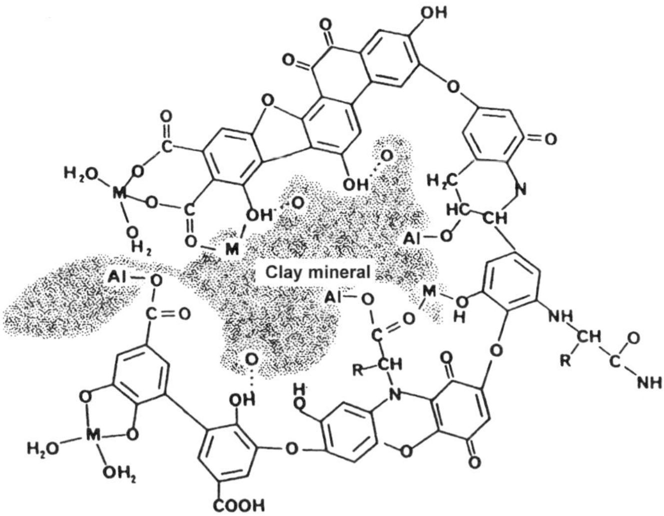

Fig. 3.7 Interaction of a clay particle and an organic molecule (Koskinen & Harper, 1990, p. 53. Reproduced with kind permission from American Society of Agronomy, Crop Science Society of America & Soil Science Society of America)

(Cheshire & Hayes, 1990; Cheshire et al., 2000). Formation of organomineral complexes results in stabilization of organic matter in terrestrial ecosystems.

Organomineral complexes are generally separated into silt and clay size fractions (<53 µm) and micro- (53–250 µm) and macro- (>250 µm) aggregates. Some workers consider the boundary for silt plus clay class to be <20 µm instead of <53 µm. Christensen (2001) classified organomineral complexes into primary and secondary organomineral complexes. The primary organomineral complexes considered as functional analogues to soil texture were divided into sand (20–2,000 µm), silt (2–20 µm) and clay (<2 µm) sized fractions. Whereas secondary organomineral complexes were divided into micro- (<250 µm) and macro- (>250 µm) aggregate sized complexes with further subdivisions into small microaggregates (<20 µm) and large microaggregates (20–250 µm). Macroaggregates have also been differentiated into small- (250–2,000 µm) and large (>2,000 µm) size classes (Degryze et al., 2004). The characteristics of primary organomineral complexes from temperate arable sandy soils are summarized in Table 3.7 (Christensen, 2001). Clay sized complexes have the highest concentration of organic matter (50–75% of the SOM) followed by silt- (20–40%) and sand- sized fraction has the least (<10%) concentration. The OM complexed with clay is dominated by microbial products, whereas the silt appears to be rich in aromatic residues derived from plants. The C/N ratio declines with decreasing particle size. Soil organic matter composition of the sand fraction is largely affected by the land use changes whereas silt- and clay- bound organic matter is more influenced by the chemical and physical environment (Guggenberger et al., 1994).

The formation of organomineral complexes results in chemical stabilization of SOM. The process of SOM stabilization is of greater importance in tropical soils

Table 3.7 Some characteristics of primary organomineral complexes isolated from temperate arable sandy soils (Adapted from Christensen, 2001)

Characteristic	Sand-associated OM	Silt-associated OM	Clay-associated OM
Composition	Enriched in plant polymers	Enriched in plant-derived aromatics	Enriched in microbial products and depleted in plant residue components
C/N ratio	Large	Medium	Small
Proportion of total SOM (%)	<10	20–40	50–70
C enrichment factor (E_c)[a]	<0.1	1–5	2–15
Cation exchange capacity (mmol kg^{-1})	10–150	60–350	300–900
Surface area[b] (m^2 g^{-1})	<10	10–50	25–100

[a]E_c = mg C g^{-1} fraction/mg C g^{-1} whole soil
[b]Surface area calculated by the BET equation using N_2 gas as adsorbent

than in temperate soils as the tropical climates favor decomposition of organic matter. In the absence of stabilization process, the tropical soils will be poor in SOM. The degree of stabilization depends on the silt plus clay content and the type of clay (i.e. 2:1 vs. 1:1 vs. allophanic clay minerals) (Sorensen, 1972; Chantigny et al., 1997; Guggenberger et al., 1999; Puget et al., 1999; Six et al., 2002). Amount of silt- and clay- associated C is related to silt plus clay content of soil (Hassink, 1997). However, the relationship differs depending on land use, clay type, and the size range defined for the silt plus clay fraction (Fig. 3.8; Six et al. 2002). Results from various studies suggest that the chemical composition of clay- and silt- sized organomineral complexes is little influenced by changes in management, while some changes can occur in sand-sized uncomplexed OM (Christensen, 2001).

In most soils, primary organomineral particles can exist as differently sized aggregates. Microaggregates can occur as free or within the macroaggregates. The potential of a soil to form aggregates depends on the size distribution of the primary complexes and their characteristics which in turn depends on fundamental soil properties, such as clay mineralogy and type and quantity of polyvalent cations and sesquioxides. The main agents in stabilization of these aggregates are microbial products, root exudates, polyvalent cations and other persistent binding agents (Christensen, 2001). Aggregation results in increased SOM accumulation in soils and provides physical protection to organic matter against decomposition. While the macroaggregates provide little physical protection (Beare et al., 1994a; Elliott, 1986), the microaggregates both in the free form (Balesdent et al., 2000) and within the macroaggregates (Denf et al., 2001) provide greater degree of protection to SOM decomposition. Jastrow et al. (1996) using ^{13}C natural abundance technique, calculated that the average turnover time of C in free microaggregates was 412 years, whereas the average turnover

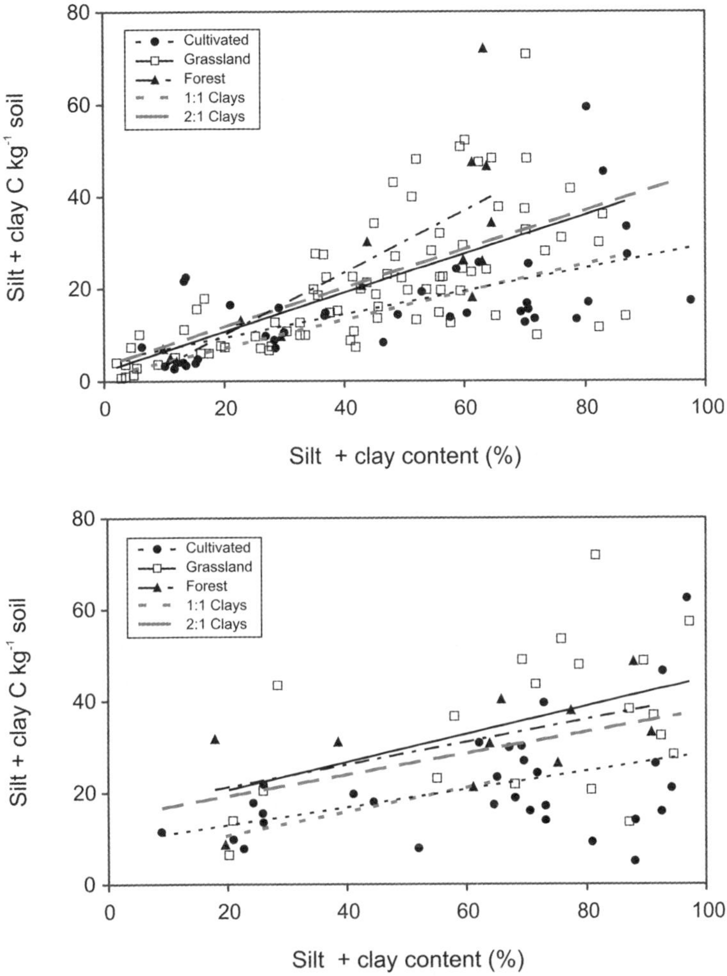

Fig. 3.8 Relationship between silt + clay content (%) and silt + clay associated C (g silt + clay C kg^{-1} soil) for grassland, forest and cultivated ecosystems. A differentiation between 1:1 and 2:1 clay dominated soils is also made. Figures (A) and (B) are for two silt + clay size boundaries viz. 0–20 μm and 0–50 μm, respectively (Six et al., 2002, p. 158. Reproduced with kind permission from Springer)

time for macroaggregates associated C was only 140 years in the surface 10 cm. The physical protection exerted by macro- and/or microaggregates on POM C is attributed (Six et al., 2002) to: (i) the compartmentalization of substrate and microbial biomass (van Veen & Kuikman, 1990; Killham et al., 1993), (ii) the reduced diffusion of oxygen into macro and especially microaggregates (Sexstone et al., 1985) and (iii) compartmentalization of microbial grazers (Elliott et al., 1980).

Various fractions of SOM respond differentially to management and stock variable amounts of soil organic carbon and nitrogen. Bayer et al. (2002) observed that in the mineral-associated SOM C and N stocks were higher by 4.6 and 16.8 times, respectively than in the particulate SOM. As compared to conventional tillage, no-tillage resulted in the largest increase of C and N stocks in the mineral-associated SOM. Degryze et al. (2004) found that afforestation on a former crop land resulted in the largest C sequestration in the fine intraaggregate POM (53–250 μm), whereas in the successional soils, C was preferentially sequestered in the mineral-associated and fine intra-aggregate POM C pools. Obviously, the identification of different fractions of SOM has implications for understanding C sequestration in terrestrial ecosystems.

3.3 Morphological Characterization of Soil Organic Matter

Morphological characteristics of SOM vary greatly between soil humus forms which are the interface between plants, soil animals and soil microbes. They are the sphere of important biological processes taking place in numerous terrestrial ecosystems. Morphological characteristics are commonly used to distinguish between different humus forms. The main humus forms are mull, moder and mor. Humus forms can be characterized as a defined combination of soil horizons containing significant amounts of organic carbon and nitrogen. Their diversity is attributed to numerous environmental factors. They are result of microbial and animal life in the soil and the locations of most biological and biochemical transformations taking place in terrestrial ecosystems. The organic material can be accumulated on the soil surface forming organic layers and in the upper mineral soil horizon. The properties of the humus forms are a suitable tool for surveying changes in ecosystems including the storage of carbon and nitrogen.

3.3.1 Classification of Terrestrial Humus Forms

In the FAO guidelines for soil profile description (FAO, 1990), organic horizons are referred to as either O or H horizons. The O horizons are those that develop upon mineral soils, described as mull, moder or mor (or raw humus according to Ulrich, 1987). According to AG Boden (2005), the O horizon (FAO, 1990) can be divided into different organic horizons, an L horizon (L: litter), an Of horizon (f: fermented) and an Oh horizon (h: humified). These forms of SOM are only saturated for a few days at a time during the year, and contain variably decomposed material with more than 20% organic carbon. Humus forms condition the development of terrestrial plant, animal and microbial communities. The L, Of and Oh horizons (AG Boden, 2005) are subjected to faster morphological alterations than the underlying Ah horizon. The H horizon is formed by an organic accumulation that is saturated for prolonged periods or is permanently saturated unless artificially drained. The H horizon should have a thickness of more than 20 cm but less than 40 cm and contain

18% or more organic carbon if the mineral fraction contains more than 60% of clay. Lesser amounts of organic carbon are permitted at lower clay contents. The H horizon may be between 40 and 60 cm thick if it consists mainly of sphagnum, or has a bulk density when moist of 1 g cm^{-3}.

The presence of the main terrestrial humus forms is linked to distinct patterns of biological activity, gaseous exchange with the atmosphere, and water and matter (particularly carbon and nitrogen) dynamics. The L horizon is an accumulation of slightly decomposed plant materials on the soil surface which corresponds to the O_i of the Soil Taxonomy (Soil Survey Staff, 1995). The Of horizon consists of <70 vol.% fine substance and the tissue of the original plant materials can be still identified. The Oh horizon consists of more than 70% none-structured organic matter.

On the basis of the characteristics of diagnostic A horizons, Schlichting et al. (1995) introduced a classification of humus forms on cultivated land, which was later completed by Beyer (1996b; Table 3.8). This classification is based on soil aggregation, SOM content and base saturation. Worm mull has been derived from the mollic epipedon (Soil Survey Staff, 1995). In contrast, crypto mull exerts an abiotic aggregation and contains less SOM. The base saturation (> or <50%, respectively) is the relevant soil property for distinguishing between crypto mull and crypto moder. In contrast to worm mull there is no formation of biotic aggregates in sand mull. Particularly the activity of soil arthropods leads to the formation of fine coagulates (Schlichting et al., 1995). Albic moder is the humus form of aric Anthrosols derived from podzols or stagnic soils. Sand grains are left with a greyish white appearance as they are stripped clean of iron hydroxides by downward or lateral migrating solutions. Thus the surface horizons of the former Podzol are depleted of iron (Bridges & Mukhopadhyay, 2003).

Table 3.8 Diagnostic properties of humus forms on cultivated land (Compiled from Beyer, 1996b)

Humus form	Diagnostic properties (A horizon)
Worm mull	Organo-mineral aggregates (biotic) SOM: >2% BS: >50%
Crypto mull	Subpolyeders, polyeders, prisma (abiotic) SOM: <2% BS: >50%
Crypto moder	BS: <50% Rest like crypto mull
Sand mull	No aggregates, only fine coagulates SOM: >2% BS: >50%
Moder of arable land	No aggregates, only fine coagulates SOM: > 2% BS: <50%
Albic moder	Greyish-white sand Rest like moder of arable land

SOM: soil organic matter; BS: base saturation

The humus forms presented previously can only be found under forests, heath and cultivated land on terrestrial soils (aeromorphic humus forms). Under the influence of groundwater or under stagnic conditions hydromorphic (semi-terrestrial) humus forms are common. In saturated conditions of bogs, organic material will accumulate as peat. Peat occurs extensively in arctic areas of North America and Eurasia and on upland areas in cool humid environments where heavy rainfall and continual saturation leads to blanket bog development. Alternatively, accumulation of peat can occur in saturated conditions associated with valleys in lowland situations. These fen peat deposits may be either acid or slightly alkaline depending upon the surrounding geology and supply of calcium-rich drainage waters. When drained these peats form excellent arable and horticultural soils, but as a result of drainage and cultivation, most of the peat becomes oxidized so that eventually only a dark-colored mineral soil remains. Limited areas of peat occur in tropical coastal areas where exploitation for agriculture has led in many cases to the development of acid sulfate conditions as pyrite oxidizes in soils and organic deposits are influenced by salt or brackish water. Subhydric humus forms develop at the bottom of lakes and are classified as Dy, Gyttja and Sapropel (AG Boden, 1994).

3.3.2 Characterization of Terrestrial Humus Forms

3.3.2.1 Mull

In biologically active soils with high earthworm population, Mull humus develops in L-Ah horizons. The A horizon (Fig. 3.9) is the place where most soil organisms are living, plant roots included (Bornebusch, 1930), and where humified organic matter is homogenized with mineral particles within organomineral aggregates (Bernier, 1998). Earthworms bury litter under their casts, thereby accelerating the incorporation of organic matter into the soil. Mull is also characterized by a rapid cycling of nutrients effected by a variety of organisms coexisting in the topsoil. Mull is typical of grassland and deciduous forest ecosystems of the temperate climate (Green et al., 1993) but is also common in forest soils of the tropical humid lowlands (Zech et al., 1997), and in semievergreen tropical forests (Loranger et al., 2003) due to favorable conditions for litter decomposition. Under deciduous forests (e.g. *Tectonia grandis*) of the subhumid and semiarid climates, the A horizons are often covered by thick L horizons during the dry season. Subsequently, during the rainy season, these litter layers are mineralized. Litter accumulation is also observed under *Acacia mangium* and *Acacia Auriculiformis* (Zech et al., 1997).

3.3.2.2 Moder

Moder humus forms are mainly found in deciduous (oak, beech) and coniferous forests of the temperate climate zone with nutrient-poor litter (Howard & Howard,

Fig. 3.9 L-Mull under beech (*Fagus sylvatica)* forest, Northrhine-Westphalia, Germany (soil: Calcaric Regosol). In Mull, the organic starting materials (L) are rapidly incorporated into the upper mineral soil horizon. Humified organic matter is homogenized with mineral particles form organo-mineral complexes (Photo: E. von Zezschwitz)

1990), in strongly acid tropical forest soils and in tropical mountains, where in comparison to tropical lowlands, organic matter turnover rates are lowered due to reduced temperatures and/or high water saturation (Zech et al., 1997). In Moder, macrofauna are smaller and reduced in abundance and diversity compared to Mull (Schäfer, 1991). Because of an absence of earthworms, organic matter is not incorporated into the mineral soil and, therefore, accumulates in the form of three organic horizons, L, Of and Oh. The transition between the O horizons are not sharp, and the cementation of organic matter in the A horizon by mineral particles is poor or nil, due to the scarcity of adhesive substances like mucoproteins or bacterial and root polysachharides (Bernier & Ponge, 1994). Most microbial biomass is fungal due to mere acid conditions than in mull. Fungi produce antibiotics and they further acidify the soil by excreting acids (Takao, 1965). Nutrients in Moder are sequestered in decaying plant debris, feces and epigeic fauna and fungi, which form the bulk Of horizon where most organisms are living (Wolters, 1988). The colonization of Of horizons by fine tree roots and their associated mycelia allows the vegetation to take up nutrients at the place where they have been released by detrital fungi and animals. Due to the conservation of organic matter in the form of animal feces and fungal biomass, nitrogen is abundant but mainly present as recalcitrant protein-N (dead fungal walls, animal cuticles, tannin-protein-complexes; Ponge, 2003). The soil atmosphere is characterized by pockets of high CO_2 and CH_4 contents (Sextone & Mains, 1990) due to the absence of carbonate buffering and poor redox values (Lee, 1999).

3.3.2.3 Mor

Mor originates from harsh climate conditions, very poor parent rocks (Ulrich, 1987), the production of acidic substances produced by mor vegetation, and the recalcitrant nature of the litter (Northup et al., 1998). As conditions are too acid for earthworms, incorporation of organic material is minimal. The Mor type of humus is strongly linked to the process of podzolization. In soils of regions where precipitation exceeds evaporation, and also where deep quartzitic sands occur from arctic climates to the humid tropical regions, the process of podzolization operates (Blume et al., 1997). Podzolization has almost always been preceded by strong leaching, which has decalcified the parent material and acidified the soil to the extent that only acidophyllous plants such as heath or coniferous forest will grow satisfactorily. These plants produce an extremely acid litter that breaks down slowly.

The faunal population is restricted, and organic breakdown mostly takes place through fungal activity. As a result, Mor develops with its L, Of and Oh horizons (Fig. 3.10). Decomposition products from plant debris are capable of chelating iron and aluminum sesquioxides and carrying them downwards in solution from the soil surface (Farmer et al., 1980, 1985; Anderson et al., 1982; Wiechmann, 1978). Acid weathering breaks down clay minerals, commonly mica, chlorite or kaolinite, aluminum is released. Several processes are involved in the precipitation of iron, aluminum and organic matter in the B horizon including bacteria that intercept and breakdown the organic linkages, the weakening of chelation bonds as they age, effects of soil wetting and drying, or the result of a pH rise in lower horizons. The results of the podzolization process are to produce an albic E horizon and a spodic B horizon.

In contrast to Moder where the transition with the mineral horizon is gradual, the Oh horizon of Mor is compacted and shows a sharp transition between the O horizons and with the underlying mineral horizon impoverished in macroaggregates (Nielsen et al., 1987). The poor nutritional value of Mor restricts the production of the vegetation. Typically the ground vegetation is made of lichens, mosses and ericaceous shrubs (Bonan & Shugart, 1989). Mor is present under coniferous trees, such as pine and (Northup et al., 1998). In mor, vegetation takes a prominent place to the detriment of animals and microbes, which are at their lowest level of abundance and diversity (Davis, 1981). Compared to Mull and Moder, the conservation of organic matter in Mor is at its optimum and the release of nutrients at its minimum (Aerts, 1995).

Tangelmor is a specific form of Mor (Zech et al., 1997), which is characterized by dark acidic to strongly acidic organic layers (Fig. 3.11). Tangelmor occurs on calcareous rock as well as on silicate rocks in mountain regions and develops under harsh climatic conditions (high precipitation, low mean temperatures). For example,

Fig. 3.10 Mor under spruce (*Picea abies*) forest, Northrhine-Westphalia, Germany, showing distinct (linear) transitions from the L to the Of and the Of to the Oh horizon (soil: Haplic Podzol). The Oh horizon is compacted (Photo: E. von Zezschwitz)

Fig. 3.11 Tangelmor under Alpine pine (*Pinus mugo*), Hohe Tauern (elevation: 1,950 m above sea level), Austrian Alps (soil: Umbric Leptosol) (Photo: R. Nieder)

in the Alps Tangelmor is present in the subalpine forest zone (*Piceetum subalpinum*) at elevations from about 1,000 m to about 1,650–1,850 m and in the *Pinus mugo* zone from 1,650–1,850 m to 2,000–2,300 m above sea level (Gansen, 1972). The cover of the organic layer with L material is frequently incomplete, whereas the Of and Oh horizons together commonly are more than 30 cm thick and thus exert a significant reservoir for organic C and N.

3.3.3 Humus Form Development in a Forest Succession

The development of humus forms can be associated with a forest succession. Mull is generally associated with early development stages of forest stands, Moder with phases of intense growth of trees before they reach maturity. During maturity of the forest stand the internal recycling of nutrients and the slower growth of trees yield more nutrients to the decomposer system (Nilsson et al., 1982). Earthworms are sensitive to environmental changes occurring during tree stand development. In soils with moderate to high base saturation, there is a global trend of decreasing abundance of earthworms during the phase of intense growth of trees, followed by progressive recovering as trees reach maturity (Bernier & Ponge, 1994). This process can be considered as a driving force for the observed changes in humus forms. In old-growth forests, functional diversity may recover during maturity through colonization by herbs and Mull-forming organisms which locally disappear during the phase of intense nutrient uptake by trees (Bernier & Ponge, 1994), and Mull may recover during maturity and senescence of the trees. In contrast, a reverse from Moder to Mull during maturity is not common in sandy acidic soils under coniferous forest where earthworms are absent (Nieder et al., 2003b).

3.3.4 Ecological Features of Humus Forms

Table 3.9 summarizes the main features of humus forms at the ecosystem level. The high plant, animal and microbial productivity and biodiversity observed in Mull can be explained by milder climate, higher richness of the parent rock, and by the permanent bioturbation created by the activity of organisms. A high level of biodiversity and productivity can be achieved only through a rapid cycling of nutrients which allows organisms to adapt themselves to a constantly changing environment and to a high level of competition. Mull organisms have a low nutrient use efficiency, which means that they use more carbon per unit nutrient taken up because they have high energy costs for capturing space and nutrients at a high level of competition (Grime & Hodgson, 1987).

In Moder, the number of functional processes is restricted compared to mull but the level of biological activity is still high. In Mor, dearth of nutrient resources and harsh climate conditions impose a stricter use of energy and nutrients. Fewer species

Table 3.9 Main morphological and ecological features of humus forms (Compiled from AG Boden, 2005; Ponge, 2003)

	Mull	Moder	Mor
Ecosystem	Grassland, deciduous forests, Mediterranean scrublands	Deciduous and coniferous forests	Coniferous forests, heathlands, alpine meadows
Horizons[a]	L-Ah	L-Of-Oh-Ah/AE	L-Of-Oh-AE
Transitions between O horizons		Not sharp	Distinct (linear)
Soil units[a]	Chernozems, Luvisols, Rendzic Leptosols, Cambisols, Kastanozems	Cambisols, Luvisols	Podzols
Productivity	High	Medium	Low
Biodiversity	High	Medium	Low
Humification	Rapid	Slow	Very slow
Humified organic matter	Organo-mineral complexes	Holorganic fecal pellets	Fecal pellets, plant debris
Nutrient availability	High	Medium	Low
Nutrient use efficiency	Low	Medium	High
Faunal group dominant in biomass	Earthworms	Enchytraeids	None
Microbial group dominant in biomass	Bacteria	Fungi	Fungi

[a] According to AG Boden (2005)

are present and functions like burying of organic matter are absent. Mor plants exhibit an intense production of recalcitrant and sometimes toxic secondary metabolites (Northup et al., 1998). The low productivity and the absence of internal disturbance due to a poor number of competitors allows a better nutrient use efficiency in Mor than in Mull. The stability of such ecosystems is not ensured by the fast adaptation of a variety of organisms to constantly changing conditions, but rather to the durability of environmental conditions created by a few Mor-forming organisms, especially through the accumulation of recalcitrant organic matter which isolates the ecosystem from outer poor environmental conditions.

Chapter 4
Organic Matter and Soil Quality

Soil, water and air quality are strongly interdependent. Soil quality has different meanings according to the perspective of a person. There is as yet not a well-defined universal methodology to characterize soil quality and to define a clear set of indicators (see Parr et al., 1992; Karlen et al., 1997; Shaxson, 1998; Bouma, 2002; Nortcliff, 2002; Schjønning et al., 2002; Gil-Sotres et al., 2005). For a farmer soil quality may mean sustaining or increasing productivity and sustaining the soil resource for future generations, for consumers it may mean healthful and inexpensive food production, and for an environmentalist it may mean a holistic view of the soil in ecosystems with respect to air and water quality, nutrient cycling, biomass production and biodiversity.

Soil organic matter is a key attribute of soil and environmental quality because it is an important sink and source of main plant and microbial nutrients and more over exerts a profound influence on physical, chemical and biological functions despite its often minor contribution to the total mass of mineral soils. It is commonly recognized that SOM imparts desirable physical condition to soil. Organic matter incorporated into the soil can affect its structure, as denoted by porosity, aggregation, and bulk density, as well as causing an impact as expressed in terms of content and transmission of water, air and heat, and of soil strength. Organic matter to a significant part exerts an influence on chemical properties of soils. During organic matter decomposition, nutrients (particularly N, P, and S) are released into the mineral nutrient pool. Reactive carbon and nitrogen compounds can also be released to the atmosphere (CO_2, N_2O, NO_x) and to surface and ground water (NO_2, NO_3^-, NH_4^+, dissolved organic C and N).

Ion Exchange capacity and the retention of metals increase following organic matter additions. Other soil chemical properties such as pH, electrical conductivity, and redox potential are determined greatly by the quality and content of SOM. One of the most fundamental functions of SOM is the provision of metabolic energy which drives soil biological processes. Surface active particles like clay minerals concentrate biological molecules at the solid-liquid interface. They can support the growth of microorganisms adsorbed on these surfaces.

Strong dependencies exist between the different groups of functions. For example, the ability of organic matter to chelate multivalent cations can affect its potential to stabilize soil structure as well as its biodegradability. The effects of organic matter on soil properties often involve interactions with the soil mineral fraction with the consequence that variations in soil properties across different soils may be a consequence of both variations in the SOM quality and quantity and in the mineral soil components.

4.1 Soil Quality

Soil quality has been a topic of greatest interest in soil science during the last few years. At present, more than 1,500 publications are available that use the term "soil quality" as a key word (Gil-Sotres et al., 2005). This interest has been focused on searching for reliable ways for evaluating soil quality. However, papers dealing with the estimation and quantification of the level of soil quality are rare, mainly because the search for quantitative indexes for soil quality is difficult. In soils, many changes can take place over the long term, and thus a change in soil quality can only be perceived when all the effects are combined over a period of time. Many soil properties are indicators of soil quality, but there is still no consensus as to how they could be used. In this chapter, the major indicators of soil quality and the trends in their use are discussed.

4.1.1 Definition and Concept

For decades, soils have increasingly been subjected to degradation and pollution. Soil may be lost in a relatively short period of time if used inappropriately or mismanaged with very limited opportunity for regeneration or replacement, indicating that the soil is not an inexhaustible resource. During the late 20th century there has been a recognition about the key involvement of the soil in crop production and in water and atmospheric purification, emphasizing the role of soil both for crop production and for environmental quality. Protecting soil and preserving its overall quality has become a key international goal (Filip, 2002). Early concepts of soil quality dealt mainly with soil properties that contribute to soil productivity, with little consideration for a definition for soil quality itself. During the last 10 years many definitions of soil quality have been proposed (e.g. Carter et al., 1997; Karlen et al., 1997; Arshad & Martin, 2002; Filip, 2002), and a recent one proposed by the Soil Science Society of America (cited by Arshad & Martin, 2002) is: "The fitness of a specific kind of soil, to function within its capacity and within natural or managed ecosystem boundaries, to sustain plant and animal productivity, maintain or enhance water and air quality, and support human health and habitation." While most countries have national criteria for water and air quality, criteria against which the quality of soil may be judged and interrelationships between soil, water and air quality are still discussed (Nortcliff, 2002). Due to the lack of agreement on the definition of soil quality, there is presently no consensus regarding a maximum quality reference soil. The different approaches can be summarized in two directions. The first considers a soil in an equilibrium with all compartments of the environment, i.e. a climax soil with climax vegetation. The second direction considers a maximum quality soil as capable of maintaining high productivity and causing the minimum environmental distortion. Climax soils have not been frequently used for soil quality studies because in most developed countries these soils have practically disappeared due to anthropogenic influences. The use of the second option regarding a maximum quality soil has been justified by the fact that the productive function of a soil is still seen as being the most important one. From a cropping system outlook in a temperate climate there is the belief that soils with deep, mollic and base-rich A horizons such as Chernozems may be the best quality soils. As the productive function of soils has been considered as the most important factor, most of the publications in recent years deal with the estimation of soil quality in agro-ecosystems and their modification as a consequence of soil tillage, fertilization and crop rotations.

Present approaches to quantify soil quality are concerned with either directly characterizing soil properties or identifying specific functions that can represent the attribute in question (Gregorich et al., 1994). The term soil quality can be seen as an integral value of compositional structures and functions of soil. Commonly accepted soil functions are listed in Table 4.1.

Carter et al. (1997) distinguished between inherent and dynamic soil quality. Inherent soil quality depends on the soil's mineral composition, soil texture and depth. It is mainly viewed as almost static and usually shows little change over time, assumed that soils are not degraded by human activities. In some cases, soils that possessed originally good inherent quality are affected by erosion, deposition or desertification. Inherent soil quality for crop production cannot be evaluated independently of extrinsic factors (Janzen et al., 1992). These include physical parameters like climatic, topographic, and hydrologic factors which can be viewed as attributes of landscape quality. In general, physical factors are of less use than biological and biochemical parameters as they alter only when the soils undergo a really drastic change (Filip, 2002). In contrast, biological parameters are sensitive to slight modifications that the soils can undergo in the presence of any degrading agent (Yakovchenko et al., 1996). Thus, attributes of dynamic soil quality are subject to change over relatively short time periods. For example, microbial biomass and population, soil respiration, C and N mineralization rates can change within a few hours or days. In comparison, labile organic matter fractions may change over a period of months to years, and stable organic matter ("inert fraction") may remain almost unchanged within a period of years or decades. Attributes of dynamic soil quality are strongly influenced by soil management.

Whenever sustainability of soil natural functions have to be evaluated, key indicators must include biological and biochemical parameters. The majority of biological and biochemical soil properties used as soil quality indicators include the activity of microbial processes and that of hydrolytic soil enzymes (Nannipieri et al., 1995). Considering the wide number of these parameters, Visser & Parkinson (1992) pointed out to consider different levels of study which includes the use of specific

Table 4.1 Functions commonly used for the assessment of soil quality

Function	Source
- Production function	FAO (1995)
- Biotic environmental function	
- Climate regulation	
- Hydrologic function	
- Storage function	
- Waste and pollution control	
- Living space function	
- Archive or heritage function	
- Connective space function	
- Recycling of organic materials to release nutrients and energy	Warkentin (1995)
- Partitioning rainfall at soil surface	
- Maintaining stable structure to resist water and wind erosion	
- Buffering against rapid changes in temperature, moisture, and chemical elements	
- Storing and gradually releasing nutrients and water	
- Partitioning energy at the soil surface	
- Sustaining biological activity, diversity, and productivity	Soil Science Society of America (1995)
- Regulation and partitioning water and solute flow	
- Filtering, buffering, degrading, immobilizing, and detoxifying organic and inorganic materials	
- Storing and cycling nutrients and other elements within the earth's biosphere	
- Provide a physical, chemical and biological setting for living organisms	Nortcliff (2002)
- Regulate and partition water flow, storage and recycling of nutrients and other elements	
- Support biological activity and diversity for plant growth and animal productivity	
- Filter, buffer, degrade, immobilize and detoxify organic and inorganic substances	
- Provide mechanical support for living organisms and their structures	

groups of properties. One level is that of the biotic community, which implies the composition and distribution of different functional groups of soil microorganisms. A second level involves population studies, considering the dynamics of specific organisms or microbial communities. A third level considers the use of properties that are related to the biogeochemical cycling of C, N, P and S. The structure of microbial population is commonly determined with molecular techniques such as fatty acid profiles and DNA characterization (Nannipieri et al., 2002). In the following section, the beneficial effects of SOM on physical, biological and biochemical properties are discussed.

4.2 Impact of SOM on Soil Physical, Chemical and Biological Properties

Soil organic matter exerts a profound influence on physical, chemical and biological properties of soil (Table 4.2). The effects of SOM on soil quality often involve the mineral fraction. Thus, variations across different soils are also a consequence of variations in the soil inorganic component.

Table 4.2 Soil functions and processes that are mainly influenced by soil organic matter (Adapted from Baldock & Nelson, 2000)

Property/attribute	Function/process
Physical	
Aggregate stability	Aggregation of soil particles by formation of bonds with the reactive surfaces of soil mineral particles
Water retention	Alteration of water storage capacity through increase in water absorption (up to 20 times its mass) and modification of pore size distribution
Soil color	Alteration of soil thermal properties by dark color
Chemical	
Exchange capacity	Adsorption of cations and anions from the soil solution
Interactions with metals	Retention or mobilization (by chelation) of metals
Interactions with organics	Alteration of biodegradability, activity and persistence of organics (e.g. agrochemicals) added to soils
Buffering capacity	Proton buffer in slightly acidic to alkaline soils
Biological and biochemical	
Energy source	Metabolic energy provision for biological processes
Reservoir of nutrients	Mineralization of organic matter provides macronutrients (N, P and S)
Microbial biomass activity	Stimulation or inhibition of microbial biomass and activity by the presence of organic materials
Enzyme activity	Indication of the oxidative activities of the soil
Ecosystem resilience	Significant pools of organic matter and associated nutrients can enhance the ability of an ecosystem to recover after natural or anthropogenic disturbances

4.2.1 Physical Properties

4.2.1.1 Bulk Density

Bulk density (BD), defined as the weight of oven-dry soil per unit volume, is an indicator of the soil's physical condition. BD is related to a soil's porosity, texture, hydraulic conductivity, aggregation, compaction, and organic matter content. Continuous cultivation tends to raise the bulk density, i.e., to compact the soil and thus reduce infiltration, aeration or root growth and to raise the energy needed for cultivation. In general, soil BD values decrease as the rate of organic residue applications and the SOM content increase regardless of the soil texture or the residue type used. The reduction of soil BD appears to be influenced primarily by the rate of organic residue addition. Khaleel et al. (1981) surveyed results of 42 field experiments dealing with the effects of manures and composts on soil properties. A highly significant correlation was found between the increase in SOM induced by manure application (ΔC) and the lowering (in percent) of soil bulk density (ΔBD):

$$\Delta BD = 3.99 + 6.62\ \Delta C\ (R^2 = 0.69;\ p \leq 0.01) \tag{4.1}$$

The data related to the beneficial effect of organic matter on the soil bulk density include a report from the Rothamsted plots where the bulk density of the soil in plots receiving only mineral fertilizers since 1852 was 1.52 Mg m^{-3} compared to a density of 1.29 Mg m^{-3} in plots amended with manure (Jenkinson & Johnston, 1977). Pettersson & Von Wistinghausen (1979) reported that the subsoil was compacted in plots receiving only inorganic fertilizers for a period of 20 years. The subsoil in the manured plots had a better structure and a lower bulk density. Such an effect on the deep subsoil layers would indicate that organic substrates move downward and are active below the plow layer. Such migration could be due to leaching of DOM or via bioturbation mechanisms such as the movement of earthworms. Reduction in soil BD is probably due not only to the dilution effect of adding less dense organic matter to the more dense mineral matter, but also to increased soil aggregation.

4.2.1.2 Aggregate Stability

Aggregation, or the binding together of individual soil particles, gives rise to what is known as soil structure. Typically, a well-structured soil has greater resistance to the forces of erosion and has improved air-water relationships. In general, hydraulic conductivity, infiltration rate, air diffusivity, surface drainage, and ease of root penetration will increase with increasing aggregation. Soil organic matter is considered important to the maintenance of aggregate stability of a wide range of soil types including Chernozems, Castanozems, Luvisols and Cambisols. Its importance tends to be less in Ferralsols and Andosols where hydrous oxides play an important stabilizing role, and in self-mulching soils like some Vertisols that contain clays with a high shrink/swell potential. Improving or increasing aggregation is more desirable on finer textured soils (silt loams, clay loams and clays). A fine textured soil will behave much like a coarse one with respect to water infiltration and drainage, if clay and silt particles are bound together into aggregates. Management techniques, such as conventional tillage, row cropping, and complete removal of vegetation can decrease the degree of aggregation and increase soil bulk density. On more sensitive soils (e.g. silty and sandy loams), surface soil structure can be completely destroyed. The need to increase soil aggregation in many situations is apparent. The aggregates formed must be resistant to degradation by the forces of water (e.g. rain drop and irrigation impact) and tillage operations.

In soils where organic matter is an important agent binding mineral particles together, a hierarchical arrangement of soil aggregates exists in which aggregates break down in a stepwise manner as the magnitude of an applied disruptive force increases (Oades, 1993; Oades & Waters, 1991). The existence of three levels of aggregation was proposed by Golchin et al. (1998): (i) the binding together of clay plates into packets <20 µm, (ii) the binding of clay packets into stable microaggregates (20–250 µm), and (iii) the binding of stable microaggregates into macroaggregates (>250 µm).

The importance and nature of the organic materials associated with each level of aggregation vary. At the scale of packets of clays, aggregation is primarily dictated by soil mineralogical and chemical properties important in controlling the extent of dispersion, and is often a function of pedological processes. The increase in aggregate stabilization is mainly due to the production of organic macromelecules by microorganisms which bind primary particles and microaggregates to macroaggregates (Becher, 1996). The binding together of clay packets to form microaggregates occurs via a range of mechanisms. The dominant mechanism is proposed to involve polysaccharide-based glues (mucilages or mucigels) produced by plant roots and soil microorganisms (Ladd et al., 1996). Emerson et al. (1986) presented transition electron micrographs showing mucilage located between packets of clay plates. Small microaggregates (<53 µm) held together by humified organic matter and biologically processed materials are bound together around a particle organic core (Oades, 1984; Beare et al., 1994b) to produce larger microaggregates and small macroaggregates <2,000 µm. Macroaggregates >2,000 µm are stabilized by the presence of roots, fungal hyphae and larger fragments of plant residues, which interconnect soil aggregates via bonding to aggregate surfaces, penetration into or through aggregates, and physical enmeshment (Churchman & Foster, 1994; Tisdall & Oades, 1982).

The addition of organic residues to soils has been found to be an effective method to not only increase total aggregation, but also increase the proportion of water-stable aggregates (Benbi et al., 1998). The effect of organic amendments on water-stable aggregate formation is less on soils with existing high levels of aggregate stability. After five annual additions of sewage sludge totalling 61 Mg ha^{-1} to a clay soil with high aggregate stability, and cropped to a grass/clover mix, aggregate stability increased by only 2% (Furrer & Stauffer, 1983). In contrast, 3 years after a single sewage sludge addition of 50 Mg ha^{-1} to a heavy, initially non-stable clay soil the soil's percentage of water-stable aggregates more than doubled (Vigerust, 1983).

The type of crop growth or rotational sequence modifies the effect of organic amendments on aggregate stability. Using various rotational sequences (grass/clover, corn, and wheat) on three different soil substrates (clay, sandy loam and loamy sand) and five annual sewage sludge additions, it was found that the longer the time period from the last grass/clover crop in the rotation, the less effect sludge had on soil aggregation (Furrer & Stauffer, 1983). The degree of improvement of structure particularly depends on particle size distribution. Sandy soils with low SOM contents lack substantial structure. Adding organic materials will increase microbial activity which in some cases lead to the buildup of SOM and the formation of macro- and micro-aggregates (Sparling et al., 1992). Also heavy clay soils are often characterized by poor structure and aeration, but they can be improved through the addition of organic amendments. In summary, the positive effect of SOM on soil structure will be more pronounced for sandy and clay soils than for silty soils.

In order to stabilize an aggregate it is necessary for the organic material to penetrate into the aggregate. In the cases of organic molecules in soil solution it is important that these diffuse to internal sorption sites. It is evident that the larger the molecule the more likely it is to be able to bridge two or more soil particles. The most effective molecules for stabilizing soil aggregates would have linear or nonlinear helix conformations in solution (see Chapter 3). Such conformations would span the longest distances, but even relatively short macromolecules could be effective binders of soil inorganic colloids. Molecules with random coil structures would span shorter distances than linear molecules with the same molecular weights. The organic molecules in soil which are most likely to be water soluble and mobile are fulvic acids and polysaccharides. The latter would arise as exudates from plant roots or as exocellular microbial polysaccharides, or contained in bacterial or fungal cell walls. Many of these polysaccharides have the properties of gums, are highly viscous, and will generally stay at the location where they were exuded or secreted.

The total soil volume occupied by pore space is inversely proportional to a soil's bulk density. As with BD, porosity is generally viewed as an index of other soil physical properties such as compaction, air-water relationships, and root penetration. Pore size distribution is an important soil physical property. Classification schemes consider pores smaller than 50 μm in diameter to be storage pores. Pores larger than this size are drained at field capacity. Storage pores are important water and nutrient reservoirs for microorganisms and plants. Pore diameters ranging from 50 to 500 μm are considered transmission pores and important for the movement of water and the exchange of gases through soil. These pores are important for root penetration. Most roots need pores in the range of 100–200 μm in which to grow. Pores greater than 500 μm in diameter are termed fissures.

Numerous studies have investigated the effect of organic amendments on soil porosity and pore size distribution. The addition of sewage sludge increased total porosity (Clapp et al., 1986). These increases occurred in various soil substrates, ranging from sandy loams to silt loams. Increases in total soil porosity were evident for as long as 12 months after sewage sludge addition. However, increases in porosity appeared to be most apparent in the first cropping season after organic material application. Porosity decreased after the winter period, but was still greater than in soils receiving no sewage sludge. Pore size distribution was observed to shift from a relatively high percentage of fissures before sludge application to greater percentages of storage and transmission pores after addition, which are most important from an agronomic as well an ecological point of view.

4.2.1.3 Moisture Retention

The addition of organic materials to soil increase water retention at both field capacity (FC: −33 kPa) and permanent wilting point (WP: −1,500 kPa) which is due to the higher water absorption capacity of SOM. Soil organic matter can absorb and hold substantial quantities of water, up to twenty times its mass (Stevenson, 1994). The greatest increases in water retention are for treatments on coarse textured soils or using high application rates of organic materials. Sludge application of 56 Mg ha^{-1} to a silt loam caused an increase in both FC and WP of 14.9 and 14.7%, respectively. The same application on a sandy loam soil increased these parameters by 17.1 and 51.7%, respectively (Kladivko & Nelson, 1979). A literature review by Khaleel et al. (1981) yielded that approximately 80% of the observed variation in increase in water retention at both FC and WP, could be explained by soil texture (% sand) and increase in SOM contents.

Organic materials can influence soil water retention directly and indirectly. The direct effect depends on the morphological structure of the organic materials and will not impart any beneficial effect to the soil unless it serves to enhance the ability of soil to hold water at potentials within the plant available range. Organic matter in the form of surface residues can also influence water retention directly by reducing evaporation and increasing the infiltration of water. The indirect effect of organic materials on water retention arises from its impact on soil aggregation and pore size distribution, and thus on the plant available water holding capacity (WHC) of soil (the difference between volumetric water content at field capacity and at permanent wilting point). This effect is best exemplified by the inclusion of SOM content as a significant parameter in pedotransfer functions which predict pore size distribution (Vereecken et al., 1989; Kay et al., 1997). Equation 4.2 presents the pedotransfer function derived by da Silva & Kay (1997):

$$\begin{aligned}\Delta\theta_w &= \text{Exp}\,[4.15 + 0.68 \ln CL + 0.42 \ln OC + 0.27 \ln BD]\,\Psi_m \\ &\quad [-0.54 + 0.11 \ln CL] + 0.002 \ln OC + 0.10 \ln BD \\ &\quad (R^2 = 0.94;\ p \leq 0.01)\end{aligned} \tag{4.2}$$

It describes the relationship between volumetric water content θ_w ($m^3\ m^{-3}$) and matric potential, Ψ_m (MPa); clay content, CL (%); organic C content, OC (%);

and bulk density, BD (Mg m^{-3}). Using this equation Kay et al. (1997) calculated predicted changes in WHC for soils ranging in clay content from 7% to 35% when organic C content was increased by 0.01 kg kg^{-1}. Increases in WHC of 0.039 and 0.020 (m^3 m^{-3}) were obtained for the soils with 7% and 35% clay, respectively, at a relative bulk density of 0.75. These results demonstrate that the presence of additional SOM enhances WHC of soils. However, the magnitude of the increase decreases with increasing clay content.

4.2.1.4 Thermal Properties

The dark color of humus contributes to the dark color of surface organic and mineral soils and can enhance soil warming and thus promote biological processes depending on temperature such as plant growth and mineralization of C, N, and S. The presence of litter layers or organic horizons (Histosols, see Chapter 1) can protect a soil against fluctuations in air temperature and solar heating (Baldock & Nelson, 2000). On Canadian forest soils subject to cold winters and cool springs, average soil temperature and the growth of fertilized seedlings were greater where the litter layers were removed compared to where they were left intact (Burgess et al., 1995). Similar effects were observed in field experiments that left different amounts of crop residue on the surface of an arable soil (Fortin, 1993).

4.2.2 Chemical Properties

4.2.2.1 Nutrient Source

Organic matter provides a large pool of macronutrients in the soil. McGill & Cole (1981) proposed that the mineralization of C, N, P, and S followed a dichotomous system involving both biological and biochemical mineralization. Biological mineralization is driven by the need of decomposer organisms for C as an energy source and accounts for the mineralization of N and C bonded S. Biochemical mineralization refers to the release of phosphate and sulfate from the P and S ester pool via enzymatic hydrolysis outside of the cell membrane. In contrast to organic N, organic P and S accumulation and mineralization in soils can occur independently of C and N dynamics. This leads to the potential for large variations in C:N:P:S ratios in SOM.

Mineralization of organic P is important to soil fertility. Soil organic P accounts for a variable proportion of the total soil P. Halstead & McKercher (1975) and Uriyo & Kessaba (1975) presented soil organic P values ranging from 4 to 1,400 μg g^{-1} soil which accounted for 3–90% of the total soil P. Organic C/P ratios in a study in Finland ranged from 61–526 (Kiala, 1963). The contribution of organic P in relation to inorganic P tends to be greater on highly weathered soils (Duxbury et al., 1989). The principal organic P-containing compounds in soils and their approximate proportions include inositol phosphates (2–50%), phospholipids

(1–5%), nucleic acids (0.2–2.5%), trace amounts of phosphoproteins, and metabolic phosphates (Stevenson, 1994).

Sulfur-containing organic compounds in soils are generally grouped into two pools. In the first pool, S can be reduced to H_2S by hydroiodic acid. In the second pool, S is directly bound to C. The hydroiodic acid-reducible fraction consists mainly of ester sulfates (C-O-S bonds) and some ester sulfamates (C-N-S bonds). The C-bonded S fraction includes amino acid S (C-S bonds) or sulfonates ($C\text{-}SO_3^-$ bonds). The ester sulfates and sulfamates are typically associated with aliphatic side chains of soil organic compounds (Bettany et al., 1979). In contrast, the C-bonded S is incorporated along with C and N into the matrix of SOM and is therefore less biologically accessible (Steward & Cole, 1989). Except for saline (Solonetz, Solonchak) and tidal (e.g., Thionic Fluvisols) soils, organic S typically accounts for >90% of the total S found in soils (Stevenson, 1994).

4.2.2.2 Exchange Capacity

The ability of SOM to adsorb both cations and anions from the soil solution is one of the most important benefits derived from its presence in soil. The release of ions to the plant root occurs by ion replacement. Cation exchange capacity (CEC) is of great importance, involving most of the cations in the soil solution. Organic matter contributes 25–90% of the CEC of surface layers of mineral soils and practically all of the CEC of peats and forest floor layers (Stevenson, 1994). The contribution is greatest for soils with low clay content or where the clay fraction is dominated by minerals with a low charge density, such as kaolinites, and is lowest for soils with highly charged minerals such as vermiculite or smectite. In sandy soils, organic matter plays a critical role in providing CEC. Very few anions are adsorbed by SOM with sufficient strength to make this a major factor in soil productivity. Charges on SOM arise from the ionization of various functional groups including carboxyl, phenolic hydroxyl, enolic hydroxyl, methoxyl, amino, and possibly other groups. However, because of potential organomineral interactions in soils and the ability to complex cations, the contribution of SOM to soil CEC values is often less than could be expected based on its carboxyl content. In acidic soils, the protonation of carboxylate groups, interaction with positively charged sites on inorganic colloids, and complexation of Al^{3+} and Fe^{3+} can significantly reduce the contribution of SOM towards CEC. The dissociation of the functional groups is pH dependent. Therefore, pH is an important factor influencing the CEC of SOM. Cation exchange capacities for humus are in the range of 60–300 (on average about 200) $cmol_c\ kg^{-1}$ (Van Dijk, 1966; Leinweber et al., 1993; Stevenson, 1994). Highly decomposed SOM has a higher exchange capacity than recent additions of organic matter to soils.

Mineralization of SOM decreases the total retention capacity of available cations in coarse textured and many tropical soils where SOM is often the major source of negative charge. The influence of organic matter amendments on the CEC depends on soil texture, initial soil CEC, length of time from the last application and nature of the organic material. The benefit of SOM in retaining nutrients depends strongly on soil texture. Sandy soils devoid of any SOM have a low CEC because sand particles have low densities of negative charge that give rise to the CEC. If the SOM content of such a sandy soil were increased slightly, soil CEC would increase proportionately. In contrast, 2:1 clay minerals are often characterized by high CEC values, and the importance of SOM to buffer nutrients in loamy or clay soils becomes less of a critical factor.

Maintenance of SOM to provide CEC is more important in highly weathered soils in the tropics than in temperate regions. However, the amount of CEC contributed to tropical soils by SOM is lower than in cooler regions because of the extended degree of blockage of negatively charged sites at low pH by Al and Fe. Attempts to increase the CEC of acid soils by increasing SOM levels may lead to short-term benefits, but in the long-term the soil system will equilibrate to block newly formed exchange sites unless soil pH is raised by liming.

4.2.2.3 Reactions with Metals

Soil organic matter is an important factor for the retention or mobilization of metals. The possible interactions can take the form of simple cation exchange reactions, such as those between negatively charged carboxyl groups and monovalent ions, or more complex interactions where coordinate linkages with organic ligands are formed. Humic substances can form complexes with cations like Ca^{2+}, Mg^{2+}, Pb^{2+}, Cu^{2+}, Ni^{2+}, Co^{2+}, Zn^{2+}, Cd^{2+}, Fe^{2+}, and Mg^{2+}. Humic and fulvic acids contain relatively large numbers of functional groups (COOH, C=O, phenolic OH) capable for combining with metals. As pH is increased, a proton is displaced from an acidic OH group of the humic or fulvic acid, allowing complexation of the metal to the organic molecule (Van Dijk, 1971). At a higher pH, a proton dissociates from water bonded covalently to the metal ion, thus forming a hydroxy complex. Metal ions and hydrous oxides have been shown to react with one acidic COOH and one phenolic OH or two acidic COOH groups (Van Dijk, 1971). These structures and the soil pH are responsible for the stability of metal-humic complexes. Their solubility differs widely. While complexes formed with fulvic acids are generally more soluble, and can be regarded as natural chelates able to reach plant roots, insoluble humic acids can withhold a considerable amount of microelements.

The influence that the complexation of inorganic cations by soil organic materials has on soil properties and processes includes the following: (i) increased availability of insoluble mineral P through complexation of Fe^{3+} and Al^{3+} in acid soils and Ca^{2+} in calcareous soils, competition for P adsorption sites, and displacement of adsorbed P (Stevenson, 1994); (ii) the release of plant nutrients through the weathering of soil parent materials by the removal of structural cations from silicate minerals (Tan, 1986); (iii) enhanced availability of trace elements in the upper part of the soil profile as a result of upward translocation by plant roots and subsequent deposition on the soil surface and complexation during residue decomposition (Stevenson, 1994); (iv) facilitated adsorption of organic materials to soil minerals which ends in the generation and stabilization of soil structure (Emerson et al., 1986); (v) buffering of excessive concentrations of otherwise toxic levels of metal cations (e.g., Al^{3+}, Cd^{2+}, Pb^{2+}; Anderson, 1995); and (vi) pedogenic translocation of metal cations to deeper soil horizons (McKeague et al., 1986).

4.2.2.4 Proton Buffer

Soil organic matter provides much of the pH buffering in most surface soils and acts over a wide range of soil pH because of the presence of weakly acidic chemical functional groups (e.g., carboxylic, phenolic, acidic alcoholic, amine, amide). With the exception of silicate clay minerals and secondary minerals (e.g., Fe and Al oxides and hydroxides, allophanes, imogolite) the buffer range of SOM is equal or even wider than that of other soil components (Table 4.3).

James & Riha (1986) reported buffer capacities of 18–36 and 1.5–3.5 $cmol_c\ kg^{-1}$ (pH unit)$^{-1}$ for the organic and mineral horizons of forest soils, respectively. Starr et al. (1996) obtained a significant correlation between acid buffer capacity and organic matter content for 29 organic and 87 mineral (E, B and C horizons) soil horizons, exhibiting buffering capacities of 9.8–40.8 and 0.1–5.2 $cmol_c\ kg^{-1}$ (pH unit)$^{-1}$, respectively. For 59 agricultural soil samples taken from the 0–15 cm layer of arable land, Curtin et al. (1996) described titratable acidity by Equation 4.3; where OC and Clay represent the soil organic C and the clay contents expressed in units of kg kg^{-1} soil and ΔpH is the reference pH minus the initial pH.

Table 4.3 Proton buffers, pH range and H^+ related reactions (Adapted from Bloom, 2001)

Proton buffer	pH range	H^+-related reactions
Soil organic matter	Whole range	Ionization of carboxyl and phenolic groups
Irreversible weathering of primary silicate minerals	Whole range	Buffering of H^+ upon dissolution of Ca^{2+}, Mg^{2+}, K^+, Na^+
Fe and Al oxides and hydroxides, allophane, imogolite, silicate clay edges	Whole range	Ionization and protonation of surface hydroxyl groups
$MgCO_3$	>9.5	Precipitation and dissolution
$CaCO_3$	7–9.5	Precipitation and dissolution
Permanent charge sites on silicate clay minerals	3.5–5	Ion exchange of H^+ and Al^{3+} with base cations
$Al(OH)_3$ – soil organic matter	5–8	Precipitation or dissolution of organic bound Al^{3+} as $Al(OH)_3$
Interlayer Al in 2:1 clays	5–7	Precipitation or dissolution of interlayer $Al(OH)_3$
Al – soil organic matter	<5	Carboxyl H^+ exchange with Al^{3+}
Dissolution of high activity silicate clays and allophane	<3.5	Buffering of H^+ upon dissolution of Al^{3+}

$$\text{Titratable acidity to pH 8} = 0.02 + 59\ \text{OC}\ \Delta\text{pH} + 3.0\ \text{Clay}\ \Delta\text{pH} \quad (R^2 = 0.95;\ p \leq 0.01) \tag{4.3}$$

Assuming that the organic C content of SOM is 58%, Equation 4.3 indicates that the buffering capacity determined by organic matter was approximately 34 (58% of 59) $cmol_c\ kg^{-1}$ (pH unit)$^{-1}$ and is an order of magnitude greater than that offered by clay (34 vs. 3 $cmol_c\ kg^{-1}$ (pH unit)$^{-1}$). The average clay:organic C ratio for the soils used by Curtin et al. (1996) was 7.9:1. Accordingly, organic C in these soils accounted for about two thirds of the soil buffering capacity.

Addition of organic matter to soils may result in increase or decrease in soil pH, depending on the dominant processes that either consume or release protons. Factors that need to be considered include the chemical nature of SOM and of the added organic materials as well as environmental factors like soil water content and leaching. Processes that lead to an increase in pH due to organic matter addition are decomplexation of metal cations, mineralization of organic N and denitrification. For an acid soil, Pocknee & Sumner (1997) found that the extent of the increase in pH was controlled by the N content to basic cation content ratio. In contrast, the addition of organic material tends to acidify soils especially under waterlogged and leaching conditions (Nelson & Oades, 1998). The main processes involved in the acidification of alkaline soils on addition of organic materials include: (i) mineralization of organic P and S, (ii) N mineralization followed by nitrification, (iii) dissociation of organic ligands, and (iv) formation of CO_2 during mineralization of organic matter.

4.2.3 Biological Properties

The provision of metabolic energy driving biological processes is the most important function of SOM. The C fixed by photosynthesis into different organic compounds (e.g., cellulose, hemicellulose, lignin, lipids, proteins) is deposited on or in the soil during and after plant growth, providing C substrates for decomposer organisms. These assimilate organic C either into body tissues or respire it to CO_2. The majority of organic matter processing is thought to be completed by the soil microbial biomass. An estimate of the C flow through a grassland soil yielded a primary production via photosynthesis of 10 t C ha^{-1} $year^{-1}$ (Oades, 1989). Of the 3 t C ha^{-1} $year^{-1}$ that was added to the soil, the soil fauna were estimated to utilize 0.3–0.45 t C ha^{-1} $year^{-1}$, while the microbial biomass was estimated to utilize 2.4 t C ha^{-1} $year^{-1}$. Although less important for the decomposition of residues, the soil fauna enhance the ability of soil microbial decomposers to utilize organic residues by fragmentation of plant debris, and by distributing organic materials throughout the soil matrix.

4.2.3.1 Nutrient Mineralization

The nutrient mineralization capacity refers to the potential of soils to transform nutrients in organic forms into mineral nutrient forms (e.g. phosphate and sulfate anions, nitrate and ammonium). The N mineralization potential has been used to assess the influence of soil management on soil quality. It increases with the addition of organic material (Pankhurst et al. (1995), intensity of pasture management (Banerjee et al., 2000), conventional plowing (Kandeler et al., 1999b), and with increasing the plowing depth (Nieder et al., 2003c). In SOM, carbon and nutrients (N, P and S) are present as part of complex organic polymers or as organically complexed cations, such as most transition metal nutrients (e.g. Mn, Cu, Fe, Zn). Generally, SOM contains up to 95% or more of the N and S and between 20% and 75% of the P in surface soils (Duxbury et al., 1989). A definite relationship can be observed between organic C, total N, organic P and total S in soils. The mean value of the C/N/S ratio of natural soils, both grassland and forest, is fairly constant at about 200:10:1, whereas the average proportion of C/N/P/S in arable soils is about 140:10:1.3:1.3 (Stevenson, 1986). These differences are ascribed to preferential mineralization of the C relative to N and S and of N relative to S in cropped soils, the generally higher nutrient contribution from agricultural crop residues, and differences in the retention capacity for the various elements in the soil-plant system after mineralization. The gradual transformation of raw plant material into stable SOM leads to the establishment of a more consistent relationship between C, N, P and S, approaching that of native humus. The mean P content in SOM is close to that for S, but is more variable than the other elements, especially in tropical soils. Thus, C/P and N/P ratios of SOM can vary widely on dependence of several factors including the relatively greater independence in the cycling of SOM-P relative to organic C, N and S. The greater constancy of soil N/S ratios indicates coupling of their cycling in the soil-plant system.

Only about 40–50% of the organic N in soils can be identified in components of definite chemical classes, with amino acids usually dominating over amino sugars (Stevenson, 1986). Although newly added and immobilized N has a different chemical composition than native soil N (e.g. more amino N and unidentified hydrolyzable N, and less heterocyclic N), it is rapidly converted to chemical forms similar to those present in native SOM. The more rapid mineralization of recently added N should be ascribed more to differences in substrate lability and availability or protection than to differences in the structural composition of added and native N.

According to NMR measurements, most of the organic P extractable from soils is in the form of monoesters of orthophosphoric acid and phosphate esters of inositols are the most abundant class of identifiable compounds, ranging from 5% to 80% of SOM-P (Hawkes et al., 1984). Small amounts of diesters and phosphates containing C-P bonds have also been detected by NMR, together with minor contributions of nucleotides and phospholipids. The capacity of inositol phosphates to form insoluble precipitates with Fe, Al and Ca, and to strongly adsorb on amorphous Fe and Al oxide surfaces, may explain their accumulation in some soils.

The most abundant forms of organic S in soil are sulfur-containing amino acids, mainly including methionine, cysteine and cystine, which also account for up to 30% of the SOM-S. Between 30% and 70% of the SOM-S is in the form of hydroiodic acid- (HI-) reducible ester sulfates (C-O-S) and C-N-S and C-S bonded sulfur. C-bonded S is not reduced by HI and comprises a smaller proportion of total SOM-S. Carbon, nitrogen, phosphorus, and sulfur are generally released in ratios different from those in which they occur in SOM. Differences in apparent mineralization-immobilization rates for N, P, and S may be due in part to their occurrence in different organic compounds and SOM fractions, and to variations in the N, P, and S content of applied crop residues (Duxbury et al., 1989).

4.2.3.2 Microbial and Plant Growth

The effect of SOM on plant and microbial growth involves the absorption or adsorption of the humic species and their impacts on biochemical properties at cell walls, cell membranes, and in the cytoplasm. The influence of humic substances on the growth of soil microorganisms also involves penetration and alteration of cell membranes. Addition of humic substances at concentrations $<30\,mg\,L^{-1}$ to a nutrient solution increased growth rates in microbial cultures (Visser, 1995), and *in vitro* growth and activity of nitrifying bacteria have been increased by the addition of humic acid (Vallini et al., 1997). The effects of humic materials on plant growth was reviewed by Chen & Aviad (1990). Favorable effects on plant growth included (i) increased uptake of water and germination rate of seeds, (ii) enhanced growth of shoots and roots, and (iii) increased root elongation, number of lateral roots, and root initiation. These results were drawn back to increased permeability of cell membranes, increased chlorophyll content and rates of photosynthesis and respiration, enhanced protein synthesis resulting from a stimulation of ribonucleic acid synthesis, and enhanced enzyme activity (Vaughan & Malcolm, 1985).

4.2.3.3 Interactions of SOM with Biological Molecules

The adsorption of important biological molecules by SOM is an important peculiarity of soil as a biological system. Enzymes can be entrapped by humic molecules or adsorbed by clay minerals, and can maintain their activity, being protected against proteolysis, thermal and pH denaturation (Nannipieri et al., 1996). In this way stabilized extracellular enzymes can be still active under conditions unfavorable for the activity of soil microorganisms (Nannipieri et al., 2002). Ladd & Butler (1975) concluded that the effect of humic acids on the activity of proteolytic enzymes varied and that the mechanism of humic acid-enzyme interactions is primarily determined by carboxyl groups of humic acids. Müller-Wegener (1988) indicated that possible humic acid-enzyme interactions that could impact enzyme activity include: (i) a direct interaction of the humid acid with the enzyme resulting in a modification of enzyme structure or changes in the functioning of active sites, (ii) interference in the equilibrium of the enzyme reaction via the humic substances acting as analog substrates, and (iii) a reduction in the availability of cations, which often act as cofactors required for enzyme catalysis.

Adsorption and binding of DNA on humic molecules can protect the adsorbed DNA against degradation by nucleases without inhibiting its transforming ability (Lorenz & Wackernagel, 1987; Khanna & Stotzky, 1992; Pietramellara et al., 1997). Transformation is a mechanism of gene transfer in soil, involving a competent bacterial cell, which can take up one of the two strands of extracellular DNA and insert it in the bacterial chromosomal DNA. In such a way the competent cells acquire all or part of the genes associated with the extracellular DNA. The fact that DNA can be adsorbed by soil colloids and protected against microbial degradation has important implications on the use of genetically modified organisms in the terrestrial ecosystems. The recombinant DNA may survive longer than expected if adsorbed by soil colloids or englobated in the genome of soil microbiota. It has been hypothesized that surface active particles concentrate biological molecules at the solid-liquid interface, which can support the growth of microorganisms adsorbed on these surfaces. Particle-size fractionation of dispersed soil commonly reveals a concentration of soil organic matter and microbial biomass in fine silt-size/coarse clay-size materials (Ladd et al., 1996). However, protection of adsorbed microorganisms against predators (Ladd et al., 1996) and protection of organic substrates, such as proteins, peptides, amino acids, polysaccharides, nucleic acids and nucleotides against microbial degradation (Nannipieri et al., 1996) do not support the hypothesis that adsorbed organic substrates sustain growth of microbes adsorbed on active surface particles (Stotzky, 1997).

4.2.3.4 Ecosystem Resilience

The important role of SOM in determining the resilience of an ecosystem can be exemplified by a comparison of the contents of chemical energy and nutrients stored in the organic matter. In temperate grasslands (prairie or steppe soils), high SOM levels are built up as a result of large below-ground additions of photosynthate, limited leaching, and slow decomposition rates. Storage of C in such ecosystems is greater in the soil than in the vegetation (Szabolcs, 1994). The large pool of chemical energy and nutrients contained in the soil organic fraction offers resistance to the loss of soil fertility induced by natural or by agricultural disturbance. Prairie soils, typically Chernozems and Castanozems, have remained agriculturally productive with limited inputs for many years (Tiessen et al., 1994). Such systems can be considered at least initially resilient. However, the question arises how long such a system can be sustained. Tiessen et al. (1983) showed that rates of organic P mineralization in a grassland soil were in excess of crop requirements over the first 60 years of agricultural management. After 60 years, only the stable, low energy providing forms of organic matter remained, and organic P mineralization rates decreased below crop demands.

In temperate forests, SOM contents are less than those of temperate grasslands and more C and nutrients are stored in aboveground vegetation as compared to the SOM pool. The impact of a natural disturbance such as fire can significantly deplete ecosystem stores of energy and nutrients. As a consequence, ecosystem recoveries are slow due to low residual contents of SOM and associated nutrients. Where temperate forests are cleared and agricultural production is initiated, SOM and nutrient losses must be minimized. Production systems in which SOM and nutrients are increased (e.g., through introduction of conservation tillage, increased organic fertilizer application, introduction of cover crops) can lead to highly productive and sustainable agriculture. In tropical forest systems, the storage of energy and nutrients in vegetation dominates, and the rapid cycling of nutrients maintain ecosystem stability. This indicates a reduced importance of SOM in resilience of a tropical forest ecosystem (Anderson, 1995). The different contribution of SOM to ecosystem resilience becomes very obvious when comparing a humus-rich Chernozem of the prairie with a tropical Ferralsol poor in SOM.

4.3 Evaluation of Organic Components as Soil Quality Indicators

Organic components of soil (soil organic matter, microbial biomass, soil enzymes) have been suggested as the most important indicators of soil quality and productivity (e.g. Larson & Pierce, 1991; Doran & Parkin, 1996). Soil fauna are commonly not regarded as a part of soil organic matter but their influence on biological processes may offer valuable long-term indicators of soil quality (Wolters, 1991). The disappearance and decomposition of plant and animal residues, the release of nutrients, the development of macropores (biopores) and the mixing of organic and mineral soil components are major functions that directly involve soil fauna. Earthworms are highly involved in these processes. Their presence and abundance in soil is commonly regarded as a visible indicator of soil quality.

4.3.1 Soil Organic Matter

Soil organic matter content has been used as a measure of soil quality because high levels usually show good correlations with other desirable attributes of soil, such as high levels of microbial biomass, available plant nutrients, and a favorable soil structure (Pankhurst, 1994). Changes in soil organic matter are commonly assessed by comparing native ecosystems (grassland or forest sites) with agricultural land, managed forests or degraded land (for details see Chapter 6). Comparisons also have been made between soils under different crop rotations, fertilization regimes and tillage treatments. The annual amounts of C and N added to soils is small relative to C and N stored in SOM. Extended periods of time (commonly several decades) are therefore required to observe significant changes in SOM. The major part of SOM consists of nonliving organic matter (see Chapter 3). Any change of the SOM level in the soil concerns mostly its decomposable (labile) part, which is mainly composed of plant, animal and microbial residues in various stages of decomposition.

Microbial biomass, fungal hyphae and faunal debris make up a significant portion of the labile fraction of SOM (Spycher et al., 1983). In cultivated soils the labile fraction typically contains 20–30% C and 0.5–2.0% N (Janzen et al., 1992), making up 2–18% of the total C and 1–16% of the total N in soil (Gregorich & Janzen, 1996). The C:N ratio is commonly between that of plant tissue and humic substances (for details see Chapter 3), because the labile fraction is less humified and undergoes more release of C than N in initial stages of decomposition. Due to gradients in climate and vegetation the proportion of C and N in the labile fraction of SOM tends to increase with increasing latitude (Christensen, 1992). The amount of labile organic matter may also vary during the growing season. Spycher et al. (1983) found that it is increased by up to 100% from early spring to summer.

In contrast to plant nutrients like P, K, Ca and Mg, up to now it has not been common to classify labile C and N fractions as standard values for optimum SOM contents. Evaluations of long-term field experiments on arable land provide some indications that may be relevant for practical purposes. For example, for sandy and loamy arable soils, optimum SOM ranges could be derived for different fractions (Table 4.4).

Optimum contents of carbon and nitrogen in soils may be within a relatively narrow range. Below the optimum ranges, yields are insufficient. Above the ranges, emissions of reactive C and N compounds to the atmosphere (CO_2, N_2O, NO_x) as well as to surface and ground water (NO_2, NO_3^-, NH_4^+, dissolved organic C and N) occur, which underlines the complex interdependence of soil, water and air quality. Effects of the SOM on the yield may reach 10% in sandy soils and up to 5% in clayey soils, whereby the percentage is the result of a comparison between an optimal mineral fertilization and an optimum combination of organic and mineral fertilization (Nieder et al., 2003a). In order to uphold the optimum SOM conditions according to Table 4.4, it is necessary to add about 2 t organic material ha^{-1} year^{-1}. This corresponds to 2 t dry matter (dry farmyard manure or solid matter from liquid manure), equivalent to 10 t fresh farmyard manure or the yearly amount of excreta produced by 1 gross weight or life weight unit of 500 kg (Isermann & Isermann, 1999). If land conservation is considered in cultivation processes combined with a reduction of farming intensity, the addition of organic substances can be reduced. Unfortunately, at present there is

Table 4.4 Optimum SOM conditions of sandy and loamy arable soils derived from seven long-term field experiments under common management in Eastern Germany (average yearly temperature: 6–10°C, average precipitation: 400–800 mm year^{-1} (Adapted from Nieder et al., 2003a)

Total SOC	0.7% (sandy soil with 3% clay) to 2.5% (loamy soil with 21% clay)
Total SON	0.07% (sandy soil with 3% clay) to 0.25% (loamy soil with 21% clay)
Hot water-soluble C (C_{hwl})	250–300 mg kg^{-1} soil
Decomposable SOC ($SOC_{dec.}$)	0.4 (0.2–0.6)%
Decomposable SON ($SON_{dec.}$)	0.04 (0.02–0.06)%
Mineralized SOC ($SOC_{min.}$)	4% of SOC_{dec}
Mineralized SON ($SON_{min.}$)	4% of SON_{dec}

SOC = soil organic carbon; SON = soil organic nitrogen

no comparable knowledge for optimizing SOM contents of grassland and forest soils, but these could be obtained by similar procedures introduced here.

4.3.2 Soil Microbial Biomass

There is a widespread use of soil microbial biomass as a soil quality indicator because it plays a key role in SOM dynamics (Jenkinson, 1987). It controls the transformation of organic matter and influences C and N storage in soil. The amounts of microbial biomass C and N are commonly closely related to amounts in the mineralizable fraction of SOM (Paul & Voroney, 1984). Global carbon and nitrogen pools in soil microbial biomass are given in Chapter 1. The microbial biomass shows some seasonal variation associated with climatic factors (Wardle, 1998). It can be lost quickly when soil becomes dry, but recovers upon rewetting (Sparling et al., 1986). Another factor, which may induce microbial biomass cycling is fluctuations of available C concentrations in soil (see Chapter 5). Available C concentration in soil will increase through rhizodepositions, farmyard manure and organic residue application. Microbial biomass is expected to be low after mineralization of these C sources. During such cycles, microbial biomass C and N are temporarily released, which may have impacts on crop production, water and air quality.

Because of its high turnover rate, soil microbial biomass responds quickly to changes in soil resulting from management. The microbial biomass C to total C ratio has been used to elucidate changes in organic matter under different cropping (Anderson & Domsch, 1989) or tillage (Carter, 1991) systems and for soil polluted by heavy metals (Brookes, 1995). An "ideal" microbial biomass content for a soil up to now has not been defined. For deriving a "target" value, it may be possible to use a soil-specific baseline resulting from comparisons of sites with the same soil type but under different management (Sparling, 1997). The results about microbial biomass in context with management obtained by different authors are partly contradictory. Microbial biomass increased with the intensity of grazing in meadow soils (Banerjee et al., 2000) and with the cereal-pasture rotation (Dalal, 1998). It decreased in cultivated versus uncultivated soils (Kandeler et al., 1999a). The effect of zero-tillage is unclear (Dalal, 1998). Microbial biomass increased in soils supplied with organic fertilizers but did not react coherently to the application of mineral N fertilizers (Ladd et al., 1994) and to the presence of herbicides (Voos & Groffman, 1997). Microbial biomass could also not be considered as a good indicator of heavy metal toxicity in soils (Dalal, 1998). On the basis of these problems, the use of microbial biomass as a reliable soil quality indicator is questionable.

4.3.3 Soil Enzymes

Dehydrogenase, phosphatase, β-glucosidase and urease activities are the most frequently used soil enzyme parameters (Gil-Sotres et al., 2005). Similar to soil organic matter and microbial biomass, the dehydrogenase activity has been used to assess the influence of management on soil quality. The addition of organic fertilizers increased the dehydrogenase activity (Langer & Gunther, 2001), while it did not react coherently to soil tillage (Bergstrom et al., 2000). Unless in very high doses, dehydrogenase activity was not affected by the presence of heavy metals (Filip, 2002). Dehydrogenase activity has been used successfully as a parameter for the evaluation of degree of recovery of degraded soils (Margesin et al., 2000; Gil-Sotres et al., 2005).

Phosphomonoesterase has been used as an index of the quality and the quantity of SOM (Bergstrom et al., 2000). The activity of this enzyme has been observed to increase during recovery of degraded soils (Gil-Sotres et al., 1992). Several studies have shown that this enzymatic activity rises as a consequence of organic fertilizer application (Chakrabarti et al., 2000), and decreases after application of mineral phosphate fertilizer (Olander & Vitousek, 2000) and by the presence of heavy metals (Kandeler et al., 1996) and pesticides (Schäffer, 1993). The enzyme β-glucosidase has been used for the evaluation of soil quality under different land use. Its activity was significantly lower in arable soils as compared to meadow and forest soils (Saviozzi et al., 2001). Organic fertilization increased the activity of this enzyme (Bandick & Dick, 1999). The activity of urease has been observed to increase due to organic fertilization (Chakrabarti et al., 2000) and to decrease as a consequence of plowing (Saviozzi et al., 2001).

4.4 Use of Combined Biological Parameters for Soil Quality Estimation

4.4.1 Indexes Developed from Two Measured Parameters

The combined effects of different management measures on individual biological functions observed in the cited literature (see section 4.3) show that the value of single properties for soil quality estimation is very limited. Simple indexes include a combination of two different parameters into a single criterion (Gil-Sotres et al., 2005). Frequently used examples are the metabolic quotient (qCO_2), the death rate quotient (qD), the ratio between biological parameters and the SOC and SON content, and the ratio of enzyme activity to microbial biomass. The metabolic quotient represents the ratio of organic material which is mineralized per unit of microbial biomass and time (Anderson & Domsch, 1985a). Any change in qCO_2 may indicate an alteration in the ratios zymogene: autochtonous flora or bacteria: fungi (Dilly & Munch, 1998). The qCO_2 value was observed to be higher in stressed (e.g. heavy metal contaminated, acidified) as compared to unstressed ecosystems (Brookes, 1995; Haynes, 1999; Filip, 2002), except for biologically highly active soils. The latter may have values of qCO_2 similar to contaminated soils (Sparling, 1997). Another point of criticism is that values that have been suggested as indicators of a stable soil system do not seem to have universal significance (Gil-Sotres et al., 2005). Similar observations can be applied to the death rate quotient (*q*D) and the ratio between biological parameters and the SOC and SON content.

An increasing ratio of enzyme activity to microbial biomass may indicate either increasing enzyme production and enzyme release by microorganisms or an enhanced release of enzymes immobilized on clay minerals or humic colloids to the soil solution. For enzymes involved in the nitrogen cycle, the ratios increased in soils with high slurry application rates (Kandeler & Eder, 1993). It was also shown that cadmium reduced the dehydrogenase and phosphomonoesterase to microbial biomass ratios (Landi et al., 2000). Nevertheless, up to now, a broad database is missing that confirms the usefulness of the ratios between enzyme activity and microbial biomass for evaluating soil quality. Summarizing, the use of simple indexes is still of little value for soil quality evaluation, mainly because of the lack of (i) reference levels, (ii) coherent behavior of the same ratio in different studies, and (iii) criteria for relating a variation in the index.

4.4.2 Indexes Developed from More than Two Measured Parameters

Indexes including more than two biological parameters, so-called complex indexes, were first used in the 1980s. The enzymatic activity number introduced by Beck (1984) includes several enzymes (dehydrogenase, catalase, phosphatase, protease, β-glucosidase). Stefanic (1994) used a so-called biological fertility index developed from polynomal formulae in which the values of several enzymatic activities are multiplied by empirical factors. Dilly & Blume (1998) proposed to select a set of ten biological properties that consider microbial biomass, microbial activity and enzyme activity. Yakovchenko et al. (1996) used parameters derived from the relationship between crop N uptake and the net N mineralization during the crop growth period. Bentham et al. (1992), using a cluster method which involved dehydrogenase activity, ATP and ergosterol content, evaluated the degree of recovery of mine soils.

Sinsabaugh et al. (1994) in their approach focused on the degradation of plant residues in soil involving a great number of enzymes. The latter authors obtained a so-called lignocellulosic factor from different enzyme activities which was strongly correlated with the weight loss of plant residues over time. Nannipieri et al. (2002) preferred this approach to others due to the accurate selection of enzyme activities. Nevertheless, the main criticism lies in the fact that it only considers carbon-related enzymes (e.g. β-glucosidase, endocellulase, exocellulase, β-xylosidase, phenol oxidase), and ignores enzymes related to the N, P and S dynamics (Gil-Sotres et al., 2005). Other authors followed similar approaches but they considered enzyme activities different from Sinsabaugh's option (Sinsabaugh et al., 1994). Fioretto et al. (2001) concentrated on the enzymes α-amylase and β-amylase. Bergstrom et al. (2000), focusing on the enzymes β-glucosidase, dehydrogenase and

L-glutaminase, obtained a decomposition factor that was capable to explain more than 90% of variation in enzymatic activity as a consequence of tillage and crop rotation. However, up to now there is only little experience about the behavior of decomposition factors in the presence of contaminants or to other soil disturbances (Trasar-Cepeda et al., 2000).

In summary, the use of complex indexes in which various biological properties are involved, seems to be a promising line of work. The inclusion of various biological properties better reflects the complexity of the soil system. However, the lack of analytical standard methods accepted by most of the research groups is a fundamental problem for the application of biological parameters for soil quality estimation. Comparisons of data from different studies are complicated by a high degree of variability of biological parameters due to both seasonal and spatial factors. Uncertainties are also due to different sampling and sample preparation, as well as lacking protocols for enzymatic activity determination from which information about incubation temperature and time, substrate concentration, etc. could be drawn. As Gil-Sotres et al. (2005) pointed out, these methodological problems contribute significantly to often contradictory interpretation and conclusion reached by different scientists. Future efforts should therefore be concentrated on the search of complex biological parameters using standard methods under standard conditions. Moreover, it will be necessary to test the methods under a wide range of conditions (soil, climate, management) in order to verify if they are globally applicable. It is necessary that soil biologists should make greater efforts in joint research programmes to increase the knowledge about a broad group of soil biological properties and to assess their value for the evaluation of their role in context to special factors (e.g. site factors, management, pollution), and finally for soil quality determination.

Chapter 5
Carbon and Nitrogen Transformations in Soils

Soil organic matter supplies major nutrients via mineralization and binds them via immobilization. The fate of C and N in the SOM is dependent on processes affecting organic matter decomposition and formation. The decomposition process is controlled predominantly by bacteria and fungi. Fauna that graze on microbes such as protozoa, nematodes and earthworms also are involved in the mineralization of C and N. Management strategies that target SOM accumulation for sustained nutrient availability particularly must provide a favorable environment to soil microflora because of their dominating role in mineralization-immobilization processes. In general, it is assumed that 1.5–3.5% of the organically bound N is mineralized annually in temperate climate ecosystems (Brady & Weil, 1999). The actual rate at which C and N are mineralized is highly variable and depends on vegetation type, SOM level, pH, soil texture, soil moisture, soil aeration, soil temperature, and management practices such as tillage and fertility amendments. Vegetation type and associated quality of residue inputs directly affect the availability of nutrients by influencing microbial C and N use efficiency. Plant residues having high C:N ratios can lead to immobilization of N. In contrast, residues containing low C:N ratios can cause N mineralization rates far more than crops need.

Inorganic carbon constitutes an important carbon pool in soils of arid and semi-arid regions with annual precipitation <500 mm. Soil inorganic carbon (SIC) exists as primary or lithogenic and secondary or pedogenic carbonates. The primary carbonates may occur as parent material in the transition zone between bed-rock and soil profile (Scharpenseel et al., 2000). The carbonates exist in several forms but the dominant biogenic forms are calcite and aragonite, both of which are stoichometerically $CaCO_3$. Calcite has six coordinated Ca atoms and is capable of incorporating several percent Mg in its lattice. Aragonite has nine coordinated Ca atoms and can take up several percent Sr into its lattice. Both forms can precipitate depending on the Ca/Mg ratio in the solution. Calcite precipitates from solutions with Ca/Mg ratios greater than 1. With smaller Ca/Mg ratios, first Mg-calcite and then aragonite are precipitated (Kempe, 1979). Dolomite ($Ca\ Mg(CO_3)_2$), the third carbonate mineral of wide importance, is formed by diagenetic disintegration of Mg-rich calcites. For the formation of dolomite the Ca/Mg ratio has to be lower than 0.13, which is normally found only in evaporative areas. The other form of carbonate, Siderite ($FeCO_3$) is formed in the absence of oxygen.

R. Nieder, D.K. Benbi, *Carbon and Nitrogen in the Terrestrial Environment,*

Nitrogen in form of NH_4^+ (radius: 0.143 nm) or potassium in form of K^+ (radius: 0.133 nm) can be fixed in the hexagonal cavity (radius 0.149 nm) of silicate tetrahedral sheets of the tetrahedral-substituted or octahedral-substituted layer charges (McBride, 1994). Entrapment of NH_4^+ occurs through interlayer bonding, accompanied by contraction of the interlayer spaces. Soils rich in 2:1-type vermiculite and montmorillonite clay minerals can fix substantial amounts of N as non-exchangeable NH_4^+.

5.1 Transformations of Organic Components

The soil microbial biomass is the principle component of the system which is responsible for the transformation of organic components, for nutrient cycling and energy flow, and represents an important reservoir for carbon and potentially available plant nutrients. Estimates of soil microbial biomass in different ecosystems have been presented in Chapter 1. Jenkinson (1990) considered soil microbial biomass as the eye of the needle through which virtually all nitrogen must pass. Saffigna et al. (1989) emphasized the value of measuring C in the soil microbial biomass, defined as that living part of the SOM excluding plant roots and fauna larger than amoeba i.e., >5,000 μm^3 (Jenkinson & Ladd, 1981), as a sensitive indicator of changes in soil organic matter following contrasting cultivation practices. The size and activity of the soil microbial biomass depend on soil organic matter quality, quantity, and distribution, and has been related to soil properties (Kaiser et al., 1992), climatic conditions (Grisi et al., 1998), and crop management (Kandeler et al., 1999b). Positive effects on soil microbial biomass have been observed after the introduction of organic farming (Mäder et al., 2002), reduced tillage (Dilly et al., 2003), and due to slurry application (Bol et al., 2003). In contrast, microbial biomass decreased due to salinization (Sardinha et al., 2003), degradation of peatland (Baum et al., 2003), the influence of antibiotics (Thiele-Bruhn & Beck, 2005), pesticides (Harden et al., 1993) and heavy metals (Suhaldoc et al., 2004). Larger organisms such as earthworms and beetles also play a vital role in nutrient cycling and can also serve as indicators of the degree of soil disturbance (Persson, 1989).

Decaying organic matter in soil is an important source of nitrogen, phosphorus and sulfur. After incorporation of easily decomposable plant material (e.g. legumes) having a small C:N ratio into the soil, nitrogen is rapidly released. In contrast, the presence of wheat straw or corn stalks with a low N content and, consequently, a wide C:N ratio commonly results in an immobilization of plant-available nitrogen by soil microorganisms to meet their N demand. Once the wheat straw or corn stalks have become highly decayed and the C:N ratio of the decomposing organic material has decreased, re-mineralization of the immobilized nitrogen begins. Net N mineralization is the outcome of two oppositely directed processes: gross N mineralization and gross N immobilization (mineralization-immobilization turnover: MIT). These processes have been discussed in detail in Chapter 9. Prediction of net N mineralization is required for the synchronization of N supply with plant

N demand. An improved understanding of MIT can potentially improve our capability to predict net N mineralization patterns.

5.1.1 Methods of Mineralization-Immobilization Measurement

A range of direct and indirect methods have been used to monitor decomposition and the release of organically bound C and N. The most important are presented in the following sections.

5.1.1.1 Soil Microbial Biomass

Direct measurement of soil microbial biomass population involves counting numbers and sizes of organisms and is exceptionally laborious. Indirect methods are therefore preferred, one of the most frequently used being the chloroform fumigation-extraction technique (Brookes et al., 1985). This involves fumigating a soil sample with $CHCl_3$ and comparing the C and N mineralized in the fumigated soil with the C and N mineralized in an unfumigated control. The flush in mineralization typically observed following fumigation is due to the recolonizing microbial population decomposing the cells killed by the fumigant. Assuming that 68% of the N in the original microbial biomass is mineralized (Shen et al., 1984), the difference in C and N mineralized between fumigated and unfumigated soils provides a measure of soil microbial biomass C and N, presupposed that the decomposability of other SOM fractions is little affected by the $CHCl_3$ fumigation (Jenkinson & Powlson, 1976). In any case, an unknown amount of microbial biomass in the rhizosphere soil attached to the roots remains on the sieve during sample pretreatment, i.e., soil material with a much higher microbial biomass availability than the bulk soil.

A similar method can be used as a measure of microbial biomass C from CO_2 release following fumigation and inoculation (Jenkinson & Powlson, 1976). However, these methods have sometimes failed to identify changes in microbial biomass C or N concentrations in spite of contrasting management regimes and the techniques at best provide only a crude assessment of biomass C and N, and hence some qualitative assessment of mineralization processes.

Researchers are becoming increasingly interested in biomass community structures, and the tools for studying the effects of perturbations on such structures are now available. Biomarkers such as sterols can be used to monitor fungal biomass and lipid phosphorus can monitor bacterial biomass (O'Donnell, 1997). For example, it is possible to detect at least the fungal part of the microbial biomass by measuring the cell membrane component ergosterol (Jörgensen, 2000), which is a specific biomarker for fungi. In contrast to the fumigation methods, ergosterol is not affected by the presence of roots and can be easily detected by a variety of methods in solid substrates (Stahl & Parkin, 1996). Ergosterol has been successfully used to differentiate between fungal and plant tissue (Newell, 1996; Wallander et al., 1997). Such novel techniques can help to increase our understanding of the soil biomass and how factors such as cultivation can affect it.

5.1.1.2 Microbial Respiration

Soil respiration is the sum of all respiratory activity within the biologically active soil layers, with the primary sources of CO_2 evolution being microbial and root respiration. The traditional manner of following the decomposition of plant residues in soil has been to measure the evolution of CO_2 from soil and plant residues incubated together and to subtract from this the loss of C from soil incubated in the absence of residues. Measurements of CO_2 efflux from soil have traditionally been made using alkali (e.g., NaOH, KOH) traps to quantify the cumulative gas respired in a closed chamber. However, such chemical absorption techniques can underestimate the gas efflux and are only capable of providing a single integrated measurement.

A novel technique, substrate induced respiration (SIR), uses patterns of utilization of contrasting C substrates to assess the functional biodiversity and activity of soil organisms (Garland, 1996). Recent research has found that differences in SIR responses between substrates gradually decline with increasing soil disturbance from pasture through ley to arable soils (Degens & Harris, 1997), with higher topsoil SIR rates (and greater microbial biomass) under minimum tillage compared with conventionally tilled soils (Kandeler & Böhm, 1996). Results suggest that differences in SIR between management regimes reflect the smaller microbial biomass in arable compared with grassland soils and arise from differences in the composition of mineralizable organic materials (Degens & Harris, 1997).

A variety of closed or open chamber methods are available for use in the field (King, 1997), with the most widely used method of measuring CO_2 concentrations being infrared gas analysis. However, this approach has limitations with respect to separation of CO_2 evolved from soil and CO_2 evolved from plant residues, including analytical errors in measuring long-time release of CO_2. This is particularly when the amount of plant residues is kept small relative to the amount of native SOM. Also, the assumption is made that addition of plant residues to the soil does not alter the decomposition rate of the native organic matter. Furthermore, there is evidence that short-term CO_2 flux from tilled soils is influenced more by mass flow processes related to a tillage-induced change in porosity than to changes in current soil microbial activity (Reicosky et al., 1997).

5.1.1.3 Litter Bag Experiments

Part of the problems related to soil respiration measurements may be avoided by the use of litter bags, in which organic residues can be placed on or into the soil and can be removed for determining weight loss and for elemental analysis after defined time intervals. By this method, the behavior of carbon and of other elements (e.g. N, P, and S), and the changes in C:N, C:P, and C:S ratios in residues during the decomposition process can be determined. The litter bag experiment presented in Fig. 5.1 shows that within 1 year, about 70% of the initial straw C was mineralized.

5.1.1.4 Isotope Labeling

Labeling substrates with ^{14}C has made it possible to follow the decomposition of added residues with high accuracy, even in the presence of relatively large amounts of native organic matter. It has also been possible to identify the plant C as it becomes incorporated into fractions of the soil humus.

The use of ^{15}N has proved a useful technique for monitoring the mineralization of organic N. The isotope dilution technique (Barraclough & Puri, 1995) involves quantifying the dilution of labeled ammonium (or nitrate) solution injected into the soil as the proportion of labeled N present in the soil mineral nitrogen pool decreases over time due to the immobilization of ^{15}N and simultaneous mineralization of unlabeled N from organic materials. This method allows field or laboratory measurement of gross rather than net mineralization and immobilization and presents the opportunity to study gross N mineralization, ^{15}N immobilization (e.g. incorporation into soil microbial biomass, plant residues or SOM) and $^{15}NH_4^+$ fixation (into clay minerals)

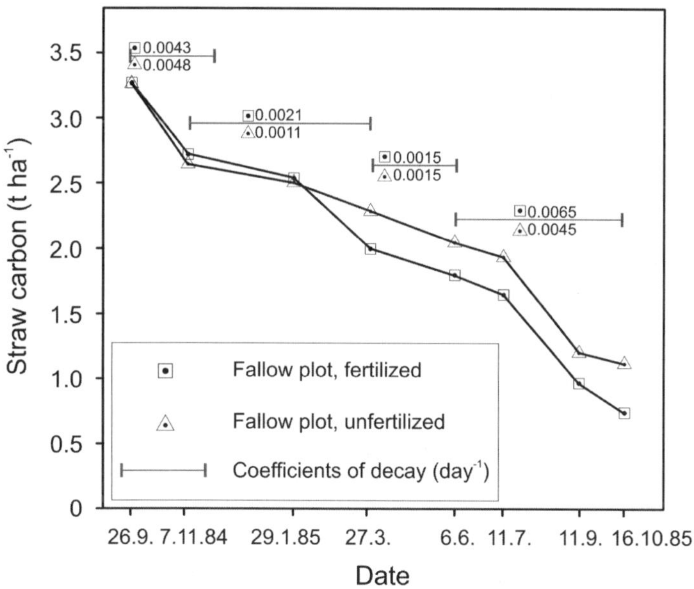

Fig. 5.1 Undecomposed straw carbon and decay coefficients (day^{-1}) for the decomposition phases. The straw (8t dry matter ha^{-1}; C:N: 103) was incorporated into a loess soil (Luvisol near Hannover, North Germany) within litter bags into a 35cm Ap horizon. The initial C mass in straw was 3.2t ha^{-1} (Nieder & Richter, 1989; p. 417. Reproduced with kind permission from Wiley-VCH)

dynamics unconfounded by processes such as nitrification and plant N uptake, which can consume (un)labeled N.

Alternatively, crop residues or fertilizers can be enriched with ^{15}N and its movement through the soil-plant system can be monitored. This is particularly useful as part of an N budget approach where major losses are quantified in addition to changes within the soil mineral N pool. This can be achieved by measuring test crop recovery of labeled N. However, when denitrification and volatilization losses are high and these fluxes are not measured, this results in an incomplete recovery of ^{15}N in measured soil and plant components (Nieder et al., 1989).

5.1.1.5 Temporal Mass Change in Soil Mineral N

An alternative method involves monitoring the change in soil mineral N (NH_4^+-N and NO_3^--N) and the change in plant N uptake of a test crop over time following contrasting management techniques like application or no application of straw or different tillage practices. This widely used method assumes that denitrification, volatilization or leaching of N are negligible during the study period. Another development has been the use of soil cores incubated *in situ* under field conditions to determine net N mineralization (Hatch et al., 1991).

5.1.2 Mineralization-Immobilization Measurements in the Field

5.1.2.1 Mass Change of Microbial Biomass in the Field

Several studies have measured microbial biomass at different intervals for a given location in the field and have attempted to interpret the trends observed in relation to environmental factors (e.g. Garcia & Rice, 1994; Nieder et al., 1996). A stimulatory effect of rhizosphere on the microbial biomass has often been observed (e.g. von Lützow & Ottow, 1994; Martens, 1990). Evidence for the microbial uptake of rhizodepositions has been provided by ^{14}C-studies (e.g. Jensen, 1993; Xu & Juma, 1994), and the microbial biomass is usually enlarged only around the rapidly growing section of the root (Newman, 1985). The temporal variability of soil microbes can be seen as a measure of its lack of stability (Tilman, 1996). Organisms that are relatively resistant to soil disturbance (e.g. soil tillage, fertilization, pesticide application, drying and wetting) tend to fluctuate less in response to changes in environmental conditions (Wardle, 1998). The factors which influence stability of the soil microbial biomass, therefore, affect nutrient conservation in soil (Wardle & Nicholson, 1996). The ratio of microbial biomass C: SOC is a possible indicator for the degree of disturbance of soil carbon cycling (Anderson & Domsch, 1989). A low biomass C: SOC ratio indicates a reduced pool of available carbon in soil (Klose et al., 2004).

A review of literature on temporal dynamics of microbial biomass data on a global scale (Wardle, 1992, 1998) showed that the trends were often contradictory, with different studies reporting peaks in different seasons, and observing positive and negative responses to temporal trends of soil moisture and plant productivity. Only a few studies from ecosystems of the temperate climate zone indicated maximum microbial biomass in spring or summer (Wardle, 1998). A significant proportion of reasons for biomass fluctuations are still unknown, since possibly each investigated site has its own controlling factor. Due to a partly high spatial variability of soil microbial biomass, it is more or less not clear if there is a uniform pattern of microbial mass change during a crop growth period on a field plot or even in a region.

Part of the variability in microbial biomass could be due to several artifacts. For example, it has been observed that not only microbial components are extracted from fumigated soils (Sparling et al., 1985), but also membranes of roots are attacked by $CHCl_3$. Microbial biomass is usually enlarged around the roots (particularly around their rapidly growing parts) (Newman, 1985). Jörgensen (2000) observed that in grassland soils the concentration of microbial biomass C was greater by 80% in the rhizosphere as compared to the bulk soil. The risk of underestimating microbial biomass in densely rooted soils is, therefore large. Pretreatment of the soil samples remains to be a critical point, particularly if the microbial biomass is determined in grassland soils or during the growth period of annual crops. Apart from the pretreatment and the analytical problem of determining soil microbial biomass in samples (with different moisture contents) taken at different times of the year, sampling problems and spatial variability may also influence the results.

Under conventional agricultural management the only carbon source for the vast majority of soil microorganisms, the heterotrophs, are the rhizodepositions of plants. During plant growth these are mainly organic compounds of low and high molecular weight (amino acids, organic acids, mono-and polysaccharides) exuded by the active sections of growing roots. Among the factors affecting C fluxes to the rhizosphere are plant species and age, size and structure of microbial populations, soil texture, mineral nitrogen in soil, and atmospheric CO_2 concentration (Nguyen, 2003). During plant growth and especially during ripening and after harvest the dieing roots themselves are the second carbon component of the rhizodepositions. Amounts of below-ground translocated C (roots and exudates) have been presented in Chapter 2. In a recent review Kuzyakov & Domanski (2000) calculated an average input of 1,500 kg C ha^{-1} for cereals and 2,200 kg C ha^{-1} for pasture plants grown longer than 100 days. This corresponds to one third to a half of total assimilated C (net assimilated plus shoot respiration). Between 10–15% of below-ground allocated C is respired by roots and 15–25% is exuded by the roots into soil (Kuzyakov, 2002).

A review by Nguyen (2003) on rhizodeposition of organic C by plants including a wide spectrum of species on the basis of 43 articles yielded that 17% of the net C fixed by photosynthesis is lost by roots and recovered as rhizosphere respiration (12%) and soil residues (5%). According to Kuzyakov & Schneckenberger (2004), cereals (wheat and barley) transfer 20–30% of the total assimilated C into the soil. Half of this amount is subsequently found in the roots and about one third in CO_2

evolved from the soil by root respiration and microbial utilization of root-born organic substances. The remaining part of below-ground translocated C is incorporated into the soil microorganisms and SOM. Pasture plants translocate about 30–50% of assimilates below-ground. For understanding dynamics of soil microbial biomass it is of interest to assess how much of the carbon translocated by the plants is used by the microorganisms for their maintenance and growth. Several reports provide information about the efficiency by which soil microorganisms use added carbon sources, mainly applied as ^{14}C-labeled substances, for their own proliferation. For glucose, which could serve as a representative compound of the low molecular and easily degradable exudates, it was found that 25–40% of added C was converted into biomass carbon (Saggar et al., 1994; Chander & Joergensen, 2001). However, less than 10% of added plant material carbon (straw, roots) was incorporated by soil microorganisms into cell carbon at a specific sampling date (Ladd et al., 1981; Martens, 1985). These estimates can vary from soil to soil as the pH and the clay content influence the efficiency of conversion (Sörensen, 1983).

However, these efficiencies of soil microbial populations in converting added carbon sources into cell carbon can not be applied for rhizodepositions as shown by growth of plants in an atmosphere with $^{14}CO_2$. Martens (1990) cultivated wheat and maize in soil columns to ripeness or late flowering in a continuous $^{14}CO_2$ atmosphere and estimated microbial biomass ^{12}C and ^{14}C at different stages of plant development. These concomitant quantifications revealed an initial decrease of biomass by 10% during the early stage of plant development. Thereafter, with increasing supply of rhizodepositions loss of biomass ^{12}C was first equalized out by the formation of biomass ^{14}C and later the initial value (100 µg C g^{-1} soil) of total biomass C was exceeded by 10%. At the end of the experiments ^{14}C in the microbial biomass represented only 0.4–0.5% of the net assimilated ^{14}C. During the growth (110 days) of Lolium perenne Kuzyakov et al. (2001) found in a $^{14}CO_2$ pulse labeling experiment that 0.2–0.8% of totally assimilated ^{14}C was in the microbial biomass.

In addition to carbon supply the dynamics of microbial biomass in the field are also determined by climatic factors, namely water potential and temperature. It has long been recognized that drying of soils has mainly two effects. It partly kills the microbial biomass. The reported extent of this effect varies considerably in the different investigations but often a reduction of 25–50% was found (Bottner, 1985; Sparling et al., 1986; Shen Shan Min et al., 1987; Kieft et al., 1987; Van Gestel et al., 1993; de Nobili et al., 2006). In addition, drying physically disrupts soil aggregates and makes additional carbon accessible to degradation (Birch, 1958). After rewetting, the degradation of died organisms and the additional carbon from soil organic matter leads to an immediate and pronounced flush of CO_2 and a rapid but limited recovery of microbial biomass. Wu & Brookes (2005) found that in a grassland soil organic matter contributed 60–70% to that flush. After rewetting the microbial biomass recovers and reached about 70–80% of its initial size. The rapid proliferation of the surviving biomass after rewetting and in addition the increased carbon availability from soil organic matter after drying raises the question about the reliability of microbial biomass C data when these are to be determined by the fumigation techniques at the time of sampling. Sparling & West (1989) investigated the consequences for carbon release different times after rewetting of dry soils for both, the fumigated and control samples. They concluded from their results that microbial biomass C by the CFE method would best be calculated when carbon in the extracts from air-dry control soil was subtracted from the carbon in the extracts of rewetted, fumigated soil and a k_{EC} value of 0.35 was applied to convert the organic C flush to microbial C. An adjustment of water contents of originally dry soil samples to 50–60% of WHC just before fumigation was also recommended by Ross (1989), and he found the water content of control samples before extraction with 0.5 M K_2SO_4 less important. For the CFI method Shen Shan Min et al. (1987) found that only internally consistent values of microbial biomass C were obtained when the original recommended procedure (Jenkinson & Powlson, 1976) was immediately applied after rewetting of air-dry soils and the carbon flush was also calculated as usual (CO_2-C from fumigated soil, 0–10 days – CO_2-C from unfumigated soil, 0–10 days). Other incubation times and their use for calculations, as proposed in the literature, were found to be inadequate.

Regarding these considerations it can be expected that the seasonal variations of microbial biomass contents in the field in a temperate climate are relatively small. Exceptions will occur when soils are exposed to extreme climatic and management events, like extreme dryness, plowing or addition of large amounts of plant residues after harvest. Corresponding investigations confirm this assessment. Patra et al. (1990) found no remarkable variations of C_{mic} throughout a year in a field with wheat from Rothamsted and they stated that possible seasonal changes will have largely been masked by experimental (s. above) and sampling errors. Nieder et al. (in press) have used results from several field studies involving numerous measurements to describe the change of soil microbial biomass C and N during the growth period of annual crops (years 1988–1992, 1994, 1995) under the temperate climate conditions of central Europe (Fig. 5.2). All of the field experiments were conducted in cash crop production farms in northern (Lower Saxony) and Central (Hessia) Germany.

The evaluation of these results through regression analysis demonstrated that microbial biomass C and N from the beginning of a year increased only slightly until summer and subsequently decreased until autumn to their initial levels. This increase on an average corresponded to a C assimilation of ~100 kg C ha^{-1} 30 cm^{-1} and an N immobilization of ~20 kg N ha^{-1} 30 cm^{-1} in the microbial biomass. This supports other studies from ecosystems of the temperate climate zone which indicated a microbial growth in spring or summer where a stimulatory effect of an active rhizosphere was made responsible for this increase in microbial biomass.

5.1.3 Results from ^{15}N Field Studies

Detected mass changes of microbial biomass can not explain N immobilization rates frequently observed in different studies using ^{15}N labeled fertilizers. A review of 16 ^{15}N balance studies (including 21 treatments) conducted on loess soils in Germany in the 1970s and the 1980s (Becker, 1990) yielded that 20–55% of the applied

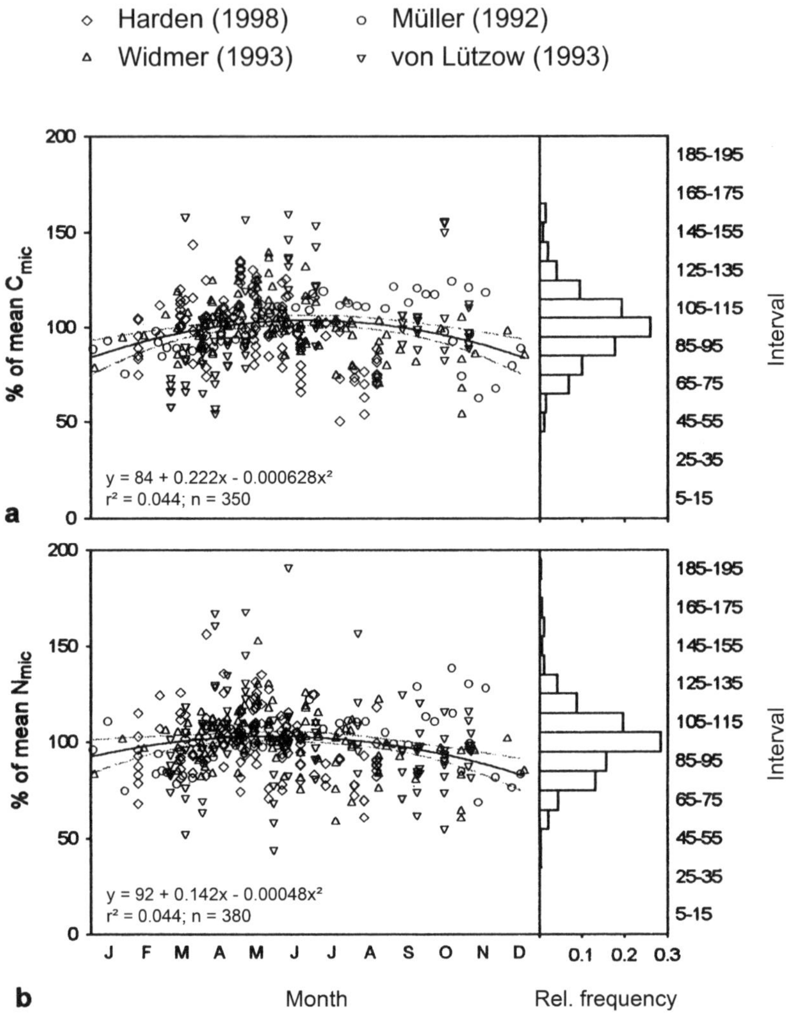

Fig. 5.2 Seasonal microbial biomass (a) C (C_{mic}) and (b) N (N_{mic}) dynamics drawn from four field studies (15 field plots) in Germany. Single measurements include different sites, replicates and treatments (e.g., different amounts and kinds of N fertilizers). They are given in percent of mean microbial biomass C and N contents. Solid lines indicate the regression line, the dashed line the 95% confidence interval of the regression. Horizontal bars indicate the relative frequency (Nieder et al., in press. Reproduced with kind permission from Wiley-VCH)

fertilizer-N (up to more than 50 kg N ha^{-1}) was immobilized during the growth period of annual crops. Most of the labeled N immobilized might have accumulated in stabilized products of decaying plant residues (mainly straw) and SOM.

As an example for the fate of ^{15}N-labeled fertilizer applied to an annual crop (winter wheat), Fig. 5.3 shows the results from a microplot experiment on a loess soil (Luvisol) near Hannover (North Germany) from September 1984 until December 1985. Before

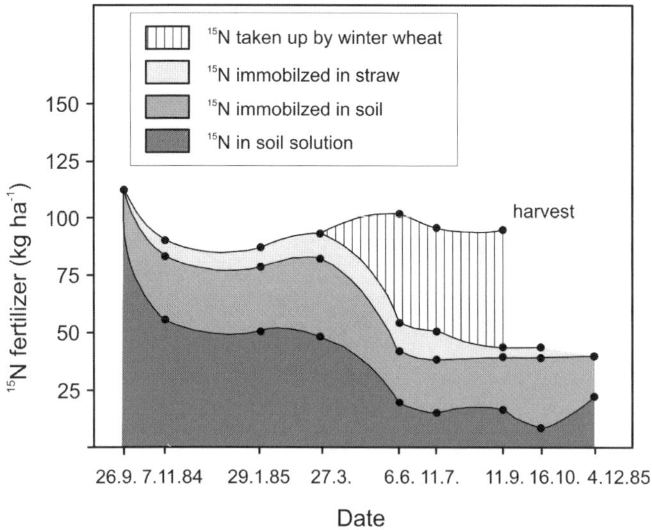

Fig. 5.3 Distribution of ^{15}N fertilizer in different pools (soil solution, soil matrix, incorporated straw, N uptake by wheat) of a microplot experiment on a loess soil (Luvisol) near Hannover, Germany) during the vegetation period of winter wheat (Nieder et al., 1993; p. 298. Reproduced with kind permission from Wiley-VCH)

sowing of the winter wheat, straw (8 t ha^{-1}) from the previous crop (winter wheat) was incorporated homogeniously into the plow layer. ^{15}N fertilizer was applied shortly after the winter wheat was sown by the end of September 1984. The mass change of ^{15}N in the straw material was quantified using a separate plot where the straw was incorporated into the soil within litter bags. At the beginning of the study (until November 1984), about 40 kg fertilizer-N ha^{-1} was immobilized in the soil matrix (rhizosphere) and the straw. During the vegetation period of winter wheat, only a small amount of fertilizer-N was remobilized. At the end of the vegetation period, about 20 kg fertilizer-N was still immobilized in the soil. Most of the labeled fertilizer has been incorporated into SOM because only small amounts of ^{15}N were found in the separately determined pool of mineral fixed NH_4^+ (up to 6.1 kg N ha^{-1} (0–90 cm) until December 1985; Nieder et al., 1993). In summary, the changes in microbial biomass (section 5.1.2.1) may be of less importance for changes in soil N storage compared to interactions with the soil matrix (SOM) and incorporated plant residues.

The above presented results from ^{15}N experiments do not completely reflect the net immobilization pattern during the vegetation period. For a complete picture, the mass changes in soil-born mineral N, using the N difference method (section 5.1.1.5), has to be considered as well. In Fig. 5.4, besides the results from the plot where ^{15}N-labeled fertilizer was applied (Fig. 5.3), results from a reference winter wheat plot without straw incorporation (same amount of N fertilizer) were included. According to this pattern, the net N immobilization (i.e., soil-born plus ^{15}N-labeled N) at maximum amounted to 37 kg N ha^{-1} (March 1985). The remobilization of N during

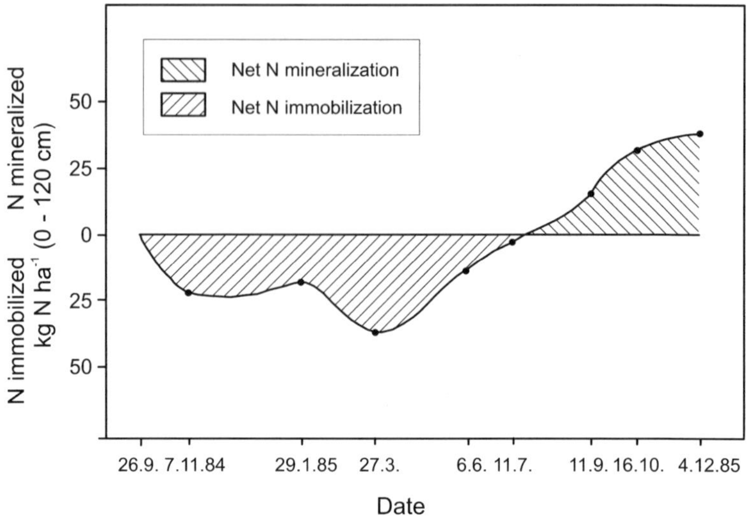

Fig. 5.4 Course of immobilization and remobilization of fertilizer-N and native soil-N in a microplot experiment on a loess soil (Luvisol, near Hannover, Germany) during the vegetation period of winter wheat (Nieder & Richter, 1989; p. 419. Reproduced with kind permission from Wiley-VCH)

the second half of the experiment can be drawn back to a decrease of the C substrate concentration (straw), causing an increased mineralization of the microbial biomass (Paul & van Veen, 1978).

5.1.4 Long-Term C and N Mineralization and Accumulation

Long-term mineralization or accumulation of C and N in soils are related to crop management or land use changes. The most important changes and their influence on the long-term C and N dynamics have been reviewed in Chapter 6. Table 5.1 gives an overview of the N mineralization and immobilization potentials following land use changes. According to Table 5.1, land use changes can be classified as sink or source system, each having a time limit for their potential action.

5.2 Transformations of Inorganic Components

5.2.1 Formation of Secondary Carbonates

Secondary carbonates in soils are formed due to the dissolution of primary ones by water plus atmospheric and/or respiratory CO_2, whereby the formed bicarbonates are reprecipitated to carbonates. Air within soils is the major source of CO_2 and

Table 5.1 Evaluation of land use changes in view of N mineralization and immobilization potentials in soils (Adapted from Gäth et al., 1997)

Kind of land use change	System type	N mineralization potential (kg N ha^{-1} 30 cm^{-1} $year^{-1}$)	N immobilization potential (kg N ha^{-1} 30 cm^{-1} $year^{-1}$)	Total duration after introduction (year)
Conversion of arable land to grassland	Sink system	–	50–100	up to 50
Sand mixed culture	Sink system	–	~200	~40
Permanent green fallow	Sink system	–	50–100	up to 50
	Equilibrium system[a]		±0	±0
Conversion of grassland to arable land (years 4–20)	Source system	100–300	–	~20
Ploughing of green fallow	Source system	~40	–	?
Conversion of grassland to arable land (years 4–20)	Source system	1,000–5,000	–	~20
Drainage of humic-rich wetlands	Source system	<500	–	?
Low moor cultivation (grassland)	Source system	~500	–	up to 100
Low moor cultivation (arable land)	Source system	~1,000	–	up to 100

[a] System in which management is kept constant in the long term

decomposition of organic material together with root respiration maintain high partial pressures of CO_2 in soil air. This can lead to dissolution and weathering of crust material and ultimately to the formation of secondary carbonates. The source of the carbonate is generally dust-borne material and/or carbonates, which are already present in the soil. Calcareous dust or Ca^{2+} from silicate weathering, dissolved by the rainwater is carried down the soil profile. But as the movement of water downward loses momentum and eventually stops and drying begins carbonates are precipitated. In loess soils, nodules similar to the size of children's heads (German term "Loesskindl") can be formed. Generally, four mechanisms of formation of secondary carbonates have been suggested (Monger, 2002) viz.: (i) dissolution of existing carbonates in the upper layers, translocation to the subsoil, and precipitation with cations added from outside the ecosystem (Marion et al., 1985), (ii) capillary rise of Ca^{2+} from shallow water table and subsequent precipitation in the surface layer through reaction with carbonic acid formed through dissolution of CO_2 in soil air (Sobecki & Wilding, 1983), (iii) carbonate dissolution and reprecipitation *in situ* with addition of cations from elsewhere (Rabenhorst & Wilding, 1986), and (iv) carbonate formation through activity of soil organisms such as termites and microorganisms (Bouquet et al., 1973;

Monger et al., 1991; Zavarzin, 2002). The rate of formation of secondary carbonates in Boreal grassland and forest regions of Canada has been reported to be 10–15 kg carbonate-C ha^{-1} $year^{-1}$ (Lundi et al., 2003).

Carbon, in the form of carbonic acid, is especially responsible for the weathering of crust materials. The chemical weathering can both add and withdraw carbon from the atmosphere. For example, dissolution of carbonates is associated with uptake of CO_2:

$$CaCO_3(s) + CO_2(g) + H_2O \rightarrow Ca^{2+} + 2HCO_3^- \quad (5.1)$$

$$Ca(Mg)CO_3 + CO_2(g) + H_2O \rightarrow Ca^{2+}(Mg^{2+}) + 2HCO_3^- \quad (5.2)$$

The equations show that for every mole of atmospheric CO_2 consumed, one mole of rock-derived CO_3^{2-} is transported. After reprecipitation of the carbonate, one mole of CO_2 is released again.

$$Ca^{2+} + HCO_3^- \rightarrow CaCO_3 + H_2O + CO_2 \quad (5.3)$$

According to the equation the precipitation of secondary or pedogenic carbonates is favored with decreasing water content, decreasing partial pressure of CO_2 or increasing Ca^{2+} or HCO_3^-. Climatic conditions regulate the long-term fate of SIC by influencing whether the equation shifts to the left or right. In arid to semiarid regions, secondary carbonates accumulate as evapotranspiration exceeds precipitation. The depth of formation of secondary carbonates increases with increase in mean annual precipitation, being shallow in arid and deep in semiarid and subhumid climates. In arid and semiarid regions of the world, dissolution of carbonates in the upper part of the soil profile and their subsequent precipitation in the lower part leads to the formation of calcic and petrocalcic horizons (Fig. 5.5). In such non-leaching environments there is no net storage of atmospheric CO_2 in soils developed from carbonates (Schlesinger, 1982; Berner, 1992). In contrast, in humid environments most of the SIC is lost in leachate to groundwater as precipitation exceeds evapotranspiration. In soils of subhumid to humid climates, leached HCO_3^- contains a mole of modern atmospheric CO_2 that is transferred to the hydrosphere in caves, lakes, and aquifers or from the continents to the oceans by river transport. Of the total bicarbonate-C (0.42 Pg C $year^{-1}$) that is transported to rivers, approximately two-third is derived from modern soil CO_2 sources and one third from ancient CO_2 sources (e.g. limestone). On an annual basis there is a sequestration of 0.14 Pg atmospheric CO_2-C (Nordt et al., 2000). This suggests that leaching soils of humid and subhumid regions play a role in the global C cycle by sequestering atmospheric CO_2, including that derived from both silicate and carbonate sources. Leaching of carbonates is also relevant to irrigated croplands in arid and semiarid regions, particularly if irrigation water is not saturated with carbonates.

Weathering of silicates consumes 2 moles of atmospheric CO_2 for every mole released during the precipitation of pedogenic carbonate (Schlesinger, 1982; Berner, 1992). For example, weathering of a non-aluminum silicate, such as Mg-olivine forestreite, can be expressed as:

Fig. 5.5 Calcisol exhibiting a compact calcrete horizon (cementation by calcite) (Photo: origin unknown)

$$Mg_2SiO_4(s) + 4CO_2 + 4H_2O \rightarrow 2Mg^{2+} + 4HCO_3^- + H_4SiO_4 \qquad (5.4)$$

This process results in the consumption of 4 moles of CO_2, but only 2 moles are returned to the atmosphere when carbonate precipitates. Therefore, this process sequesters modern atmospheric CO_2. Weathering of an aluminosilicate such as feldspar albite to montmorillonite may be written as:

$$2NaAlSi_3O_8 + 2CO_2 + 2H_2O \rightarrow Al_2Si_4O_{10}(OH)_2 + 2Na^+ + 2HCO_3^- + 4H_4SiO_4 \qquad (5.5)$$

The solution becomes alkaline, and carbonates will precipitate only if $Na_2(HCO_3)_2$ is produced by strong evaporation. Weathering of anorthite ($CaAl_2Si_2O_8$) produces one mole of Ca and 2 moles of HCO_3^- and only half of the CO_2 is returned to the atmosphere upon precipitation of calcite. Since weathering of silicate mineral requires CO_2, it is an effective mechanism for the removal of CO_2 from the atmosphere. However, the magnitude of soil inorganic carbon flux with the atmosphere is difficult to estimate at local, regional or global scales.

The silicate weathering reaction, which in nature takes place on a geological time scale is being considered as a CO_2 mitigation option by accelerating the process (IPCC, 2005). The process, termed mineral carbonation or mineral sequestration is based on the reaction of CO_2 with Ca and Mg oxides to form insoluble carbonates. In mineral carbonation high concentrations of CO_2 captured from an anthropogenic large point source (such as major CO_2 emitting industries, large fossil fuel or biomass energy facilities) is reacted with metal oxide bearing materials with the purpose of fixing the CO_2 as carbonates (Seifritz, 1990; Dunsmore, 1992; Lackner et al., 1995). When CO_2 reacts with metal oxides the corresponding carbonate is formed and heat is released. The amount of heat released depends on the specific metal and on the material containing the metal oxide. For example, mineral carbonation of Olivine (Mg_2SiO_4), Serpentine ($Mg_3Si_2O_5(OH)_4$) or Wollastonite ($CaSiO_3$) releases 89, 64 and 90 kJ heat mol^{-1} CO_2 at 25°C and 0.1 MPa (Robie et al., 1978). Because of the exothermic nature of the reaction, the formation of carbonates is thermodynamically favored at low temperature. Whereas at high temperature (above 900°C for calcium carbonate and above 300°C for magnesium carbonate, at a CO_2 partial pressure of 0.1 MPa) the reverse reaction, that is calcination is favored. Interest in mineral carbonation arises due to the abundance of metal oxide bearing materials, particularly natural silicates, and the permanence of storage of CO_2 in a stable solid form. However, mineral carbonation is still an immature technology and is in the research phase. Some processes using industrial wastes are in the demonstration phase (IPCC, 2005).

Another inorganic source of C, which needs mention here is the agricultural lime. Application of lime, such as crushed limestone or dolomite, for management of croplands and grasslands on acidic soils is a common practice to reduce soil acidity. The application results in emission of CO_2 primarily due to dissolution of lime under acidic conditions. As per IPCC methodology (Houghton et al., 1997) all the mass fraction of C in the rock is assumed to be released as CO_2 to the atmosphere. West & McBride (2005) contended that emissions may be much smaller as a significant portion of dissolved lime constituents may leach through the soil and be transported by rivers to the ocean. Much of the fraction transported to the ocean will precipitate as $CaCO_3$. Their analysis showed that net CO_2 emission from the application of agricultural lime is 0.059 Mg C per Mg limestone and 0.064 Mg C per Mg dolomite as compared to IPCC emission coefficients of 0.12 and 0.13 Mg C per Mg crushed rock, respectively. In view of the widely divergent estimates, there is a need for further research to better quantify SIC dynamics in agricultural soils.

5.2.2 *Nitrification*

Nitrification is the process of conversion of NH_4^+ to NO_3^- under aerobic conditions. The process is mediated by specific genera of aerobic chemoautotrophic microorganisms, namely *Nitrosomonas* and *Nitrobacter* (Alexander, 1977). These organisms use CO_2 as a carbon source, with *Nitrosomonas* oxidizing NH_4^+ to nitrite (NO_2^-) and *Nitrobacter* converting NO_2^- to NO_3^-.

$$NH_4^+ + 1\tfrac{1}{2}\,O_2 \rightarrow NO_2^- + 2H^+ + H_2O + \text{Energy} \quad (5.6)$$

$$NO_2^- + \tfrac{1}{2}\,O_2 \rightarrow NO_3^- + \text{Energy} \quad (5.7)$$

Since there is production of $2H^+$ per N during nitrification, this may cause acidification in some environments. The energy released during the formation of nitrite (272 kJ) and nitrate (79 kJ) is used by *Nitrosomonas* and *Nitrobacter* organisms for carrying out their life functions. Commonly, nitrite (NO_2^-) is oxidized rapidly to NO_3^- but NO_2^- could accumulate in the soils with high pH and high concentration of ammoniacal N ($NH_4^+ + NH_3$). Concentration of NH_4-N, soil aeration status (expressed as oxygen or moisture content), temperature, soil pH and soil texture arc the major factors that govern nitrification in soils. As oxygen is strictly required in the production of NO_3^-, a lack of oxygen in soils inhibits nitrification. Oxygen supply is moderated by soil moisture, and in the normal soil moisture range, the effect of soil water content on nitrification probably reflects its effect on oxygen diffusion. Both excessively dry and waterlogged soil conditions could severely inhibit the process of nitrification. However, nitrification may not be completely inhibited in wet or flooded rice soils because NH_4 could be oxidized to NO_3 in the thin O_2-containing surface soil layer and in the overlying water phase of flooded soil (Engler & Patrick, 1974). In wetlands nitrification is primarily confined to the water column and a very thin (<1 cm) surface layer of soil. Very little nitrification also occurs within the rhizosphere of some wetland plants that create an aerobic microclimate. Under flooded soil conditions, nitrification rate is influenced by alkalinity of the flood water, inorganic C source, microbial population and NH_4-N concentration besides temperature and pH.

Nitrification seems to be restricted by low temperatures to a greater extent than ammonification. Nitrification is low at temperatures below 5°C. The optimum temperature for nitrification ranges from 30–35°C. *Nitrobacter* is more sensitive to low temperature than is *Nitrosomonas*. In cold conditions this may lead to a NO_2^- accumulation in the soil, which may have a toxic effect on plants. Nitrification is sensitive to soil pH, with significant reduction in the process below pH 6.0 and above pH 8.0. Nitrification is negligible below pH 4.5. Weber & Gainey (1962) found that the nitrification stopped at soil pH 3.9–4.1. Chemolitho-autotrophic bacteria are the main nitrifying agents in most acid soils, whereas heterotrophs in these soils are generally thought to make only a small contribution to NH_4^+ oxidation. However, in certain acid soils, heterotrophs may also play an important role in nitrification. The optimum pH for nitrification is 7–8. At pH >8 the conversion of NO_2^- to NO_3^- is inhibited to a greater extent than the conversion of NH_4^+ to NO_2^-. This is because *Nitrobacter* is sensitive to pH, mainly because free ammonia and nitrous acid are both toxic to it. *Nitrosomonas* is relatively less sensitive to high pH or ammonia. Thus, NO_2^- can accumulate in alkaline soils. Low temperatures have a similar effect and promote accumulation of NO_2^-. A significant interaction between pH and temperature has been found and in some cases it leads to accumulation of NO_2 as *Nitrobacter* is affected more than *Nitrosomonas*. At extremes of pH, the effect of temperature is more pronounced on nitrite than ammonium-oxidizing organisms, while at optimum pH ammonium-oxidizing organisms are more temperature sensitive. Wong-Chong & Loehr (1975) showed that the optimal temperatures for the two processes vary with pH.

Crop residue management could also influence the rate of nitrification indirectly by controlling the mineralization-immobilization turnover depending on their C:N ratio. The rate of nitrification is especially important in grassland soils, where much recycled N, either from animal excreta or mineralized from soil organic matter occurs as NH_4^+. Though it is usually assumed that nitrification rates exceed the rate of mineralization in many grassland and natural systems, significant quantities of NH_4^+ may accumulate in the soil profile (Jarvis & Barraclough, 1991). High levels of heavy metals such as copper, cadmium and zinc in soils are known to depress N mineralization, and the inhibitory effect of heavy metals is more pronounced on nitrification which lead to the accumulation of NH_4-N in the soil (Benbi & Richter, 1996; Fig. 5.6). In general, mineralization process is less sensitive to the effects of heavy metals than nitrification probably because mineralization can be carried out by a large diversity of microorganisms. Nitrification instead is performed by rather few specialized organisms only (Benbi & Richter, 1996).

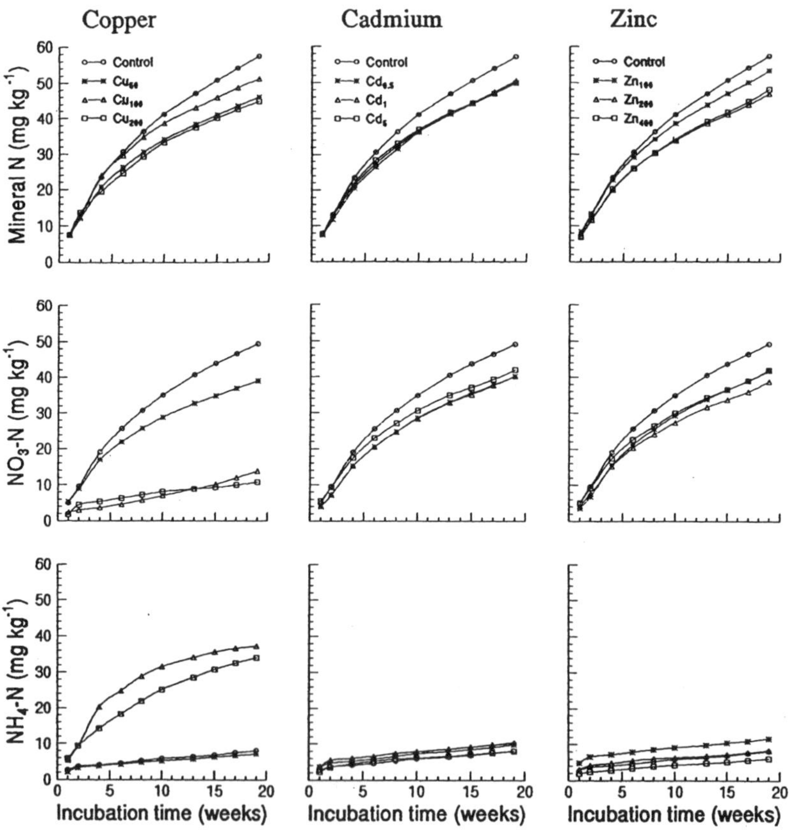

Fig. 5.6 Effect of heavy metal (Cu, Cd and Zn) additions on N mineralization and nitrification in a sandy soil. The numbers as subscripts to the metals indicate the rate of application (mg kg^{-1} soil) (Benbi & Richter, 1996; p. 20. Reproduced with kind permission from Springer)

Nitrification process is central to the flows, transfers, losses or utilization of N. Despite its importance as a rate limiting process and a reasonable knowledge of the ecology and environmental demands of the organisms involved, it is a poorly defined process in many soils. It influences fertilizer-use efficiency (FUE) by crops as well as N losses from soils. Since a major portion of total fertilizer N used in world agriculture is urea and other ammoniacal fertilizers, the rate of nitrification is a key determinant of N losses. Nitrification rates in excess of NO_3^- utilization by plants can result in increased levels of NO_3^--N in runoff and groundwater rendering it unsafe for human consumption. On the other hand slow rates of nitrification could result in accumulation of NH_4^+-N which may enhance FUE by reducing denitrification and leaching losses. However, reduced nitrification could also increase N losses via NH_3 volatilization. There is an increasing evidence that nitrification may be a major source of the N_2O and NO emitted by soils. Nitrification influences the amount of N_2O and NO released to the atmosphere during nitrification itself (Fig. 5.7). In soils, there may occur conditions that inhibit the second stage of nitrification, thus N_2O and NO concentrations may increase in the soil and emitted to the atmosphere. In well-aerated systems, where NO can be swept quickly from the soil, NO evolution is usually 10–100 times larger than that of N_2O (Hutchinson & Davidson, 1992). Nitrous oxide evolution becomes relatively more important as the moisture content of the soil increases. This is probably because the residence time of NO in soil is longer due to decreased diffusion, which increases the potential for NO to be converted to other products; and secondly the activity of enzymes within *Nitrosomonas* spp., which reduce NO_2^- to N_2O are favored in less oxic conditions (Poth & Focht, 1985). Nitrification also influences the rates of N_2O and NO released as result of denitrification and chemodenitrification, by providing substrate for these reactions (Fig. 5.7). Under oxygen limited conditions nitrifiers can use NO_2^- as a terminal electron acceptor and result in the production of NO and N_2O or dinitrogen (N_2). In many rice growing areas, continuous flooding cannot be maintained and there are periods of alternate drainage and flooding. Ammonium accumulated during flooding can be nitrified rapidly during the aerobic drainage period and nitrates thus produced are subsequently reduced to N_2O or N_2 through denitrification when soils are reflooded. The emission of oxides of nitrogen is influenced by a number of soil, environmental and management factors. These are discussed in detail in Chapter 8.

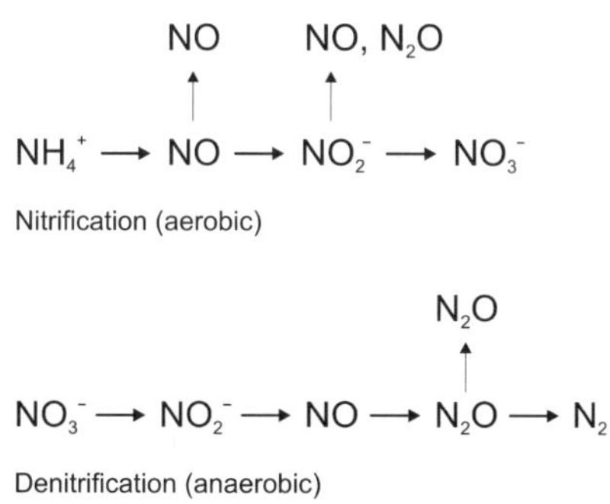

Fig. 5.7 Emission of oxides of nitrogen through nitrification and denitrification processes in soil

In view of the increasing CO_2 concentration in the atmosphere, the microbial mediated processes of the N cycle are expected to be modified but the impact is poorly understood and the reports are often contradictory. For example, Zak et al. (1993) found increased rates of net N mineralization under elevated CO_2 in a short term laboratory incubation. Other studies reported decrease in net mineralization due to greater N immobilization (Rouhier et al., 1994; Torbert et al., 1995). However, there is paucity of information on the effect of elevated CO_2 on nitrification and denitrification processes. Autotrophic nitrifiers are not likely to be directly affected by increased C flow under elevated CO_2 concentration as these organisms utilize inorganic C instead of preformed organic C as a C source (Paterson et al., 1997). Nitrifiers may be indirectly influenced by increased microbial activity under elevated CO_2. Nitrifying population may also be influenced by small changes in soil moisture content, pH or nutrient status due to increased plant growth under elevated CO_2. In a study on four European grasslands, Barnard et al. (2004) did not found any effect of elevated atmospheric CO_2 on nitrifying and denitrifying enzyme activity.

5.2.3 Fixation and Defixation of Ammonium

A part of inorganic N as NH_4^+ in soil may undergo fixation reactions which result in entrapment of ammonium ions in interlayer spaces of phyllosilicates, in sites similar to K^+ in micas. The fixed ammonium, also called non-exchangeable or interlayer NH_4^+, cannot be extracted with neutral normal potassium salt solutions (SSSA, 1997). While smectites, illites and vermiculites all can fix ammonium, the capacity to fix applied ammonium decreases in the order vermiculite, illite and smectite (Allison et al., 1953). The amount of NH_4^+ fixation also depends on the degree of K-saturation of the interlayers of the 2:1 clay minerals. Soils rich in exchangeable potassium have low contents of non-exchangeable ammonium. As a result of competition for fixation sites, the presence of NH_4^+ or K^+ may alter both fixation and release of these cations. Addition of K^+ prior to NH_4^+ has been reported to depress NH_4^+ fixation (Nommik & Vahtras, 1982), and addition of NH_4^+ prior to or at the same time as K^+ reduces K^+ fixation (Aquaye & MacLean, 1966; Bartlett & Simpson, 1967). However, there are reports, which indicate that K^+ pre-addition either did not block subsequent NH_4^+ fixation (Drury et al., 1989) or the presence of K^+ induced greater NH_4^+ fixation (Chen et al., 1989).

Studies with ^{15}N labeled NH_4^+ have provided direct evidence for the fixation of applied ammonium wherein 18–23% of $^{15}NH_4^+$ was fixed after a 15-day incubation in soils with high vermiculitic contents (Drury et al., 1989). Fixation is usually fast and occurs within the first few days after fertilizer application. For example, Nieder et al. (1995b) found that in a loess soil most of the NH_4^+ fixation had occurred within 1 h after application of NH_4^+-acetate solution and the fixation reactions were virtually over within 10 h after NH_4^+ application. Similarly, in some soils of eastern Canada, more than 40% of the added NH_4^+ was fixed in less than 2 h and further fixation was negligible (Sowden et al., 1978). The amount of added NH_4^+ fixed depends on NH_4^+-fixing capacity of the soil (see Chapter 1). Fischer et al. (1981) found that in a gley soil with high NH_4^+ fixing capacity, 47% of added ^{15}N was present in the fixed ammonium fraction, whereas in a histosol with very low NH_4^+ fixation capacity only 7% of the added NH_4^+ was fixed. This suggests that depending upon the NH_4^+ fixation capacity of the soil, the level of fixed NH_4^+ can be increased upon addition of an NH_4^+ source of fertilizer but this recently fixed NH_4^+ may be subsequently released for uptake by plants as the NH_4^+ concentration around plant roots decreases.

While the process of ammonium fixation is fairly well understood there are conflicting reports on its role in N dynamics in the soil-plant system. The results differ because of differences in methodology, soil type, mineralogical composition and agroclimatic conditions. Most investigators have determined fixed NH_4^+ using strong oxidizing agents (KOBr or KOH) to remove organic N. The suitability of these methods to recover quantitatively the recently fixed fraction of non-exchangeable NH_4^+ is less certain and thus do not necessarily reflect the fraction that is truly unavailable to plants (Chen et al., 1989). The published reports indicate that the native fixed NH_4^+ may be unavailable to plants as it could be released only on geological weathering (Smith et al., 1994; Liu et al., 1997). However, the recently fixed NH_4^+ that originates from fertilizer application may be released (defixed) during crop growth period (Haas et al., 1993; Nieder & Benbi, 1996). Presumably native fixed NH_4^+ is trapped more in the center of the interlayers and thus tightly bound, while recently fixed NH_4^+ may be less strongly bound in the peripheral zone of the interlayers (Scheffer & Schachtschabel, 1984) and thus released during the crop growth period (Fig. 5.8).

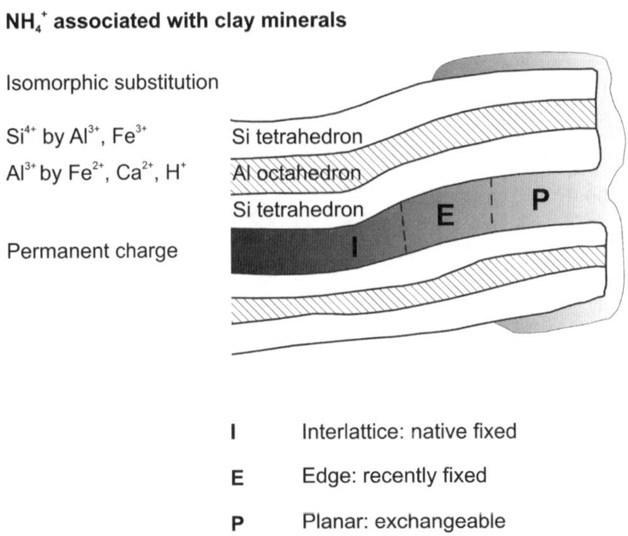

Fig. 5.8 Schematic representation of fixation of ammonium in interlayers of clay minerals

Differences in clay mineralogy and K-saturation of the minerals also influence the release of clay-fixed NH_4^+. Dou & Steffens (1995) showed that 90–95% of newly fixed $^{15}NH_4^+$ was released during a 14-week period when perennial ryegrass (*Lolium perenne* L.) was grown under greenhouse conditions. Under field conditions, 66% of the recently fixed NH_4^+-N was released in the first 86 days and the remaining being strongly fixed over the next 426 days (Kowalenko, 1978). Amounts ranging from 35% to 72% of fertilizer derived recently fixed NH_4^+-N were released for uptake by ryegrass (Norman & Gilmour, 1987). As percentage of total non-exchangeable NH_4^+-N, the release of this fraction ranged from 4% to 25% in different soils (Osborne, 1976; Smith et al., 1994; Steffens & Sparks, 1997). In field experiments on loess soils, Nieder et al. (1996) observed that the time course of fixed NH_4^+ in a Phaeozem Ap horizon was significantly correlated ($r = 0.63$) with that of mineral N during winter wheat growth (Fig. 5.9).

Apparently, the amounts of fixed NH_4^+ increased after fertilizer application in spring before it is depleted by the growing plant root system (NH_4^+ release in summer). The NH_4^+ refixation in autumn may be mainly due to increased mineralization of organic N after harvest of annual crops. This relationship suggest that in addition to fertilizer N, the plant cover and the microbial biomass exert a great influence on the dynamics of non-exchangeable NH_4^+. Some studies have suggested that heterotrophic microorganisms can rapidly assimilate NH_4^+ from clay-fixed fraction primarily under a C-supplying plant cover (Drury & Beauchamp, 1991; Nieder et al., 1996). The magnitude of NH_4^+ fixation and the release of fixed NH_4^+ are dependent upon concentration gradients existing between the amounts of clay-fixed NH_4^+ and the exchangeable and soil solution NH_4^+ (Nommik & Vahtras,

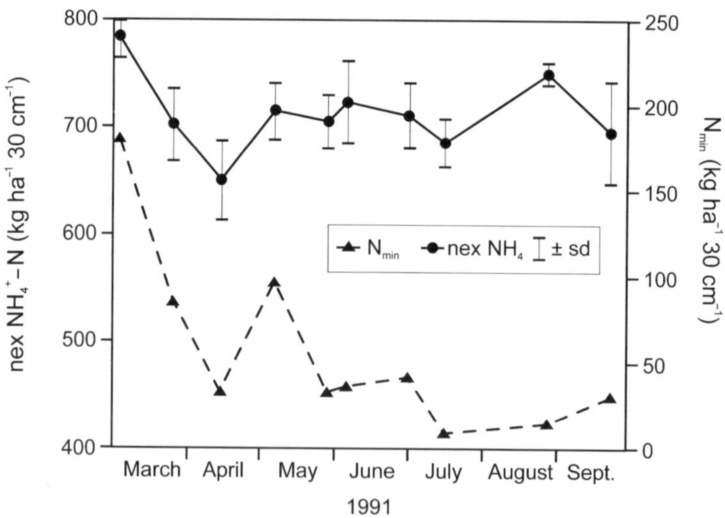

Fig. 5.9 Dynamics of non-exchangeable (nex) NH_4^+ and mineral N (Nmin) in a phaeozem A_p horizon cultivated to winter wheat with conventional N application (Nieder et al., 1996; p. 182. Reproduced with kind permission from Springer)

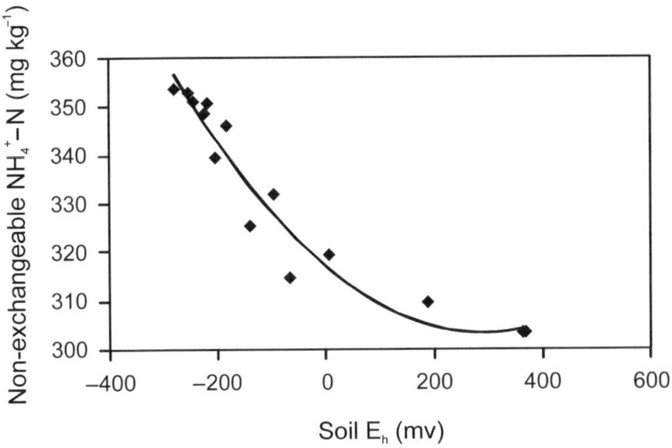

Fig. 5.10 Relationship between soil redox potential (E_h) and non-exchangeable NH_4^+-N in soil (Zhang & Scherer 2000; p. 519. Reproduced with kind permission from Springer)

1982). With the lowering of NH_4^+ concentration in soil solution, NH_4^+ ions diffuse out of interlayers. Therefore, factors such as fertilizer N application, plant cover, soil organic matter, microbial biomass, clay content and clay mineral composition that affect concentration of NH_4^+ in soil solution may promote either release or fixation of NH_4^+. Soil submergence or flooding (rice culture) has been shown to increase NH_4^+ fixation (Keerthisinghe et al., 1984) mainly due to increased availability of NH_4^+ for fixation and low redox potential (Zhang & Scherer, 2000). A close relationship has been found to exist between soil redox potential (E_h) and non-exchangeable NH_4^+ (Fig. 5.10).

The results available so far indicate that native fixed NH_4^+ has no significance in the soil N dynamics, however, the temporal changes in the content of recently fixed NH_4^+ show that this fraction is actively involved in the nitrogen dynamics during the crop growth period. The fixation and release of added NH_4^+-N may help reduce N losses from soils by acting as a temporary sink for fertilizer N that is subsequently made available for plant uptake.

Chapter 6
Anthropogenic Activities and Soil Carbon and Nitrogen

When humans cultivate land, they impact matter storages in soils and alter their fluxes in the soil-plant-atmosphere system. The most pronounced impacts are exerted through land use and especially land use changes, which almost certainly change the fluxes between soils and the atmosphere (Bolin & Sukumar, 2000). Land use and land cover are linked to climate in complex ways. Form and intensity of land use influence the constellation of the atmosphere and the whole global climate system. Key links include the exchange of greenhouse gases (GHGs) between the land surface and the atmosphere, the radiation balance of the land surface, the exchange of sensible heat between the land surface and the atmosphere (USGCRP, 2006). Further interaction occurs between the roughness of the land surface and its uptake of deposits from the atmosphere (Fowler, 2002). Because of these strong links, changes in land use and land cover can be important contributors to climate change. Since there is a close relationship between the atmosphere, soils and climate, the global climate change will strongly impact all soil processes (Rosenzweig & Hillel, 1995), particularly carbon and nitrogen dynamics. Most land use changes have been and still are a significant source for the release of former plant and soil C and N from SOM into both the atmosphere and the drainage zone. In this chapter, the extent of land use changes and their influence on C and N fluxes are presented on regional and global scales and the impacts of land use changes on the global climate are discussed.

6.1 Land Use Changes

6.1.1 Land Use Area Distribution and Its Global Change

In the past years, the area of natural vegetation cover has steadily diminished in favor of agricultural landforms or urbanization (Adger & Brown, 1995; Field & Raupach, 2004). One exception is the tundra-covered area where no intensive agricultural production can be practiced due to unfavorable climatic conditions. Most natural land cover forms, especially forests, grasslands and wetlands, have been

R. Nieder, D.K. Benbi, *Carbon and Nitrogen in the Terrestrial Environment,*

subject of anthropogenic land use changes (Bork et al., 2006). The causes and rates of land use changes vary by region and scale but the major reason for the ever-increasing demand of areas for human activities is population growth (Apps et al., 2001).

The temperate zone is the most populated zone of the world. For thousands of years, forest area has diminished particularly in the temperate zone where forests were cleared for agriculture and pasture. Clearing of the southern European Mediterranean region began about 5,000 years ago. In Central Europe and in China, deforestation occurred in early Medieval times. In parts of Russia and Mongolia, deforestation occurred in the late Medieval times, and in North America, deforestation occurred particularly in the 19th century.

The demand for area which is suitable for agricultural production and other human activities has increased significantly since the beginning of industrialization in the second half of the 18th century. Concern about SOM losses from soil as a result of plowing up grassland has been voiced since about 1850. In the second half of the 19th century, North American prairies have been increasingly converted after the arrival of settlers from Europe (Bork et al., 2006). At present, grassland areas cover approximately 41% of the terrestrial land surface (White & Vanasselt, 2000). Until the 1940s, conversion of natural landforms was located in the temperate zones of the northern hemisphere, but since the 1950s, the areas of severe land conversion have shifted from developed to developing countries in the southern hemisphere. These trends can be observed in Fig. 6.1.

African forest and woodland has decreased consistently during the observed period, whereas cropland and pasture have increased accordingly. The trend to increasing deforestation in recent years can also be observed for Latin America, Asia and other pacific countries. In contrast, after previous decline, forest and woodland area since the mid 20th century increased in North America and Europe (Kauppi et al., 1997). The cropland area in both regions decreased accordingly. In the former USSR, a significant decline in the deforestation rate was observed until 1980. According to Apps et al. (2001), this trend has turned to an increase in forest area since about 1980. Forest area in the temperate zone is increasing partly because agricultural yields have improved or because the profitability of marginal agriculture has declined.

6.1.1.1 Changes in the World Forest Area

Until the 20th century most of the conversion of forest to cultivated land area occurred in the (now a days) developed countries (Forster et al., 2007). During the past few decades, the major deforestation has occurred in tropical forests (Field & Raupach, 2004; Forster et al., 2007). In the tropics, economics still focus primarily on natural resource extraction rather than sustainable land management. Improper and incompatible methods of deforestation and subsequent land development lead to soil degradation and a rapid decline in crop yields (Lal, 1995a). Tropical forests were largely intact until the 1920s (Apps et al., 2001). Loss of tropical forests esca-

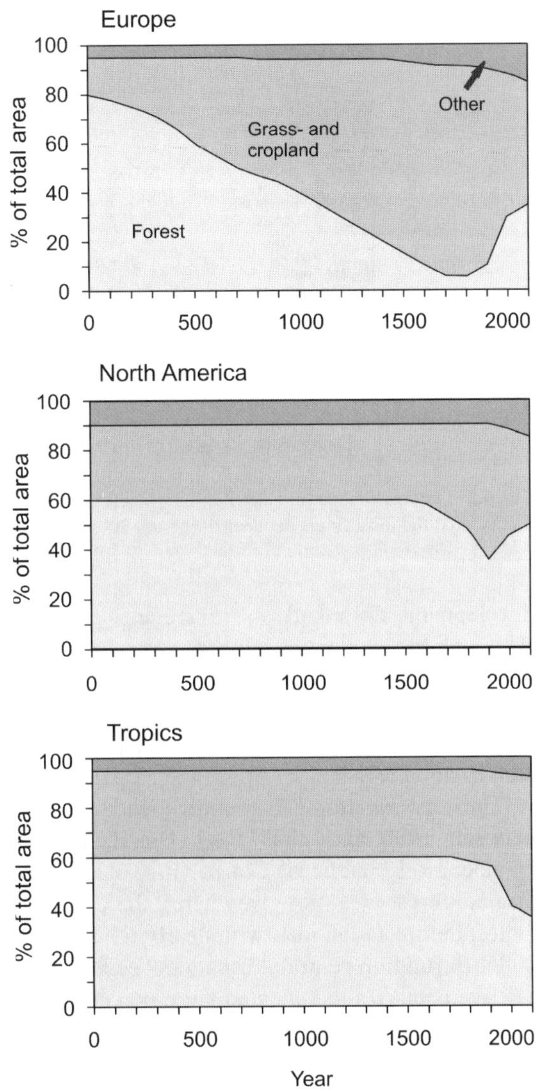

Fig. 6.1 Schematic figure indicating historical changes in land use in three world regions (Apps et al., 2001; p. 310. Reproduced with kind permission from Technical Support Unit Working Group III, IPCC)

lated in the second half of the 20th century. Roughly one fifth of the world's tropical rainforest has been destroyed between 1960 and 1990 in only 3 decades. According to the FAO (cited in Apps et al., 2001), about 15.4 million hectares of natural tropical forests are presently lost every year. The largest part of this loss occurs in Latin America (42%), Africa (31%) and Asia (27%) (Apps et al., 2001). Similar values are presented by the NASA Earth Observatory (2007; Fig. 6.2).

Deforestation of tropical forest in the 1990s ranged from about 55,600–120,000 km^2 every year (NRDC, 2004), with an average rate of 19.1 million hectares year^{-1} during the period 1990–1995 (Apps et al., 2001). Exceptions of the

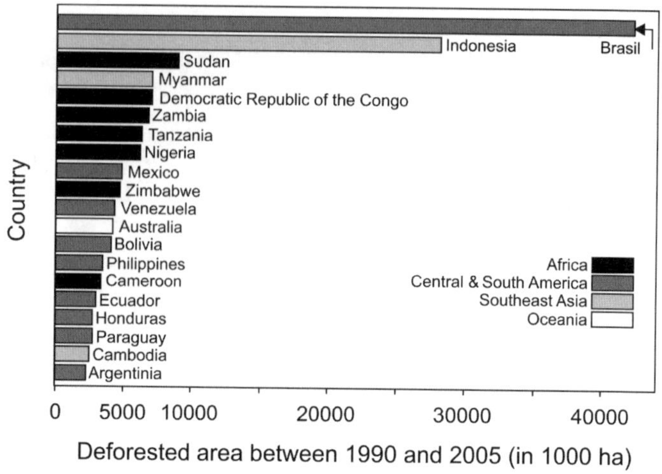

Fig. 6.2 Deforested area (in 1,000 ha) in tropical countries represented in a list of the top 20 countries that cleared the most forest between 1990 and 2005 (Adapted from NASA Earth Observatory, 2007)

development described above are e.g. Australia and some countries in Asia. Although the economic status in Australia resembles that of the USA and Europe, the Australian biosphere still acts as a source of CO_2, due to deforestation in some parts of the country (Field & Raupach, 2004). In India, in contrast, deforestation rates have declined since 1980, despite population growth, owing to effective forest conservation legislation (Apps et al., 2001).

The top five tropical countries with the greatest total area of deforestation between 1990 and 2005 were Brazil, Indonesia, Sudan, Myanmar, and the Democratic Republic of Congo (Fig. 6.2). In terms of percentage of the loss of original forest area (not given here), the Comoros (north of Madagascar) show the largest deforestation rate, with nearly 60% of its forests cleared between 1990 and 2005. Burundi in central Africa was second, clearing 47% of its forests. The other top five countries that cleared large percentages of their forests were Togo (44%), Honduras (37%), and Mauritania (36%). Thirteen other tropical countries and island territories cleared at least 20% of their forests from 1990–2005 (NASA Earth Observatory, 2007).

Despite the decreases in deforestation in temperate and boreal regions, the global wood consumption is projected to increase by 50% until the year 2050 (NRDC, 2004). Apps et al. (2001), however, reports that rates of tropical deforestation have declined slightly in the last decade. Nevertheless, if the present rate of deforestation is maintained, all tropical forests may be gone by the year 2090.

The most common reasons for deforestation are land conversion for agriculture and cattle grazing, the need of timber, fiber, burning materials or space for urbanization, mining and petroleum extraction. The world marked, i.e. the consumers in developed countries, is of increasing importance. The NRDC (2004) report that North Americans use 27% of the wood commercially harvested worldwide, although only 5% of the world's population lives in the United States. Each US citizen consumes averaged 75 m^3 of solid wood per year, corresponding to one ancient tree. The reasons for deforestation vary for different regions of the world. Non-sustainable logging has been the leading factor in parts of Southeast Asia, whereas excessive harvest of wood fuel has been important only in specific subcountry regions and in some African countries. Population and livestock density and external debt are key factors responsible for deforestation. In Africa, key factors were extraction of fuel wood, production of charcoal and demand for cropland. In Asia and Latin America, key factors were increase of cropland and increase of livestock density, respectively (Apps et al., 2001).

The method mostly applied to clear forest in the tropics is slash and burn by which trees are cut down, and the trunks and litter are burned. This method is increasingly used by shifting cultivators to gain profit from short-term yields. Shifting cultivation systems represent a broad group of land use systems based on a few years of crop production. The short-term consequences are release of C and N to the atmosphere and loss of a C sink in form of living biomass. The decrease in SOM as a consequence of cultivation is of the same order as that for temperate regions, averaging 25–30% of the initial SOM contents (Davidson & Ackermann, 1993). Shifting cultivation systems include a recovery fallow period of a few years (i.e. bush fallow) or for longer periods (secondary forest succession). In South East Asia, systems based on secondary forests can be an early stage of "agroforest" development.

Forests in 2005 were estimated to cover 3,952 million hectares, or 30% of the total land area of the world (Table 6.1). These estimates include undisturbed forests, forests modified by humans through use and management (so called seminatural forests) and human made forests (i.e., forest plantations) created artificially by afforestation or reforestation.

Table 6.1 Forest area estimates, changes in forest area (negative numbers indicate forest area decrease), and carbon in biomass (Adapted from FAO, 2006)

Region	Forest area (million hectares) 2005	Annual change (million hectares $year^{-1}$) 1 990–2000	Annual change (million hectares $year^{-1}$) 2000–2005	Total biomass (Tg)	C in biomass (Tg)
Africa	635.412	–4.4	–4.0	120,139	59,923
Asia[a]	571.577	–0.8	1.0	65,396	32,458
Europe[b]	1001.394	0.9	0.7	87,509	43,614
North and Central America	705.849	–0.3	–0.3	43,176	21,566
Oceania[c]	206.254	–0.4	–0.4	18,660	8,414
South America	831.540	–3.8	–4.3	151,464	74,464
World	3,952.026	–8.9	–7.3	486,344	240,439

[a]Without the Russian Federation
[b]Including the Russian Federation
[c]Including Australia

Most relevant for the C and N cycles is that between 2000 and 2005, deforestation continued at a rate of 12.9 million hectares year^{-1}, which is mainly a result of converting forests to agricultural land, but also due to expansion of settlements, infrastructure, and unsustainable logging practices (FAO, 2006). During the 1990s, deforestation was slightly higher, at 13.1 million hectares year^{-1}. Of this, 42% occurred in Latin America, 31% in Africa, and 27% in Asia (FAO, 1996). Due to afforestation, landscape restoration and natural expansion of forests, the net loss was about 8.9 million hectares year^{-1} in the 1990s and 7.3 million hectares year^{-1} from 2000 to 2005 (see Table 6.1). The loss is largest in South America, Africa and Southeast Asia. Thus, the C stock is still decreasing in these regions, whereas it is increasing in other regions.

Particularly forest plantations have increased forest growth in many temperate (mainly poplars) and (sub)tropical (mainly eucalypts) regions. The area of forest plantation was about 45–60 million hectares in 1980, 80–100 million hectares in 1995, 140 million hectares in 2005, and increased by 2.8 million hectares year^{-1} between 2000 and 2005 (FAO, 2006). Forest plantations are tree crops which are in many ways analogous to agricultural crops. They have a simple structure and are commonly restricted to one or a few species chosen for their fast growth, yield of commercial products and ease of management which results in higher productivity than for natural forests.

According to the Millennium Ecosystem Assessment scenarios (MEA, 2005), forest area in industrialized regions are projected to increase between 2000 and 2050 by about 60–230 million hectares. However, these estimates have to be regarded with care because of increasing demand for areas on which crops are grown for bioenergy. The forest area in developing regions will decrease by about 200–490 million hectares. In addition to the decreasing forest area globally, forests are increasingly affected by fires, insects, diseases, and extreme climatic events including drought, storm and floods. Such disturbances annually affect about 100 million hectares (FAO, 2006).

Increasing the area of forests by reforestation and afforestation may significantly increase the C storages in both plant biomass and soils, and thus mitigate global warming. Reforestation or afforestation of bare land, redundant arable land or degraded pastures has been proposed as a good measure for sequestering atmospheric CO_2 in growing trees and increasing SOM (IPCC, 1995). Reforestation and afforestation are different terms. Congruence is given as far as both terms describe the plantation or regrowth of trees on a previously non-forested area. Both practices aim at reestablishing forest that is comparable to natural forest. The difference is that reforestation describes the growing of trees on an area that had originally been covered by forest at some time in the past. Afforestation, in contrary, describes the growing of wood on an area that has not been forest-covered before. Another practice is tree farming, or tree plantation. In tree plantations, trees are grown mostly in monoculture, for the only purpose of logging. Here, no reestablishment of forest is intended.

There has been a large increase in area devoted to forest plantations. By 1990, there were 61.3 million hectares under plantations and the rate of establishment is now about 3.2 million hectares year^{-1} (Apps et al., 2001). For example, in Australia the area of new plantations established in 1999 was about 95,000 hectares, and 150,000 hectares were projected for 2000 (Polglase et al., 2000). Another example for vast reforestation is China. The third largest land-use change in China has been reforestation of previously cultivated land (Heilig, 1999). Some 970 million hectares were reforested between 1988 and 1995, especially in the provinces of Shaanxi, Inner Mongolia, and Yunnan (Heilig, 1999). While these losses of cropland may have somewhat diminished China's agricultural land resources, they will certainly help to prevent or reduce environmental disasters in the future, such as desertification or flooding. In temperate regions, a large-scale afforestation of agricultural land is only possible if adequate supplies of food, fiber, and energy can be obtained from the remaining area. This is currently possible in the E.U. and the USA through intensive farming systems.

6.1.1.2 Changes in the World Grassland Area

The largest total area with grassland is present in sub-Saharan Africa and Asia, amounting to 14.5 and 8.9 million square kilometers, respectively. The five countries with the largest total grassland area are Australia, the Russian Federation, China, the US, and Canada. Grasslands are found most commonly in semiarid zones (28% of the world's grasslands), followed by humid (23%), cold (20%), and arid zones (19%). Temperate grasslands, savannas, and shrub lands have experienced heavy conversion to agriculture, more so than other grassland types including tropical and subtropical grasslands, savannas, and woodlands. The extent of these changes can be linked to technological developments in the agricultural sector and population growth.

Various grassland types can be defined according to different climatic regions and growth conditions. Large coherent areas of natural grasslands are the tropical savannas with approximately 15%, and temperate grasslands (steppes) with 8% (Apps et al., 2001). Further (natural and artificial) grasslands can be found in nearly all climatic regions, e.g. the Boreal or Chaparral zone, where humans have cleared large areas from the natural (forest or shrub) vegetation. There is a wide variation in the extension of grassland even between countries of similar latitudes. For example, grassland occupies 21% of the agricultural area in Denmark and about 90% in Ireland (Whitehead, 1995). In western and northern Europe and in the former USSR, grassland occupies 50–60% of the agricultural area (Whitehead, 1995). A global overview of the area occupied by grassland in 1998 is given in Table 6.2.

Savanna and shrub land are the most widespread grassland types, followed by nonwoody grassland and tundra of the cooler climates. The loss of grassland in favor of other landforms is globally maintained up to today. However, North American forests and grasslands recently are not being converted to other land use forms on a large scale (EIA, 1998). Forestland in North America increased slightly in extent, and the grassland area remained almost constant between 1987 and 1992. Net gains in some grassland regions in South America are mostly derived from destruction of

Table 6.2 Extension of different grassland types by area and percentage of total land area (White et al., 2000)

Grassland type	Area (million square kilometers)	Percent of total land area
Savanna	17.9	13.8
Shrubland	16.5	12.7
Nonwoody grassland	10.7	8.3
Tundra	7.4	5.7
World (total)	52.5	40.5

Table 6.3 Conversion of historical grassland into other forms of land use. The converted area is given in percent of original grassland area (White et al., 2000)

Grassland type	Remaining grassland (%)	Agricultural land (%)	Urbanization (%)	Other (%)
Tropical and subtropical grasslands, savannas, and shrublands	71.3	15.4	0.8	11.8
Temperate grasslands, savannas, and shrublands	43.4	41.4	6.1	7.4
Flooded grasslands and savannahs	48.2	21.7	2.9	24.4
Alpine grasslands and shrublands	70.6	7.7	1.4	18.7
Mediterranean shrublands	48.0	11.9	4.4	34.9
Tundra	71.2	0.1	0.1	23.7

the tropical rain forest. The major reasons for grassland conversion are urbanization and agriculture (Adger & Brown, 1995; White et al., 2000). The figures of these changes in the different grassland types are presented in Table 6.3.

The figures in Table 6.3 suggest that temperate grasslands underwent the most widespread conversion, followed by the Mediterranean shrub land (over 50%). Almost similarly high are the losses of flooded grassland. The losses of tropical, subtropical and alpine grassland as well as tundra amount to only 30%. Table 6.4 provides figures of grassland conversion regarding different continents.

Conversion of grassland area has been highest in North (almost 90%) and South (76%) America. The loss of grassland areas in Oceania (including Australia) accounts for 39%. The proportion of converted grassland areas in Asia and Africa (approximately 20%) are comparatively smaller. Figures for grassland conversion in Europe are not available from the same source. However, development patterns can be assumed to be comparable to those of North America.

Table 6.4 Percentages of estimated remaining and converted grassland (White & Vanasselt, 2000)

Continent/ region	Remaining grasslands (%)	Converted to croplands (%)	Converted to urban areas (%)	Total converted (%)
North America/ Tallgrass Prairie	9.4	71.2	18.7	89.9
South America/ Cerrado Woodland and Savanna in Brazil, Paraguay and Bolivia	21.0	71.0	5.0	76.0
Asia/Daurian Steppe in Mongolia, Russia and China	71.7	19.9	1.5	21.4
Africa/Central and eastern Mopane and Miombo in Tanzania, Rwanda, Burundi, Dom. Rep., Congo, Zambia, Bots-wana, Zimbabwe and Mozambique	73.3	19.1	0.4	19.5
Oceania/Southwest Australian Shrublands and woodlands	56.7	37.2	1.8	39.0

6.1.1.3 Changes in Total Agricultural Area

Agricultural land in 2002 was estimated to cover 5,023 million hectares of which 3,488 million hectares (or 69%) was under pasture and 1,405 million hectares (28%) under cropland (Table 6.5). From 1961 to 2002, agricultural land gained almost 500 million hectares from other land uses. During this period, on average 6 million hectares of forest land and 7 million hectares of other land were converted to agricultural land, particularly in the developing countries. This trend will probably continue in the future (Rosegrant, 2001; Green et al., 2005).

Despite a decline in per capita agricultural land (FAOSTAT, 2006), per capita food availability has increased significantly during the last 4 decades (Table 6.6). The consumption of animal products has increased significantly in the developing countries, particularly in East and Southeast Asia. Meat demand in developing countries during the period from 1967 to 1997 increased from 11 to 24 kg per capita

Table 6.5 Agricultural area estimation, changes in agricultural area (negative numbers indicate agricultural area decrease) (Adapted from FAOSTAT, 2006)

	Agricultural area (million hectares)					Change 2000s/1960s	
Region	1961–1970	1971–1980	1981–1990	1990–2000	2001–2002	%	million hectares
Developing countries							
Agricultural land	2,682	2,801	2,955	3,119	3,184	19	502
Arable land	650	682	724	760	792	22	142
Permanent crops	59	68	80	–	99	81	48
Permanent pasture	1,973	2,051	2,152	2,260	2,286	16	313
Developed countries							
Agricultural land	1,879	1,883	1,877	1,866	1,838	–2	–41
Arable land	648	649	652	633	613	–5	–35
Permanent crops	23	24	24	24	24	4	1
Permanent pasture	1,209	1,210	1,201	1,209	1,202	–1	–7
World							
Agricultural land	4,562	4,684	4,832	4,985	5,023	10	461
Arable land	1,297	1,331	1,376	1,393	1,405	8	107
Permanent crops	82	92	104	123	130	59	49
Permanent pasture	3,182	3,261	3,353	3,469	3,488	10	306

Table 6.6 Per capita food supply in developing and developed countries (negative numbers indicate decrease in Cal day^{-1} or g day^{-1}) (Adapted from FAOSTAT, 2006)

Region	1961–1970	1971–1980	1981–1990	1990–2000	2001–2002	Change 2000s/1960s %	Change 2000s/1960s Cal day^{-1} or g day^{-1}
Developing countries							
Energy, all sources (Cal day^{-1})	2,032	2,183	2,443	2,600	2,657	31	625
% from animal sources	8	8	9	12	13	77	–
Protein, all sources (g day^{-1})	9	11	13	18	21	123	48
% from animal sources	18	20	22	28	30	67	–
Developed countries							
Energy, all sources (Cal day^{-1})	3,049	3,181	3,269	3,223	3,309	9	261
% from animal sources	27	28	28	27	26	−2	–
Protein, all sources (g day^{-1})	92	97	101	99	100	10	8
% from animal sources	50	55	57	56	56	12	–

year^{-1}, with an annual growth rate of more than 5% by the end of that period. A further increase in global meat demand by 57% may be expected until 2020 (Rosegrant et al., 2001) with the greatest increases in demand for poultry of 83% (Roy et al., 2002).

6.1.1.4 Changes in the Global Wetland Area

The global area of wetlands is estimated by various authors between 5.3 and 12.8 million square kilometers, which corresponds up to 9% of the total land area (Zedler & Kercher, 2005). A number of factors contribute to the destruction of wetlands. It becomes obvious, that human activities constitute more threats for wetlands than natural processes. The main reasons for loss of wetlands are commonly land reclamation for agriculture, urbanization and infrastructure. Worldwide, the total loss in wetlands has been estimated at 50% of the wetlands that existed in 1900 (Moser et al., 1998). Much of this loss in the northern hemisphere occurred during the first 50 years of the 19th century. Since the 1950s, increasing pressure for conversion to alternative land use has been put on tropical and subtropical wetlands. The observed trend of increasing wetland reclamation goes hand in hand with the industrial, technological and population development in industrializing countries.

In North America, the destruction of wetlands through land use had major effects on the C fluxes (Bridgham et al., 2006). In the past, wetlands were mostly considered wastelands. As the US was settled and people moved west, swamps and marshes were obstructions along the way. Wetlands were drained to be replaced by farmland, railroads and road construction. In the 1960s, more than 88 million hectares of wetlands existed in 48 states of the USA. Since then, more than 50% of the original wetland area was drained and converted to farmland (MBG net, 2002). In the densely populated regions of southern and eastern Asia, wetland loss has been occurring for thousands of years due to rice cultivation. Lowland rice cultivation began in Southeast Asia about 6,500 years ago, and sophisticated drainage and irrigation systems had been developed in parts of the Middle East by the fourth millennium BC (Moser et al., 1998).

Over the centuries, vast areas of wetland in southern and eastern Asia have been converted into rice fields or drained for other forms of agriculture and human settlement. In some areas, this conversion of wetlands has been complete. For example, the natural floodplain wetlands of the Red River delta in Vietnam, which originally covered 1.75 million hectares, were completely converted. Likewise, there is nothing left of the 1 million hectares of natural floodplain vegetation, which once covered most of the Sylhet Basin in Bangladesh or the 6 million hectares of floodplain wetlands in the lowlands of central Myanmar (former Burma). Much of the 40 million hectares of rice cultivation in the central plains of India were developed at the expense of natural wetlands, and the same is true of the 1.9 million hectares of rice cultivation in the central plains of Thailand. An exemplary overview of wetland losses is provided in Table 6.7.

Wetlands vary widely in their genesis, geographical location, water regime, chemistry, and plant communities. Nevertheless, they have in common the (periodically) water-saturated soil conditions. These conditions establish oxygen-excluded environments, which allow certain anaerobic processes, e.g. methanogenesis and immense accumulation of organic materials (Apps et al., 2001). Microorganisms responsible for organic matter decomposition are inhibited under anaerobic conditions. However, other species are obliged to the anaerobic conditions, the so-called *methanogens*. In general, intact wetlands are sinks for CO_2 and N compounds in soil and plant tissues, but sources for CH_4 (Van den Bos, 2000; Bridgham, 2006).

6.1.2 Change in SOC and SON Following Land Conversion

Under natural conditions, SOM contents are constant in the long term due to balanced input of organic residues and microbial decomposition. The quantity and quality of organic matter inputs and their rate of decomposition are determined by the combined interaction of climate, soil properties, and land use

Table 6.7 Estimated losses of wetland until the 1980s (Compiled from Moser et al., 1998 and others)

Continent/region or state	Wetland loss (% of original wetland area)	Author
North America/ Conterminous US	56–65	OECD (1996)
California	53	Dahl & Allord (1999)
Florida	91	Dahl & Allord (1999)
Ohio	46	Dahl & Allord (1999)
New Hampshire	90	Dahl & Allord (1999)
Rhode Island	9	Dahl & Allord (1999)
Alaska	37	Dahl & Allord (1999)
Hawaii	1	Dahl & Allord (1999)
Canadian Pacific coast estuarine wetlands	12	Dahl & Allord (1999)
Canadian Atlantic tidal and salt marshes	80	National Wetlands Working Group (1988)
St. Lawrence River shoreline marshes and swamps	65	National Wetlands Working Group (1988)
Minnedosa pothole region of SW	71	National Wetlands Working Group (1988)
Manitoba	71	
Mexico	35	Moser et al. (1998)
South America/	6	OECD (1996)
Cauca River Valley system (Colombia)	88	Restrepo & Naranjo (1987)
Magdalena River delta (Colombia)	80	Restrepo & Naranjo (1987)
Europe/	56–65	OECD (1996)
Germany	>50	Jones & Hughes (1993)
Spain	>50	Jones & Hughes (1993)
Greece	>50	Jones & Hughes (1993)
Italy	>50	Jones & Hughes (1993)
France	>50	Jones & Hughes (1993)
Portugal	>50	Jones & Hughes (1993)
UK estuaries	23	Davidson et al. (1991)
UK saltmarshes	50	Davidson et al. (1991)
UK wet grasslands	40	RSPB (1993)
Asia/	27	OECD (1996)
Indonesia	31	Scott (1993)
Israel	100	Immirzi et al. (1992)
Thailand	82	Immirzi et al. (1992)
West Malaysia	71	Immirzi et al. (1992)
Sarawak and East Malaysia	11	Immirzi et al. (1992)
Indonesia	18	Immirzi et al. (1992)
China	13	Immirzi et al. (1992)
Africa/	2	OECD (1996)
Tunisia	15	Hollis (1992)
Medjerdah catchment	84	Hollis (1992)

(continued)

Table 6.7 (continued)

Continent/region or state	Wetland loss (% of original wetland area)	Author
Tugela Basin	>90	Taylor et al. (1995)
Mfolozi catchment	58	Taylor et al. (1995)
Australia and New Zealand/		
Victoria	27	Usback & James (1993)
Southeast Australia	89	Usback & James (1993)
New Zealand	90	Cromarty (1996)
World	26	OECD (1996)

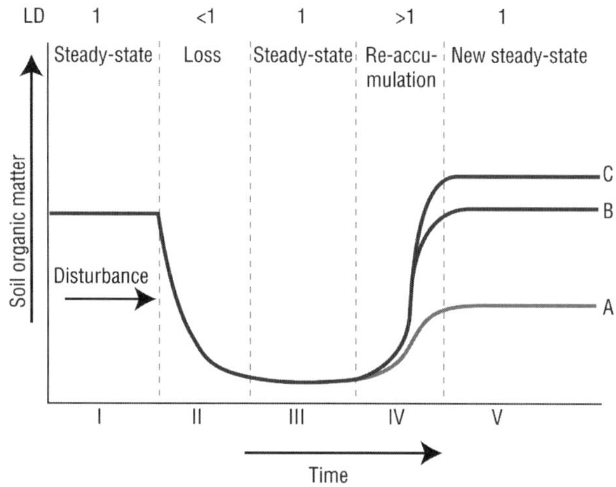

Fig. 6.3 Conceptual scheme of SOM decomposition following disturbance and reaccumulation. Under steady-state conditions (I), C input ("In") from plant residues equal C losses ("Out") via SOM decomposition (In/Out = 1). Due to disturbance, "Out" often exceeds "In", resulting in a loss of SOC (II), until a new but lower steady state is reached (III). Introduction of management systems where "In" exceeds "Out" results in a re-accumulation of SOC (IV) until a new but higher steady state is reached (V). The level of the new steady state (A, B or C) depends on the management that was introduced (Apps et al., 2001; p. 313, adapted from Johnson, 1995; IPCC, 2000c. Reproduced with kind permission from Technical Support Unit Working Group III, IPCC)

(EPA, 2006). Contents of SOM decrease in the sequence forest, grassland, arable land (Zbell & Höhn, 1995). Ecosystem disturbances are driving forces that determine the transition of landscapes from C and N sink to source and vice versa. For total C and N stocks in soils, the effects of disturbances may be detectable for decades or centuries (Fig. 6.3).

Each soil has an "equilibrium" C and N content depending on the nature of vegetation, precipitation and temperature (Gupta & Rao, 1994). The equilibrium C and N content is the result of a balance between inflows and outflows to the pool

(Fearnside & Barbosa, 1998). Typical well-drained mineral soils contain 1–6% SOC by weight, although some mineral soils that experience long-term water saturation may contain significantly more. Disturbances of natural ecosystems affect the C and N stocks in vegetation and in SOM. These stocks vary over time as a function of the history of disturbances (MacLaren, 1996; Kurz & Apps, 1999). Following a disturbance, C and N are transferred from living material to the dead organic matter pools. Some studies have reviewed the effects of land use changes on SOM stocks, such as forest clearing (Allen, 1985), tropical forest clearing (Detwiler, 1986), disturbance and recovery (Schlesinger, 1986), cultivation (Davidson & Ackerman, 1993) deforestation for pasture (Neil & Davidson, 2000) and from cultivation and native vegetation into grasslands (Conant et al., 2001). In case of forest or steppe fires (natural or artificial), part of the C and N is released immediately into the atmosphere. Following cultivation, part of the SOC and SON is released until a new steady state is reached (Bhatti et al., 2001).The largest changes in SOC occur with the conversion of natural ecosystems to arable land (Apps et al., 2001). The rate and ultimate magnitude of C loss depends on pre-conversion conditions, conversion method and subsequent management practices, climate, and soil type (Post & Mann, 1990; Lal, 2004a). In the tropics, 40–60% of the SOC loss generally occurs within the first 10 years following conversion. Carbon stocks continue to decline after 10 years but at a much slower rate (EPA, 2006). In temperate regions, C loss can continue for several decades, reducing stocks by up to 40% of native C levels (Mann, 2004). On average, most studies after cultivation showed a decline in SOC of about 30%. Globally, conversion to arable land has resulted in a SOC loss of about 50 Pg C and total emissions of C from land use change, including that from biomass loss are estimated at about 122 ± 40 Pg C (Apps et al., 2001). If regrowth of forests or grassland follows, the corresponding re-sequestration of C and N may last 50–200 years or more. Afforestation of arable land leads to a build up of SOM and thus promotes CO_2 mitigation (Zbell & Höhn, 1995).

Published results on SOC and SON changes in many cases are limited because the authors did not adjust their data for soil bulk density changes following land conversion. According to Fig. 6.5, which shows results from numerous studies, bulk density generally increases upon conversion from forest to agricultural land. There is also little recognition in the literature that soil bulk density also changes upon conversion of grassland to cultivated land. Because the bulk density of a surface grassland soil (0–10 cm) is often <1.0 g cm^{-3}, and the density of plow layers (0–>25 cm) usually is 1.2–1.5 g cm^{-3}, C losses calculated on a mass basis are less than those reported on a concentration basis.

Cultivation tends to break soil aggregates and thus compacts the soil. The average increase in bulk density was 12.9 ± 1.5% (n = 78; Fig. 6.4), with changes after conversion to cultivated land use of 16.9 ± 2.2% (n = 36) and pasture of 9.5 ± 2.1% (n = 42).

The results confirm that bulk density changes are likely to have confounded the results of many studies and that SOC and SON changes reported in the studies, which do not deal with this problem are likely to have been under- or overestimated.

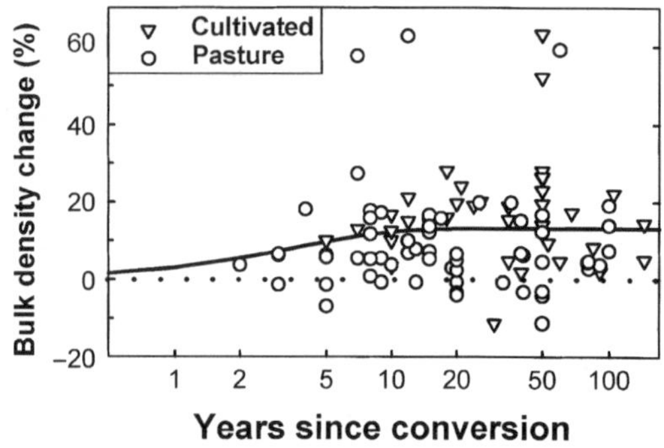

Fig. 6.4 Changes in soil bulk density after land conversion for pasture or cultivated soils (Murty et al., 2002; p. 107. Reproduced with kind permission from Wiley-Blackwell)

Besides soil bulk density, the depth to which soil samples are taken can significantly influence estimates of total SOC and SON. To make reliable assessments of the impacts of soil conversion on SOC and SON reserves, comparable sampling depths as well as bulk density measurements at each sampling depth are required.

Guo & Gifford (2002) have reviewed the literature for the influence of land use changes on SOC stocks on the basis of a meta analysis (Fig. 6.5). For studies where soil bulk densities (BD) were not available, these were estimated according to an equation developed by Post & Kwon (2000):

$$BD = \frac{100}{\%OM/_{0.244} + {}^{100-\%OM}/_{1.64}} \tag{6.1}$$

As an overall average across all of the given conversion categories, land use change reduced SOC by 9%. The stocks of SOC declined after the conversion from pasture to plantation (–10%), forest to plantation (–13%), and particularly from forest to crop (–42%) and pasture to crop (–59%). Soil C increased after the conversion from forest to pasture (8%), crop to plantation (18%), crop to secondary forest (53%) and crop to pasture (19%).

With respect to different sampling depths, the SOC results were influenced drastically after land use change from forest to crop (Fig. 6.6). In all other land use change categories, the factor soil sampling depth had no significant influence on the SOC results. The stocks of SOC decreased by about 50% if the sampling depth was less than 60 cm, but there was no significant change below 60 cm.

6.1.2.1 Conversion of Grassland to Arable Land

The SOM contents under natural and artificial grassland are generally higher than under other land use forms. Compared to arable land, there are several factors which promote the build up of these high storages of C and N under grassland vegetation. Among these are a high density of roots, high root exudation rates and

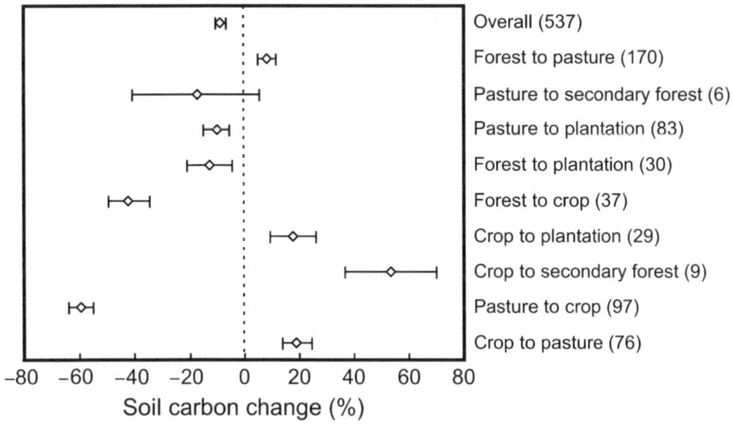

Fig. 6.5 Response of SOC to various land use changes including 537 observations from 74 publications (95% confidence intervals are shown and numbers of observations are in parenthesis) (Guo & Gifford, 2002; p. 374. Reproduced with kind permission from Wiley-Blackwell)

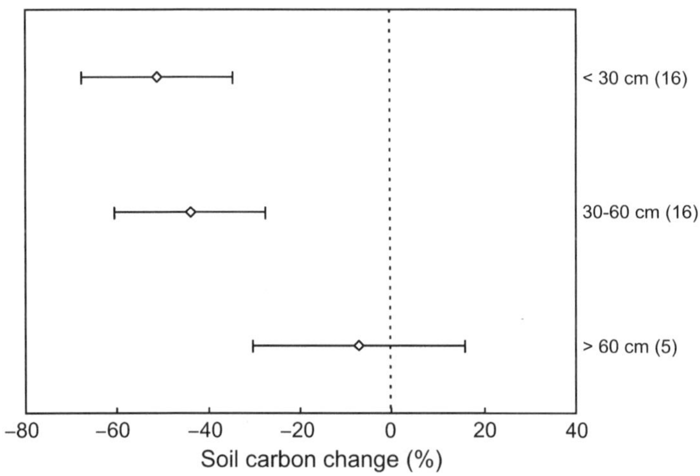

Fig. 6.6 Changes of SOC in different sampling depths after land use change from forest to crop (95% confidence intervals are shown and numbers of observations are in parenthesis) (Guo & Gifford, 2002; p. 351. Reproduced with kind permission from Wiley-Blackwell)

limited soil aeration because of the absence of tillage. Developing grasslands are regarded as C and N sinks. Under maintained grassland, C and N storage and release are balanced after an equilibrium state is reached. The equilibrium SOM content under grassland varies according to climatic region, soil type and grass species. Estimates for SOC of different grassland types range from 120 to 400 Mg C ha^{-1} (Adger & Brown, 1995).

Conversion of grassland to arable land causes changes in the soil properties and SOM storages which may promote the release of GHGs. The extent of these changes depends on the grassland type, climatic and environmental conditions and initial soil properties (White et al., 2000). In the past, temperate grasslands, savannas, and shrublands have experienced heavy conversion to agriculture, more so than other grassland types including tropical and subtropical grasslands, savannas, and woodlands. For example, almost 90% of the North American Tallgrass Prairie has been converted to croplands (White et al., 2000). Reinstallation of grassland on former agricultural soils combined with sustainable grassland management provide potential sinks for atmospheric C and N compounds and may help to mitigate global warming.

One of the most important process that is influenced by grassland conversion is the long-term mineralization of SOM. Plowing of grasslands increases the amount of CO_2 released and N mineralized. The age of the grassland and the initial amount of SOM are the main factor determining the mineralization rates. Nieder et al. (2003a) reported higher losses for soils with higher initial SOM pools. Old grasslands which are converted to arable land may release up to 400 kg N ha^{-1} $year^{-1}$ in the first year after plowing (Whitehead, 1995). Losses of SOM generally occur from all particle-size separates, although loss rate constants increase as particle size increases. Lobe et al. (2001) for sandy soils in South Africa found that the concentrations of SOC reached an equilibrium 34 years after conversion for the bulk soil and after 55 years for clay-size separates. Organic matter attached to silt continued to be lost due to wind erosion as long as the cropping continued. Similar observations were made by Römkens et al. (1985) where changes were highest in the coarse soil fraction. In humid temperate regions, the highest rates of SOC loss were observed in the first 20 years following cultivation. Soils with high C content were found to loose at least 20% of the initial SOC due to cultivation (Mann, 2004). In contrast, in soils with a very low initial SOC content, amounts of organic matter increased slightly after cultivation.

Haas et al. (1957) reported SOM losses up to 50% (C concentration basis) at many Great Plain sites (semi-arid climate) after 40 years of cultivation in wheat-fallow systems with conventional tillage. Simulation of cultivation effects across the Great Plains showed that the greatest loss in SOM would be in the wetter and warmer part (south-west) of the region, which reflects the greater initial amounts of SOM found in this area (Cole et al., 1989). There is no clear relationship between soil texture and SOM loss. In the north of the Great Plains, percent loss in soil C on fine-textured soils is 46% whereas on sandy soils the loss is 42–48%. In the south, losses range from 38% to 54% on the fine-textured soils and from 44% to 45% on the sandy soils area (Cole et al., 1989). Peterson & Vetter (1971), using total soil N as an index, reported that wheat-fallow systems in the Great Plains had organic N losses of 20–30% (mass basis).

Decreasing the tillage intensity results in less SOM losses after cultivation of grassland. Carbon losses from 1970 until 1990 for no-till, stubble mulch and plow tillage systems were 4,500, 6,900 and 11,200 kg C ha^{-1}, respectively (Peterson et al., 1998). For the first 12 years, C losses from no-till managed soil were 11% of that for the native sod with no additional C loss from 1982 to 1990. In the same time period, stubble mulch and plow tillage systems lost 1,000 and 2,400 kg C ha^{-1}, respectively. These data suggest that reducing tillage (no-till relative to stubble mulch and plow tillage) promotes C retention after sod breaking.

Losses of SON occur in a similar way compared to SOC. Extreme N mineralization rates following plowing of old grassland in the UK were reported to be 70% of the original N content during a 20–30 year cultivation period (Whitehead, 1995). The highest N loss rates were observed during the first few years of cultivation. More than half of the initial N of a 25 cm plow horizon was lost within the first 5.5 years, and more than 90% of the total N loss occurred within 18 years. A comparative study in the USA showed that the N content was 31–56% lower under fields that were under at least 40 years of continuous cultivation compared to natural grassland (Whitehead, 1995).

Besides other factors, climate has a major influence on loss of SOM after grassland conversion. Guo and Gifford (2002) reported that after conversion, more SOC was lost from land with 400–500 mm precipitation (−75%) compared to lands with 300–400 mm (−54%) and >500 mm.

6.1.2.2 Conversion of Arable Land to Grassland

The reverse process of the conversion from pasture to crop, i.e. of arable land to grassland significantly sequesters carbon from the atmosphere (Freibauer & Schrumpf, 2006). Reestablishment of pasture commonly results in a rapid recovery of the total SOM content. The high root production by grasses may explain why pastures accumulate large amounts of SOM (Cerri et al., 1991). Guo & Gifford (2002) indicated that besides organic C (C_{org}), the microbial C (C_{mic}) and the C_{mic}:C_{org} ratio were consistently higher in pasture soils than in equivalent soils under arable land. Episodic grazing or cutting of pastures may enhance SOM accumulation due to the rapid death of roots following each defoliation event followed by root regrowth as the pasture sward reestablishes. Compared to ungrazed areas, controlled grazing can lead to increased annual net primary production (Conant et al., 2001). Most pasture plants (about 80%) are perennial and have well developed root systems. The relative belowground translocation of assimilated C by pasture plants can reach up to 80% (including C respired by roots) but up to only 60% by trees (Kuzyakov & Domanski, 2000).

According to Römkens et al. (1985), almost 90% of the pasture-derived C that was mineralized during intensive maize cropping was replaced within 9 years. Soil texture has a strong influence on the SOC accumulation after reestablishment of grassland. Due to initially lower contents, particularly the medium and coarse size fractions (>150 µm) accumulated C rather quickly after pasture reestablishment. Both fractions were almost completely regenerated by the input of root-derived SOM.

The time necessary to reach a new SOM equilibrium depends on soil type, climate, vegetation cover and grassland management. Data obtained at Rothamsted

(UK) from plots on a silty clay loam indicated that more than 100 years may be necessary to restore equilibrium in SOC and SON contents if soils are turned to grassland from preceding long-term cultivation. The average accumulation rate for nitrogen was about 56 kg N ha^{-1} $year^{-1}$ during the first 40 years. In another experiment in the UK, the increase in soil N under a grazed grass-clover sward sown on a previously arable soil was about 75 kg N ha^{-1} $year^{-1}$ during the first 10 years (Whitehead, 1995). In New Zealand, the average increase in soil N under grazed grass cover swards was estimated to be 112 kg N ha^{-1} $year^{-1}$. Larger increases occurred in soils with lower initial N contents. The influence of climate on long-term grassland soils at equilibrium has been shown in the USA. Along a transect from the East to the West, the SOM content was positively correlated with moisture (Whitehead, 1995).

6.1.2.3 Conversion of Forest to Agricultural Land

The C stocks in forests can be divided into two different pools: the biomass and the SOC pool. About two thirds of C is stored in soils, and one third in vegetation biomass. The percentage of the soil C pool is especially high in boreal forests (80%), while it is only 50% in the tropics (Kasang, 2004). The soil C pool reacts more slowly to environmental impacts (e.g. fire or deforestation) than the living biomass, though both pools are closely interrelated. Tropical forests contain the largest pool of terrestrial biota and NPP (Field & Raupach, 2004). Vegetation and soils of tropical forests store 460–575 Pg C (NASA Earth Observatory, 2007). Globally, the C reservoir of forests amounts up to about 1,000 Pg C (Kasang, 2004). Temperate and tropical forests together account for approximately 75% of the world's plant C and 40% of the world's SOC (Field & Raupach, 2004). Because of their high soil C storages the boreal forests of Canada, Russia and Alaska alone hold about 50% of the C that is fixed in forests worldwide and the total boreal forests were estimated to contain 61 Pg C at the end of the 1990s (Kasang, 2004).

In the tropics, SOC pools are small and react quickly to changes in the ecosystem (Nieder et al., 2003a). Depending on the form of land clearing, and the subsequent use of the wood product, the release of CO_2 from the plant material can be immense. If the wood is burned on the site, or cut as fuel wood, almost all the C sequestered in vegetation will be released to the atmosphere (Apps et al., 2001). If the wood is used as timber, less C is emitted instantly from the wood products. Apart from the form of forest clearing, the change in plant C is mainly influenced by the subsequent form of land use (Lal, 1995a). The decrease in plant C storage is most severe if the land is left bare after deforestation or cultivated. If a new forest develops on the site, under certain conditions the C in plant biomass may be restored almost completely within a century (Schlesinger, 1997). The second C storage that is affected by deforestation is the soil C storage. In tropical regions, the SOC stock is relatively low (Fig. 6.3) due to rather high average temperatures and soil moisture contributing to high microbial decomposition rates. In contrast, in temperate and especially boreal forests the storage of SOM can be enormous.

Since the beginning of agriculture, 750 million hectares forests have been converted to agricultural land. This conversion has caused loss of 121 Pg C from soils and biomass worldwide (Kasang, 2004). However, in many regions like Western Europe and North America, C pools have now stabilized and are recovering. In most countries in temperate and boreal regions forests are expanding, although current C pools are still smaller than those in preindustrial or prehistoric times (Apps et al., 2001). While complete recovery of prehistoric C pools is unlikely, there is potential for substantial increases in C stocks (Field & Raupach, 2004). Carbon stocks in tree biomass during the last few decades may have increased by 0.17 Pg C $year^{-1}$ in the USA, and by 0.11 Pg C $year^{-1}$ in Western Europe, which means that about 10% of the global fossil CO_2-C emitted might have been absorbed during that period. In some tropical countries, however, the average net loss of forest C stocks continues, though rates of deforestation may have declined slightly during the 1990s (Apps et al., 2001).

Thuille et al. (2000) examined below- and above-ground C stocks on one site under meadow and Norway spruce, respectively, in the southern Alps. The original forest vegetation was cleared 260 years ago in order to create grazing land. Due to this deforestation C was lost from the organic layer (53 Mg C ha^{-1}) and from the upper mineral soil horizon (12 Mg C ha^{-1}). During the following 200 years of grassland use, the new Ah horizon sequestered 29 Mg C ha^{-1}. After the abandonment of the meadows, a spruce stand was established. The C stocks in tree stems increased exponentially with stand age. Thuille et al. (2000) estimated the C stocks at about 190 Mg C ha^{-1} in both the regrown 62 year old Norway spruce, and in a 130 year old Norway spruce-white fir control site. During reforestation, C stocks in the organic soil layer increased linearly at a rate of 0.36 Mg C ha^{-1} $year^{-1}$. The continuous soil C sequestration during forest succession was attributed to increasing litter inputs by forest vegetation, and constantly low decomposition rates of coniferous litter. Carbon accumulation in woody biomass seemed to slow down after 60–80 years, but continued in the organic soil layer.

Harms et al. (2004) investigated changes in soil C after tree clearing in semiarid rangelands in Queensland (Australia). The original SOC stocks (excluding surface litter, extractable roots and coarse charcoal) at uncleared sites were 29.5 Mg ha^{-1} for 0–0.3 m soil depth, and 62.5 Mg ha^{-1} for 0–1.0 m depth. Soil C decreased by 8% for 0–0.3 m soil depth (2.5 Mg C ha^{-1}) and by 5.4% (3.5 Mg C ha^{-1}) in 0–100 cm soil depth due to clearing. Changes in soil C after tree clearing were strongly correlated to initial soil C contents, and were associated with particular vegetation groups and soil types (Harms et al., 2004). Changes in soil N were strongly correlated with changes in soil C. Figure 6.7 shows results from numerous studies where forest was cultivated to arable land. The observed sites correspond to those shown in Fig. 6.4.

In all but 11 of the observations, SOC decreased following land conversion. The mean percentage change in SOC 10 or more years after conversion was –30.3 ± 2.4% (n = 75). The largest change in SOC was –72% after 75 years of cultivation of various crops in Georgia, USA (Giddens, 1957). The largest increase in SOC was 49% on a site where banana was included as part of a regular crop rotation

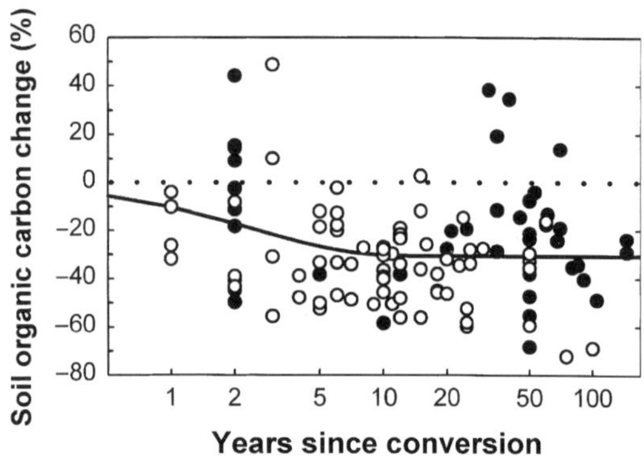

Fig. 6.7 Changes in SOC (% of original C content) after conversion of forest to cultivated soil. Closed circles (•) show data where bulk density effects have been considered by the authors, and open circles (○) show data with remaining uncertainty about the procedures (Murty et al., 2002; p. 108. Reproduced with kind permission from Wiley-Blackwell)

sequence (Nye & Greenland, 1960). This build-up of SOC was attributed to the large litter input resulting from banana cultivation.

Pasture established following clearing of forests has a greater potential SOC stock than it has following crop. Well-managed pastures maintain or even increase SOM levels compared with native forest. Tate et al. (2000) reported that total SOC stock was 13% higher in the grassland than in the forest they studied (199 vs. 167 Mg C ha^{-1}).

In comparing more than 25 sets of paired pasture and mature forest sites, Lugo & Brown (1993) found that soil C stocks under pasture varied from 60% to 140% of those under forests and that on average, SOM under pasture was not significantly different than under mature forest. In Colombia, Fisher et al. (1994) reported very large belowground C increases of 25–70 Mg ha^{-1} within 5–10 years after establishing pastures of deep-rooting African grasses. In the semiarid tropics, pasture land is the predominant land use system. The degree to which improved pasture management is practised (i.e. introduction of grasses and legumes, soil fertility maintenance) has a major impact on SOM levels. In many regions, poor management has resulted in overgrazing and nutrient deficiencies, leading to soil erosion and SOM losses (Eden et al., 1991). The sequestration potential for C and N in moist tropical pastures can be significant under favorable conditions. Fisher et al. (1995) suggested that improved pastures which replace native savannas throughout South America could account for an additional sequestration of 100–500 Tg C $year^{-1}$ in these tropical soils.

Substantial soil C inputs may be attributed to the deep-rootedness of grasses in improved tropical pastures. Indeed, 75% of the claimed increased C sequestration was found below 20 cm soil depth and is thus likely to be due to root inputs. Fisher et al. (1995) found that the large increase in SOM under improved tropical pastures (up to 70 Mg C ha^{-1}) was associated with a substantial increase in the C:N ratio, giving ratios in the SOM of 33:1 compared with usual SOM values of ~12:1. It is thus likely that only partial decomposition of roots occurred leading to the increased SOM content. Extrapolation from a fitted double exponential decay model to laboratory incubation data of tropical pasture materials suggested that between 43% and 47% of legume root C and 54–62% of grass roots was theoretically "non-decomposable".

Similarly, after incubation of root materials for 1 year, <50% of C had been evolved through respiration, whereas 80% of C from leaves of the tropical grass *Brachiaria humidicola* had been lost (Cadish & Giller, 2001). *Brachiaria* and legume roots had a higher C:N ratio and a higher lignin content than leaves. The lignin:N ratio is widely known to govern residue decomposition of many plant materials and is also used to allocate residue fractions to the slow decomposing structural pool in many models.

According to the study by Guo & Gifford (2002), clearing forest for pasture increased the SOC stocks by 24% in areas with 2,000–3,000 mm precipitation, but had no effects in areas with rainfall less than 2,000 mm and more than 3,000 mm. It is possible that in areas with annual rainfall >3,000 mm the rainfall led to initial topsoil erosion and associated loss of SOC.

6.1.2.4 Afforestation and Reforestation

In Europe, afforestation of 30% of (surplus) arable land (total arable land area: 40.6 million hectares) would increase total soil organic C stocks by 3.58 Pg over 100 years (Smith et al., 1997a). Tree growth additionally sequesters C in wood. Jenkinson (1971) estimated the C in standing woody biomass to be three times that found in soil on natural woodland regeneration experiments. Standing woody biomass would therefore accumulate 10.74 Pg C over 100 years (Smith et al., 1997a). This is sequestered only temporarily, unless converted to durable bioproducts. However, if management intensity decreases because of environmental concerns or changes in policy (Enquete Commission, 1995), this option may no longer be available. In 2004, the United States forests sequestered 637 Tg C, which corresponds to about 10.6% of all the CO_2 released by fossil fuel combustion (US EPA, 2007). If reestablishment of forest is possible in the tropics, 1 million planted trees might fix 0.9 Tg C during their typical 40 year lifetime (US EPA, 2007).

Over time periods <5 years, SOC generally decreases following afforestation. In contrast, in afforested sites more than 10 years old, SOC increased in the surface soil layer. During the early stages of stand development, little detrital matter is produced due to the small biomass, and low rates of litter fall return. Therefore, directly following agricultural abandonment, the decline in C is attributable to the greater loss of C through decomposition than gain through litter production. The subsequent accumulation of C indicates that annual inputs of C through NPP exceeded the amount of C lost by decomposition. Although changes in soil C following afforestation are not well documented, it is generally assumed that over decades, the C content increases following afforestation.

In the long term, the equilibrium SOC level is generally slightly above that of the preceding agricultural soil which is due to both, increase in SOC stock in mineral soil as well as in forest floor (US EPA, 2007). This trend can be observed especially on highly eroded former arable land. The Rothamsted long-term experiments on average demonstrated a 0.34–0.55 Mg C m^{-2} $year^{-1}$ accumulation over a 100-year period (Polglase et al., 2000). In contrast to the majority of the afforestation studies reviewed, some workers found that soil C initially increased or did not change with time (Polglase et al., 2000). Reductions in SOC stocks are possible in systems with large deep drainage, or where erosion may continue even in afforested systems. Different time periods are needed to reach a new equilibrium (or near equilibrium) of SOM in different ecosystems. According to Polglase et al. (2000), the time taken for the equilibrium state was calculated to be 10 years following natural forest succession in Nigeria (0–10 cm), 30 years under eucalypt and pine plantations in Congo (0–5 cm), 40–60 years under pine-oak stands in Massachusetts (0–15 cm), 45–60 years under conifer forests in Wisconsin (0–15 cm), and more than 60 years following natural forest succession in Minnesota (0–10 cm).

Previous land use has a significant effect on change in SOM after start of forest regrowth. Considering the first 10 years, change in soil C on ex-pasture was −9.75 g C m^{-2} $year^{-1}$, and on ex-crop (arable) land it was +142.3 g C m^{-2} $year^{-1}$ (Polglase et al., 2000). If land was cropped, the rate of decrease in soil C may be limited because the soil C largely consists of stable humus resistant to breakdown. In contrast, if the land was formerly improved pasture, soil may have a relatively high C content that is susceptible to loss after plantation. In addition to changing aboveground C stocks, afforestation of grasslands also influences SOC stocks by changing soil structure, soil moisture, magnitude and dynamics of soil C and N inputs, and other factors. Model simulations (CGR, 2006) revealed an initial decrease in soil C after reforestation, followed by an increase in soil C stocks as the forest stands became more established. These dynamics are strongly influenced by the initial soil C content and, to a lesser extent, forest type. Residues such as dead roots from the preexisting grass decompose rapidly. Fine (<3 mm) and medium (3–10 mm) tree roots also decompose rapidly, but decomposition of large woody roots (>10 mm diameter) is slower (Polglase et al., 2000). Tree roots may also add C deeper in the soil profile than pasture roots.

With the change in litter quality from grass litter (which is low in lignin) to needle litter with higher lignin content, soil C:N ratios are likely to increase. This is likely to become even more pronounced when branch and bark litter contributes to the litter pool. Simulations by Kirschbaum (2004) suggested that it could take many decades for the change in the C:N ratio of the mineral soil, because the most pronounced changes occur in the organic layer.

Soil C:N ratios due to the change from more easily decomposable grass litter to more resistant litter from pine needles and woody plant components increased from about 16.5–20 (Kirschbaum, 2004). However, in some cases afforestation of rangelands may lead to a net increase in GHG emissions. For example, when grasslands, which are strong sinks of C, are replaced by very low-productivity forests, there may be a net increase in trace gas and CO_2 emissions (CGR, 2006).

Following the conversion of arable land to plantations, changes inevitably occur in the quality, quantity, timing, and spatial distribution of soil C inputs. These changes, together with the changes in the soil microenvironment, affect decomposition rates. For example, decomposition rates may decrease because of afforestation due to the cooler soil surface under the canopy and litter layer. Decomposition of soil C added via tree roots may decrease because of the lower soil temperatures and reduced microbial activity at greater depth (Polglase et al., 2000). Furthermore, the soil C dynamics significantly depend on stand age.

Accumulation of SOM in tree plantations is affected by species as some species produce more litter than others. According to Guo & Gifford (2002), planting conifer trees (mainly *Pinus radiata*) onto pasture land significantly reduced SOC stocks. Planting broadleaf trees had little effect on SOC. The difference between conifer and broadleaf trees may be related to their individually inherent strategy to allocate assimilated C below ground. Root biomass may be more important determinants for the accumulation of SOC in mineral soils of forests than aboveground litter input. During their 3 year study, Coleman et al. (2000) found that pine fine root production was only 2.9% of that of poplar. Fertilization or planting N_2-fixing species can increase biomass production in plantation forest and potentially increase C input into the soil, but it may also enhance decomposition. These may be the reasons why fertilizers do not necessarily increase SOC stocks.

Soil organic N after planting trees may be moderately lost for a few years, but the changes are negligible thereafter. Up to a few hundred kg N ha^{-1} may be transferred from the soil into the growing plant pools in the longer-term. In Australian replanted forests, total SON losses or gains did not exceed a few percent of the total initial N content (Kirschbaum, 2004). Due to the increase in biomass pools, changes in litter pools occurred, with residues comprising initially N-rich foliage litter, but changing thereafter to nutrient-poorer and longer-lasting branch litter. The amount of carbon in litter increased linearly with time, whereas the N amount in litter increased strongly during the first 20 years and more slowly thereafter. After 5 and 50 years, about 100 kg N ha^{-1} and 180 kg N ha^{-1}, respectively, were stored in plant biomass and aboveground litter (Kirschbaum, 2004). The forest system gained N only through atmospheric deposition which was estimated to be <5 kg N ha^{-1} $year^{-1}$. The increase in plant and litter N was accompanied by slightly decreasing SON stocks of the mineral soil during the first few years which was drawn back to disturbance of the soil due to tree planting. After the initial N loss, N stocks increased again. In summary, the system's total N stocks remained almost constant. Leaching and gaseous losses of N during 50 years in this N-limited system on average comprised only about 2% of the N which was turned over. Models of litter decay assume that about 0.04–0.5 of C in litter becomes humus (Polglase et al., 2000). Decomposition of litter is faster under eucalypts than conifer plantations. Needle litter tends to remain as semi-decomposed residues on the soil surface. Litter pools show decomposition rates between 0.30 and 0.65 $year^{-1}$ under northern hemisphere pines, and between 0.08 and 0.47 in northern hardwoods (Polglase et al., 2000). Depending on the climate and substrate quality of litter, it may take 1–6 years before it can be considered to be intimately mixed with soil.

6.1.2.5 Future C Storage Potential Through Reforestation and Afforestation

Reforestation and afforestation may be of significant importance in future C management (Apps et al., 2001). Reforestation and afforestation are likely to continue to generate C sinks throughout the 21st century. Global estimates by Field and Raupach (2004) for the C sequestration potential through forest (re-) growth are in the range of some Pg C to a few tens of Pg for the entire 21st century. The IPCC Second Assessment Report estimate that about 60–87 Pg additional C could be conserved or sequestered in forests by the year 2050.

6.1.2.6 Wetland Reclamation

Wetlands provide a potential sink for atmospheric C and N (Mitra et al., 2005; Nieder et al., 2003a). When wetlands are drained or degraded, several processes are initiated that promote the decomposition of the stored organic materials. If not managed properly, wetlands become sources of greenhouse gases (Mitra et al., 2005). In context with the global warming potential, the role of wetlands is discussed controversially. Bridgham et al. (2006) put emphasis on the CH_4 emissions, whereas the gross of authors value the sink function for CO_2 and N compounds (e.g. Schlesinger, 1997; Röhricht et al., 2006; Freibauer & Schrumpf, 2006).

Of the estimated 1,500 Pg C stored in global soils, one third is supposed to be contained in peat soils but the share of the area covered by peat soils is far less than one third of the land mass (Adger & Brown, 1995). This ratio indicates the especially high storage of C in peat soils. Globally, wetlands contain about 300–600 Pg C (Apps et al., 2001). A major portion of this C is found in peat-forming wetlands in both northern (302 Mha, 397 Pg C) and tropical (50 Mha, 144 Pg C) biomes, often in association with forest vegetation (Apps et al., 2001). According to Adger and Brown (1995) under blanket bog up to 1,200 Mg C ha^{-1} can be found. This is about ten times more than under neglected grassland with only 120 Mg C ha^{-1}. This extremely high storage of C is a consequence of the special conditions under wetlands. Besides soil temperature, the position of the water table decides on the rate of organic matter decomposition. In non-drained peat profiles, the cool temperatures and anoxic conditions retard the rate at which organic matter decomposes. The plants growing on bogs deliver a coarse fibred, acid litter, which even under optimized conditions would undergo slow turnover rates. The combination of these unfavorable factors promotes the accumulation of organic residues. The storage form in wetland is long-term sequestration in form of peat (Röhricht et al., 2006). Turunen et al. (2000) found average long-term C accumulation rates of 17.1 g C m^{-2} $year^{-1}$ in central West Siberia. Thus, positive C and N gains are common and immense storages can be found in wetland soils. For example, Bleuten et al. (2000) estimated that West Siberian peats, which represent 50% of the global peat area, store 51 Pg C.

Drainage of organic soil deposits results in a decrease in surface elevation (subsidence). Drained organic soils are subsiding at the rate of several centimeters per year (Netherlands: 1.75 cm $year^{-1}$; Quebec in Canada: 2.07 cm $year^{-1}$; Everglades in the USA: 3 cm $year^{-1}$; San Joaquin Delta in the USA: 7.6 cm $year^{-1}$; Hula Valley in Israel: 10 cm $year^{-1}$; Terry, 1986). Among the reasons for subsidence are shrinkage due to drying, loss of the buoyant force of groundwater, compaction, wind erosion, burning, and microbial oxidation. Microbial oxidation is the predominant cause of Histosol subsidence. Approximately 73% of the loss of surface elevation in Everglades Histosols accounted for microbial oxidation (Terry, 1986). An assumed C release from drained wetlands by oxidation of the organic material of 10 Mg C ha^{-1} $year^{-1}$ yields a global annual C release of 0.05–0.35 Pg C. Drainage of 10^7 km^2 of Gleysols causes an extra release of 0.01 Pg C $year^{-1}$. The resulting global release from Histosols and Gleysols ranges between 0.03 and 0.37 Pg C $year^{-1}$.

Based on an average storage rate of 200 kg C ha^{-1} in wetlands of the cool climate and assuming an area of about 350 × 10^{10} m^2, the annual accumulation before disturbance was 0.06–0.08 Pg C $year^{-1}$ (Armentado & Menges, 1986). The total area drained in the period 1795–1980 was 8.2 × 10^{10} m^2 for crops, 5.5 × 10^{10} m^2 for pasture and 9.4 × 10^{10} m^2 for forests. In the tropics about 4% of the wetland area has been reclaimed in the period 1795–1980. The annual shift (loss of sink strength and gain of source strength) in the global C balance is 0.063–0.085 Pg C due to reclamation of histosols in cool regions. Including tropical Histosols, the global shift would be 0.15–0.184 Pg C $year^{-1}$ (Bouwman, 1990). The potential to increase C levels in soils under cultivation will be largely restricted to upland soils. Restoring C sinks in artificially drained wetland soils is unlikely unless they are taken out of agricultural production and reverted to natural wetlands.

6.1.3 Land Use Changes and Greenhouse Gas Emissions

6.1.3.1 Greenhouse Gas Emission from Upland Soil Conversion

The consequence of land use changes is an increasing release of the greenhouse gases CO_2, CH_4 and N_2O from soils to the atmosphere (NEIC, 1998; EIA, 2000). While the majority of global CO_2 emissions are from the burning of fossil fuels, roughly a quarter of the C entering the atmosphere stems from land use changes (WRI, 2007). Expansion of cropland, pasture and infrastructure have increased CO_2 emissions significantly in the past two centuries (Zbell & Höhn, 1995; Field & Raupach, 2004). Emissions of CO_2 from land use changes vary widely according to the soil and environmental conditions (Bouwman, 1990; Lal, 2004b). The GHG effect from land use change depends on factors like the history of land use before conversion, the original SOM content, the kind of land use change, climate, and other effects (EIA, 2000; Lal, 2004a). The total C fluxes from 1950 to 2000 for different countries are given in Fig. 6.8.

The positive values in Fig. 6.8 indicate a net flux of CO_2 from soils to the atmosphere due to land use changes. The data include emissions from living and dead vegetation disturbed at the time of clearing or harvest, emissions from wood products (including fuel wood), and emissions from the oxidation of SOM in the years following initial cultivation. Negative values indicate a net flux of CO_2 from the atmosphere into the soil-plant system. The data should be treated as order-of-magnitude estimates (WRI, 2007). The WRI (2007) states that yearly flux estimates are

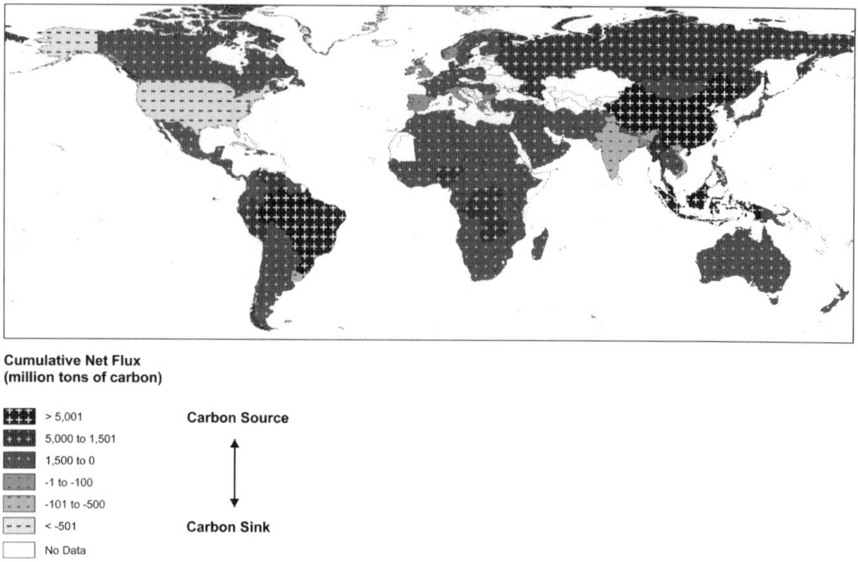

Fig. 6.8 Net C flux to the atmosphere 1950–2000 from land use changes (WRI, 2007. Reproduced with kind permission from World Resources Institute)

uncertain in the order of ±150% for large fluxes, and ±50 Tg C year^{-1} for estimates near zero.

Forest soils can be a significant source or sink of important GHGs, such as NO, N_2O, CH_4, and ammonia (NH_3). Among the main factors influencing the source or sink function of newly established forests are initial soil C and N content, climatic and environmental conditions, previous land use, tree species, stand age and soil properties (Polglase et al., 2000). In general, areas with high SOC content prior to afforestation show higher soil C losses, higher rates of methane oxidation, and higher N_2O emissions (CGR, 2006). Differences in forest floor dynamics and soil C and N fluxes are substantially reduced as the forest system approaches equilibrium. Hence, reforestation and afforestation do not only affect the net changes in aboveground C stocks, but also belowground C storage and non-CO_2 GHG emissions.

The amount of N_2O emitted from forest soils depends on N inputs and availability, organic C availability, O_2 partial pressure, soil moisture content, pH, temperature, and tree planting/harvesting cycles (EPA, 2006). The effect of the combined interaction of these factors on N_2O flux is complex and highly uncertain. In the short term, soil disturbance after tree planting increases the availability of nutrients and microbial activities. More N is available to be nitrified and denitrified. Reforestation may also increase root biomass and root exudation which is a major source of DOC causing increased soil microbial activities including nitrous oxide (N_2O) or methane (CH_4) production/consumption (CGR, 2004). In consequence, trace gas emissions such as NO and N_2O usually increase after forest planting. The amount and quality of fresh litter which varies from forest type to forest type strongly affects the trace gas emissions. Fresh litter has a low C:N ratio, larger litter fractions will be stored into the more labile pools, resulting in faster decomposition rates. Differences in organic contents as well as trace gas emissions can be expected for the same site with different afforestation species. For example, reforestation with oak compared to pine or fir will potentially lead to higher N_2O emissions due to oak's higher foliar N content (CGR, 2006). The presence of ground vegetation contributes to lowered denitrification rates by depleting the mineral N pool. Competition of nitrifying organisms with plant uptake and microbial N immobilization has been observed to be responsible for rather low rates of net nitrification in the Harvard Forest soils in the USA (CGR, 2006). A developing forest takes up more nitrogen compared to a steady-state forest. The development of tree vegetation decreases the N available for microbial nitrification and the potential for trace gas production. In the long term, establishment of forest vegetation reduces trace gas emissions.

Loss of carbon from mineral soils also varies during the first 50 years following afforestation, with little variation across forest types. NPP influences litter inputs and C allocation, which in turn influence soil C dynamics and trace gas emissions. Among other factors, the rate of NPP is strongly influenced by tree species. Growth rates for Fir, Pine and Oak classes varied slightly, with Pine exhibiting the highest growth rates (CGR, 2006). After 50 years, C accumulation in the forest floor may account for 6.5–7.5 Mg C ha^{-1}, or 15–20% of the tree C stocks (CGR, 2006). However, estimates of C storage in litter layers vary for different forest types (Polglase et al., 2000). Litter C storage in Australia was found to vary from 600 to 1,610 g C m^{-2} under pine plantations, and from 380 to 2,200 g C m^{-2} under various eucalypt plantations (Polglase et al., 2000).

Deforestation increases the amount of CO_2 and other trace gases in the atmosphere. When a forest is cut and burned to establish cropland and pastures, the C that was stored in the tree trunks (wood consists to 50% of C) is released into the atmosphere as CO_2 (Roper, 2003). Carbon emissions from deforestation in 2004 were estimated at 1.6 Pg C $year^{-1}$, or 20–25% of total anthropogenic emissions (Field & Raupach, 2004). In comparison, fossil fuel burning (coal, oil, and gas) releases about 6 Pg $year^{-1}$ (NASA Earth Observatory, 2007). Tropical deforestation presently accounts for about 20% of all human-caused CO_2 emissions. Besides CO_2, gaseous N forms can be emitted. Depending on fire intensity and other factors, such as moisture content, a variable fraction of C and N in biomass and the soil will be released as CH_4, CO, N_2O, and NO_x (Apps et al., 2001).

The loss of forests has a profound effect on the global C cycle. From 1850 to 1990, deforestation worldwide released 122 Pg C into the atmosphere (NASA Earth Observatory, 2007). Within the past 20 years, deforestation has contributed 30% of the present anthropogenic increase of atmospheric CO_2 (Field & Raupach, 2004). Currently, the most deforestation occurs in developing countries. Deforestation in the Brazilian Amazon is the largest single source of CO_2 emission from deforestation (NASA Earth Observatory, 2007). The deforestation rate in the Amazon region was estimated to be 1.38 million hectares $year^{-1}$ in 1990, corresponding to emission of 251 Tg C $year^{-1}$ (Apps et al., 2001). The deforestation rate has increased in recent years, to 2.91 Mha $year^{-1}$ in 1995 and 1.82 Mha $year^{-1}$ in 1996. In Brazil, about 90% of the original forest is still intact which also means that this country remains a large potential source of future emissions. Reducing the deforestation rate by 50% would conserve 125 Tg C year^{-1} (Apps et al., 2001).

6.1.3.2 Greenhouse Gas Emission from Intact Wetlands

Wetlands play an important role in the sequestration and release of climatically relevant gases. Of special importance in the context with GHG emissions from wetlands are CO_2, CH_4 and N_2O (Röhricht et al., 2006). These gases can be emitted by intact wetlands, though to various extents. Emission rates of carbon dioxide from intact peats are relatively small (commonly 5–10 g CO_2 m^{-2} day^{-1}) compared with the other systems. Another factor is that plant tissues, e.g. mosses and shrubs of northern peatlands, provide a substrate that decomposes slowly, with exponential decomposition, or k values, between 0.02 and 0.25 year^{-1} (Scanlon & Moore, 2000). This means that intact wetlands sequester significant amounts of CO_2. In contrast, they release CH_4 and thus have to be considered as a sources for GHGs (Van den Bos, 2000; Röhricht et al., 2006). In natural wetlands, the level of the groundwater table is the main factor regulating the redox potential, and therefore CH_4 emissions through methanogenesis (Bouwman, 1990). Major individual sources for CH_4 are wetland rice, natural wetlands, ruminants, termites, landfills for solid waste dumping, biomass burning, coal mining, oil and gas exploitation, gas distribution and water reservoirs (Bouwman, 1990). Changes in the above land use forms alter fluxes of CH_4. All these sources are increasing at present. In natural wetlands, the water table level is the main factor regulating CH_4 emissions. Temporary water-saturated soils and wet forests in boreal regions contribute substantially to the global CH_4 budget. Emissions of CH_4 from natural wetlands may even increase in response to global warming. Since the increase in temperature will probably be strongest in northern latitudes, the boreal and tundra ecosystems are likely to show changes in CH_4 and probably as well N_2O emissions (Bouwman, 1990).

Intact wetlands are a source of constant CH_4 emissions. However, the mechanisms causing different CH_4 emission rates on different wetland types are not yet completely understood (Van Breemen & Feijtel, 1990). For example, differences in CH_4 emission rates between bogs (ombrotrophic peats; pH 3–4) and fens (minerotrophic peats; pH 6–7) were negligible (Van Breemen &d Feijtel, 1990). In contrast, in other studies higher methane production was found in fens compared to bogs (Van Breemen & Feijtel, 1990). Despite the fact that high temperatures stimulate methane production (Van den Bos, 2000) estimates for different climatic regions show almost the same values (Van Breemen & Feijtel, 1990). Methane production rates ranged from 1 to 200 g m^{-2} year^{-1} in tropical, cool temperate, and subarctic environments (Bouwman, 1990).

6.1.3.3 Emission of GHGs from Flooded Soils

The soil redox potential decreases after flooding of soils. According to Mayer & Conrad (1990), exposure of dry, aerated paddy soils to anoxic conditions resulted in a decrease of the soil redox potential and initiation of CH_4 production. The soil's capacity for initiation of CH_4 production could be regenerated by either addition of glucose or by exposure to H_2 or addition of reducing agents (Mayer & Conrad, 1990). In a greenhouse experiment, Mishra et al. (1997) found that CH_4 efflux was almost ten times higher from continuously flooded soils than from continuously non-flooded conditions. Intermittently flooded regimes (alternately flooded and non-flooded cycles of 40 or 20 days each) emitted distinctly less methane than the continuously flooded system. A significantly negative correlation was found between CH_4 emission under different water regimes and rhizosphere redox potential. Extractable Fe^{2+}, readily mineralizable C and root biomass presented a significant positive correlation with cumulative methane emission.

6.1.3.4 Emission of GHGs from Drained Wetlands

A consequence of drainage is the aeration of the upper soil horizons or peat layers which leads to the oxidation of organic material and the release of C and N compounds into the atmosphere and the drainage water. Especially intensively utilized wetland soils show significant release of CO_2 and N through peat mineralization. For example, up to 1,000 kg total N ha^{-1} $year^{-1}$ and 4 kg N_2O-N ha^{-1} $year^{-1}$ were released from degraded fens in eastern Germany (Augustin et al., 1995b). Due to the high climatic relevance of N_2O, the amounts released through wetland reclamation are significant. The use of drained wetland as forest area and arable land resulted in higher N emission rates compared to grassland. Due to higher water tables and lower aeration rates, in extensively managed wetland soils, a reduced peat mineralization can be expected (Röhricht et al., 2006).

The important effects of wetland drainage are reduction in C sequestration potential, oxidation of soil C reserves and reduction in CH_4 emissions. The emission of CH_4 decrease under increasingly aerated soil conditions because oxygen inhibits the methanogenetic archaebacteria (Van den Bos, 2000). In contrast, the emission of CO_2 increases with rising soil oxygen contents because the decomposition rate is enhanced under aerated conditions (Bleuten et al., 2000; Bridgham et al., 2006). The emission rates from drained wetlands depend on the state of drainage and the subsequent form and intensity of soil management. Extremely high CO_2 emission rates were observed on drained wetland forests, i.e. swamps, on arable land derived from fens, as well as from intensively used grassland.

According to Freibauer & Schrumpf (2006), more than 90% of German wetlands are drained. Annual emission rates from these areas are approximately 3.1 Mg C ha^{-1} $year^{-1}$, which is above the global average regarding emissions from wetland soils. The current contribution to the global climate stress through German wetlands is estimated at 6–12 Tg C $year^{-1}$. Annual CO_2 emissions from drained wetland areas in eastern Germany which are now used for agricultural production are estimated at 1.4 Tg C (Röhricht et al., 2006). Following wetland reclamation an average release of 4 kg N_2O-N ha^{-1} $year^{-1}$ can be assumed (Augustin et al., 1995b). Emissions of N_2O may also increase due to global warming.

6.1.3.5 Emission of GHGs from Restored Wetlands

Restoration of wetlands stops degradation of organic matter and even leads to new formation and accumulation of peat and is expected to reduce emissions of CO_2 and N_2O but increases the emissions of CH_4. Rising of water table reduced the emissions of N_2O from drained wetlands (Augustin et al., 1995b). Significant mitigation of GHGs emissions could be reached through vast restoration of (previously drained) wetlands. For example, restoration of all degraded German wetland soils could yield an 11–20% contribution to the German emission reduction in the Kyoto protocol (Freibauer & Schrumpf, 2006). However, the effect of CH_4 emissions negotiates only a small part of the achieved benefit through reduction of CO_2 and N_2O emissions. Contrary observations were made by Bridgham et al. (2006) who recognized the benefits of C sequestration in wetland soils and the reestablished wetland vegetation. Nevertheless, Bridgham et al. (2006) reported that CH_4 emissions from wetlands may largely offset benefits of C sequestration in terms of climate forcing. An exception was made for the restoration of estuarine wetlands, which showed throughout positive net effects regarding the influence on climate change.

6.1.3.6 Fluxes of CO_2 and CH_4 in Relation to Temperature and Groundwater Table

Several studies indicate a high sensitivity of wetland CH_4 emissions to temperature and water table (Denman et al., 2007). Model simulations by the same authors yielded an increase in CH_4 emissions by 19% due to an increase in temperature by 2°C. The combined effects of 2°C warming and 10% increase in precipitation yielded an increase in of CH_4 emissions by 21%. Temperature also influences the moisture regime of wetlands, which determines the type (aerobic or anaerobic) of decomposition. Emissions increase under a scenario where an increase in temperature is associated with increases in precipitation and NPP, but emissions decrease if elevated temperature results in either reduced precipitation or reduced NPP.

Van den Bos (2000) monitored CO_2 and CH_4 fluxes from three sites in the western Netherlands in relation to temperature and groundwater level. CO_2 fluxes varied throughout the year from about 25–2,000 mg CO_2-C m^{-2} h^{-1} and were found to be strongly correlated with temperature. Fluxes of CH_4 varied from 0.5 mg CH_4-C m^{-2} h^{-1} at sites with low groundwater tables up to 6.2 mg CH_4-C m^{-2} h^{-1} with high groundwater levels. It could be concluded that most coastal peatlands in the Netherlands act as a C source and therefore contribute to global warming (Van den Bos, 2000).

6.1.4 Fire Regimes

Fire is a natural factor in large areas of the world, thus an important part of the global C cycle, and is a major short-term source of atmospheric C, but it adds to a smaller longer-term sink (<0.1 Pg C $year^{-1}$) in form of charcoal (Field & Raupach, 2004). Savanna fires are thought to comprise one of the largest sources of CO_2 from biomass burning (Hao & Liu, 1995) which would otherwise have been incorporated into the SOM pool. In savannas, fire frequency and intensity are intimately connected to tree-shrub dynamics through a complex series of feedbacks involving fuel load, grazing pressure, and competition between tree and grass species that have different relative resistance and have a range of responses to fire (Crutzen & Goldhammer, 1993).

In many parts of the world, human activities have led to changes in fire frequency in savanna regions, some of which are probably dating from prehistoric times (Bird, 1995). In some areas such as Africa, there has been an increase in fire frequency as a result of an increase in the number of fires lit by humans utilizing savanna resources (Swaine, 1992). In other cases, such as parts of South Africa, Australia, and North America, there has been a decrease in fire frequency related partly to active fire suppression or overgrazing (Bird et al., 2000). A decrease in fire frequency generally leads to the establishment of "woody weeds" at the expense of grassland, while an increase in fire frequency generally favors the expansion of grasses (Menault et al., 1990).

Given the large area covered by savannas globally ($1.3–1.9 \times 10^8$ ha), and the importance of the SOM pool in modulating changes in the size of the atmospheric CO_2 pool, it is possible that anthropogenic changes in fire frequency may have altered the dynamics of the savanna SOM pool, with a significant consequent impact on the global carbon cycle. It is also known that savanna fires can significantly affect parameters such as microbial populations, inorganic nutrient levels, nitrogen levels, trace gas fluxes, water infiltration rates and molecular composition of the SOM (Bird et al., 2000). There are only few studies on the effect of fire on soil carbon stocks in savanna ecosystems. These suggest that complete fire suppression will increase SOM contents, while increasing fire frequency in an area previously subjected to a lower fire frequency may lead to a decline in SOM.

Fire management is a suggestion to increase C sequestration in soils. For example, in the immense extent of tropical savanna and woodland, a 20% fire suppression would yield a C storage of 1.4 Mg C ha $year^{-1}$ with associated mitigation of 0.7 Pg C $year^{-1}$ (Field & Raupach, 2004). The latter authors also estimated that additional fire suppression in Siberian boreal forest, and tropical savanna and woodland might conceivably decrease the rate of accumulation of atmospheric CO_2 by 1.3 Pg C $year^{-1}$, or about 40%. In a study, which has made use of a long-term series of fire trials in Zimbabwe, vegetation plots from both sandy and clay-rich soil types have been subjected to fire frequencies ranging from annual burn to complete protection for the last 50 years (Bird et al., 2000). Variations in the 0–5 cm SOM contents were predominantly related to soil texture, with C concentrations at the sandy sites being consistently 35–50% lower than those at comparable clay sites. Average 0–5 cm carbon densities for all the burnt plots were approximately 100 and 50 mg cm^{-2}, at the clay site and the sandy site, respectively. In both cases, lower fire frequencies had resulted in a ~10% increase, while higher fire frequencies had resulted in a ~10% decrease in these average values. Plots from which fire had been excluded experienced a 40–50% increase in carbon stocks in the 0–5 cm interval, compared with the average of the burned plots. Increasing fire frequency results in a relative increase in the fine particulate SOM.

Burning of vegetation residues in native forest or plantations may cause a slow but long-term increase in charcoal (Polglase et al., 2000). In contrast to native forest or previous plantation land, residues on former agricultural land produce little charcoal, due to the lack of woody components.

6.2 Agricultural Management

When humans developed a systematic agriculture, soil cultivation became an acceptable practice for preparation of a more suitable environment for plant growth. Pictures in ancient Egyptian tombs portray a farmer tilling his field using a plow and oxen prior to planting the seed. Tillage became almost synonymous with agriculture. Agriculture presumably began as no-tillage system where a pointed stick was used to place seed directly into untilled soil. In many parts of the tropics, no-tillage is still a part of slash and burn agriculture.

6.2.1 Soil Tillage

Tillage is one of the factor that directly impacts our environment. It affects decomposition processes through the physical disturbance and mixing of soil, by exposing soil aggregates to disruptive forces, and through the distribution of crop residues in the soil. Tillage systems have evolved over long time periods. Tillage also affects soil temperature, aeration and water relations by its impact on soil structure. By increasing the effective soil surface area and continually exposing new soil to wetting/drying and freeze/thaw cycles at the surface, tillage makes aggregates more susceptible to disruption and physically protected inter-aggregate organic material becomes more available for decomposition (Beare et al., 1994a). Among drastic tillage-induced changes in soil properties are bulk density, infiltration rate, aggregation, microbial activity, species diversity and SOM and nutrient profile (Prihar et al., 2000). Tillage also impacts carbon and nitrogen sequestration and thus the evolution of greenhouse gases such as CO_2 and N_2O. There are two general categories of tillage practices: conventional and conservation tillage. Both practices have different effects on soil properties, crop yields and GHG emissions from soils. For both practices, the initial content of C and N in the soil determines the direction of changes in soil storages after tillage activities. A general trend seems to be increased effluxes of CO_2 after conventional and increased emission of N_2O after conservation tillage practices. Most agricultural soils in temperate climates have lost significant amounts of SOC due to excessive tillage. Conservation tillage practices that include reduced and no-tillage farming and increased cropping intensity, along with reseeding of marginal croplands to permanent cover, can increase SOM and store a significant portion of C from the atmosphere.

6.2.1.1 Plow Tillage

Plow tillage systems incorporate crop residues and distribute organic matter evenly throughout the plow layer. The process of tillage may affect the SOM pool in two ways. SOM from deeper layers and from the interior of aggregates is brought to the surface (Whitehead, 1995; Lal, 2004a). Subsequently, mineralization rates increase, due to increased aeration and availability to microorganisms. However, SOM from the surface (as zone of more active mineralization) is transferred to deeper layers with lower turnover rates, thus contributing to the soil's humus pool (Nieder et al., 2003a; Lal, 2004a). Increased tillage frequency results in loss of SOM due to higher aeration and microbiological activity (Grant, 1997). In the long term, the SOM content of mineral soils with constant rotation is in a state of quasi-equilibrium if the plowing frequency and depth are approximately constant. Under such circumstances, soils attain a balance between gains and losses of C, N, S and P. Carbon and nutrients in SOM are temporarily liberated but the amounts released are compensated for by incorporation of equal amounts into newly formed humus.

In Europe (e.g. Denmark, France, The Netherlands, Belgium and Sweden; Cannell, 1985) and North America (Rasmussen et al., 1998), typical depths of plowing by established methods usually range from 15 to 25 cm. In Western Germany, increasing the plowing depth from <25 to >35 cm in the 1970s has been a feature of intensive agriculture (Nieder & Richter, 2000). In the territory of East Germany (the former GDR), depths of plowing rarely exceed 25 cm (Nieder, 2000). Increasing the plowing depth initially causes dilution of the SOC (and SON) content in the Ap due to mixing of underlying C-poor subsoil material (Fig. 6.9a).

Subsequently, SOM contents present before the onset of deeper plowing were quasi reestablished within several decades (Fig. 6.9b). From 1970 to 2000, large amounts of SOC and SON have been accumulated in the deepened plow layers. About 10 Mg C ha^{-1} and 1 Mg N ha^{-1} were accumulated in loess soils of cash crop production farms and roughly 20 Mg C ha^{-1} and 2 Mg N ha^{-1} were accumulated in sandy soils of livestock production farms (Fig. 6.10).

As a consequence, a significant part of the N surplus, which since the 1970s in Western Germany amounts to more than 100 kg N ha^{-1} $year^{-1}$ (Nieder et al., 2007), has been buffered in the deepened plow layers (about 30 kg N ha^{-1} $year^{-1}$ in cash crop production farms and roughly 70 kg N ha^{-1} $year^{-1}$ in livestock production farms) and thus been prevented from leaching. The higher SOC and SON accumulation rates in livestock production farms compared to cash crop production farms might be due to higher inputs of organic residues from animal production. In most of the arable soils, the buffering capacity for C and N has reached its limits because organic matter "equilibria" have been reestablished. Due to the recent SOM accumulation, the N mineralization potentials of arable soils have increased significantly (Fig. 6.11).

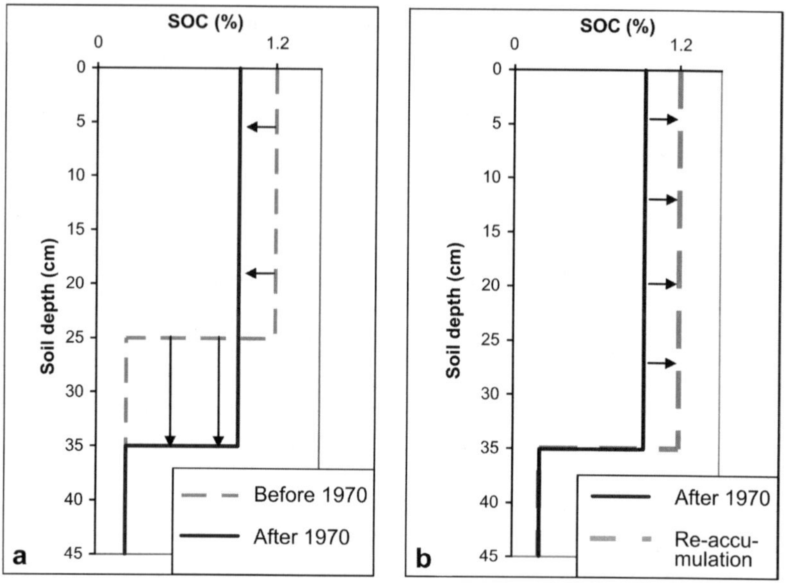

Fig. 6.9 Dilution of the SOC content due to deepening the plough layer (a) and accumulation of SOC due to quasi-reestablishment of the original SOM content (b)

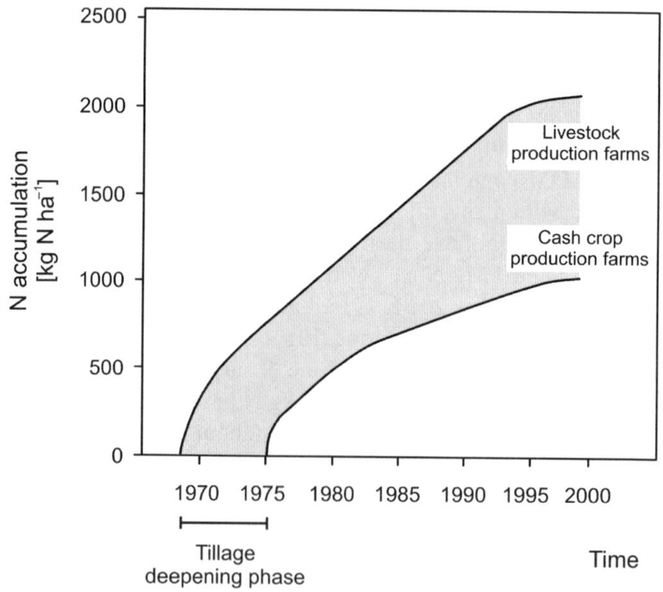

Fig. 6.10 N accumulation after deepening the tillage (16 farms, 120 plots) from <25 to >35 cm around 1970 (Southern Lower Saxony, North Germany) (Nieder et al., 2003c; p. 176. Reproduced with permission from the authors)

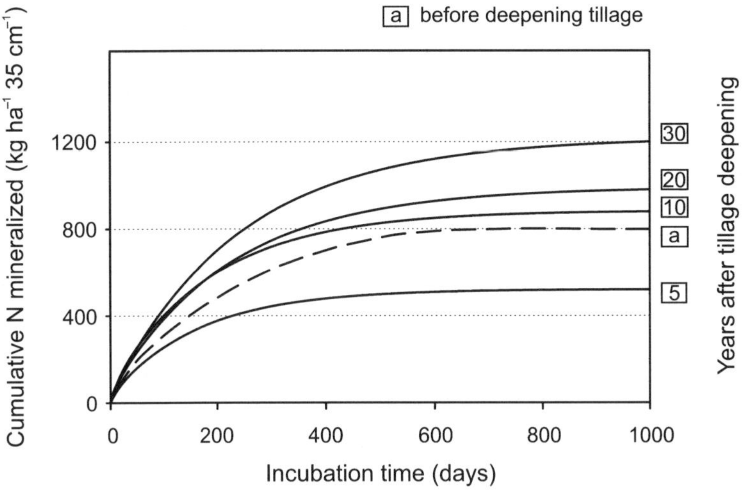

Fig. 6.11 Cumulative N mineralization (from incubation experiments) in Southern Lower Saxonian arable loess soils 5, 10, 20 and 30 years after deepening the tillage (Southern Lower Saxony, North Germany) (Nieder et al., 2003c; p. 176. Reproduced with permission from the authors)

According to optimized N mineralization parameters from long-term incubation experiments, the N mineralization potentials have increased from 600 to 800 kg N ha^{-1} 30 cm^{-1} in the late 1970s to about 1,200 kg N ha^{-1} 35 cm^{-1} in 2000 (Fig. 6.10). This means that a significant part of the newly accumulated N has become part of the "active" pool of SOM. These results suggest that there is significant potential for the storage of SOM in arable soils, but it has to be taken into account that the stored amounts of SOM can be easily released again under inappropriate management practices. The changes in the N (and C) mineralization potentials are of particular interest for N fertilization strategies as well as for assessing N and C sequestration in arable soils.

6.2.1.2 Conservation Tillage

Conservation tillage is defined as a system having at least 30% or more crop residues covering the soil at planting (CTIC, 2000). Conservation tillage practices can be subdivided in no-tillage ('pure' no-tillage or strip-tillage), ridge tillage (building ridges with in-season cultivation), minimum tillage and mulch tillage (field-wide tillage). The mulch cover is a substantial requirement in achieving the positive effects of conservation tillage practices. In the past 2 decades, conservation tillage (zero tillage and various reduced tillage management systems) have received increased attention owing to the potential of these management systems for abating soil erosion, conserving soil moisture, enhancing water quality, and cutting monetary and energy inputs of crop production systems. In Europe, less than 5% of the cropland area was in conservation tillage in the 1990s (Nieder, 1998), and presently there is only a slightly increasing tendency. In the United States in 1992, conservation tillage was used on 31% of the cropped land (Cannell & Hawes, 1994). By 2010, this portion is expected to increase to 63–82% (OTA, 1990).

The influence of tillage management on SOC and SON has been investigated intensively throughout the world. Most studies have examined changes in concentration of SOC (e.g. Dick, 1983; Karlen et al., 1994; Frede et al., 1994; Salinas-Garcia et al., 1997). Few studies have examined the changes in the mass of SOC (e.g. Campbell et al., 1995; Reicosky et al., 1995; Van den Bygaart et al., 2002). Generally, SOC and SON concentrations in the surface 15 cm of no-tilled soils are greater than in tilled soils, especially when they are moldboard plowed (Fig. 6.12).

Conservation tillage induces not only stratification of SOM and related nutrients but also enhances the size of soil microbial biomass in the upper part of the surface soil

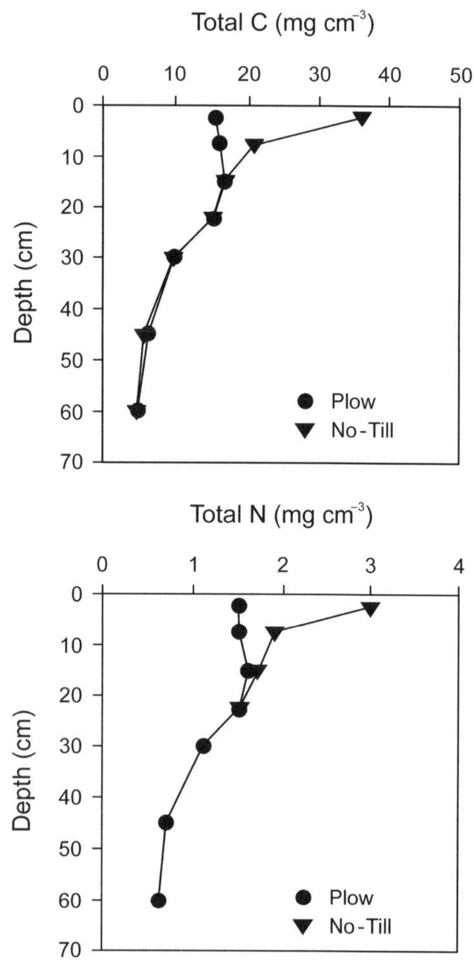

Fig. 6.12 Tillage treatment effects on SOC and SON within the upper 60 cm of a silt loam soil following 12 years of continuous corn (Karlen et al., 1994; p. 323. Reproduced with kind permission from Elsevier)

(Logan et al., 1991). The magnitude of these changes depends on soil texture, climate, and cropping system. The new equilibrium in soil properties by conversion from plow-till to no-till may be attained over a period of 10–20 years (Kern & Johnson, 1993; Frede et al., 1994). The equilibrium status may be reached more quickly in coarse-textured soils of the tropics than in heavy-textured soils of the temperate climate. According to Van den Bygaart et al. (2002), there are numerous factors that affect the dynamics of SOC under no-tillage including climate, management history, soil type and landscape processes. The initial SOC (and SON) content also play an important role for the dynamics of SOC (and SON) after the introduction of conservation tillage systems.

Rates of C accumulation in soils under no-till or conservation-till reported in the literature vary widely, ranging from below 0 to 1.3 Mg C ha^{-1} $year^{-1}$ in the upper 15 cm (Reicosky et al., 1995). In a number of long-term experiments (up to 20 years under no-till) conducted in humid regions, zero tillage increased SOC by an average of 3 Mg C ha^{-1} compared with conventional tillage with a moldboard plow (Paustian et al., 1997b). Van den Bygaart et al. (2002) found that no-till increased the storage of SOC in western Canada by 2.9–1.3 Mg C ha^{-1}. However, in eastern Canada, conversion to no-till did not increase SOC. Differences in SOM between conventional and zero-tillage systems in semiarid regions are generally small because conventional tillage is less intensive and shallower than in humid regions (Unger, 1991). In western Canada, Campbell et al. (1995) reported that over a 12-year period, a continuous spring wheat system under zero-tillage in 0–15 cm soil depth gained about 1.5 Mg C ha^{-1} compared with conventionally tilled plots. On a loamy chernozem in a spring wheat-fallow system, the gain was only 0.5 Mg C ha^{-1}. On a sandy soil, neither tillage nor fallow frequency influenced C sequestration over an 11 y period (Campbell et al., 1996).

Experiments in the subhumid and humid tropics have demonstrated the potential for no-till systems to maintain higher SOM levels compared to conventional cultivation directly after land clearing. Reduced soil erosion and lower soil temperatures with surface mulches are particularly important attributes of no-till systems in the tropics. Agboola (1981) reported organic matter losses <10% with no-till compared with 19–33% losses in tilled treatments, after 4 years of continuous maize.

Under reduced tillage practices, compared to conventional tillage, decomposition of SOM and emissions of CO_2 are reduced. Tillage also affects the conditions for N_2O emissions from soils. In some studies, higher N_2O losses were observed for no-tillage systems compared to conventional tillage (IFA/FAO, 2001) which may be particularly due to higher denitrification activity. Under reduced tillage, soil moisture contents are increased, which may thus increase denitrification activity. However, according to Duiker (2005) conversion to conservation tillage only caused increased N_2O fluxes if the soils were poorly drained. Higher rates of denitrification and N_2O fluxes were observed after application of organic residues on the soil surface as mulch compared to sites where the residues were removed (IFA/FAO, 2001). This effect may be particularly related to the effect of residues placed on the soil surface to reduce evaporation. Incorporation of residues may stimulate the mineralization of SOM and, therefore, accelerate emissions of N_2O (IFA/FAO, 2001). Six et al. (2001) investigated the effects of tillage on the fluxes of CH_4 and N_2O. On average, in soils of the temperate climate under no-tillage, compared to conventional tillage, CH_4 uptake by soils increased by 0.42 ± 0.10 kg CH_4-C ha^{-1} $year^{-1}$. In contrast, emissions of N_2O increased by 1.95 ± 0.45 kg N_2O-N ha^{-1} $year^{-1}$. The increased N_2O emissions lead to a negative global warming potential when expressed on a CO_2 equivalent basis.

6.2.2 *Fertilization*

Fertilizers are the most widely used form of chemicals in agriculture. The use of fertilizers is not without controversy. For example, inappropriate or excessive fertilizer application can lead to increased losses of N from soils. Losses can occur through runoff, leaching or as gaseous N compounds. Losses in gaseous forms can have negative impacts on the global climate, especially as N_2O (Harrison, 2003).

Legumes can supplement N by fixing N_2 from the atmosphere. Forage legumes contain 3–4% N that can originate from both the soil and air (Evers, 2001). When legumes are incorporated into the soil, the decaying residues release N, which becomes available for subsequent crops. In grasslands, the decay of legume roots provides additional N for the grass. In many agricultural systems, legumes such as clover, soybeans and alfalfa deliver important inputs of N (IFA/FAO, 2001). Estimates of the amount of fixed N range from 58 to 120 kg N ha^{-1} for annuals and about 228 kg N ha^{-1} for alfalfa (Evers, 2001). Fixation of N from the atmosphere is inhibited by fertilizer N application (Whitehead, 1995).

6.2.2.1 Synthetic Fertilizers

Mineral fertilization enhances crop growth and C inputs from increased organic residue production. There exist numerous forms of synthetic fertilizers (Chapter 2). They vary in their behavior during processes causing leaching and volatilization of nitrogen. Positive effects on the soil C balance particularly have been observed due to increased synthetic N application rates (Table 6.8). With every 1 Mg ha^{-1} increase in SOC pool in the root zone, crop yields can be increased by 20–70 kg ha^{-1} for wheat, 10–50 kg ha^{-1} for rice, and 30–300 kg ha^{-1} for maize (Lal, 2005). Recently, Benbi & Chand (2007) showed that contribution of 1 Mg SOC ha^{-1} to wheat productivity in subtropical India ranged from 15 to 33 kg ha^{-1}. The wheat productivity per Mg of SOC declined curvilinearly as the native SOC concentration increased. Adoption of recommended management practices on agricultural lands and degraded soils enhance soil quality including water holding capacity, cation exchange capacity, soil aggregation, and susceptibility to crusting and erosion. One possibility to enhance SOC is the application of organic fertilizers that do not only contain N and other nutrients, but also C in their organic tissues. In addition to direct effects of fertilization on C inputs, high N availability may also enhance the formation of recalcitrant humic substances in soils (Fog, 1988).

The application of fertilizer N has often little or no effect on SOC (Table 6.10), which may be mostly due to the fact that N fertilization may stimulate soil microbes

Table 6.8 Soil organic C response to fertilizer N application in long-term field experiments. SOC levels are given as % increase over unfertilized control (Adapted from Nieder et al., 2003a)

Rotation	Period (year)	N application (kg ha^{-1})	SOC (% increase over control)	Country	Reference
Mixed rotation	20	112–128	7	Germany	Körschens & Müller (1996)
Continuous rye	112	40	13	Germany	Garz (1995)
Mixed rotation	81	30–50	11	Germany	Welte & Timmermann (1976)
Continuous wheat	135	144	23	United Kingdom	Jenkinson et al. (1994)
Cereal/root crops	30	80		Sweden	Paustian et al. (1992)
-Straw removed			18		
-Straw added			16		
-Sawdust added			15		
Continuous maize	12	200	7	USA	Barber (1979)
Mixed rotation	22	320		India	Benbi & Biswas (1997)
-Mineral N			15		
-Mineral N + FYM			62		

to mineralize SOM. Nitrogen fertilization has been found to promote SOC and SON accumulation particularly in soils that were originally poor in SOM (Whitehead, 1995). A positive effect of fertilizer N on SOM accumulation is also likely in poorly-drained clay soils, in which the mineralization of plant residues is slow due to restricted aeration. N fertilization has been found to increase C sequestration primarily because of the higher plant productivity, leading to greater return of plant residues to the soil.

6.2.2.2 Organic Fertilizers

Application of organic fertilizers is an efficient measure to increase the SOM pool. The advantage of organic fertilizers compared to inorganic ones is the slow and continuous release of N. Organic fertilizers are mainly applied as manure and sewage in solid and liquid form, and crop residues. The feces of farm animals consists mostly of undigested food that has escaped bacterial action during passage through the body. This undigested food is mostly cellulose or lignin fibers (Nieder et al., 2003a). Animal wastes are more concentrated than the original feed in lignins and minerals.

The feces also contain the cells of microorganisms. Nitrogen in manure solids occurs largely in organic forms (undigested proteins and the bodies of microorganisms). The C:N ratio of farmyard manure is usually 15–30. Liquid manure may also contain significant amounts of NH_4^+ which has been formed from urea through hydrolysis. The manure applied to cropland varies greatly in nutrient content, depending on animal type, ration fed, amount and type of bedding material, and storage condition. Both N content and availability of the N to plants decreases with losses of NH_3 through volatilization and NO_3^- through leaching. Manures aged by cycles of wetting and drying and subjected to leaching with rainwater may have lost so much N that very little will be available to the crop in the year of application.

Sewage sludge from biological treatment of domestic sewage, is a stabilized product with an earthy odor and which does not contain raw, undigested solids. Liquid sewage sludge is blackish and contains colloidal and suspended solids. Most sludges, as produced in a sewage treatment of roughly equal parts of organic and inorganic material. Heavy metals like Zn, Cu, Pb, Cd, Hg, Cr, Ni may occur in quantities sufficient to adversely affect plants and soils. The availability of any given metal in soil will be influenced by pH, SOM content, type and amount of clay, content of other metals, cation exchange capacity, variety of crops grown, and others. The organic component is a complex mixture consisting of digested constituents that are resistant to anaerobic decomposition, dead and live microbial cells, and compounds synthesized by microbes during the digestion process. The organic material is rather rich in N, P, and S, and the C:N ratio of digested sludge ranges from 7–12. N availability in sludges decreases as the content of NH_4^+ and NO_3^- decreases and as the organic N becomes more stable as a result of digestion during biological waste treatment. Conservation of the N that often volatilizes as NH_3 would greatly increase the value of sewage sludge as an N source.

Manure applications have been observed to increase C sequestration (Table 6.9). However, changes in SOM contents due to altered organic fertilizer application run very slow. If the changes are in dimensions relevant to practice, it may take more than 10 years until they can be proved. With the extension of the experimental question of the Static Fertilization Experiment Bad Lauchstaedt (near Halle, central Germany), there was the chance to quantify the changes in the C and N contents in a Chernozem following extreme changes in fertilization (Körschens & Müller, 1996). At a high starting level the annual decrease is 0.013% C equaling to 0.5 Mg C ha^{-1} $year^{-1}$ and 0.0011% N corresponding to 44 kg N ha^{-1} $year^{-1}$. This amount corresponds roughly with the difference in the N uptake comparing the nil treatment and the previously completely fertilized treatment. As concerns the previously unfertilized treatment, the annual C increase is lower, reaching 0.0081% $year^{-1}$, the amount of N increased by 0.0012% $year^{-1}$. It becomes apparent that it will take several decades to reach new equilibrium SOM levels.

Numerous long-term field experiments show that with realistic and practically feasible amendments the SOM levels, compared with unfertilized controls, can hardly be increased by more than 30%. Assuming a practically achievable increase of about 0.3% C in the 690 million hectares of arable mineral soils in the temperate climatic zone, this overall C sequestration would be in the order of about 6 Pg C

Table 6.9 Nitrogen surplus calculations for different world regions (Adapted from Van Drecht et al., 2005)

Region	Area (10^6 km^2)	Bouwman et al. (2005) (Tg N year^{-1})	Green et al. (2004) (Tg N year^{-1})	Siebert (2005) (Tg N year^{-1})
World	128.1	232.2	210.8	252.8
Canada	9	7.6	5.2	4.9
USA	9.1	20.4	16.9	19.4
Central America	2.6	5.6	4.9	8.2
South America	17.4	40	25.3	44.7
North Africa	5.7	5.5	4.9	2.9
Western Africa	11.2	22	17.2	24.9
Eastern Africa	5.8	10.1	7.8	14.1
Southern Africa	6.7	10.6	10.8	15.0
Western Europe	3.5	12.2	12.7	13.3
Eastern Europe	1.1	3.7	4.4	2.5
Former USSR	21.6	20.1	16.8	15.1
Middle East	5.8	6.5	6.6	6
South Asia	5	21.8	24.2	27.3
East Asia	11	24.3	33.4	29.5
Southeast Asia	4.1	12.2	9.6	11.5
Oceania	8.1	8.5	9.3	12.4
Japan	0.3	1.1	0.9	1.3
Australia	7.5	6.9	7.8	10
Brazil	8.4	24.1	13.2	24.2
India	3.2	16.3	17.9	20.5
China	9.2	23.3	31.2	26.9

within a period of 50 years (Sauerbeck, 1993). Even though this applies only to the temperate climate zone, this figure is probably more realistic than the global 20–30 Pg C estimate in the IPCC Report (Cole et al., 1996), since the practical chances to sequester more carbon in tropical arable soils are comparatively small (Batjes, 1998; Batjes & Sombroek, 1997).

6.2.2.3 Nutrient Management

Nitrogen applied in fertilizers, manures, biosolids and other sources is not always used efficiently by crops (Galloway et al., 2003). The surplus N is susceptible to leaching of NO_3^- and emission of N_2O. Table 6.9 shows the total N surplus, calculated as the difference between inputs (atmospheric N deposition, biological N_2 fixation, mineral fertilizer and animal manure application) and outputs (crop and grass harvest and grazing export, ammonia volatilization) for different world regions.

There are significant differences in the N surplus estimates between different authors. The most uncertain aspects of these calculations are the input terms biological N fixation and atmospheric N deposition, and the output terms (export of N from fields in harvested crops and grazing, and NH_3 volatilization (Van Drecht et al., 2005). In Table 6.10, the N surplus per unit agricultural area is given for 15 European countries.

Table 6.10 Nitrogen surplus in European countries (Adapted from Nieder et al., 2007)

Country	N surplus (kg N ha^{-1} $year^{-1}$)
The Netherlands	>250
Belgium	140
Denmark	110
Luxemburg	95
Germany	85
Ireland	>50
Finland	>50
Greece	<50
France	<50
Italy	<50
Spain	<50
Portugal	<50
UK	<50
Austria	<50
Sweden	<50
Mean of the 15 countries	~50

More than 80% of the anthropogenic N_2O emission stems from agriculture. Crop production is responsible for about 50% of N_2O emissions from the agricultural sector. N_2O is also emitted from manure, soil-borne N (especially in fallow years), legumes, plant residues and compost. Based on statistical models, the global annual emissions from fertilized arable land was estimated to 3.3 Tg N_2O-N $year^{-1}$, and to 1.4 NO-N $year^{-1}$ (Stehfest, 2006). Fertilizer induced N_2O emissions, which are currently estimated by the IPCC to be 1.25 ± 1% of the N applied, range between 0.77% (rice) and 2.76% (maize). In the 1990s, simulated N_2O emissions from agricultural soils amounted up to 2.1 Tg N_2O-N $year^{-1}$ (Stehfest, 2006). Emission rates of N_2O from agricultural soils are significantly affected by fertilization rate, SOM content, soil pH, texture, crop type, and fertilizer type. NO emissions are significantly determined by fertilization rate, soil N content, and climate.

Improving N use efficiency can reduce NO_3^- leaching and N_2O emissions and indirectly GHG emissions from N fertilizer manufactures (Schlesinger, 1999). By reducing leaching and volatilization losses, improved efficiency of N use can also reduce off-site N_2O emissions. Practices that reduce N balance surpluses include: (i) adjusting application rates based on precise estimation of crop needs (e.g., precision farming, agricultural system models), (ii) use of slow-release fertilizers or nitrification inhibitors, (iii) applying N when least susceptible to loss, and (iv) placing the N more precisely into the soil to make it more accessible to crop roots (e.g., Robertson, 2004; Monteney et al., 2006; Kersebaum et al., 2007).

Animal manures can release significant amounts of N_2O and CH_4 during storage. Emissions of CH_4 from manure stored in lagoons or tanks can be reduced by cooling, use of solid covers, mechanically separating solids from slurry, or by capturing the CH_4 emitted (Amon et al., 2001; Clemens & Ahlgrimm, 2001). The manures can also be digested anaerobically to maximize CH_4 retrieval as a renewable energy source (Clemens et al., 2006). Handling manures in solid form (e.g., compost) rather than liquid form can suppress CH_4 emissions, but may increase N_2O emissions (Paustian et al., 2004). For most countries there are limitations for manure management, treatment or storage. However, the limitations are often insufficient from an ecological point of view. To some extent, emissions from manure might be curtailed by altering feeding practices (Kreuzer & Hindrichsen, 2006) or by composting manure (Pattey et al., 2005), but if aeration is inadequate, CH_4 emissions during composting can still be substantial (Xu et al., 2004).

Manures also release GHGs after application to cropland or deposition on grazing lands. Practices that tailor nutrient additions to plant uptake can reduce N_2O emissions (Dalal et al., 2003). However, deposition of feces and urine from livestock complicates management of nutrients. Compared to synthetic fertilizers, manures (and the release of nutrients from manures) are not as easily controlled nor as uniformly applied (Oenema et al., 2005).

6.2.3 Introduction of Fallow Systems

Fallow systems are introduced to "give the soil a break" from the tiresome job of producing crops. The soil is left bare (bare fallow) or planted with vegetation (green fallow) that is plowed in the soil before the next crop-growing period. In regions with food surpluses (e.g. EU, USA, Canada), about 25 million hectares (15% of total cropland) has been taken out of production by government set aside programs since the 1980s. Enrollments in the US Conservation Reserve Program are for 10 year periods after which time the land may be returned to annual crop production. The EU agricultural set aside programs are for 1–5 years and can include annual cropping for non-food production (e.g., oil seed for fuel). Under certain conditions, fallowed soils may cause depletion of SOM through degradation and can be a significant source of GHGs. Reasonably managed fallow systems can improve soil properties and turn cultivated soils in sinks for C. Recent studies indicate that introduction of bare fallow depletes SOC and may thus increase the source function (Larionova et al., 2003), while introduction of green fallow systems is considered to improve the sink function of cultivated soils (e.g. Apps et al., 2001). In sub-Saharan Africa introduction of fallow systems generally has the highest potential for SOC sequestration with estimated rates up to 28.5 Tg C $year^{-1}$ (Vågen et al., 2004).

6.2.3.1 Bare Fallow

In the temperate zone, traditional fallow meant that land was plowed and tilled, but left unsown, usually for a year. This practice was undertaken in order to allow the soil to recover from more or less intensive cropping. In modern agriculture, improvements in crop rotations and manure application have diminished the necessity of the bare fallow in temperate zones. Bare fallow is uneconomical because the land is left unproductive. Moreover, the risk of nitrate leaching is increased. Presently, bare fallowing may only be advised for heavy soils and in dry climates. Bare fallow is therefore used extensively in semiarid areas of the world (e.g. Canada, USA, Australia, former Soviet Union) to offset rainfall variability and increase soil water storage. In a cereal-fallow rotation, there may be only up to 6 months of crop cover during 24 months. Increased aeration of a soil under tilled bare fallow increases the decomposition of SOM (Larionova et al., 2004). During bare fallow periods, compared to cultivated areas, mineralization of SOM is generally enhanced due to increased soil moisture. This effect is further promoted by decreased or no plant residue inputs.

In tropical regions, in the course of shifting cultivation, traditional shifting cultivators after 3–4 years of cultivation kept sites fallow for periods of 10–20 years (Lal, 1995a). In the fallow period, vegetation from the surrounding tropical forest is introduced again to cleared site, which is burned again before new cultivation. These management practices allowed the infertile tropical soils to recover, and the crops could profit from the nutrients transferred to the soil from burning of the succession vegetation. Tropical areas are an important source of CO_2 not only because of widespread clearing of new lands but also due to the introduction of fallow periods in shifting agriculture systems (Apps et al., 2001). Recently, traditional shifting cultivation systems have been increasingly replaced by more sustainable systems one of which is the introduction of minimum tillage with mulching of crop residues, weeds or legume fallow (Roose & Barthes, 2001).

Soils kept bare are susceptible to all forms of soil degradation. SOM and especially SOC are easily lost from bare soils, thus contributing to the soil's susceptibility towards degradation and erosion. Moreover, soils bare from vegetation lack any protection against affecting erosive forces in form of wind and water. As a consequence, large amounts of soil material and adhering nutrients may be lost from bare sites (Armstrong, 1990; Raffaelle et al., 1997; Sonder, 2004). It can be expected that the decomposition of SOM under bare fallow contributes to increased concentrations of atmospheric CO_2 (Larionova et al., 2003) and N_2O (AF, 2000). In addition, soils kept bare lack an active C sink in form of growing biomass. Recent observations in Alberta (Canada) show that CO_2 emissions from agricultural soils have declined due to reduced summer fallow areas. Reducing the number of fallow years in Alberta may reduce CO_2 emissions by up to 0.17 Mg CO_2–C ha^{-1} $year^{-1}$ (AF, 2000). In Saskatchewan (also Canada), during the last decade, the area of summer-fallow decreased from 43% to 20%, which in this province lead to an increase in SOC up to about 3.8 Tg C $year^{-1}$ (Van den Bygaart et al., 2002).

6.2.3.2 Green Fallow

In green fallow systems, usually species are cultivated which are favorable for soil properties such as legumes. Green fallow systems combine positive aspects from bare fallow, while avoiding the negative effects. Under green fallow, a vegetation cover reduces the risk of erosion and soil degradation by mitigating the impacts of wind and water. There are two different types of green fallow systems. In the one type, soils are planted with grasses, often in combination with clover. In the other form, a cover crop is planted after harvest, which covers the soil during the following winter. The main purpose of green fallow is to cover soils, in order to reduce nutrient losses as observed under bare fallow. The plants grown under green fallow are generally plowed into the soil as green manure before the next cropping period. Green manures increase SOM and improve the soil structure, thus further reducing the risk of erosion. Green manures are also recommended measures for increasing N availability. In case of legumes, the soil N storage may be increased significantly (Whitehead, 1995; Bognonkpe, 2004), especially if initial N contents are low. However, in the semiarid regions of North America wheat productivity on soils was observed to decrease after green fallow. This decrease in productivity, however, seemed to be related to a depletion of the soil water content by the fallow legumes, rather than negative effects on soil properties and nutrients (Schlegel & Havlin, 1997; Vigil & Nielsen, 1998). However, legumes commonly have a positive influence on the yields of following main crops in regions where the water supply is not restricted.

Estimates for the US cornbelt suggest an additional C accumulation of 4,000 kg C ha^{-1} in 100 years from the use of winter cover crops (Lee et al., 1993). The use of perennial forage crops can significantly increase SOM contents, due to high root C production, lack of tillage disturbance, and protection from erosion. If arable soils revert to grassland, SOM contents in upper soil horizons could reach levels comparable to their precultivation condition. Increases up to 550 (Jenkinson, 1971) and to1,000 (Jastrow, 1996) kg C ha^{-1} $year^{-1}$ have been documented for cultivated land planted to grassland. SOM accumulation would continue only until soils will reach a new equilibrium value, most of which would be realized over a 50–100 years period. Considering the area of cropland with real or potential surpluses (about 640 million hectares in Europe, USA, Canada, former Soviet Union, Australia, Argentina) and assuming recovery of the SOM originally lost due to cultivation (25–30%), a permanent set aside of 15% of this land area (about 96 million hectares) might accumulate 1.5–3 Pg C in SOM (Paustian et al., 1998).

There exists a high potential for increasing SOC through establishment of natural or improved fallow systems (agroforestry) with attainable rates of C sequestration in the range of 0.1–5.3 Mg C ha^{-1} $year^{-1}$ (Vågen et al., 2004). In consequence, the soil's sink function for CO_2 is increased, and net CO_2 and N_2O emissions from decomposition of SOM are reduced. Compared to other green fallow species, legumes may increase N_2O emissions because growing of legumes increases soil N contents (Whitehead, 1995), which increases the potential for denitrification, thus promoting emissions of N_2O. Growing of legumes on the other hand reduces the amounts of mineral N fertilizer application, and thus the potential for emissions of N_2O from this source (AF, 2000). In the whole, the positive effects of green fallow on soil properties may be assumed favorable for the global climate.

6.2.4 Crop Rotation Effects

Among annual crops, cereals generally produce the most residues while crops such as grain legumes, dry beans and root crops produce less. Thus, SOM levels tend to be lower under maize-soybean rotations compared with continuous maize (Paustian et al., 1997b). Changes in SOM for six rotations including maize, sugar beet, navy bean, oats and lucerne were directly correlated to amounts of residue returned and the frequency of maize in the rotation (Zielke & Christensen, 1986). The inclusion of perennial forages (i.e. leys) in rotations increases SOM levels relative to rotations with annual crops alone. Experiments in Europe with 3 or more years of ley within annual crop rotations had up to 25% more SOM compared to rotations with only annual crops (Van Dijk, 1982; Nilsson, 1986). Carbon accumulation by perennials is attributed to the high relative allocation of organic materials below-ground, greater transpiration leading to drier soils, the formation of stable aggregates within the network of grass roots, and the absence of soil disturbance by tillage (Paustian et al., 1997a).

6.2.4.1 Rice-Based Systems

Intensive irrigated rice systems belong to the most important food production systems. Worldwide, about 80 million hectares of irrigated rice are harvested annually and the intensive lowlands will remain the major source of rice production in future (Dobermann & Witt, 2000). Intensive rice-based cropping systems are rice-rice, rice-rice-rice, rice-rice-pulses, rice-wheat, and rice-rice-maize. Lowland rice systems differ markedly from upland crop systems in C and N cycling processes. Irrigated rice is concentrated on alluvial floodplains, terraces, inland valleys, and deltas in the humid and subhumid subtropics and the humid tropics of Asia. Acidity, salinity, alkalinity and Al toxicity are usually less severe in rice soils due to the long periods of submergence. Flooding and intensive rice cultivation create distinct micro-environments that differ in physical and chemical properties. Components of the soil-floodwater system from the upper to the lower include floodwater, surface-oxidized soil, reduced soil, rhizosphere-oxidized soil, plow pan and oxidized or reduced subsoil. Another specific feature of irrigated rice is the presence of a photosynthetic aquatic biomass. The daily primary production of the floodwater community was estimated to be 0.2–1.0 g C m^{-2} day^{-1}, or on average about 700 kg C ha^{-1} per rice crop, with an annual turnover rate of 70–80% (Roger, 1996).

Under irrigated rice double-cropping, SOM levels tend to be stable or to increase (Nambiar, 1994). Organic matter content is generally lower in rice-upland crop rotations such as rice-wheat or rice-maize. When crop residues are not incorporated in these systems, the SOM amounts can decrease to the point of reducing the supply of N through mineralization-immobilization turnover, which could lead to low grain yields. Economic constraints often promote removal of straw from the field. In South Asia and China, rice residues are used for fuel, animal feed, roofing, and other uses. Only a small part is composted with animal waste and recycled to the fields (Flinn & Marciano, 1984). In South and Southeast Asia where combine harvesting is common, rice straw is burned. In general, straw management offers significant potential benefits, but optimal management is not straightforward and agronomic objectives often clash with economic objectives.

6.3 Ecosystem Disturbance

Degradation can occur due to natural and anthropogenic influences. Natural phenomena that lead to soil degradation are land slides, dune development, glacier retreat and vulcanism. These phenomena are local and may therefore not contribute to significant changes in global C cycling. The area of ecosystems disturbed by human activities was estimated to occupy worldwide 1,216 Mha of which 130 Mha are located in South Asia (Lal, 2004b). Land degrading processes caused mainly by unsuitable land use and management practices include soil erosion and deposition, surface mining, salinization and acidification and SOM and nutrient depletion. Most of the degradation is caused by accelerated erosion, which is a serious problem especially the humid tropics. Global hotspots of soil degradation are sub-Saharan Africa, South Asia, the Himalayan-Tibetan region, the Andean region, Central America and the Caribbean.

Severe water erosion is extensive in the humid regions of southeast Asia, including Myanmar, Thailand, Malaysia, and Indonesia, numerous islands in the Pacific and Oceania; along mountain regions of the Pacific coast in Central America, including south-eastern Mexico, Honduras, Nicaragua, and Costa Rica, and in some regions of the Amazon Basin (Lal, 1995a). High sediment yields are observed from river basins draining humid tropical regions (Lal, 2004b). Similarly, high sediment yields are reported from humid regions of Costa Rica, Java, Malaysia, Panama, Papua New Guinea, Australia, Philippines, and Thailand (Lal, 1995a). A serious and long-running water erosion problem exists in China, on the middle reaches of the Yellow River and the upper reaches of the Yangtze River. From the Yellow River, over 1.6 billion tons of sediment flows each year into the ocean. The sediment originates primarily from water erosion in the Loess Plateau region of northwest China. Examples of erosion rates in humid tropics are presented in Table 6.11.

On the Chinese loess plateau, horticulturists cleared the vegetation from the slopes and plateau (Bork et al., 2006). The prevailing Cambisols were completely eroded and fertile soils were irreversibly lost. In the tropics, the loss of trees anchoring the soil with their root system causes widespread erosion. The rate of soil loss after forest clearing is therefore enormous. A study in Ivory Coast (Côte d'Ivoire) showed that forested slope areas lost 0.03 Mg soil ha^{-1} $year^{-1}$. In contrast, cultivated slopes annually lost 90 Mg soil ha^{-1} $year^{-1}$, and bare slopes lost 138 Mg soil ha^{-1} $year^{-1}$. In Costa Rica, about 860 million tons soil are lost every year. In disturbed systems, biomass and SOM are depleted relative to native ecosystems and well-managed agroecosystems. Reducing land degradation and restoring existing degraded land are significant options to conserve SOM.

6.3.1 Erosion and Deposition Effects

Soil erosion and deposition may play important roles in balancing the global C budget through their impacts on the net exchange of C between terrestrial ecosystems

Table 6.11 Magnitude of soil erosion by water observed in some countries of the humid tropics (Lal, 1995b)

Country	Soil erosion rate (Mg ha^{-1} $year^{-1}$)
Brazil	18–20
Ecuador	200–600
Peru	15
Guatemala	5–35
Jamaica	90
Guinea	18–25
Madagascar	25–250
Nigeria	15–300
Côte d'Ivoire	60–600
Papua New Guinea	6–300

and the atmosphere (Liu et al., 2003). The net effect of soil erosion on atmospheric CO_2 is still uncertain as the C removed may be deposited elsewhere and at least partially stabilized (Apps et al., 2001; Lal, 2004b; Renwick et al., 2004). Soil erosion is an intrinsic natural process, but in many places, it is increased by human land use. During erosive processes, the soil is disturbed and SOC may be depleted, which causes the release of CO_2.

Erosion and deposition also redistribute considerable amounts of SOM within a toposequence or a field which drastically alter the mineralization process in landscapes. SOM buried on deposition sides is withdrawn from the active C and N pool. Whereas erosion and deposition only redistribute SOM, mineralization results in a net loss of C from the soil system to the atmosphere. The decrease of SOM content on the eroded sites affects soil quality in a negative way. The permanent erodic output of Ap material induces SOM dilution in cultivated soils. The SOM content decreases due to annual plowing at constant depth and mixing with underlying, C-poor subsoil material. Erosion results in decreased primary productivity, which in turn adversely affects SOM storage because of the reduced quantity of organic C returned to the soil as plant residues. The annual SOM supply in the eroded soils are roots, crop residues and organic manures containing proteins, lipids, polysaccharides and lignins (Beyer et al., 1999). This causes a relative enrichment of litter compounds in eroded soils. In contrast, the Ap of colluvic soils (FAO, 1998a: Colluvi-cumulic Anthrosol) receive SOM from the annual supply of the eroded topsoil materials as well. These are rich in humus and lead to a dominance of humic compounds (Beyer et al., 1999). In addition, SOM in the Ap of colluvic soils is not diluted by tillage because of the colluvic materials underlying the Ap. Colluvic soils usually contain a larger proportion of SOM in labile fractions because this material can be easily transported. If the accumulation of soil material in depositional areas is extensive, the net result of the burial of this active SOM would be increased SOM storage because decomposition is substantially slowed.

Erosion and deposition of soil material have significant influences on soil properties, processes and fertility (Lal, 2004b). Not only the areas where soil is removed are affected by these changes but also the deposition sites. Particularly areas of the tropics are prone to water erosion due to the high intensity rainfall, and C displaced by erosion is estimated to be about 1.6 Pg C $year^{-1}$ for the tropics as a whole (Lal, 1995b). Annual C transport from tropical soils to oceans may make 0.16 Pg C or about 10% of the total C displaced. Soil erosion by water may be an important source of atmospheric CO_2. Lal (2004b) estimated that on a global scale about 1.14 Pg of C may be emitted annually into the atmosphere through erosion induced processes.

6.3.1.1 Soil Erosion and Depletion of SOC

Due to its low weight, SOM is especially susceptible to transport, thus it is among the first constituents to be removed. Bouwman (1990) reported that SOM was five times higher in eroded material than in the original soil before erosion. Together with mineralization, erosion can locally be an important cause of SOM decline in cropping systems. This is especially so on sites with poor soil cover, steep slopes and erosive rain conditions (Roose & Barthes, 2001).

Knowledge of the impact of erosive processes on SOC dynamics, and understanding the fate of C translocated by erosive processes is crucial to assessing the role of erosion on emissions of GHGs into the atmosphere. Severely eroded soils may have lost one-half to two-thirds of their original C pool (Lal, 2004b). Loss of SOC is higher in soils with higher initial C pools, and higher in the tropics than in temperate regions. Preliminary results from runoff plots on hill slopes in West Africa indicated that C losses by erosion and leaching ranged between 10 and 100 kg C ha^{-1} $year^{-1}$, depending on annual rainfall and vegetation cover (Roose & Barthes, 2001). According to Bouwman (1990) runoff erosion may cause soil losses of 5–10 Mg ha^{-1} $year^{-1}$. An average SOC content of 2% would result in 100–200 kg C loss ha^{-1} $year^{-1}$.

There is a growing recognition that soil erosion and deposition play an important role in the C cycle, from site to global scales. However, it is still uncertain if erosion creates an atmospheric CO_2 sink or source. There are two competing theories about the impacts of erosion on the availability of C (Izaurralde et al., 2007). Some authors assume that C from eroded fields is sequestered or stored in depressions (e.g. Stallard 1998; Renwick et al., 2004). This C sequestration through burying of SOC withdraws C from active cycling and renders it unavailable for release as CO_2. On the other hand, erosion events cause aggregate breakdown of physically protected C, thus making it accessible for oxidation and emission of CO_2 (Izaurralde et al., 2006). Other authors, therefore, assumed an additional release of CO_2 through erosion and deposition. Liu et al. (2003), for example, reported about significant CO_2 release from deposition sites. Several experiments have shown on-site depletion of the SOC pool by accelerated erosion (Lal, 2004b). However, on-site depletion does not necessarily imply emission of GHGs into the atmosphere. The fate of the eroded SOC depends on the circumstances of deposition. Soil organic carbon may be partly transported into aquatic ecosystems and depressions, where it may be mineralized and released as CO_2. To another part, SOC may be buried and sequestered. However, mineralization of organic matter during transport is not the only process that should be considered in quantifying the impacts of soil erosion and deposition on SOC. The gradual exposure of C-poor subsoil at eroding sites leads to continuous C sequestration through root and litter input into the soil, whereas the C that is buried at depositional sites may be only slowly mineralized. Thus, soil erosion and deposition may lead to C sequestration at the watershed to global scales, possibly explaining part of the missing sink of atmospheric CO_2.

Liu et al. (2003) developed a general ecosystem model to simulate the influences of rainfall-induced soil erosion and deposition on SOC dynamics in soils of Mississippi including ridge top (without erosion or deposition), eroding hill slopes, and depositional sites that had been converted from native forests to croplands in 1870. Changes in SOC storages were compared to a control site. The SOC storage was reduced at the eroded sites, while the SOC storage increased at the depositional areas. In the long term, the results indicated that soils were consistently C sources from 1870 to 1950. The source strength was lowest at the eroded sites (13–24 g C m^{-2} $year^{-1}$), intermediate at the ridge top (34 g C m^{-2} $year^{-1}$), and highest at the depositional sites (42–49 g C m^{-2} $year^{-1}$). During the observed period, C emissions were reduced via dynamically replacing surface soil with subsurface soil. The subsurface soil showed lower SOC contents (quantity change) and higher passive SOC fractions (quality change). From 1950 to 1997 soils at all landscape positions became C sinks due to changes in management practices (e.g. intensification of fertilization and crop genetic improvement). The sink strengths were highest at the eroding sites (42–44 g C m^{-2} $year^{-1}$), intermediate at the ridge top (35 g C m^{-2} $year^{-1}$), and lowest at the depositional sites (26–29 g C m^{-2} $year^{-1}$). The enhanced C uptake at the eroded sites was attributed to the continuous SOC loss through erosion and replenishment with enhanced plant residue input. Overall, soil erosion and deposition reduced CO_2 emissions by exposing C-poor soil at the eroded sites and by burying SOC at depositional sites (Liu et al., 2003). The results suggest that failing to account for the impact of soil erosion and deposition may potentially contribute to an overestimation of both the total historical C released from soils owing to land use change and the contemporary C sequestration rates at the eroding sites.

6.3.2 Mine Spoil Reclamation

Man's search for mineral resources leads to severe impacts on the land surface. Open cast, i.e. surface mining activities result in a drastic disturbance of large land areas, even entire landscapes. A common characteristic of reclamation areas is the lack of vegetation and SOM. Since production of SOM and its decomposition is considered a key component for carbon and nutrient cycling in terrestrial ecosystems, the course of SOM development has received considerable attention in reclamation and restoration research (Waschkies & Huettl, 1999; Rumpel et al., 1999; Wali, 1999).

Chemical properties of spoils provide clear indications of changes both over time and within the spoil by depth. In naturally revegetated 45 year old chronosequences located in the mixed grass prairie of North Dakota (USA), organic C showed a rate of increase of 131 kg C ha^{-1} $year^{-1}$ (Wali, 1999). This accumulation rate was about one half that reported for mine spoils in Saskatchewan (Canada; 282 kg C ha^{-1} $year^{-1}$) by Anderson (1977) and for eastern Montana (USA; 256 kg C ha^{-1} $year^{-1}$) by Schafer & Nielsen (1979). In the study by Wali (1999), incremental rates of N accumulation between successive age groups were calculated between 1 and 17 years (rate of accumulation: 24 kg N ha^{-1} $year^{-1}$), between 17 and 30 years (36 kg N ha^{-1} $year^{-1}$) and between 30 and 45 years (16 kg N ha^{-1} $year^{-1}$). If these N accumulation rates were linearly extrapolated into N values at unmined areas, the mined sites would take nearly 240 years to reach N equivalence with the unmined sites. As these data show, the rate of N accumulation decreases with time, and the process is likely to take much longer. C:N ratios indicate a trend toward rehabilitation of mined soils. In young spoil materials, C:N ratios of SOM are high due to production of organic material by pioneer plant communities coupled with a delay in decomposition. When C:N ratios are high, nearly all mineralized N will be used by microorganisms and no inorganic N will accumulate. The lack of mineral N in spoils may cause the successional stagnation of some plant communities. Considerable moderation of C:N ratios takes place only after relatively long time periods, i.e. decades. In the 1 year old sites of the 45 year study (Wali, 1999), C:N values showed a wide range (5–40), but 70% of the 45 year old sites showed values below 15, comparable to unmined sites.

In the lower Lusatian mining district (eastern part of Germany), topsoiling, i.e. spreading of former soil material on the surface of reclamation sites, is not practised in the reclamation process as natural soils (coarse pleistocene soils) are mostly of poor quality. In mine spoils originating from Tertiary strata of the overburden sequence, the percentage of geogenic organic C ranges from 0.5% to more than 5.0% in highly carboniferous substrates (Haubold et al., 1998). Large quantities of this organic C are in inert forms that are resistant to microbial utilization. After amelioration with lignite-derived ash and NPK fertilizer, the spoils were reforested with coniferous and deciduous trees. A chronosequence of young mine soils under planted pine forest yielded high C accumulation rates. At 11 and 17 year old Scots pine sites, carbon accumulated mainly in the forest floor (L and Of horizon), whereas in the forest soil of the oldest, 32 years site, most of the carbon was accumulated in the upper mineral horizon (Ai horizon). Carbon accumulation measurements showed a cumulative C production of 50.2 t over 32 years (Rumpel et al., 1999) corresponding to a mean rate of increase of 1,570 kg C ha^{-1} $year^{-1}$. Similar C accumulation was observed in natural pine forest soils of the Lusatian mining district. The degree of humification of 32 year old spoil sites was at the same level as in natural forest soils.

In the Rhineland lignite mining area (western part of Germany), soils reclaimed for agriculture are mainly loess-derived and very poor in SOM contents (<0.5% SOM) due to their high raw loess content. The process of SOM accumulation begins rapidly with cultivation of the reclaimed loess soils. The dimensions and patterns of the SOM accumulation process are essentially influenced by the addition of organic fertilizers. The oldest SOM accumulation experiment in that area began in 1969. Measurements until 1995 on different fertilizer treatments on a basis of 30 cm plow horizon yielded a C accumulation of 13.3 Mg ha^{-1} (crop residues removed), 16 Mg ha^{-1} (crop residues not removed), 25.1 Mg ha^{-1} (3.6 Mg ha^{-1} $year^{-1}$ manure), 24.9 Mg ha^{-1} (3.7 Mg ha^{-1} $year^{-1}$ sewage sludge), 24.1 Mg ha^{-1} (3.7 Mg ha^{-1} $year^{-1}$ waste compost), 31.4 Mg ha^{-1} (7.2 Mg ha^{-1} $year^{-1}$ waste compost), 18.3 Mg ha^{-1} (3.7 Mg ha^{-1} $year^{-1}$ straw), and 23.8 Mg ha^{-1} (3.8 Mg ha^{-1} $year^{-1}$ peat; calculated from Delschen, 1999). The fertilizer type appears to be of secondary importance. More important to the SOM accumulation is the amount of organic fertilizer and the organic matter contained in it.

Depending on the treatment, accumulation rates were between 0.49 and 1.16 Mg C ha^{-1} $year^{-1}$ with values decreasing with time (Delschen, 1999). From these results, it is estimated that reclaimed soils receiving organic fertilizers will take much longer than the previously assumed 25 years. It is important to determine whether the young SOM of reclaimed soils will have properties similar to those of older SOM in undisturbed topsoils.

6.3.3 Salinization

Salt and alkali-affected soils worldwide cover about 427 million hectares, including salt marsh soils, inland desert soils and agricultural soils that have undergone secondary salinization due to irrigation practices (Szabolcs, 1989). In comparison with many non-saline systems, they are depleted in SOM because of a reduction in vegetation density. Attempts are being made to convert or reclaim some of these soils for agricultural production of halophyte crops using saline water (including seawater) for irrigation (Olson et al., 1996). Many of these soils support halophyte communities with high rates of primary production (Le Houerou, 1993). With supplemental or tidal irrigation, halophytes may yield as large a biomass as many agronomic crops (O'Leary et al., 1985).

In saline systems, organic matter decomposition rates may be lower and potential soil carbon storage higher than in fresh water systems. In a high saline environment, C losses from the soil via microbial respiration and leaching may be much less than from systems with lower salinity. Reduction in soil microbial respiration due to increased salinity has been reported for various salts (Malik & Haider, 1977). Saline soils could store significant quantities of C particularly if halophyte crop residues were plowed annually into the soil (Glenn et al., 1993). It has been suggested that by reclamation of the 35 million hectares of salt-affected wasteland in India, up to 2 Pg soil organic carbon could be sequestered (Gupta & Rao, 1994).

6.3.4 Soil Acidification

In humid regions, soils become acid as precipitation exceeds evapotranspiraion. "Natural" rainfall is acidic (pH ~ 5.6) and adds H_2CO_3 to soils. This acidification results in a gradual leaching of exchangeable basic cations, such as Ca^{++} and Mg^{++}

from soils. Weathering of silicates results in a release of Al^{3+}, which replaces the basic cations at the exchange complex. Exchangeable Al^{3+} is in equilibrium with soil solution Al^{3+}, which can react with H_2O to produce protons, which further acidify the soil. Most of the acidity in soils between 4.0 and 5.5 is due to hydrolysis of Al^{3+}. Other natural processes that contribute to soil acidification include soil respiration to produce CO_2 which can react with water to form H_2CO_3, and mineralization of nitrogen with subsequent nitrification of NH_4^+ to NO_3^-. Soil acidification can also be caused by acidic deposition, fertilization, leaching, fallowing, increase in organic matter and N contents, and deforestation (Nilsson, 2003). Acidic deposition is known to influence the growth of plants. Direct impacts concern the aboveground parts of a plant, whereas indirect impacts are a consequence of a decrease in soil quality due to acidification. Acid soils are characterized by high exchangeable Al, multiple nutrient deficiencies (e.g. P, Ca, Mg, and Zn), and high P fixation (Prihar et al., 2000).

Increased acidic deposition has been a major research focus for the last 30 years. It results in the addition of strong acids to soils. The primary origin is from the emission of sulfur dioxide (SO_2) and N oxides (NO_x). Typical sources of acidic deposition include coal-, gas- and oil-burning power plants, the transport sector, industrial operations (e.g. cement production and smelters), and land use. When nitrogen and sulfur gases enter the atmosphere, they react with water to nitric (HNO_3) and sulfuric (H_2SO_4) acids. In North America and Europe, pH of rain averaged 4.0, and in extreme cases, it was as low as 2.4 (Likens et al., 1979). Up to now, increased N and S depositions have mainly been considered a problem of the industrialized countries of Europe and North America, but the problem of high N and S depositions is rapidly extending to other parts of the world such as East Asia (Kuylenstierna et al., 1995).

The quality of the acidic deposition, i.e. the relative percentages of HNO_3 and H_2SO_4, is controlled by the spatial distribution of the sources of NO_x and SO_2. While S depositions in Europe and North America decreased in the 1990s, inputs of NH_4^+ and NO_3^- remained unchanged or even increased (Matzner & Murach, 1995). Ammonium, which is mainly produced in livestock farming regions causes further acidification through the subsequent nitrification process which commonly occurs in soil. In some regions of the world, emissions of NH_y-N exceed NO_x-N emissions. For example, in Germany, about 60% of the gaseous N emissions originate from agriculture (mainly NH_y-N; Umweltbundesamt, 1999). The transport and energy sectors are sources for another 40% (mainly NO_x-N). The soil acidification beyond that produced by natural processes depends on the buffering capacity of the soil and on the type of management.

Until the 1950s, nitrogen acted as a growth-limiting element in most coniferous forests of the Northern Hemisphere (Berg & Verhoef, 1998). Since about 5 decades, northern forests (i.e. forests of the Boreal and the Temperate Zones) are subjected to increased acidic deposition (Nilsson, 2003). Compared to other land uses, forests additionally take up N and S via forest canopy. Northern forest soils are believed to be efficient at retaining deposited nitrogen, and up to now, only certain forest types show significant losses of nitrate to streams or groundwater (Dise et al., 1998)

Nitrogen deposition in many forests of the temperate climate may amount to 20–80 kg N ha^{-1} year^{-1} (Isermann & Isermann, 1995). While the nitrate N deposition rates are distributed relatively homogeneously over temperate regions, a great proportion of the NH_y-N is deposited in the vicinity (up to several kilometers) of its source. In regions with a high livestock density like in North-West Germany and the Netherlands, agriculture may contribute to >80% of the total N deposition (Barunke, 2002). In forests that were in vicinity of livestock production farms, Heinsdorf and Krauss (1991) observed total N depositions that exceeded 100 kg ha^{-1} year^{-1}. In Germany, at present the "critical" N input to forest ecosystems of about 10 kg ha^{-1} year^{-1} (Nagel & Gregor, 1999) is exceeded on 99.7% of the whole forest area (Umweltbundesamt, 1999). Besides increased N and S deposition, rising atmospheric carbon dioxide (CO_2) concentration and associated global warming during the last 50 years could also impact forest ecosystems (Scharpenseel et al., 1990; Houghton et al., 1995; van Breemen et al., 1998).

The most visible issue related to acidic deposition in forests has been widespread forest decline (German expression: "Waldsterben") in Europe and in North America. However, evidence that forest decline is caused solely by acidic deposition is lacking and complicated by the interactions between acidification and other environmental or biotic factors that influence growth of trees. Similar to agricultural systems, elevated concentrations of O_3 can also cause damage to forest vegetation. In Central European forest soils, soil pH has decreased since the 1950s by up to 2 pH and by 0.5 pH on average (Nieder et al., 2000). Long-term acidification of forest soils has lead to the liberation of ionic aluminum from the soil minerals into the soil solution. The presence of aluminum in the soil solution leads to an inhibition of the repolymerization of organic substances in the humus cycle, while the breakdown is not affected. This process leads to a long-term increase in DOM concentration and an accumulation of inorganic nitrogen in the mineral soil (Eichhorn & Huettermann, 1999). In contrast, decrease in soil pH on many sites causes an increase in the thickness of the forest floor. This is because the roots avoid to penetrate the acid mineral soil and organic residues are slower decomposed on the soil surface than in the soil. Long-term experiments (1966–1995) in mature forests of the Solling mountains have shown that the SOM pool in the forest floor on average has increased by 700 kg C ha^{-1} year^{-1} under a 150 year old beech forest and by 1,400 kg C ha^{-1} year^{-1} under a 110 year old spruce forest (Meesenburg et al., 1999). Elevated N inputs to forest may also enhance the accumulation of C and N in SOM through increased biomass production (Aber et al., 1998).

Short-term acidification can occur as a consequence of climate influences. For example, the temperate zone and the boreal zone are characterized by a cool and wet climate, where temperature is a factor that limits microbial activity and mineralization in soils. A dry and warm year influences an ecosystem not only by drought, but also by increased mineralization. If the amount of nitrate that is formed by N mineralization cannot be taken up by plants, the result will be a temporary accumulation of nitric acid. These climatically induced changes of soil acidity as a result of a temporary lack of equilibrium between the metabolism of the soil microorganisms and plant roots were first postulated by Ulrich (1980). They were confirmed by Murach (1983) on the basis of data collected in spruce forest soils of the Solling mountains (Northern Germany) in the cool and wet summer of 1981 and the warm and dry summer of 1982. During the summer of 1981, the mass of living roots exceeded the root necromass. This relation was changed considerably during 1982. A steady increase in nitrate was observed in the soil solution, accompanied by a steady decrease of pH. This acidification push resulted in a significant change in the relationship of living roots and dead fine roots. In the Fichtelgebirge (Southern Germany), Guggenberger & Beudert (1989) observed that the DOM concentrations below the forest floor undergo high seasonal variations. The highest concentrations of DOM were found when periods of drought were followed by heavy rains. Besides N mineralization, the C mineralization push caused an increase in DOM and a pronounced decrease in pH (production of CO_2 with subsequent reaction in soil solution to H_2CO_3).

In contrast to forests, agricultural impacts related to acidic deposition are less of concern because of commonly high buffering capacities for these ecosystems. High atmospheric ozone concentration is often more responsible for crop damage than the presence of acid substances. Acid rain contains N and S which are important plant nutrients. Artificial foliar application of acid rain to some crops (e.g. alfalfa, tomato, corn, lettuce) at critical growth stages was shown to be beneficial (Pierzynski et al., 2000). Negative responses to acid rain have been identified with broccoli, carrots, mustard and radish. Agricultural lands with silty, loamy and clay-rich textures are commonly maintained at pH levels between 6.0 and 7.0. Soils with coarser texture such as sandy soils and organic soils are maintained at pH 5.0–5.5. Acidity in agricultural soils is commonly neutralized by application of limestone ($CaCO_3$) or dolomite ($CaMg[CO_3]_2$). Further management options include selection of crop and pasture species, crop sowing time, crop varieties, and stock management. Due to the reduction in atmospheric S depositions in Europe and North America since the 1980s, fertilizer S consumption has grown. Atmospheric N depositions in some areas and for some agricultural management systems may be a significant source compared to the 100–300 kg N ha^{-1} $year^{-1}$ required by agricultural crops.

Nitrification of NH_4^+ containing fertilizers produces H^+ ions that decrease soil pH. Leaching of NO_3^- from the root zone promotes acidification by uncoupling the proton balancing system. The acidifying effect of fertilizers follows the order ammonium sulfate > ammonium nitrate > anhydrous ammonia > urea > calcium nitrate (Bouman et al., 1995). The degree of acidity caused by a fertilizer is modified by soil characteristics, cropping systems, and environmental variables. Fertilization may also cause acidification by the export of basic cations (Bolan et al., 1991). Acidification is accelerated when the harvested crop removes an excess of basic cations (Ca^{2+}, Mg^{2+}, K^+, Na^+) over anions (Cl^-, SO_4^{2-}, NO_3^-). In winter rainfall regions, such as southern Australia, soil acidification is associated with extended periods of legume pasture leys (Helyar & Porter, 1989). Substantial increases in SOM and total soil N contents are believed to cause acidification by increasing levels of carboxylates as well as leaching during summer when autumn and winter active pastures are dormant. In pasture-cropping systems, where little or no N fertilizer is applied, acidification may occur during the N building phase.

6.3.4.1 Impact of Acidification on Soil C and N in Forest Ecosystems

Local transformations of humus forms and modifications of important soil functional processes have been described as a consequence of atmospheric depositions of acids and nutrients (especially N and S species) (Nilsson, 2003), forest liming (Deleporte & Tillier, 1999), forest transformation (Fischer et al., 2002), clear-cutting and afforestation (Pastor & Post, 1986). Deposition of acids and nitrogen in wide areas led to changes in humus forms during the last few decades (Beyer, 1996b). Global increases in temperature are also discussed as a possible reason for changes in humus forms (Scharpenseel et al., 1990). According to studies conducted in numerous old-growth forests of Central Germany during the 1970s (von Zezschwitz, 1976, 1980) it was still possible to distinguish between humus forms (Mull, Moder, Mor and their transient humus forms) according to C:N ratios of the organic layer materials. "Standard" C:N ratios were published by Arbeitskreis Standortskartierung (1996) referring to studies which were conducted about 20 years ago. They range from C:N 29–38 for Mor, 25–31 for Moder, and 14–17 for Mull. In forest floors of the North German lowlands, present mean values for Mor (28) and Moder (25) are lower than the cited "standard" C:N ratios (Nieder, 2004). This indicates a relative increase in the concentration of total N in these forest floors. Compared to the organic layers of coniferous forests, the amounts of organic matter stored in the underlying upper mineral soil (Ah horizon) change much less with time (Böttcher & Springob, 2001; Romanya et al., 2000).

Due to the decrease in C:N ratio, proton concentrations in soils have further increased (von Zezschwitz, 1985). In contrast to high available N contents in forest soils, available Ca, Mg and K concentrations have become deficient (Beyer, 1996b). The influence of N depositions on the morphology of humus forms is discussed controversially. Belotti (1989) and Belotti & Babel (1993) detected no changes in humus forms due to increased depositions of acids and nitrogen in forests of southern Germany. In contrast, Beyer (1996b) in North Germany (luvisol on glacial till under old-growth beech) observed a transformation from Mull to Moder within only 25 years which was drawn back to increased soil acidification. The pH in the A horizon within that time period decreased from 4.0 to 3.2, and the base saturation from 40% to 13%. On pleistocene sands of the North German lowlands, the typical humus form of Podzols under >60 year old pine (*Pinus sylvestris*) stands for long time periods was Mor (Hofmann, 1997). In many parts of this region, Moder nowadays is the dominating humus form which is a consequence of elevated N depositions (Nieder, 2004). According to the latter study, the transformation of Mor to Moder has drastic consequences for the C and N dynamics of the forest ecosystem, because in mature pine stands of this region, the organic layers of Moder (~50 Mg C ha^{-1} and 2.2 Mg N ha^{-1}) store significantly less carbon and nitrogen as compared to Mor (up to 90 Mg C ha^{-1} and 3.2 Mg N ha^{-1}). However, the lack of historic data (Scholten, 1990) and the partly high spatial variability in humus forms within even one forest stand (Belotti, 1989) makes it difficult to detect long-term changes in humus morphology and C and N dynamics, particularly on a large scale.

Chapter 7
Leaching Losses and Groundwater Pollution

Large amounts of nitrogen fertilizers and organic manures are added to soil under intensive agriculture, but their use efficiency is generally low and varies greatly under different cropping and ecosystems. The unutilized N may accumulate in the soil profile and become a contaminant in streams or ground water. Nitrate drained into surface water bodies, e.g. rivers, lakes, or estuaries, can cause deterioration of surface water quality, resulting in eutrophication, algal bloom, and fish poisoning. High concentrations of NO_3^- in drinking water is deemed harmful to human health. As per World Health Organization (WHO) standards, groundwater having more than 10 mg NO_3^--N L^{-1} is unfit for drinking. Reports from some developed countries show that the critical limit has been exceeded in a significant proportion of water samples. For example, in the US, nitrate levels are higher than 10 mg N L^{-1} in approximately 20% of wells in farmland areas, between 2–10 mg L^{-1} in 35% of wells, and below 2 mg L^{-1} in only 40% of wells (Galloway et al., 2004). Indications of increasing NO_3^--N concentration in water are also emanating from countries with emerging economies, particularly China. Results of the studies from North China with intensive vegetable production showed that in half of the 110 locations investigated, nitrate contents in ground and drinking water exceeded the critical WHO value for drinking water (Zhang et al., 1998). The bulk of the nitrate comes from mineral fertilizers and manure applied to crops and grasslands. Further intensification of fertilizer use may aggravate the problem. Besides NO_3^--N, dissolved organic carbon (DOC) and nitrogen (DON) are important constituents of the soil solution. Export of DOC through leaching is being implicated in the loss of organic matter from soils and transport of DON from surface soils to groundwater and streams is of concern from a nitrogen balance and ecological point of view. Estimates of the role of DOC in terrestrial carbon balance are generally based on river DOC fluxes that range from 1 to 10 g C m^{-2} $year^{-1}$. Although these fluxes are small compared to primary productivity and heterotrophic respiration, but the production and transport of DOC influences many biological and chemical processes in soils, and transfer of nutrients from terrestrial to aquatic ecosystems. On regional and global scales, examination of the retention and turnover of DOC is useful in characterizing and quantifying the C storage capacity of soils. In this chapter we examine nitrate leaching losses in soils and the concentration and fluxes of DOC and DON in different ecosystems. The influence of environmental conditions,

R. Nieder, D.K. Benbi, *Carbon and Nitrogen in the Terrestrial Environment,*

agricultural land use, and soil factors on NO_3-N leaching and the management strategies for mitigating the problem are also discussed.

7.1 Dissolved Organic Carbon

Dissolved organic matter (DOM), which includes dissolved organic forms of carbon and nitrogen comprises a continuum of organic molecules of different sizes and structures that pass through a filter of 0.45 ∝m pore size. It represents the most reactive and mobile form of organic matter in soils and plays an important role in the biogeochemistry of C, N, and P, the transport of soil pollutants and in the pedogenesis. It is mainly composed of high molecular weight complex humic substances. Only small proportions of DOM, mostly low molecular weight substances such as organic acids, sugars, and amino acids can be identified chemically (Herbert & Bertsch, 1995). Fractionation and structural analysis of DOM in soil solutions have shown that microbial metabolites constitute a significant proportion of DOM (Kalbitz et al., 2000). Fungi are the most important agents in the process of DOM production, probably because of incomplete degradation of organic matter by this group of the decomposer community. The carbohydrate fraction of DOM is chemically different from that of plant residues or bulk humus in that DOM carbohydrates have a higher proportion of hexose- and deoxy-sugars than pentose sugars. Dissolved organic matter from agricultural soils has a higher proportion of hydrophobic compounds when compared with extracts from grassland and forest soils (Raber et al., 1998).

The main source of DOC is plant residues accumulated in the uppermost soil layer, although the mineral horizon is also thought to produce dissolved carbon (Kaiser et al., 1997). Dissolved organic carbon may be transported to groundwater or surface waters, utilized by microbes or retained in the soil by abiotic mechanisms. The bioavailability of DOC depends on its origin and chemical characteristics. The labile fraction of DOC in soil solution is easily decomposable, whereas recalcitrant C such as humic substances is biologically inert (Yano et al., 1998). The flux of DOC in soil facilitates transport of nutrients and contaminants in soil. Generally, DOC concentrations in soil solution decline with depth in mineral soils as a result of DOC retention by soil surfaces. Qualls & Haines (1992) suggested that abiotic retention of DOC via adsorption to soil surfaces was primarily responsible for reduction in DOC concentrations; however decomposers may facilitate adsorption processes by removing organic compounds held on soil exchange complex, thereby opening more sites for additional adsorption. Laboratory studies (McCracken et al., 2002) show that microbial decomposition is a significant factor regulating organic carbon concentrations in soils.

The soluble fluxes of organic compounds from throughfall and out of the litter layer can amount to 1–19% of the total litterfall carbon flux and 1–5% of net primary production (Gosz et al., 1973; McDowell & Likens, 1988; Qualls et al., 1991). A review of the published studies across a range of forest ecosystems (Neff & Asner, 2001) showed that surface soil fluxes of DOC range from 10 to 85 g C m^{-2} $year^{-1}$ and these decline to 2–40 g C m^{-2} $year^{-1}$ below the surface horizons. DOC fluxes vary from 1 to 10 g C m^{-2} $year^{-1}$ in streams, but a few substantially higher fluxes can occur in drainages containing sandy or highly organic soils (Moore & Jackson 1989; Hope et al., 1994). Mean annual concentration of DOC in temperate forest ecosystems of North America and Europe have been reported to range between 3 and 35 mg L^{-1} in throughfall solution, 20–90 mg L^{-1} in solutions of humic layer (Oa; L according to AG Boden, 2005), and 2–35 mg L^{-1} in B horizons (mineral soil) (Michalzik et al., 2001). The DOC fluxes, which are largest in humic layers range from 100 to 400 kg ha^{-1} $year^{-1}$. The fluxes with throughfall and seepage fluxes in the B horizon are relatively small and range from 40 to 160 and 10 to 200 kg DOC ha^{-1} $year^{-1}$, respectively (Fig. 7.1). As is evident from the figure concentrations of DOC decrease from the A horizon to the B horizon, whereas fluxes of DOC decrease

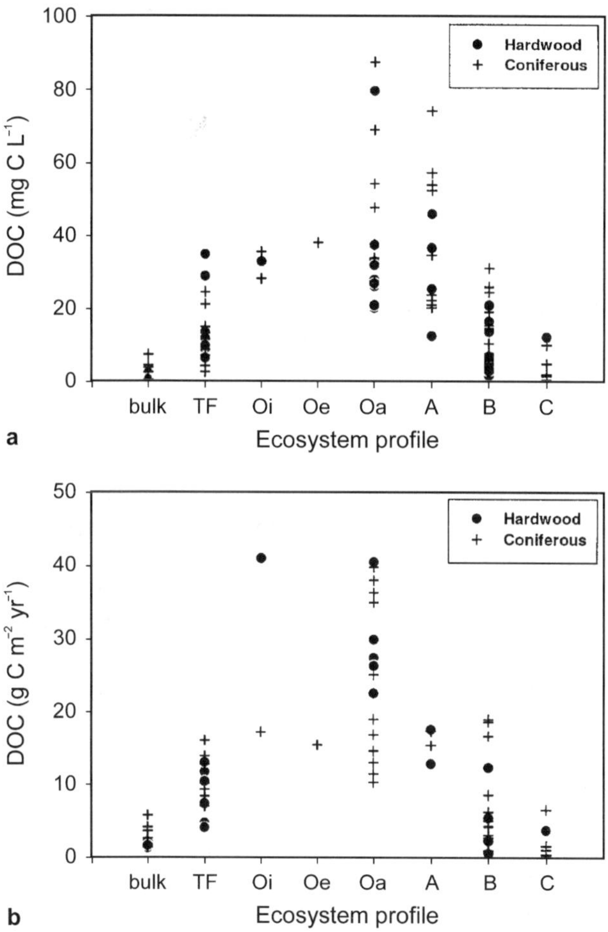

Fig. 7.1 (a) Mean annual concentration of DOC and (b) annual fluxes of DOC along a vertical profile in forest ecosystems [bulk: bulk precipitation; TF: throughfall precipitation; Oi: litter; Oe: fermented; Oa: humic layer (L, Of and Oh, respectively according to AG Boden, 2005); A, B and C: horizons of the mineral soil] (Michalzik et al., 2001; p. 187. Reproduced with kind permission from Springer)

rapidly from the forest floor to the A horizon. Water flux and velocity have been found to be the important parameters influencing concentrations and fluxes of DOC (Mertens et al., 2007). Khomutova et al. (2000) studied the mobilization of organic carbon in undisturbed soil monoliths of a deciduous forest, a pine plantation, and a pasture. After 20 weeks of leaching, the amounts of DOC removed constituted 6.4%, 3.8% and 6.2% of initial soil organic carbon in soil monoliths of deciduous forest, pasture and coniferous forest, respectively. Cumulative values of DOC production decreased in the sequence coniferous forest > deciduous forest > pasture. The variability in DOC concentration and flux measurements apart from depending on the type of ecosystems also depends on the measurement and sample collection method, each of which may yield different results.

Dynamics of DOC and DON appear to be controlled by biological, chemical and physical processes, which interact by antagonistic and synergistic mechanisms. However, the relevance of each process for the description of DOM dynamics under field conditions is unclear (Kalbitz et al., 2000). Results of laboratory studies indicate that the release of DOC from forest floors generally increased with temperature and soil moisture, decreasing ionic strength, increasing sulfate concentrations, increasing C:N ratio of the solid phase, increasing leaching frequency and decreasing metal saturation of DOC (Michalzik et al., 2001). The results with regard to the influence of pH are inconsistent. While some studies reported a positive relationship between the release of DOC and increasing pH of the extraction solution (Chang & Alexander, 1984; Vance & David, 1989), others (Cronan, 1985) reported no difference in the amounts of mobilized DOC within a pH range between 3.5 and 5.7. The effect of most of these factors is yet to be confirmed in the field.

Results from modeling studies indicate the importance of representing both root carbon inputs and soluble carbon fluxes to predict the quantity and distribution of soil carbon in soil layers. For a test case in a temperate forest, DOC contributed 25% of the total soil profile carbon, whereas roots provided the remainder. The analysis showed that physical factors- most notably, sorption dynamics and hydrology play the dominant role in regulating DOC losses from terrestrial ecosystems but that interactions between hydrology and microbial-DOC relationships are important in regulating the fluxes of DOC in the litter and surface soil horizons (Neff & Asner, 2001).

Most of the published studies on leaching of DOC have been reported from forested ecosystems and there is very little information on the loss of DOC and DON under cropped soils and grazed pastures. Forested ecosystems apparently support larger DOC fluxes than grazed pastures (Ghani et al., 2006). The quantity of DOC leached from grazed pastoral soils will depend on inputs from sources including animal (urine and faeces), pasture (grass and clover) residues, fertilizer and native organic matter. Ghani et al. (2006) examined effects of these inputs on the leaching of DON and DOC from soils using intact soil cores containing resident perennial grass/clover pasture. Root senescence (caused by glyphosate application) resulted in greatest leaching of DOC (15.2 kg C ha^{-1}) followed by urine application (10.5 kg C ha^{-1}). Application of dung, grass or clover litter and fertilizer N did not influence significantly the leaching of DOC, which ranged from 5.9 to 7.4 kg C ha^{-1}. Under rice agriculture, Lu et al. (2000b) found that DOC in the root zone increased with plant growth reaching maximum between rice flowering and maturation. Their results suggested that DOC in the root zone of rice plants is enriched by root-derived C and the rates of CH_4 emission are positively correlated with the dynamics of DOC in the root zone. Not only the quantity but also the quality of DOC pool differed between root zone and non-root zone soils with non-root zone being more recalcitrant and stable as compared to root-zone DOC (Lu et al., 2000)

7.2 Dissolved Organic Nitrogen

The existence of soluble organic forms of N in rain and drainage waters has been known for many years, but these have not been generally regarded as significant pools of N in agricultural soils (Murphy et al., 2000). This is probably because of the difficulties with measurement of dissolved organic nitrogen (DON). With the availability of advanced analytical techniques, studies conducted in the last over a decade indicate that DON represents a significant pool of soluble N in soils and there is a need to include it in ecosystems budgets and N cycling studies. DON is usually composed of a wide range of compounds ranging from low molecular weight (LMW) amino acids and amino sugars to high molecular weight (HMW) polyphenol-bound N (Stevenson, 1982; Antia et al., 1991). In arable soils, free amino acids only make up 3% of DON, amino sugars and heterocyclic-N bases, on average 15%; the remainder of the hydrolyzable fraction of soluble organic nitrogen is present in amino compounds (Murphy et al., 2000). Jones et al. (2004) hypothesized that there are two distinct DON pools in soil. The first pool comprises mainly free amino acids and proteins and is turned over very rapidly by the microbial community, so it does not accumulate in the soil. The second pool is a high molecular weight pool rich in humic substances, which turns over slowly and represents the major DON losses to freshwaters. The LMW pool may directly regulate the rate of ammonification and nitrification in soil as it provides the initial substrate for these N transformation pathways. The influence of HMW may be indirect, through nonspecific inhibition of enzymes such as proteases.

Large pools of DON have been measured in leachate from forest floors and it is recognized as a major contributor of nitrogen to surface water in forested watersheds. Even in areas with large anthropogenic inputs of dissolved inorganic nitrogen, DON constitutes the majority of total dissolved N in stream exports (Campbell et al., 2000b). Qualls et al. (1991) observed that 94% of the dissolved N leaching through a deciduous forest soils was present in organic form. Similarly, Yu et al. (2002) observed that DON accounted for 77–99% of the total dissolved nitrogen in Oa horizon leachates of forest soils. Proteins and peptides were the main contributor to DON in Oa horizon leachates and combined amino acids released by acid hydrolysis accounted for 59–74% of the DON. Most of the DON was found in the hydrophobic fraction, which suggests the presence of protein/peptide-polyphenol complexes or amino compounds associated with humic substances.

In relation to the input of C and N to soil by annual above-ground litter fall, the annual transport of DOC and DON from the forest floor into the mineral soil amounts to an average of 17% (range 6–30%) of the annual litter input of C and to 26% (range 1–53%) of the litter N input (Michalzik et al., 2001). Throughfall is an important source of DON to the forest floor and the fluxes from the forest floor into the mineral soil are largely dependent on the water flux (Solinger et al., 2001). A survey of the published studies (Michalzik et al., 2001) on concentration and fluxes of DON in forest ecosystems of the temperate zone shows that mean annual concentration of DON range from 0.25–1.11 mg L^{-1} in throughfall, from 0.4 to 2.45 mg L^{-1} in the forest floor and from 0.2 to 1.1 mg L^{-1} in the mineral soil horizons. Fluxes of DON with throughfall range from 1.2 to 11.5 kg ha^{-1} $year^{-1}$. In the Oa layer and the B horizon, DON fluxes vary considerably between 0.2 and 18 kg ha^{-1} $year^{-1}$ and 0.1 and 9.4 kg ha^{-1} $year^{-1}$, respectively. The fluxes of DOC and DON in forest floor leachates increased with increasing annual precipitation and were also positively related to DOC and DON fluxes with throughfall. Concentrations of DOC in forest floor leachates were positively correlated to the pH of the forest floor.

Solinger et al. (2001) measured the concentration and fluxes of DON and DOC in bulk precipitation, throughfall, forest floor leachate and soil solution of a deciduous stand in Germany. The DOC and DON concentration and fluxes were highest in leachates originating from the Oa layer of the forest floor (73 mg C L^{-1}, 2.3 mg N L^{-1} and about 200–350 kg C, 8–10 kg N ha^{-1} $year^{-1}$). The DOC and DON concentrations in throughfall were positively correlated with temperature. Borken et al. (2004) studied the effect of compost application on leaching of DOC in six nutrient depleted forest soils in Germany. Compost treatment significantly increased cumulative DOC outputs by 31–69 g C m^{-2} at 10 cm depth and by 0.3–6.6 g C m^{-2} at 100 cm. The mineral soils between the 10 and 100 cm depths acted as significant sinks for DOC, as shown by much lower cumulative outputs at 100 cm of 3–11 g C m^{-2} in the control and 6–16 g C m^{-2} in the compost plots (Fig. 7.2).

Compared to seminatural systems, little is known about the form and functions of DON and the role that it plays in soil N cycling in agricultural soils. Murphy et al. (2000) found that soluble organic N (SON) extracted from soils (by water, KCl, etc.) is of the same order of magnitude as mineral N. In a wide range of agricultural soils from England, SON has been found to vary between 20 to 30 kg N ha^{-1}. Its dynamics are affected by mineralization, immobilization, leaching and plant uptake in the same way as those of mineral N, but its pool size is more constant than that of mineral N. Across different soils, crops and extractants, the SON has been reported to range between 7 to 45 kg N ha^{-1}, 23–55% of which is hydrolysable (Murphy et al., 2000). Results from Broadbalk continuous wheat experiment at Rothamsted show that approximately 10% of the N leached from drains is likely to be leached in organic form. More total N and DON is leached from plots receiving FYM compared to inorganic N (Table 7.1). Very large amounts of DON (up to 20% of total N lost) have been found in drainage waters leaving grassland lysimeters in Devon, UK (Hawkins et al., 1997). It is not clear whether DON leaving soils can be transformed to NO_3^--N in surface- or groundwaters.

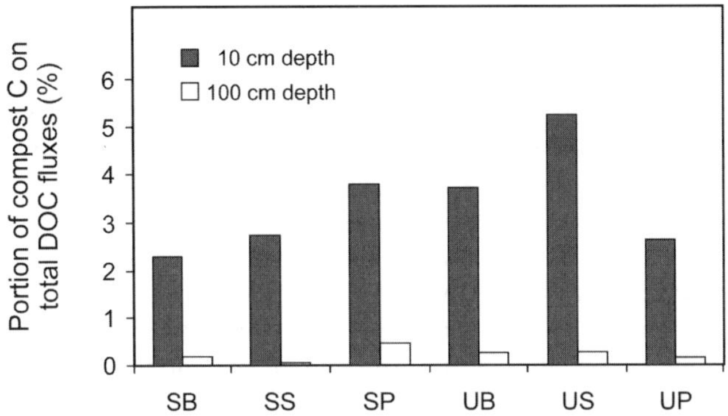

Fig. 7.2 Percent of compost C found in seepage fluxes at 10 and 100 cm depth in the Solling and Unterlüß beech (SB, UB), spruce (SS, US) and pine (SP, UP). (Borken et al., 2004; p. 95. Reproduced with kind permission from American Society of Agronomy, Crop Science Society of America & Soil Science Society of America)

Table 7.1 Mineral N and dissolved organic N (kg N ha^{-1}) in the drainage solution collected from tile drains (September–November 1998) located at 65 cm soil depth in the Broadbalk continuous wheat experiment at Rothamsted, UK. (Adapted from Murphy et al., 2000)

Treatment	Mineral N	Dissolved organic N
No N	9.9	1.2
144 (kg N ha^{-1} $year^{-1}$)	6.3	1.1
288 (kg N ha^{-1} $year^{-1}$)	29.0	2.5
FYM (~240 kg N ha^{-1} $year^{-1}$)	52.0	7.0

Leaching of DON may have several ecological consequences, such as constraining N accumulation in terrestrial ecosystems (leading to N limitations) and enhancing N bioavailability to aquatic ecosystems. The mobility of DON appears to be regulated by sorption to the mineral soil component and to a lesser degree, by biodegradation and uptake by biota (Qualls & Haines, 1992). Thus the mobility of DON is a function of its chemical composition and is strongly linked to hydrological parameters, such as variations in water flow paths through soils and dissolution kinetics of humic soil components (Hedin et al., 1995). Once in the aquatic ecosystem (streams and lakes), the chemical forms of DON will affect N bioavailability and possibly aquatic primary productivity. Leached DON may take with it nutrients, chelated or complexed metals and pesticides. Apparently, DON is an important pool in N transformations, but there are still many gaps in our understanding. The experimental evidence available indicate that DON uptake from soil may not contribute largely to N acquisition by plants but may instead be primarily involved in the recapture of DON previously lost during root exudation (Jones et al., 2005).

7.3 Nitrate Leaching

The movement of water through soil can result in the transport of N down out of the rooting zone of the plant. This process of N loss is called leaching and it usually occurs when nitrogen is in the nitrate (NO_3^-) form since nitrate being negatively charged moves freely with the soil-water unless the soils have significant anion exchange capacity (AEC). The leaching of NH_4^+ in soils may not be a problem except when applied in very large quantities on coarse textured soil having low cation exchange capacity (CEC) or in variable charge soils such as Alisols, Acrisols and Ferralsols of the (sub)tropics. When variable charge surfaces are protonated (on acidification) the soil loses its ability to retain cations in outer-sphere complexes. The cations instead remain in solution where they may be taken up by plants, heterotrophs or nitrifiers (NH_4^+), or leached from the system. If the pH continues to drop, AEC will increase and eventually exceed CEC, resulting in a soil in which NH_4^+ is more mobile than NO_3^-. More NO_3^- is retained as AEC increases, in part offsetting increased NH_4^+ loss. In practice, the NO_3^- is retained mainly in the subsoil, where low levels of SOM result in high point of zero charge (Fox, 1980).

Globally, the calculated amount of N leached from agricultural soils is 55 Tg year^{-1} (Fig. 7.3), contributing about 22 Tg year^{-1} to the river export at the river estuary (Van Drecht et al., 2003). The current contribution of deep groundwater flow, which is influenced largely by historical fertilizer use is in the order of 10%. Therefore, most of the current river export is due to recent development in fertilizer use such as Europe and Asia. However, due to scarcity of data it is difficult to validate these results, particularly because a number of environmental, soil, plant and management factors influence the rate and extent of NO_3^--N transport and leaching to ground and surface waters. Two major factors controlling leaching losses of NO_3^- are the concentration of NO_3^- in the resident soil solution and the amount of water percolating through the soil profile. High soil NO_3^- levels and

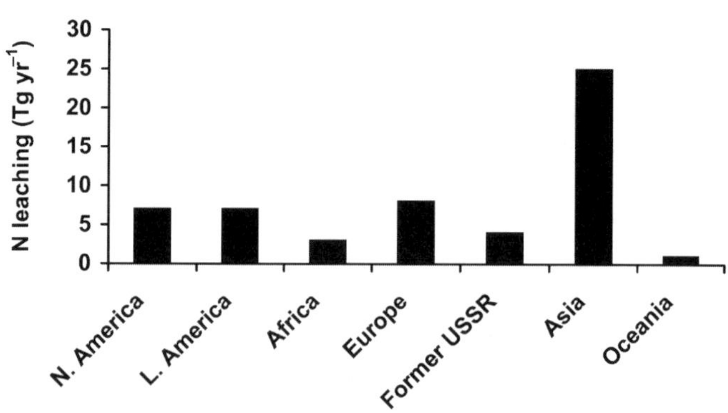

Fig. 7.3 Calculated regional leaching losses of nitrogen from agricultural soils (Drawn from data presented in Van Drecht et al., 2003)

sufficient downward movement of water to move NO_3^- below the rooting depth are often encountered in soils of the humid and subhumid zones, to a lesser extent in soils of the semiarid zone. In agroecosystems, NO_3^- originates from mineralization of SOM and crop and animal residue, fertilizer N not used by crops and to a lesser extent from atmospheric deposition. Fertilizer N application generally increases leaching losses. When high fertilizer rates are combined with heavy irrigation regimes on coarse-textured soils, leaching losses of NO_3^- can be substantial. Hence the loss is greater for high-fertility soils and under heavy N fertilization. The most important environmental factors that influence NO_3^- leaching include rainfall, evaporation and temperature. These factors may affect NO_3^- leaching directly through their effect on water drainage and indirectly through their effect on the soil NO_3^- content. For example, in temperate regions nitrate leaching losses usually increase in seasons with high amounts of drainage, e.g. during the autumn-winter period when evapotranspiration is low. However, in humid subtropical regions of India the leaching losses are likely to be more during the summer coinciding with high monsoonal rains.

Among the soil factors, texture and structure interact to influence leaching of NO_3^-. Leaching losses of N may be severe on rapidly percolating sand, gravely, and lateritic soils under conditions of heavy rainfall or excessive irrigation. Generally NO_3^- leaches more rapidly from sandy than silt and clay soils. The nature of soil determines the time taken for the leached NO_3^- to reach the groundwater. In clayey soils with a deep water table, it may take years for the downward moving NO_3^- to reach groundwater, whereas in sandy soils with a high water table, it may reach the groundwater in a few days or weeks. In a study on arable soils of northern Germany for a 3 years period (Nieder et al., 1995a), an average annual leaching rate of 63 kg NO_3^--N ha^{-1} was estimated for sandy soils as compared to 16 kg NO_3^--N ha^{-1} for heavier arable soils developed in loess. Even the low leaching figure for the heavier soils would lead to nitrate concentrations in the leaching water at or above 50 mg L^{-1} NO_3^- in this region where annual drainage is about 150 mm or less. The effect of texture is greatly modified by soil structure and the microscale distribution of NO_3^- in the soil (White, 1985). If NO_3^- is held within soil aggregates it will be protected from leaching when bypass or preferential flow occurs (Thomas & Phillips, 1979), however if NO_3^- is held on the outside of aggregates bypass flow causes it to leach faster than it would by uniform displacement (Addiscott & Cox, 1976). Beside soil texture and structure, SOM provides a critical control on catchment's susceptibility to enhanced N leaching. Nitrogen richness of organic soils, expressed as C:N ratio has been found to be an effective indicator of soil susceptibility to enhanced N leaching (Evans et al., 2006).

Several studies have reported that the leaching losses of N are related to the amount of fertilizer input with losses increasing with increasing rate of fertilizer N application (Benbi, 1990; Benbi et al., 1991a). Kolenbrander (1981) showed that for a distinct drainage rate, nitrate leaching increases with the nitrogen application rate, the coarseness of soil texture (i.e. the lighter soil leaches more) and with decreasing continuity of soil cover (i.e. arable crops leach more nitrate than grassland). In a long-term experiment, Benbi et al. (1991a) found that the amount of residual NO_3^--N in the 2.1 m soil profile after 16 cycles of corn-wheat-cowpea (fodder) cropping was directly related to the amount of fertilizer N, P and K application. The NO_3^--N content in the soil profile was minimum when half (50% NPK) the recommended amount of NPK was applied and the maximum amount accumulated under 1.5 times (150% NPK) the recommended application of fertilizers (Fig. 7.4). Evaluation of results from a number of long-term experiments under permanent grassland and arable crop rotations showed that the leaching below grassland was independent of fertilization upto 200 kg N ha^{-1}, above that the leaching increased linearly (Walther, 1989). In contrast, under arable land, the leaching increased linearly from zero with the amount of fertilization. Di & Cameron (2000) showed that there was a quadratic relationship between the annual N leaching losses and potentially leachable N (mineral and mineralizable N) in the soil.

In addition to fertilizer input, animal excreta, sewage effluent and decomposition of soil organic matter contribute substantially to NO_3^--N leaching to the groundwater bodies. The nitrate leaching is likely to be higher in agricultural systems based to a greater extent on organic inputs than those in which a larger proportion of the crop's N requirement is met from optimum and well-timed application of inorganic fertilizers. Results from Broadbalk and Hoosfield long-term experiments indicate large losses of N where FYM has been applied for a long period (Powlson et al., 1989). Nitrogen balances at Broadbalk winter wheat and Hoosfield spring barley experiments show that 124 kg NO_3^--N ha^{-1} is leached from the FYM treatment compared to 25 kg ha^{-1} from the inorganic fertilizer treatment. In a study in which equal amounts of poultry manure N and inorganic fertilizer N were applied in spring to barley, leaching of manure-derived N was much higher than fertilizer N during a 3 year period (i.e. 28 and 3.5 kg N ha^{-1}), primarily due to leaching during autumn and winter (Bergström & Kirchmann, 1999). Therefore, agricultural systems based to a greater extent on organic inputs/organic farming are likely to cause more

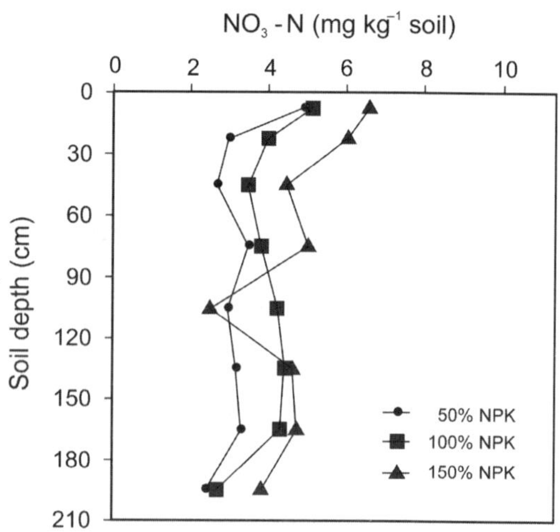

Fig. 7.4 Influence of fertilizer application rates on residual NO_3^--N in soil profile after 16 cycles of corn-wheat-cowpea (fodder) cropping (Benbi et al., 1991a; p. 176. Reproduced with kind permission from Springer)

nitrate pollution mainly due to lack of synchrony between N release from organic sources and crop demand.

Nitrate leaching losses vary considerably between different land use systems and these are generally higher under arable cropping as compared to forest ecosystems. The potential for NO_3^- leaching in different land use systems typically follow the order forest < cut grassland < grazed pastures, arable cropping < plowing of pasture < vegetables (Di & Cameron, 2002). Higher leaching potential in grazed pastures as compared to mowed pastures is attributed to the return of large proportion (60–90%) of N ingested by the grazing animals to the soil in the form of urine and dung. Urine patches are important source of potentially leachable nitrogen in soil. Based on N loading in the urine patch and the paddock area covered by animal urine, Silva et al. (1999) estimated that NO_3^--N leaching loss was about 33 kg N ha^{-1} $year^{-1}$ from an unfertilized grazed pasture and 36–60 kg N ha^{-1} $year^{-1}$ with the application of 400 kg N ha^{-1} $year^{-1}$ through urea or dairy shed effluent. Studies using ^{15}N have shown that the leaching losses from urine depends on the timing of application with relatively less losses from spring application as compared to summer and fall application (Decau et al., 2004).

Several studies have monitored input and output fluxes of nitrogen in forest ecosystems. It has been observed that elevated N deposition and the subsequent gradual N saturation of forest soils may lead to substantial NO_3^- leaching to ground and surface water (Aber et al., 1998). Higher NO_3^- concentrations are frequently found in regions with chronic N input from deposition (>8–10 kg N ha^{-1} $year^{-1}$). A compilation of regional and continental data from temperate forests indicate that a combined N flux to the soil of 50–60 kg ha^{-1} $year^{-1}$ from N deposition and litterfall may be a threshold for nitrate leaching in undisturbed forests (Gundersen et al., 2006). According to IFEF (Indicators of Forest Ecosystem Functioning) database, nitrogen deposition in throughfall in forest ecosystems in Europe ranges from <1 kg N ha^{-1} $year^{-1}$ in northern Norway and Finland to >60 kg N ha^{-1} $year^{-1}$ in the Netherlands and Czech republic (MacDonald et al., 2002). The amount of NO_3^- leached in the regional forests ranges from 1 to 40 kg N ha^{-1} $year^{-1}$ with 23% of sites (total 181) leaching between 5 and 15 kg N ha^{-1} $year^{-1}$, 13% leaching more than 15 kg N ha^{-1} $year^{-1}$ and the rest (64%) leaching less than 5 kg N ha^{-1} $year^{-1}$. The amount of NO_3^- leached is strongly related to the amount of nitrogen deposited in throughfall (input) and nitrogen status in the forest, expressed as C:N ratio (Fig. 7.5).

By stratifying the data based on C:N ratio in the organic horizon, MacDonald et al. (2002) obtained highly significant relationships between N input (kg ha^{-1} $year^{-1}$) and NO_3^- leached (kg ha^{-1} $year^{-1}$) for sites with C:N ratio ≤ 25 (Equation 7.1) and C: N ratio > 25 (Equation 7.2).

$$NO_3^- \text{ leached} = 0.65^{*}\text{N throughfall} - 3.81 \quad (7.1)$$

$$NO_3^- \text{ leached} = 0.32^{*}\text{N throughfall} - 1.05 \quad (7.2)$$

The higher slope of the relationship (Equation 7.1) for sites with C:N ratio ≤25 (nitrogen enriched sites) suggests that the risk of NO_3^- leaching is more in soils

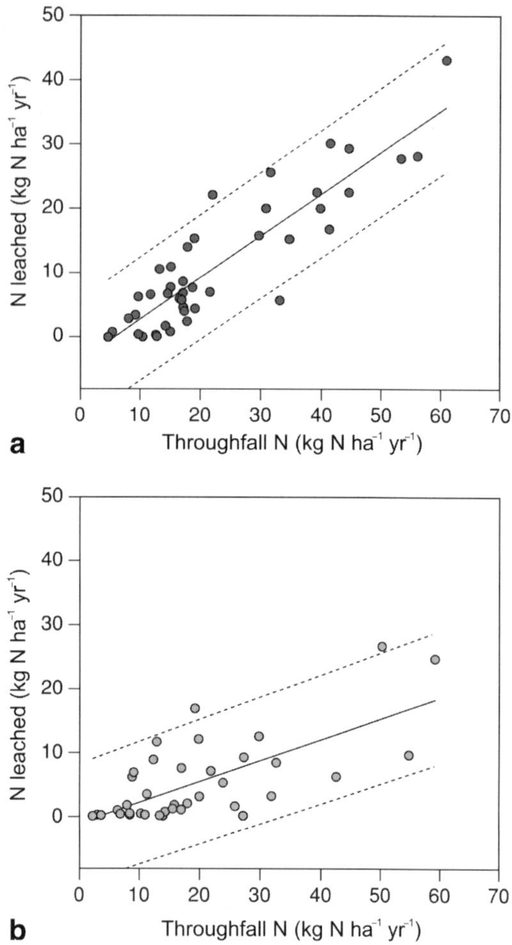

Fig. 7.5 Relationship between N input in throughfall and NO_3^--N leached at (a) sites where the organic layer C:N ratio is ≤25 and (b) sites where the organic layer C:N ratio is >25 (MacDonald et al., 2002; p. 1031. Reproduced with kind permission from Wiley-Blackwell)

with narrow C:N ratio as compared to those with wider C:N ratio (>25). Borken & Matzner (2004) also obtained similar slope values for forest stands in Germany though the relationship was relatively poor. Their analysis showed that the beech, oak and spruce forests released about 24–31% of N input to the groundwater whereas pine forests lost only 9% of throughfall N by leaching. The low leaching rates of the pine forests was attributed to low seepage fluxes at the sites under pine rather than to the (pine) tree species.

7.3.1 Reducing Leaching Losses

Since the concentration of NO_3^--N in the soil and water drainage through the soil profile are the major factors influencing NO_3^- leaching, practices that maximize N use efficiency together with improved water management help reduce the leaching losses. The measures to reduce NO_3^- leaching include optimal and balanced fertilization, synchronizing N supply to plant demand, manipulation of water applications and rooting depth, appropriate cropping sequence, use of cover crops, and the use of slow release fertilizers and nitrification inhibitors (Benbi, et al., 1991a; Prihar et al., 2000). Benbi et al. (1991a) showed that balanced application of N, P and K under intensive agriculture can significantly reduce the amount of unutilized nitrates in the soil profile by enhancing the N recovery in crop plants. The N recovery averaged 25% for annual application of N alone, 42% for N and P, and 56% for N, P and K. All N treatments increased the residual soil NO_3^--N but the plots receiving N, P and K where crop yields and N recovery were maximum, had the least residual NO_3^--N in the 2.1 m soil profile; thus reducing the risk of NO_3^- leaching.

Development of suitable irrigation schedules with respect to timing and amount so as to synchronize the NO_3^- rich zone with the moist zone of high root activity can help efficient N uptake by plant. It has been shown (Pratt, 1984) that where roots have access to the entire soil solution, nitrate is not leached unless excess fertilizer N is added or the soil is over irrigated. Soil-water and nitrogen dynamics models (e.g. Benbi et al., 1991b) could be used for efficient on-farm NO_3^- management under both irrigated and rainfed conditions.

Slow release fertilizers such as urea-formaldehyde, isobutylidene diurea, sulfur coated urea control nitrification by slowing down the rate at which NH_4^+ is made available for nitrification. However, the use of slow release fertilizers have been limited because of their high cost and possible mismatch between nitrogen availability and crop demand. Nitrification inhibitors such as Dicyandiamide (DCD), N-serve and calcium carbide, which retard the formation of nitrate by nitrifying bacteria are known to increase the fertilizer use efficiency provided that maintaining the added fertilizer N as NH_4^+ does not lead to increased ammonia volatilization. The beneficial effect of nitrification inhibitors in reducing N leaching losses depend on soil type and time and rate of N application. The effect is likely to be more in coarse-textured soils at rates of N application below optimum (Hoeft, 1984).

Vegetation retards NO_3^- leaching from the root zone by absorbing nitrate and water. Rooting habits of plants exert a profound influence on NO_3^- mobility in the root zone with high mobility under shallow rooted crops like potato as compared to deep rooted crops such as wheat. Therefore, choice of appropriate cropping sequence in which heavily fertilized shallow-rooted crops are followed by low-nutrient requiring deep-rooted crops can minimize residual NO_3^--N accumulation in the soil profile and thus reduce the risk of leaching. Use of cover crops after harvesting can be effective in reducing nitrate leaching compared to bare fallow. A number of studies have reported the beneficial effect of cover crop vis-à-vis bare fallow in reducing leaching losses of nitrogen. In a 7-year study, Shepherd (1999) found that on average winter crops decreased NO_3^--N leaching by 25 kg N ha^{-1} $year^{-1}$. The cover crops decreased the average N concentration in the drainage water from 24 to 11 mg N L^{-1}. Similarly, McLenaghen et al. (1996) showed that the N leaching loss under a ryegrass cover crop in New Zealand was only 2.5 kg N ha^{-1} compared with 33 kg N ha^{-1} for bare fallow soil. However, in some studies long-term

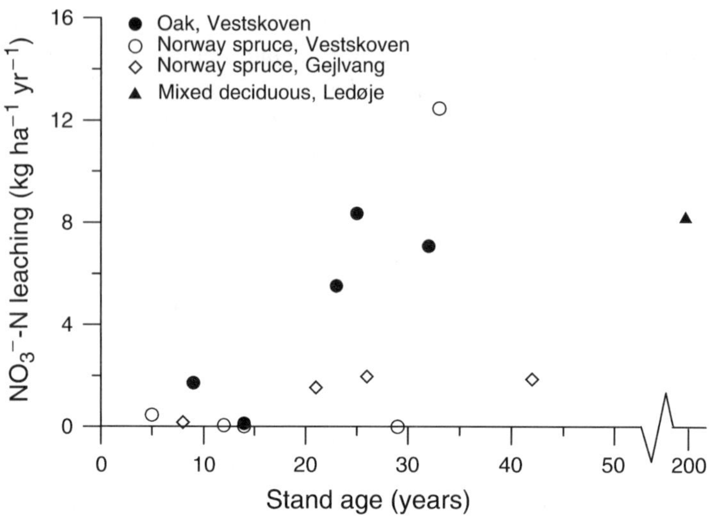

Fig. 7.6 Average annual NO_3^--N leaching below the root zone at 90 cm depth as a function of time since afforestation in three chronosequences in Denmark (Hansen et al., 2007; p. 1257. Reproduced with kind permission from Wiley-Blackwell)

use of cover crops, has been shown to increase N leaching compared with cropping systems without cover crops. Hansen et al. (2000) showed that NO_3^- leaching was 29% higher in plots with 24 year of cover crops than in plots without cover crops. This has been attributed to the large inputs of crop residues from cover crop.

Measures to reduce NO_3^- leaching from grasslands and pastures, *inter alia*, include avoiding plowing or better timing of plowing pasture leys, removing stock from the fields earlier in the grazing seasons, improved stock management, and precision farming. For pasture systems, splitting the annual fertilizer application rates into a number of applications to match the pasture N demand can result in lower NO_3^- leaching losses compared with few applications at higher rates (Di et al., 1998; Silva et al., 1999). Francis (1995) suggested that the most reliable way to minimize N leaching losses in Canterbury, New Zealand, is to delay the plowing of pasture for as long as possible in autumn or winter. Delaying the plowing of pasture until late autumn (May) reduced the leaching loss from 72–106 kg N ha^{-1} to 8–52 kg N ha^{-1}. Removing stocks from the fields earlier in the grazing season reduces the accumulation of high concentrations of potentially leachable NO_3^- in the soil of grazed pastures but increases the quantity of manure produced by housed animals and the need to recycle this effectively. Supplementing grass diets with low-nitrogen forages such as maize silage reduces the quantity of nitrogen excreted by livestock (Cuttle & Scholefield, 1995).

Since forests have low NO_3^- leaching, afforestation of abandoned cultivated land has been suggested as an effective way to reduce leaching of NO_3^- to groundwater (Iversen et al., 1998). This also has a co-benefit of enhanced carbon sequestration. Hansen et al. (2007) evaluated the effect of afforestation of former arable land on nitrate leaching, based on three afforestation chronosequences in Denmark. Afforestation of former arable land initially resulted in lower nitrate leaching than that occurring under the former agricultural land use. Nitrate concentrations became almost negligible in forest stands of 5–20 years of age (Fig.7.6).

However, after canopy closure (>20 years) nitrate concentration below the root zone and nitrate leaching tended to increase. This was attributed to increased N deposition with increasing canopy development and decreased N demand once the most N-rich biomass compartments had been built-up. In a recent study (Van der Salm et al., 2006) from the Netherlands it has been shown that conversion of arable land into oak and spruce forest could decrease NO_3^- leaching to groundwater but it reduces water recharge of ground and surface reservoirs, thus affecting the local hydrological cycle. Apparently a number of management options are available for reducing nitrate leaching but the appropriate strategy for a given ecosystem will depend on soil, environmental and cultural variables.

Chapter 8
Bidirectional Biosphere-Atmosphere Interactions

There is an increasing recognition that the emission of four principal greenhouse gases (GHGs) viz. carbon dioxide (CO_2), methane (CH_4), oxides of nitrogen (nitrous oxide, N_2O and nitric oxide, NO) and the halocarbons (a group of gases containing fluorine, chlorine and bromine), which stem from human activities are bringing about major changes to the global environment. These gases accumulate in the atmosphere and their concentration in the environment has increased significantly with time. The increased concentration of these gases cause global warming, deplete the concentration of ozone in the stratosphere that acts as a shield against excessive exposure to ultra violet (UV) rays at the Earth's surface, and also contribute to acid deposition. The biosphere plays a massive role in the global cycling of carbon dioxide and other GHGs. Although most of the anthropogenic CO_2 emissions come from the combustion of fossil fuels, there is also a substantial contribution from land use change, due to biomass burning and increased mineralization of soil organic carbon, following the conversion of forest or native grassland to agriculture. In all terrestrial ecosystems except those with bare land resulting from agricultural or forestry operations, soil emissions of CO_2 generally take place within a two-way exchange between the land and the atmosphere. This exchange is usually known as the Net Ecosystem Exchange (NEE) and involves the component exchanges of CO_2 from the plant canopy, its elements, and the ground surface. The NEE is the net sum of the gains of C in photosynthesis by the vegetation, the losses of C from respiration by above-ground plant tissues, and losses by below-ground roots, mycorrhiza and heterotrophic microorganisms (soil respiration). The most important of these fluxes are the gains by plant photosynthesis and losses through soil respiration. Atmospheric CH_4 originates from both natural and anthropogenic sources. Wetlands, rice agriculture, livestock, landfills and waste treatment have been implicated as dominant sources of methane emission. However, methane growth in atmosphere depends upon its photochemical destruction and methane oxidation in soil. Oxides of nitrogen are emitted to the atmosphere through biogenic and abiogenic processes and a part of the emitted nitrogen oxides and ammonia are deposited back on the terrestrial ecosystems.

In the last 2 decades, much experimental work has been undertaken to quantify the emission of these gases and to identify the key sources and factors that govern them. Though fluxes vary greatly between ecosystems, and are often subject to

R. Nieder, D.K. Benbi, *Carbon and Nitrogen in the Terrestrial Environment*,

great variations, both spatial and temporal, within ecosystems, attempts have been made to develop regional and global budgets. In the recent report of the Intergovernmental Panel on Climate Change (IPCC, 2007a, b) it has been shown that the emission of GHGs have increased substantially and with current climate change mitigation policies and related sustainable development practices, global GHG emissions are expected to grow over the next few decades. In this chapter, we examine the current emission scenario and the contribution of different factors governing fluxes with particular emphasis on soil-atmosphere gaseous exchange. Since the chlorofluorocarbons (CFCs), the principal form of halocarbons are entirely of industrial origin and unaffected by land-atmosphere processes, these are not considered further here. However, the abundance of CFC gases is decreasing as a result of international regulations for the protection of ozone layer (Denman et al., 2007).

8.1 Atmospheric Nitrogen Depositions

Nitrogen inputs to the terrestrial ecosystems occur through fertilizer application, atmospheric deposition and by the action of microorganisms that fix atmospheric N_2. Globally, the supply of N to terrestrial ecosystems has doubled as a consequence of anthropogenic activities, such as industrial N fixation, cultivation of N-fixing legumes and production of nitrogen oxides by fossil-fuel burning. The most important substances emitted by human activities are oxides of nitrogen and NH_3. Many different sources are responsible for their emissions. It is estimated that on an average 70–80% of the emitted N is deposited back on land and water bodies. Model simulations have shown that 50–80% of the fraction deposited on land falls on natural (non-agricultural) vegetation indicating the importance of atmospheric transport in dispersing pollution from agricultural and industrial regions to natural ecosystems (Dentener et al., 2006). The atmospheric deposition can occur as wet or dry deposition. The deposited N besides impacting a number of processes in soil and vegetation, greatly modifies the global C and N cycles.

8.1.1 Wet and Dry Deposition

Atmospheric nitrogen is deposited to the terrestrial ecosystems through rain, snow and hail (wet deposition) or dust and aerosols (dry deposition). The input originates mainly from previously emitted NH_3 and NO_x. Deposition occurs in the form of NH_3 and NH_4^+ (collectively termed NH_x) and as NO_y and its reaction products: gaseous nitric acid (HNO_3), nitrous acid (HONO) and particulate nitrate (NO_3^-). While HNO_3 usually features a rapid downward (net deposition) flux to the surface (Huebert & Robert, 1985), the exchange of NO, NH_3, HONO and NO_2 between surface and atmosphere may be bi-directional (Trebs et al., 2006). Dry deposition of NH_3 is most important close to a source and wet deposition of NH_4^+ is most important some distance downwind from the source. Far from the source the deposition of NH_4^+ is on an annual average halved approximately every 400 km (Ferm, 1998). In parts of Europe, with high NH_3 emissions, like the Netherlands, Belgium and Denmark, dry deposition of NH_3 represents the largest contribution to total NH_x deposition. In countries with low NH_3 emission densities only wet deposition of NH_4^+ from remote sources dominates the deposition (Asman et al., 1998). Dry deposition of N to wet surfaces in an agricultural region of Iowa, USA has been shown to be several times greater than to dry surfaces, suggesting that NH_x absorption by water associated with wet surfaces is an important transport mechanism (Anderson & Downing, 2006).

Though dry deposition contributes substantially to the total atmospheric deposition of nitrogen, most of the earlier studies measured wet deposition only and there is limited information available on dry deposition. This is mainly because the techniques to measure dry deposition are not as well established and there is also high variability in the observed deposition to different types of surfaces. Lovett & Lindberg (1986) measured dry deposition of N in a deciduous forest by three different methods. The flux estimates varied widely (1.8–9.1 kg N ha^{-1} $year^{-1}$) reflecting the variability in the measurements associated with methodology. Goulding and associates (1998) computed the total deposition of all N species to winter cereals at Rothamsted to be 43.3 kg N ha^{-1} $year^{-1}$, out of which 84% were oxidized species and 79% dry deposited.

In many natural and seminatural ecosystems, the atmospheric N deposition has increased over the years. As compared to estimated inputs of 1–3 kg N ha^{-1} $year^{-1}$ in the early 1900s (e.g. Galloway, 1995; Asman et al., 1998), the atmospheric N deposition rates of 20–60 kg N ha^{-1} $year^{-1}$ in non-forest ecosystems in Western Europe, and up to 100 kg N ha^{-1} $year^{-1}$ in forest stands in Europe or the USA have been observed (Bobbink et al., 2002). While there has been an increase in deposition rate across all the biomes at both temperate and tropical latitudes, the increase is greatest in the northern hemisphere temperate ecosystems (Table 8.1; Holland et al., 1999). The high deposition rates are probably driven by biomass burning, soil emissions of NO_x and NH_3 as well as lightning production of NO_x. Mixing ratios of NO_2 and water-soluble N species in the gas and aerosol phase are reported to be significantly enhanced when widespread biomass burning takes place, resulting in high N deposition rates. Trebs et al. (2006) observed that on a tropical pasture site in Brazil, dry deposition accounted for 46.5% of the depositions during the dry (biomass burning) season, whereas during the wet season (clean conditions) dry deposition accounted for 22.3% of depositions as compared to 77.6% from wet deposition.

Global estimates of total atmospheric N deposition show a tremendous increase during the last 150 years (Table 8.2) and these are projected to increase further with time. Compared to total annual NO_y deposition of 12.8 Tg N in 1860 the deposition has increased to 45.8 Tg N in the early 1990s (Galloway et al., 2004; Lelieveld & Dentener, 2000). The greatest deposition to the continents occurs in Asia (6.5 Tg N $year^{-1}$) followed by Africa and Europe (5.0 Tg N $year^{-1}$ each). In the same period, total deposition of NH_x has increased from 18.8 Tg $year^{-1}$ to 56.7 Tg N $year^{-1}$ and

Table 8.1 Preindustrial and contemporary N depositions (Tg N $year^{-1}$) onto biome types estimated by model MOGUNTIA (Dentener & Crutzen, 1994) (Adapted from Holland et al., 1999)

Biome	NH temperate latitudes	SH temperate latitudes	Tropical	NH temperate latitudes	SH temperate latitudes	Tropical
	Preindustrial			1990s		
Grasslands	0.88	0.17	0.38	6.39	0.32	2.87
Forests	1.94	1.18	4.18	9.47	0.19	4.94
Mixed	0.82	0.51	2.77	4.16	1.04	6.44
Life-forms wetland and riparian zones	0.37	0.01	0.31	6.61	0.21	1.30
Ice	0.02	0.00	–	0.02	0.00	–
Total	4.03	1.87	7.64	26.65	1.76	15.55

NH = northern hemisphere; SH = southern hemisphere

Table 8.2 Global atmospheric deposition of NO_y and NH_x (Tg N $year^{-1}$) (Adapted from Galloway et al., 2004)

	1860		1993	
	NO_y	NH_x	NO_y	NH_x
Terrestrial	6.6	10.8	24.8	38.7
Marine	6.2	8.0	21.0	18.0
Total	12.8	18.8	45.8	56.7

the largest deposition flux (16.1 Tg N $year^{-1}$) is in the Asian continent. This is obviously because of high population, expansion of industrialization, and enhanced food production. The global annual nitrogen deposition over land is expected to increase by a factor of ~2.5 by the year 2100, mostly because of the increase in nitrogen emissions. This will significantly expand the areas with annual average deposition exceeding 1,000 mg N m^{-2} $year^{-1}$ (Lamarque et al., 2005). As per the simulated estimates (Lamarque et al., 2005) the current deposition over land ranging between 25 and 40 Tg N $year^{-1}$ is expected to increase to 60–100 Tg N $year^{-1}$ by 2100. The deposition over forests is expected to increase from 10 to 20 Tg N $year^{-1}$. Simulation results presented by Dentener et al. (2006) show that NO_y depositions in 2030 generally remain unchanged except in Asia where the depositions are expected to increase by 50–100%. The NH_4^+ depositions are predicted to decrease by 20% in Europe but increase by 40–100% in Central and South America, Africa, and parts of Asia. However, these estimates need to be interpreted with caution. Comparison of mean modeled and measured wet deposition rates for the year 2000 shows (Fig. 8.1) that while 70–80% of modeled NO_y depositions in Europe, North America, Africa and East Asia were within ± 50% of the measured values, the modeled deposition rates are underestimated by a factor of 2 (~130 mg N m^{-2} $year^{-1}$) in South Asia (India). On the contrary, NH_4^+ depositions in south Asia (India) are strongly overestimated by on average 350 mg N m^{-2} $year^{-1}$. Globally, the agreement of NH_x deposition with measurements is relatively less than for NO_y with 30–60% of the modeled deposition within ±50% of the measurements.

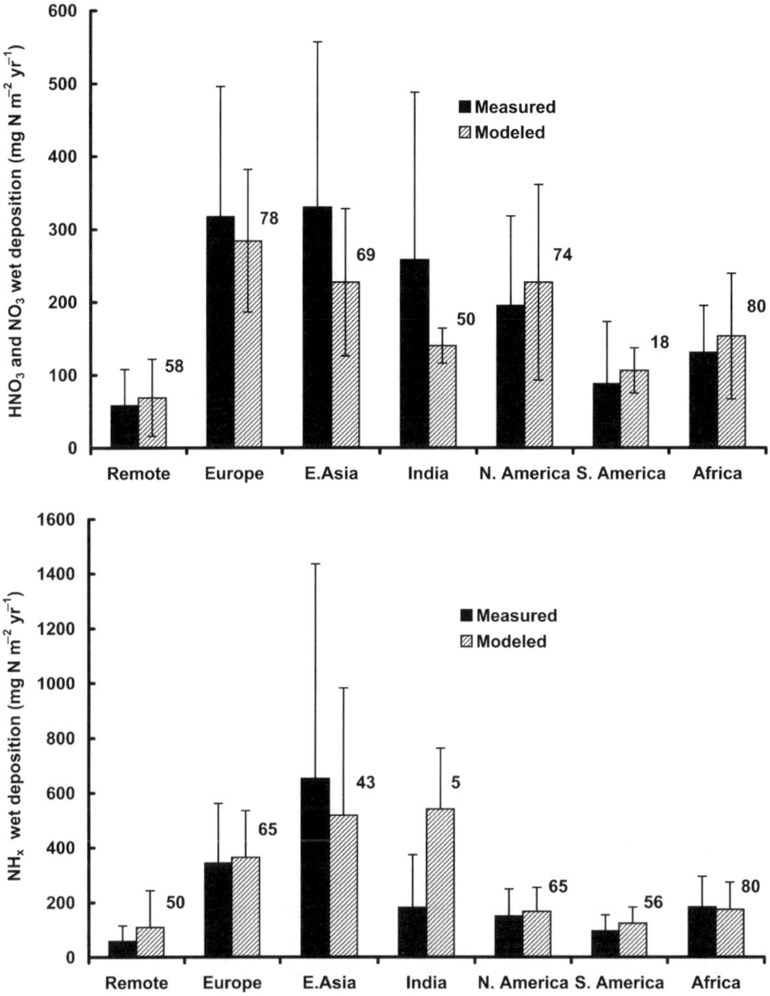

Fig. 8.1 Comparison between mean measured and modeled wet deposition of HNO_3 and aerosol NO_3^- and NH_x in different regions of the world. Vertical lines indicate ± standard deviation (SD) and numbers next to the modeled bars show the percentage of the modeled deposition within ±50% of the measurements. For the sake of clarity – SD bars for NH_x are not shown (Drawn from Dentener et al., 2006)

The atmospheric portion of the analysis of the global N cycle only addresses inorganic N species (NO_x, NH_3 and N_2O) and does not include oxidized atmospheric organic N (i.e. organic nitrates such as peroxyacetyl nitrates), reduced atmospheric organic N (e.g. aerosol amines and urea), and particulate atmospheric organic N (e.g. bacteria, dust). The emission and deposition of organic N is probably a significant component of the atmospheric N cycle. Neff et al. (2002) estimated that contribution of organic N to total N loading constitutes around a third of total N load with a median value of 30% (standard deviation 16%). Preliminary estimates indicate that the global flux of atmospheric organic N range between 10 to 50 Tg N $year^{-1}$, though there are considerable unresolved uncertainties.

8.1.2 Effect of N Deposition on Ecosystems

Increased input of N to terrestrial ecosystems impacts a number of ecological processes operating at different temporal and spatial scales. While deposition of N to agricultural or croplands could serve as a source of nutrient, it could also be lost by different pathways and thus become a contaminant. Using computer simulation model, Goulding et al. (1998) studied the fate of deposited N and estimated that out of 45 kg N deposited ha^{-1} $year^{-1}$ at Broadbalk Continuous Wheat Experiment at Rothamsted, around 5% is leached, 12% is denitrified, 30% is immobilized in the soil organic matter and 53% taken off in the crop. Several studies have reported the negative effects of excessive N deposition to natural and seminatural ecosystems. Nitrogen deposition can lead to eutrophication and acidification. Since both NH_4^+ and NO_3^- are easily available nitrogen forms that could be taken up by soil and vegetation, their total deposition affects eutrophication. The effect due to NH_x deposition in Europe has worsened. Despite the fact that NH_3 is a base, it can cause acidification of soil. In the air NH_3 neutralizes acids and forms NH_4^+. When NH_4^+ is taken up by roots, H^+ ion is released from the root that acidifies the soil. NH_4^+ can be oxidized to NO_3^- resulting in the release of two H^+ ions. When the nitrate is taken up by the roots a HCO_3^- or OH^- ion is released. As a result there will be net acidification when the NO_3^- is lost from the soil by leaching (i.e. $2H^+$ and $1OH^-$ ions are created). In temperate ecosystems, addition of excess N from the atmosphere leads to soil acidification and base cation depletion, but strong plant N uptake slows the rate of change. In tropical systems, soil acidification due to N deposition is affected by surface charge properties of the soils (Matson et al., 1999). The acidification of soils due to increased N input and its influence on forest ecosystems has been described in Chapter 6. High N deposition increases vulnerability of forests to other stress factors such as frost, drought and pest as well as the probability of trees falling during storms caused by roots growing closer to the surface. Enhanced N deposition can increase NO_3^- leaching and thus become a contaminant in groundwater. In an agricultural region of Iowa, USA, it has been reported (Pan et al., 2004) that the deposited N in a forested watershed is contributing 1.23 kg N ha^{-1} $year^{-1}$ to leaching losses at the current level of N deposition. It was estimated that if the N depositions were twice as large as the current ones, the leaching losses would increase by a factor of more than 3. Several studies have shown the effect of increased N deposition on trace gas flux from forest and grasslands (Steudler et al., 1989; Mattson, 1995; Bowden et al., 2000). But the investigators usually applied N rates substantially higher than the ambient atmospheric N depositions, which may not simulate the trace gas flux appropriately. Ambus & Robertson (2006) observed that with realistic levels of N input (1–3 g N m^2 $year^{-1}$) to unmanaged forest and grassland communities for 2 years there was no effect on trace gas flux and soil N concentration. Such results emphasize the need to use realistic levels rather than saturating levels of N input for simulating increased N deposition.

The major effects of excessive deposition of nitrogen may be summarized as: (i) reduction in biodiversity and changes in species composition from oligotrophic or mesotrophic to relatively fast growing nitrophilic ones, (ii) soil acidification (leading to nutrient imbalances and mobilizing aluminum and toxic metals), (iii) saturation of woodland ecosystems, (iv) increased nitrous oxide emissions from denitrification and nitrification in soils and reduced methane oxidation rates, (v) altering the balance of nitrification and mineralization/immobilization, (vi) increased nitrate leaching from the soil to the deeper groundwater, (vii) direct toxicity of nitrogen gases and aerosols to the above-ground parts of individual plants especially near the sources of NH_x and NO_y, (viii) increased susceptibility to secondary stress factors such as drought, frost, pathogens or herbivores, and (ix) increased carbon storage. The impact of N deposition on ecosystem processes and N losses have been studied primarily in N-limited ecosystems in the temperate zone; it is possible that tropical ecosystems may respond differently to increasing deposition. Matson et al. (1999) concluded that inputs of N into tropical forests are unlikely to increase productivity and may even decrease it due to indirect effects on acidity and the availability of phosphorus and cations.

The severity of the impacts of atmospheric nitrogen deposition depends on the duration and amount of the increased inputs, the chemical and physical form of the atmospheric nitrogen input, the sensitivity of the plant and animal species to the increased input, the abiotic condition in the ecosystem, and the land use or management. Excessive deposition of NH_x will be more harmful than nitrate. It is well known that extremely high NH_3 concentration can kill trees. Since a number of variables determine the severity of an effect of N deposition, high variations in sensitivity of different ecosystems to atmospheric nitrogen deposition have generally been observed.

Generally, the critical load approach is used to describe the vulnerability of ecosystems to N deposition. A critical load is defined as a quantitative estimate of an exposure to one or more pollutant below which significant harmful effects on specified sensitive elements of the environment do not occur according to the present knowledge (Nilsson & Grennfelt, 1988). The critical loads generally range between 10 to 20 kg N ha^{-1} $year^{-1}$for forest ecosystems, 10–30 kg N ha^{-1} $year^{-1}$ for grasslands and tall forb dominated ecosystems (except for more sensitive moss and lichens dominated mountain habitats: 5–10 kg N ha^{-1} $year^{-1}$), and 5–35 kg N ha^{-1} $year^{-1}$ for mire, bog and fen habitats (Bobbink et al., 2002). The critical loads for inland surface water, coastal and marine habitats range between 5–20, 10–25 and 30–40 kg N ha^{-1} $year^{-1}$ (Table 8.3). However, the critical loads for some ecosystems are speculative and need to be validated by studying long-term effects of increased N deposition on ecosystem processes. For example, Persson et al. (1995) predicted the long-term effect of atmospheric N deposition on Norway spruce stands in south-western Sweden. They reported that annual N deposition of 20 kg N ha^{-1} during the

Table 8.3 Empirical critical loads for nitrogen deposition to natural and seminatural ecosystems (Adapted from Bobbink et al., 2002)

Ecosystem	Critical load (kg N ha^{-1} $year^{-1}$)
Forest	10–20
Tundra	5–10
Arctic, alpine and subalpine scrub habitat	5–15
Heathland	10–25
Grasslands	5–25
Mire, bog and fen	5–35
Inland surface water	5–20
Coastal	10–25
Marine	30–40

next 30 years in southwestern Sweden would not affect the growth of Norway spruce stands.

Model estimates show that the critical loads for acidification and eutrophication are exceeded in 7–18% of the global area of natural ecosystems with serious problems in the heavily industrialized regions of eastern USA, Europe, the former Soviet Union, and large parts of Asia. Both acidification and eutrophication risks are projected to increase in Asia, Africa and South America in the near future (Bouwman et al., 2002b). But there are major uncertainties in the approach used, particularly with respect to upscaling the estimates, base cation emission and deposition fluxes. Results of 23 atmospheric chemistry transport models (Dentener et al., 2006) show that currently 11% of the world's natural vegetation receives nitrogen in excess of the critical load threshold of 1,000 mg N m^{-2} $year^{-1}$. The regions most affected are the United States (20%), Western Europe (39%), Eastern Europe (80%), South Asia (60%), East Asia (40%), Southeast Asia (30%), and Japan (50%). The global fraction of vegetation exposed to N loads in excess of 1,000 mg N m^{-2} $year^{-1}$ increases globally to 17–25% in 2030. The regions most affected by exceedingly high nitrogen loads are Europe and Asia, but also parts of Africa.

8.1.2.1 Nitrogen Deposition and Carbon Storage

Elevated N inputs to forests may enhance the accumulation of carbon and nitrogen in SOM through increased biomass production (Aber et al., 1998). In N-limited temperate ecosystems, N deposition has been shown to enhance carbon storage, which may have substantial impacts on global CO_2 concentration (e.g. Townsend et al., 1996). Estimates of global C sink induced by nitrogen enrichment range from nearly zero to 2.3 Pg C $year^{-1}$. Levy et al. (2004) estimated that cumulative change in N deposition over 100 years will change the total C content of the coniferous forest ecosystem of Sweden by ~20 kg C (kg N)$^{-1}$. However, there is considerably uncertainty in the estimates. Contrary to temperate ecosystems, higher N inputs to most tropical systems may lead to lower productivity and reduced carbon storage. The decreased productivity and consequent reduced C storage could result from losses of base cations due to increased leaching of nitrate, effect of increasing soil acidity on phosphorus availability, and increased Al mobility into soil solution that may be toxic to plant growth and microbial activity (Matson et al., 1999). As discussed in Chapter 6, enhanced N deposition and consequent acidification of the soil could cause changes in humus forms, which may have great implications for carbon and nitrogen dynamics in forest ecosystems.

8.2 Carbon Fixation via Photosynthesis

Photosynthesis (*photo = light, synthesis = putting together*) is a process in which green plants use solar energy to transform H_2O, CO_2 and minerals into oxygen (O_2) and organic compounds, mainly carbohydrates. Photosynthesis is performed by higher plants, algae, some bacteria and some protists, all of which are collectively referred to as photoautotrophs. As nearly all non-photoautotrophic life depends on the carbon compounds produced by photosynthesis, it is the most important biochemical pathway.

8.2.1 Photosynthetic Pathways

In terrestrial plants, three photosynthetic pathways exist: C_3, C_4 and CAM (Crassulacean acid metabolism) (Ehleringer & Monson, 1993). The C_3 pathway is an ancestral pathway for CO_2 fixation and occurs in all taxonomic plant groups, whereas C_4 photosynthesis is common in the more advanced plant taxa and occurs especially in monocots (i.e., grasses and sedges) but less in dicots (Sage & Monson, 1999). The CAM pathway only occurs in epiphytes and succulents from arid regions, which are limited in global distribution and C cycling. The focus in the following section will, therefore be on the C_3 and C_4 pathways.

The anatomy of C_4 leaves with so-called 'Kranz' cells differs fundamentally from that of C_3 plants. The chloroplasts of C_3 plants are of homogeneous structure, while two types of chloroplasts occur in C_4 plants. The mesophyll cells contain normal chloroplasts, that of the vascular bundle sheath have chloroplasts without grana, i.e. they are partially impaired in function. This peculiarity does not affect the *Calvin cycle*, it concerns only the light reactions of photosynthesis. Photosynthesis is a multistep pathway in which CO_2-C is fixed into stable organic molecules. A simple general equation is:

$$6\,CO_2 + 6\,H_2O + \text{light} \rightarrow C_6H_{12}O_6 + 6\,O_2 \quad (8.1)$$

In the first step, RuBP (ribulosebisphosphate) carboxylase oxygenase (Rubisco) combines RuBP (a 5C molecule) with CO_2 to form two molecules of the 3 C molecule phosphoglycerate (PGA):

$$RuBP + CO_2 \rightarrow PGA \tag{8.2}$$

The enzyme Rubisco is capable of catalyzing two different reactions. The one reaction leads to the formation of two molecules of PGA when CO_2 is the substrate and the other reaction results in one molecule of PGA and another one of phosphoglycolate (PG, 2C molecule) when O_2 is the substrate (Lorimer, 1981). The latter reaction results in less efficient CO_2 fixation and may lead to the release of CO_2 in a process named photorespiration:

$$RuBP + O_2 \rightarrow PGA + PG \tag{8.3}$$

The proportion at which Rubisco catalyzes CO_2 versus O_2 depends on the ratio of CO_2 to O_2. This relationship establishes a link between current atmospheric conditions and photosynthetic activity. The efficiency of the C_3 pathway is presently decreasing as a consequence of the Rubisco sensitivity to O_2. The C_4 pathway is a biochemical modification of the C_3 pathway. It reduces the Rubisco oxygenase activity and thereby increases the photosynthetic rate in low-CO_2 environments (Ehleringer et al., 1991).

The C_3 cycle in C_4 plants is restricted to interior cells (the bundle-sheath cells). In mesophyll cells surrounding the bundle-sheath cells, PEP (phosphoenolpyruvate) carboxylase (a much more active enzyme) fixes CO_2 as HCO_3^- into the C_4 acid oxaloacetate. The latter diffuses to the bundle-sheath cell where it is decarboxylated and refixed in the common C_3 pathway. As a consequence of the higher activity of the PEP carboxylase, CO_2 is effectively concentrated in the places where Rubisco is located, which results in a high ratio of CO_2 to O_2 and limited photorespiration.

The quantum yield for CO_2 uptake defined as the slope of the photosynthetic light response curve at low light levels (or the light use efficiency) strongly differ between C_3 and C_4 taxa. The reduced quantum yield values of C_4 taxa are temperature independent, whereas the quantum yield values of C_3 taxa are reduced with increased temperature. As a consequence, the light use efficiencies of C_3 taxa will exceed that of C_4 taxa at lower air temperatures and will fall below that of C_4 taxa at higher temperatures.

8.2.2 Global Distribution of C_3 and C_4 Pathways

The current global distribution of C_3 and C_4 photosynthetic pathways is particularly a function of temperature which has been documented by numerous monocot studies worldwide. In most of these studies, >90% of the variance in C_3/C_4 abundance in present ecosystems is explained by temperature alone. Both long-term aboveground harvest studies (Epstein et al., 1997) and belowground SOC studies (Tieszen et al., 1997) independently indicate a C_3/C_4 transition near 45° N. Collatz et al. (1998) predicted that C_4 abundance can be expected in all regions where the mean monthly temperature exceeds 22°C and monthly precipitation exceeds 25 mm.

8.2.3 Response of C_3 and C_4 Pathways to Increasing Atmospheric CO_2 Concentration

Changing atmospheric CO_2 levels may modify the geographical distribution of C_3 and C_4 pathways. The global emergence of C_4-dominated ecosystems during the late Miocene suggests that atmospheric CO_2 levels decreased across a threshold of ~500 ppmv favoring C_4 over C_3 photosynthesis in warm ecosystems (Ehleringer & Cerling, 2001). During glacial periods when atmospheric CO_2 levels decreased to approximately 180 ppmv, C_4 taxa were apparently more abundant than they are today. These changes in C_3/C_4 abundances had enormous impacts on both evolution and mammalian grazers. The basis for this impact may be feeding preferences associated with differential digestibility of C_3 versus C_4 grasses.

The C_4 photosynthesis is nearly CO_2-saturated at present atmospheric CO_2 concentration. In contrast, C_3 photosynthesis is operating well below potential CO_2 fixation. It is, therefore, often suggested that, except for dry environments, the present increase in atmospheric CO_2 concentration favors C_3 versus C_4 plants. The enhanced photosynthetic potential of C_3 plants under elevated CO_2 is of immense importance for the competition between C_3 and C_4 plants (Kirschbaum, 1994). At a location where C_3 and C_4 plants coexist, they must be competing for other limiting resources like water, nutrients or light. Increasing CO_2 concentration causes a selective advantage on the C_3 over C_4 plants. Increased C gain by C_3 plants would allow them to either increase root growth and compete more successfully with their C_4 neighbors for nutrients, or increase foliage production to compete more successfully for light. Where differences are observed within a single generation, these are likely to be further compounded over successive generations.

Some examples support the above thought concerning the effects of elevated CO_2 concentration on the C_3/C_4 balance. In a wetland mixed community of C_3 sedge and C_4 grasses, elevated CO_2 resulted in an increase in C_3 plant above ground dry mass and a concomitant decrease in C_4 plant above ground dry mass (Drake, 1992). Archer et al. (1995) attributed that woody C_3 vegetation invaded C_4-dominated grasslands in some locations during the past 200 years when global CO_2 concentration increased from 280 to 360 ppmv and further increases in CO_2 concentration might significantly influence future C_3/C_4 competitions. Although C_4 photosynthesis is almost nearly CO_2-saturated at present CO_2 concentration, C_4 plants can respond positively to elevated CO_2. The mechanism of stomata closure in C_4 plants exposed to elevated CO_2 leads to increased water use efficiency and there are direct observations that C_4 plants growth may be stimulated when at the same time growth of C_3 plants is not affected (Owensby et al., 1993). In contrast, Henderson et al. (1994) in Australia found a significant increase in the representation of C_4 grasses in the flora of southern and eastern Australia.

In summary, there is still some diverging discussion concerning relative effects of global environmental change on C_3 and C_4 plants. However, with other factors unchanged, increasing atmospheric CO_2 concentration seems to enhance the competitive advantage of C_3 over C_4 plants. Up to now it is not clear how a change in the C_3/C_4 balance *per se* would affect the global C cycle.

8.3 Biological N_2 Fixation

Nitrogen is the nutrient that is most commonly deficient, contributing to reduced agricultural yields throughout the world. Molecular nitrogen or dinitrogen (N_2) makes up four-fifths of the atmosphere but is metabolically unavailable directly to higher plants or animals. Higher plants and animals obtain nitrogen from nitrogen-fixing organisms or from nitrogen fertilizers (including nitrogen compounds formed during lightning strikes). Molecular nitrogen is available to some species of microorganism (so-called diazotrophs) through biological N_2 fixation in which atmospheric nitrogen is converted to ammonia by the enzyme nitrogenase (Kim & Rees, 1992) and a protein termed *ferredoxin* is used as electron donator. The produced NH_3 can be further converted to form organic compounds. Depending on the type of microorganism, the reduced ferredoxin is generated by photosynthesis, respiration or fermentation. Two moles of NH_3 are produced from one mole of nitrogen gas, at the expense of 16 moles of ATP and a supply of electrons and protons (Serraj et al., 1999):

$$N_2 + 8\,H^+ + 8e^- + 16\,ATP = 2\,NH_3 + H_2 + 16\,ADP + 16\,Pi \tag{8.4}$$

Bacteria that fix N_2 can be divided into free-living and symbiotic species. The free-living diazotrophs require a chemical energy source if non-photosynthetic, whereas the photosynthetic diazotrophs utilize light energy. The free-living diazotrophs contribute little fixed N_2 to agricultural crops. Associative nitrogen-fixing microorganisms are those diazotrophs that live in close proximity to plant roots (that is, in the rhizosphere or within plants) and can obtain energy materials from the plants (Cocking, 2003). The symbiotic relationship between diazotrophs called rhizobia and legumes (for example, clover and soybean) can provide large amounts of nitrogen to the plant and can have a significant impact on agriculture. The symbiosis between legumes and the nitrogen-fixing rhizobia occurs within nodules mainly on the root and in a few cases on the stem. A similar symbiosis occurs between a number of woody plant species and the diazotrophic actinomycete Frankia. The plant supplies energy materials to the diazotrophs, which in turn reduce atmospheric nitrogen to ammonia. This ammonia is transferred from the bacteria to the plant to meet the plant's nutritional nitrogen needs for the synthesis of proteins, enzymes, nucleic acids, chlorophyll, and so forth.

Legumes and N_2 fixation are very important in the developing world (Serraj et al., 1999), where much of the increases in food production must occur to accommodate increasing world population. It is essential that tropical legumes are exploited to replace fertilizer nitrogen, to avoid serious environmental problems of local and global proportions. The need for food, fuel, and building material has made deforestation an increasingly pressing problem in the developing world where legumes and other nitrogen-fixing trees offer a means of reversing this trend, especially the use of fast-growing N_2-fixing trees. By the year 2050, world population is expected to double from its current level of more than five billion. It is reasonable to expect that the need for fixed nitrogen for crop production will also double at least. If this is supplied by industrial sources, synthetic fertilizer nitrogen use will increase from presently 80 to about 160 million tons of nitrogen per year, about equal to that produced by the biological process. This amount of nitrogen fertilizer will require burning some 270 million tons of coal or its equivalent. However, expanded exploitation of biological N_2 fixation could reduce, and in the longer term substantially replace, the need for industrially produced fertilizer nitrogen.

8.3.1 N_2 Fixation by Non-symbiotic Bacteria

Non-symbiotic fixation of N_2 by soil bacteria (e.g. Azotobacter) requires a readily available energy source and is encouraged by restricted oxygen supply. This process is therefore likely to occur to greater extent in grassland than under arable land. Field studies in England and Wales showed that non-symbiotic fixation under grassland rarely amounted to more than 5 kg N ha^{-1} $year^{-1}$, and similar rates of non-symbiotic N_2 fixation (up to 8 kg N ha^{-1} $year^{-1}$) were reported for prairie soils in Ohio (Whitehead, 1995). Rates of N_2 fixation were much less in areas that had been previously treated with fertilizer N. The global mean N_2 fixation rate for grassland has been estimated to be 5 kg N ha^{-1} $year^{-1}$ (Smil, 1999). For arable crops also, non-symbiotic bacteria can make only a limited contribution to the N nutrition, because large amounts of organic nutrients are not continuously available to microbes in the rhizosphere (Table 8.4).

Non-symbiotic fixation is reduced by the presence of ammonium and nitrate, and therefore by the application of fertilizer N. When soil nitrogen is depleted, associative nitrogen fixers, for example *Azotobacter* spp. and *Azospirillum* spp., function vigorously when supplied with an energy source. However, they are considered of minor agricultural significance. Recently, two other free-living diazotrophs, *Acetobacter diazotrophicus* and *Herbaspirillum* spp., were found to live endophytically in the vascular tissue of sugarcane, where there is access to abundant sucrose as a possible source of energy for nitrogen fixation (Dšbereiner et al., 1993). This finding may explain the large positive nitrogen budgets measured with some cultivars of sugarcane in Brazil (Urquiaga et al., 1992). The fixation of nitrogen in these cultivars reduces the energy required for production of ethanol from sugarcane.

Table 8.4 Rates of asymbiotic N_2 fixation under different cropping systems (Adapted from Cocking, 2003)

Legume	N_2 fixation rate per crop: range (kg N ha^{-1} $year^{-1}$)
Rice-blue green algae	10–80
Rice-bacterial associations	10–30
Sugarcane-bacterial associations	26–160
Wheat-bacterial associations	10–30

8.3.2 N_2 Fixation by Symbiotic Bacteria

Plants and microbes form symbiotic associations in legumes, lichens, and some woody plants. The system most important for agriculture is the legume-rhizobia symbiosis (Cocking, 2003). The fixation of atmospheric N_2 occurs within root nodules after rhizobial penetration of the root. Thus, many legumes can grow vigorously and yield well under nitrogen-deficient conditions, and may contribute nitrogen to the farming system in the crop residues after grain harvest, or more significantly as green manure incorporated in the soil. Legumes are important sources of protein, mainly for feed in the developed world and for food in the developing world. They have been exploited as sources of nitrogen most notably in the agricultural systems of Australia and New Zealand. The successful introduction of legume crops necessitates the simultaneous introduction of compatible rhizobia bacteria (inoculants in various forms) (McInnes et al., 2004), which have been in use for about 100 years.

There are more than 13,000 described species of legumes. Of the approximately 3,000 species examined, more than 90% form root nodules (in which nitrogen fixation presumably occurs in symbiosis with rhizobia). Because few have been exploited for food, there is the prospect that the utilization of legumes could be expanded substantially. It is estimated that about 20% of food protein worldwide is derived from legumes. The highest consumption occurs in the former Soviet Union, South America, Central America, Mexico, India, Turkey, and Greece. The dietary use of legumes is quantitatively in the following order: dry bean (*Phaseolus vulgaris*), dry pea (*Pisum sativum*), chickpea (*Cicer arietinum*), broad bean (*Vicia faba*), pigeon pea (*Cajanus cajan*), cowpea (*Vigna unguiculata*), and lentil (*Lens culinaris*) (Agostini & Khan, 1986). Peanut (*Arachis hypogaea*) and soybean (*Glycine max*) are dominant sources of cooking oil and protein. They are also major food sources in some regions. The residual meal of soybean is an excellent and relatively inexpensive source of protein. Although a small percentage of the meal is incorporated into human foods, most of it is used for feeding livestock and pets. Symbiotic nitrogen fixation in legumes allows them to grow well without the addition of fertilizer nitrogen. However, it may be necessary to apply phosphorus and other deficient nutrients, as well as lime to alleviate soil acidity.

The importance of legumes in animal feed should not be overlooked. Alfalfa (*Medicago sativa*), clovers (*Trifolium* spp.), stylosanthes (*Stylosanthes* spp.), desmodium (*Desmodium* spp.), and other forages are grown extensively. They are either grazed or fed as hay or silage. Alfalfa silage furnishes not only roughage and high-quality protein, but also a variety of vitamins, minerals, and other nutrients. The anaerobic ensiling process supports a rapid fermentative acidification of the plant material, serving to preserve nutritional quality.

Legumes can contribute nitrogen to cropping systems in several ways (Howieson & Ballard, 2004). A gain in nitrogen will accrue to the soil if the total nitrogen in the plant residues left after harvest is greater than the total amount of nitrogen absorbed from the soil. In general terms, the less nitrogen available in the soil and the lower the nitrogen- harvest index of the legume crop, the greater will be the nitrogen gain by the system. To maximize the nitrogen contribution from a legume crop, the total crop must be incorporated in the soil as a green manure. This can be achieved by conventional means or by alley cropping (with legume shrubs) or agroforestry (with legume trees) approaches. Regular coppicing of the shrubs and trees provides foliage for incorporation into the soil as a green manure or for application as mulch. *Sesbania rostrata*, a legume shrub that is tolerant of waterlogging, forms nodules on stems, and has a high nitrogen-fixing capacity. It has been used to good effect as a green manure in paddy fields in Thailand and Senegal. Similarly, in some parts of China and Southeast Asia, Azolla is allowed to grow in paddy water (Choudhury & Kennedy, 2004) and is then incorporated into the soil as a green manure.

Legumes are often a component of intercropped systems in tropical agriculture (Baldock & Ballard, 2004), and the possibility of direct benefit to the nonlegume as a result of nitrogen excretion by the legume has been a contentious issue. Data in the literature show that nitrogen exchange does occur in certain circumstances, but it can be detected only under conditions of very low availability of soil nitrogen because it occurs only in small amounts. There is evidence that mycorrhizal connections between the intercropped components may provide a route of nitrogen transfer. Such nitrogen benefit to an intercropped cereal would be significant only under low-yielding conditions. When parts or all of the legume senesces and decomposes, the associated crop can obtain nitrogen in larger quantities.

Rates of symbiotic N_2 fixation besides species depend on site factors such as plant-available water, temperature, pH, soil mineral N content (Nieder et al., 2007), and use of rhizobial inoculants (Deaker et al., 2004). Each of these factors may cause a high variability of N_2 fixation rates (Table 8.5).

Symbiotic nitrogen fixation is highly sensitive to drought, which results in decreased N accumulation and yield of legume crops (Serraj et al., 1999). The effects of drought stress on N_2 fixation usually have been perceived as a consequence of straightforward physiological responses acting on nitrogenase activity and involving exclusively one of the three mechanisms: carbon shortage, oxygen limitation,

Table 8.5 Rates of symbiotic N_2 fixation under different species (Compiled from Whitehead et al., 1995; Cocking, 2003; Nieder et al., 2007)

Legume	N_2 fixation rate per crop: range (kg N ha^{-1} $year^{-1}$)
Peanut (*Arachis hypogaea*)	37–206
Pigeon pea (*Cajanus Cajan*)	7–235
Chickpea (*Cicer arietum*)	3–141
Soybean (*Glycine max*)	0–450
Broad bean (*Vicia faba*)	18–380 (178)
Dry pea (*Pisum sativum*)	18–334 (134)
Mungbean (*Vigna radiata*)	9–112
Acacia (*Leucaena leucocephala*)	100–300
Prostrate sesbania (*Sesbania rostrata*)	11–458
White lupine (*Lupinus albus*)	6–228 (98)
White clover (*Trifolium pratense*)	0–600
Grass clover	100–350
Rice-azolla	20–100

Figures in parenthesis are mean values

or feedback regulation by nitrogen accumulation. The sensitivity of the nodule water economy to the volumetric flow rate of the phloem into the nodule offers a common framework to understand each of these mechanism. As these processes are sensitive to volumetric phloem flow into the nodules, variations in phloem flow as a result of changes in turgor pressure in the leaves are likely to cause rapid changes in nodule activity. This could explain the special sensitivity of N_2 fixation to soil drying. It seems likely that N feedback may be especially important in explaining the response mechanism in nodules. A number of studies have indicated that nitrogenous signals, associated with N accumulation in the shoot and nodule, exist in legume plants so that N_2 fixation is inhibited early during soil drying. The existence of genetic variation in N_2 fixation response to water deficit among legume cultivars opens the possibility for enhancing N_2 fixation tolerance to drought through selection and breeding.

At the forest stand level as well, high rates of biological N_2 fixation are most often reported for actinorhizal and leguminous plants which fix nitrogen in symbiosis with procaryotes. Examples include red alder (*Alnus rubra*) with N_2 fixation rates up to 130 kg N ha^{-1} $year^{-1}$ (Binkley, 1981), *Casuarina equisetifolia* with N_2 fixation rates in the range of 12–85 kg N ha^{-1} $year^{-1}$ (Diem & Dommergues, 1990) and snowbrush (*Caenothus velutinus*) with N_2 fixation rates in the range of 20–100 kg N ha^{-1} $year^{-1}$ (McNabb & Cromack, 1983; Youngberg & Wollum, 1976; Zavitovski & Newton, 1968).

8.3.3 Global Estimates of Biological N_2 Fixation

Asymbiotic and symbiotic biological systems may fix an estimated 110–160 Tg of nitrogen annually (Table 8.6), and this probably has not changed substantially during the last century. About 40 Tg are attributed to forested ecosystems (Burns & Hardy, 1975).

Our understanding of spatial patterns and rates of biological N_2 fixation may be better for agricultural systems compared to natural ecosystems (Galloway et al., 2004).

Table 8.6 Total N inputs from biological N_2 fixation for the world and some regions (Adapted from Van Drecht et al., 2005)

Region	Biological N_2 fixation: mean of sources[a–d] (Tg N $year^{-1}$)	Biological N_2 fixation: range (Tg N $year^{-1}$)
World	135.4	110.2–160.1
Canada	4.8	2.5–7.3
South America	27.8	20.7–34.5
North Africa	2.7	1.0–4.3
Eastern Europe	1.4	1.2–2.1
Former USSR	11.2	9.0–16.4
East Asia	8.3	7.9–10.1
Oceania	8.6	7.1–10.4

Data from [a]Van Drecht et al. (2003), [b]Boyer et al. (2004), [c]Green et al. (2004) and [d]Siebert (2005)

In agricultural areas there are relatively good records of the distribution of cultivated croplands along with statistical information on agricultural management practices (Smil, 2001). In contrast, in natural ecosystems, it remains a challenge even to map the spatial distribution of natural vegetation species hosting N_2-fixing bacteria (Boyer et al., 2002). Moreover, there is a broad spectrum of N_2-fixing organisms in the natural environment, having complex distributions across the landscape. Furthermore, even in a single plant community, there exists a large degree of temporal and spatial heterogeneity in factors controlling N_2 fixation rates (Smil, 2001). In summary, there still remains huge uncertainty in understanding the magnitude of biological N_2 fixation at regional scales which is mainly due to a large degree of variability in observed rates and a relatively sparse number of measurements. This highlights the need for considerably more research in this area.

8.4 Carbon Dioxide Emission

Carbon dioxide cycles between the atmosphere, oceans and land biosphere (see Chapter 2). The atmosphere contains 762 Pg C and the total quantity of CO_2-C exchanged annually between the land and atmosphere due to natural processes such as photosynthesis, respiration, decay and sea surface gas exchange (gross primary productivity) is estimated at ~120 Pg C year^{-1}; and that between the ocean and the atmosphere at ~90 Pg C year^{-1} (Denman et al., 2007). However, there is an imbalance between emissions and uptake, caused by anthropogenic activities leading to increased concentration of CO_2 in the atmosphere. Over the last 250 years the atmospheric concentration of CO_2 has increased globally by ~100 ppm (36%) from about 275 ppm in the preindustrial era (AD 1000–1750) to 379 ppm in 2005 (Fig. 8.2;

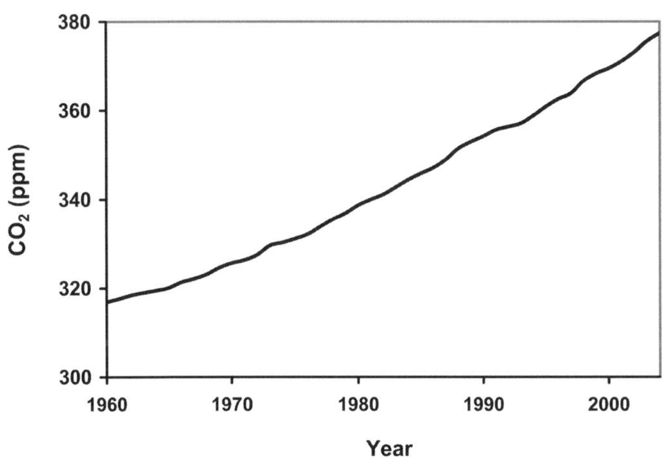

Fig. 8.2 Atmospheric CO_2 concentration derived from in situ air samples collected at Mauna Loa, Hawaii (Using data from Keeling & Whorf, 2005. Reproduced with kind permission from the Carbon Dioxide Information Analysis Center)

Denman et al., 2007). Direct instrumental measurements show that during the period 1960 to 2005 the atmospheric CO_2 concentration increased at 1.4 ppm year^{-1}. However, the highest average growth of 19 ppm occurred during the decade 1995–2005. The increase in global atmospheric CO_2 is mainly due to emissions from the combustion of fossil fuel and cement production though there is substantial contribution from land use changes and management such as deforestation, biomass burning, crop production and conversion of grassland to croplands (Andreae & Merlet, 2001; Houghton, 2003; van der Werf et al., 2004). Annual emissions of CO_2 from fossil fuel burning and cement production since 1960 has increased by a factor of more than 3, from ~2.5 Pg C year^{-1} in 1960 to ~7.8 Pg C year^{-1} in 2005 (Marland et al., 2006; Forster et al., 2007). Before 1900, emissions due to fossil fuel burning were well below 1 Pg C year^{-1}. Currently, fossil fuel combustion is responsible for more than 75% of anthropogenic CO_2 emissions and the remainder coming from land use changes (Fig. 8.3). Regional distribution of CO_2 emissions due to fossil fuel combustion, gas flaring and industrial activities for the years 1990 to 2005 shows (Table 8.7) that North American region (USA, Canada and Mexico) is the highest emitter accounting for about one-fourth of the total global emissions followed by Asia and OECD Europe (IEA, 2006). China contributes more than 50% to the total emissions from Asia and its emissions has increased from 0.69 Pg in 1990 to 0.94 Pg in 2000. Except for former USSR and non-OECD Europe, the emissions from different regions of the world have increased over the years.

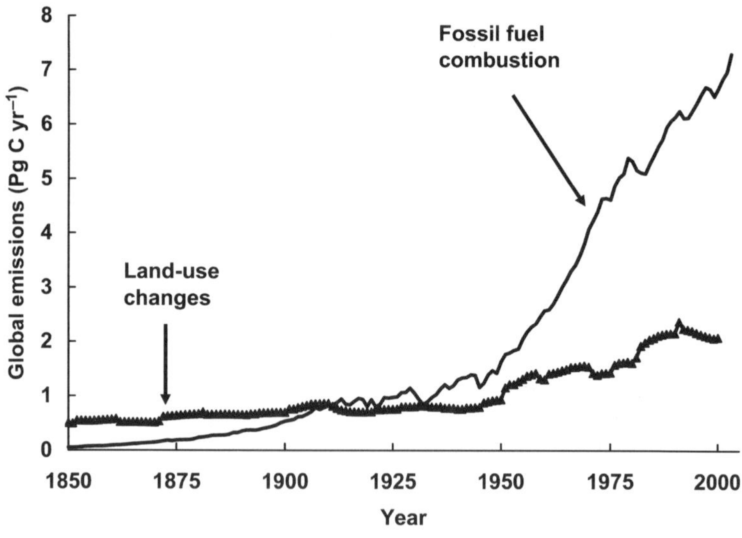

Fig. 8.3 Annual global CO_2 emission from fossil fuel burning and cement manufacture (1850–2003) and land-use changes (1850–2000) (Using data from CDIAC web site; Houghton & Hackler, 2002 Marland et al., 2006. Reproduced with kind permission from the Carbon Dioxide Information Analysis Center)

Table 8.7 Regional distribution of CO_2-C emissions (Pg C) from fossil fuel burning, gas flaring and industrial processes for the years 1990, 1995 and 2000 (Recalculated from IAE, 2006)

Region	1990	1995	2000
OECD N. America (Canada, Mexico and USA)	1.55	1.64	1.84
OECD Pacific (Australia, Japan, Korea and New Zealand)	0.45	0.52	0.56
OECD Europe[a]	1.12	1.10	1.13
Non-OECD Europe[b]	0.11	0.07	0.07
Former USSR	0.94	0.68	0.62
Africa	0.31	0.32	0.54
Middle East	0.19	0.25	0.30
Latin America	0.42	0.42	0.58
Asia (including India and China)	1.27	1.62	1.73
India	0.20	0.25	0.32
China	0.69	0.92	0.94
International bunkers	0.18	0.19	0.23
World	6.54	6.81	7.60

[a] OECD (Organization for Economic Development) – Europe includes Austria, Belgium, Czech Republic, Denmark, Finland, France, Germany, Hungary, Iceland, Ireland, Italy, Luxemburg, the Netherlands, Norway, Poland, Portugal, Slovak Republic, Spain, Sweden, Switzerland, Turkey, the United Kingdom

[b] Non-OECD Europe includes Albania, Bulgaria, Cyprus, Gibraltar, Malta, Romania, Former Yugoslavia, Bosnia-Herzegovina, Coatia, FYR of Macedonia, Serbia/Montenegro, Slovenia

The CO_2 emissions due to land use changes during the 1990s are estimated as 0.5–2.7 Pg C $year^{-1}$, contributing 6–39% of the CO_2 growth rate (Brovkin et al., 2004). In carbon cycle simulations by Brovkin et al. (2004) and Matthews et al. (2004), land use change emissions contributed 12–35 ppm of total CO_2 rise from 1850 to 2000. Until the beginning of the 20th century emissions from changes in land use and management were greater than those from fossil-fuel burning, but the latter now dominates by a factor of about 3 (Fig. 8.3). According to estimates presented by Houghton (2003), total emissions from 1850 to 2000 from land use change amounted to 156 Pg C, about 60% of which was from the tropics. During this period, the greatest regional flux was from tropical Asia (48 Pg C), while the smallest regional flux was from north Africa and Middle East (3 Pg C). Global annual flux during 1980s and 1990s averaged 2.0 and 2.2 Pg C $year^{-1}$, respectively, dominated by fluxes from tropical deforestations. Outside the tropics, the average net flux of carbon attributable to land use changes and management decreased from a source of 0.06 Pg C $year^{-1}$ during the 1980s to a sink of 0.02 Pg C $year^{-1}$ during the 1990s (Houghton, 2003).

The observed increase in atmospheric CO_2 concentration accounts for only 55% of the CO_2 released by human activity since 1959. The rest has been taken up by the balance between sources (emissions due to human activities and natural systems) and sinks (the removal of the gas from the atmosphere by conversion to a different chemical compound). The global carbon budget (Table 8.8) shows that as compared to atmospheric increase of 3.2 Pg C $year^{-1}$ in 1990s the atmospheric load increased at a rate of 4.1 Pg C $year^{-1}$ during the years 2000–2005. During

Table 8.8 The global carbon budget (Pg C $year^{-1}$) during 1990s and 2000–2005. Errors represent ± standard deviation. Positive fluxes indicate emissions to the atmosphere and negative fluxes are losses from the atmosphere (sinks) (Adapted from Denman et al., 2007)

	1990s	2000–2005
Emission from fossil fuel and cement production	+6.4 ± 0.4	+7.2 ± 0.3
Net ocean to atmosphere flux	−2.2 ± 0.4	−2.2. ± 0.5
Net land to atmosphere flux	−1.0 ± 0.6	−0.9 ± 0.6
Land use change	+1.6 (0.5–2.7)	
Residual terrestrial sink	−2.6 (−4.3 to −0.9)	
Atmospheric increase	+3.2 ± 0.1	+4.1 ± 0.1

the later period, while fossil fuel burning and cement production are the net source of ~7.2 Pg C $year^{-1}$, ocean and land partially offset these emission by ~3.1 Pg C $year^{-1}$.

8.4.1 Carbon Dioxide Emissions from Biomass Burning and Soils

Fire is a major agent for conversion of biomass and soil organic matter to CO_2. Globally, wildfires oxidize 1.7–4.1 Pg C $year^{-1}$ or about 3–8% of total terrestrial net primary productivity (Denman et al., 2007). Estimates of carbon emitted to the atmosphere due to biomass burning are highly uncertain as the combustion efficiencies and the extent of burned area are not precisely know. Mouillot et al. (2006) using a 100-year, 1° × 1° global fire map and a biogeochemical carbon cycle model estimated total direct emissions from fires as 3.3 Pg C $year^{-1}$ out of which 50% come from savannas, 38% from tropical forests, 6.2% from boreal forests and 5.6% from temperate forests. But these estimates differ considerably compared to the previous estimates of 2–2.9 Pg C $year^{-1}$.

Soils constitute the largest pool of actively cycling C in terrestrial ecosystems (see Chapter 1) and stock about 1,500–2,000 Pg C (to a depth of 1 m) in various organic forms, from recent plant litter to charcoal to very old, humified compounds (Amundson, 2001) and 800–1,000 Pg as inorganic carbon or carbonate carbon (Post et al., 1982; Eswaran et al., 1993). About a third of the soil organic C occurs in forests, another third is in grasslands and savannas, and the rest is in wetlands, croplands and other biomes. Atmospheric CO_2 enters terrestrial biomass via photosynthesis, at a rate of about 120 Pg C $year^{-1}$ (Gross Primary Productivity) and about half of it is soon released as CO_2 by plant respiration, so that net primary productivity is ~60 Pg C $year^{-1}$. Heterotrophic respiration (largely by soil microorganisms) and fire return an equivalent amount (~60 Pg C $year^{-1}$) back to the atmosphere. Averaged over total area of continents, these C fluxes amount to about 4 Mg C ha^{-1} $year^{-1}$ (Janzen, 2004). Estimates of historic loss of soil organic C from the cultivated cropland soils of the world range from 41 to 55 Pg C (Houghton & Skole, 1990; Paustian et al., 1998).

8.4.2 Carbon Dioxide Emission Mitigation Options

According to Kyoto Protocol, industrial countries are to reduce their emissions of GHGs by an average of 5% below their 1990 emissions by the first commitment period, 2008–2012. Therefore, there has been increased focus to look for options for mitigating the emission of GHGs. The approaches to mitigate or stabilize concentration of CO_2 in the atmosphere include:

- Reducing energy consumption by increasing the efficiency of energy conversion and/or utilization. According to IPCC (2001b) improvements in energy efficiency have the potential to reduce global CO_2 emissions by 30% below year-2000 levels using existing technologies at a cost of less than US\$30 t^{-1} CO_2 (US\$100 t^{-1} C).
- Decarbonizing energy supplies either by switching to less carbon intensive fuels (for example natural gas instead of coal) or using alternative, non-CO_2 emitting energy sources, such as wind, solar, or nuclear energy.
- Capturing and storing CO_2 chemically or physically in repositories such as deep ocean or geological formations (IPCC, 2005). The process of CO_2 capture and storage (CCS) involves collection and concentration of CO_2 produced in industrial and energy related sources, transporting it to a suitable storage location, and then storing it away from the atmosphere such as geological formations, in the ocean, in mineral carbonates or for use in industrial processes. As of mid-2005 there are three commercial projects linking CO_2 capture and geological storage: one each in Norway, Canada, and Algeria each of which captures and stores 1–2 Mt CO_2 $year^{-1}$. The technology is not mature enough yet and has not yet been applied at a large scale but it may become a viable option by 2015 or 2020.
- Replacing fossil fuel with biofuels that recycle recently photosynthesized atmospheric CO_2, rather than introducing new, previously dormant C into active cycling. Biomass from agricultural residues and dedicated energy crops can be an important bioenergy feedstock, but its contribution to mitigation depends on demand for bioenergy from transport and energy supply, on water availability, and on requirement of land for food and fiber production. It has been estimated that annually 0.5–1.5 Pg fossil fuel C could be substituted by dedicated biofuels (0.25–1 Pg C $year^{-1}$), shelterbelts and agroforestry (0.06–0.25 Pg C $year^{-1}$) and crop residues (0.21–0.32 Pg C $year^{-1}$). The benefits are diminished or negated if excessive fossil fuel is used to produce the biofuel, or if removal of more NPP reduces the amount of C stored in terrestrial ecosystems (Sauerbeck, 2001). Widespread use of agricultural land for biomass production for energy may compete with other land uses and can have positive and negative environmental impacts and implications for food security (IPCC, 2007b). There are already indications that the growing use of cereals, sugar, oilseed and vegetable oils for ethanol and bio-diesel are changing crop prices and animal feed costs.
- Increasing the amount of C stored in vegetation and soil (C sequestration): Any practice that increases net primary productivity or reduces the rate of heterotrophic respiration will increase C storage. Since, the Kyoto Protocol provides

for C sequestration through Clean Development Mechanisms, the option has attracted particular attention of ecologists and others probing the global C cycles. Better management of agricultural soils, restoration of degraded soils and ecosystems, restoration of former wetlands now being used for agriculture has a vast potential of C sequestration. Practices that enhance C sequestration include afforestation and reforestation, conservation tillage and mulch farming, integrated nutrient management and adopting systems with high biodiversity (Lal, 2004a). Carbon sequestration besides being a cost-effective strategy has the co-benefits of restoring soil fertility and productivity, reducing risk of soil erosion and sedimentation, and enhancing biodiversity.

- Reducing other agriculture related emissions of CO_2 such as less energy use in agricultural operations (such as through reduced tillage, optimal fertilizer use efficiency, improved irrigation techniques and enhanced solar drying) and minimizing conversion of new land to agriculture in the tropics. It has been estimated that by exploiting all the possibilities of fuel saving a 10–40% reduction in the present agricultural energy requirement equivalent to 10–50 Tg C $year^{-1}$ may be achieved (Sauerbeck, 2001).

The mitigation options associated with land use changes are strongly related to major climatic zones and the most significant opportunities appear to be in the humid tropics and in tropical wetlands (Paustian et al., 1998). Choice and effectiveness of one or more mitigation options will depend on a variety of factors such as the potential of an option to deliver emission reductions, the national resources available, the accessibility of a technology for the country concerned, national commitments to reduce emissions, the availability of finance, public acceptance, likely infrastructural changes, environmental side-effects, etc. (IPCC, 2005). Terrestrial C sequestration through biotic processes appears plausible option of reducing the rates of CO_2 emissions while abiotic processes of carbon storage and alternatives to fossil fuel take effect.

8.4.3 Role of Forests in CO_2 Mitigation

Forests are an important component of the global C cycle containing about half of the C residing in terrestrial vegetation and soil, amounting to some 1,200 Pg of C. Forests both influence and are influenced by climate change, and their management will have a significant influence on global warming in the present century. Carbon in forests is stored in living biomass, including standing timber, branches, foliage and roots, and in dead biomass, including litter, woody debris and SOM. Compared to other terrestrial ecosystems, forest vegetation has a very high density. The C stored in the soil and litter of forest ecosystems also makes up a significant portion of the terrestrial C pool. Globally, SOC represents more than half of the stock of C in forests. Boreal forests account for more than any other terrestrial ecosystem (26% of total terrestrial C stocks), while tropical and temperate forests account for

20% and 7%, respectively (Dixon et al., 1994). There are, however, considerable variations among forest types. In boreal ecosystems, 80–90% is stored in the form of SOM, whereas in tropical forests, the carbon is fairly equally distributed between vegetation and soil. The main reason for this variation is the influence of temperature on the relative rates of production and decay of organic matter. At high latitudes, SOM accumulates because it is produced faster than it can be decomposed. In contrast, at low latitudes, warmer temperatures enhance decomposition of SOM and recycling of nutrients.

Any activity that affects the amount of biomass in vegetation and soil has potential to sequester C from, or release C into, the atmosphere. Forest management can contribute towards the mitigation of global warming through emission reductions and C sequestration. Forestry measures alone will not be enough to halt the increase in atmospheric CO_2 concentration. They can only complement efforts to reduce C emissions from the burning of fossil fuels. Particularly the effects of the rise in global atmospheric CO_2 concentrations and increased N deposition rates in forests near industrial regions have lead to an increase in forest biomass in recent years. Through the combined effects of reforestation, regrowth of degraded forests and enhanced growth of existing forests, between about 1 and 3 Pg C $year^{-1}$ may be absorbed (Malhi et al., 1999).

8.4.3.1 Management of Forest Carbon

There are several strategies for the management of forest C (Table 8.9). The first is to reduce the demand for fossil fuel by increasing the use of wood for durable wood products (C substitution). The second is to reduce or prevent the rate of C release from existing C sinks (C conservation), and the third is to increase the rate of C accumulation by enhancing or establishing C sinks (C sequestration).

In contrast to the combustion of fossil fuel, the use of biofuels does not result in a net release of CO_2 into the atmosphere because the CO_2 released through combustion of biofuels is the modern C taken up by the growing biomass. If current biofuel use were to be replaced by energy from fossil fuels, an additional 1.1 Pg C $year^{-1}$

Table 8.9 Forest carbon management strategies and measures (Adapted from Bass et al., 2000)

Management strategy	Management measure
C substitution	– Conversion of forest biomass into durable wood products in place of energy-intensive materials – Use of biofuels (establishment of bioenergy plantations) – Use of harvest residues (e.g. sawdust or straw) for biofuel
C conservation	– Conservation of biomass and SOM in existing forests – Improved efficiency of wood processing – Fire protection
C sequestration	– Afforestation, reforestation and restoration of degraded land – Introduction of agroforestry systems on arable land – Improved silviculture techniques to enhance growth rates

would be released to the atmosphere (IPCC, 2000c). The C substitution of fossil fuels by biofuels will result in a reduction of C emissions, which is proportional to the mass of fossil fuel C replaced. Estimates of the future contribution of biofuels to meet the energy demand range from 59 to 145 × 10^{18} J for 2025 and to 94 to 280 × 10^{18} J for 2050 (Bass et al., 2000). Establishment of new biofuel plantations will also have a long-term C sequestration effect if they replace land use systems with a lower or zero C sequestration rate.

In terms of forestry, the conservation of existing forest carbon stocks has the greatest potential for mitigation of climate change. Reducing the present rate of deforestation will produce a more direct effect on global atmospheric CO_2 levels than the measures listed in Table 8.9 under 'C sequestration'. If deforestation were stopped immediately, 1.2–2.2 Pg C $year^{-1}$ could be conserved (Dixon et al., 1993). Brown et al. (1996) estimated that a reduction in deforestation in tropical regions could conserve 10–20 Pg C by 2050 (0.2–0.4 Pg C $year^{-1}$). The most important management practice to conserve C stocks in existing forests is the use of reduced impact logging in the tropics. Conventional logging practices lead to a high level of damage to the residual stand, with up to 50% of remaining trees killed or damaged (Kurpick et al., 1997). Extreme weather conditions caused by climate change will increase the risk of wildfires. Fire management practices have the potential to conserve C stocks in forests. However, fire prevention and fire fighting efforts are to be combined with land use policy measures to address the needs of rural population.

Carbon sequestration rates as a consequence of afforestation/reforestation depend on the site characteristics, species involved, and management. Silvicultural activities that increase the productivity of forest ecosystems, such as timely thinning, can increase forest C stocks to some extent. However, compared with afforestation/reforestation, the effect of varying silviculture systems on total C stocks is relatively low (Dixon et al., 1993).

8.4.3.2 Carbon Yield Following Forest Management Measures

Carbon sequestration rates for forest plantation, forest management and agroforestry vary within a wide range (Table 8.10). At the global level Sedjo & Solomon (1989) estimated C yields of about 6.24 Mg C ha^{-1} $year^{-1}$ while Nordhaus (1991) estimated a range of only 0.8–1.6 Mg C ha^{-1} $year^{-1}$. Typical C sequestration rates following forest plantation are 0.8–2.4 Mg C ha^{-1} $year^{-1}$ in boreal forests, about 1–10 Mg C ha^{-1} $year^{-1}$ in temperate regions and 2–19 Mg C ha^{-1} $year^{-1}$ in the tropics.

Estimates by Richards & Stokes (2004) suggest that it may be possible to sequester 0.25–0.5 Pg $year^{-1}$ in the US alone, and up to 2.0 Pg $year^{-1}$ worldwide. Assuming a global land availability of 345 million hectares for afforestation/reforestation and agroforestry activities, Brown et al. (1996) estimated that over the next 50 years at least 38 Pg C could be sequestered, i.e., 30.6 Pg by afforestation/

Table 8.10 Carbon yields following forest management measures (Adapted from Richards & Stokes, 2004)

Authors	Region	Forest plantation ($Mg\ C\ ha^{-1}\ year^{-1}$)
Sedjo & Solomon (1989)	Global	6.24
Nordhaus (1991)	Global	0.8–1.6
Brown et al. (1996)	Boreal	0.8–2.4
	Temperate	0.7–7.5
	Tropical	3.2–10
Moulton & Richards (1990)	USA	2.0–10.9 (0–7.6)[a]
Dudek & LeBlanc (1990)	USA	3.7–8.9
Adams et al. (1993)	USA	2.0–10.9
Richards et al. (1993)	USA	0–9.4
Parks & Hardie (1995)	USA	3.3–5.1
Richards (1997)	USA	0.9–9.4
New York State (1991)	New York/USA	2.1 (1.1)[a]
van Kooten et al. (1992)	Canada	0.6–0.8 (0.6–0.12)[a]
Wangwacharakul & Bowonwiwat (1995)	Thailand	2.21–18.75 (0.95–6.25)[b]
Barson & Gifford (1990)	Australia	7.5
Tasman Institute (1994)	New Zealand	7.7

[a]Carbon yield following forest management
[b]Carbon yield following agroforestry

Table 8.11 Potential contribution of forest management measures to global C sequestration, 1995–2050, based on a total C sequestration potential of 38 Pg (Adapted from Brown et al., 1996)

Management measure	Percent contribution
Temperate afforestation/deforestation	31
Temperate agroforestry	2
Boreal afforestation/reforestation	6
Tropical agroforestry	17
Tropical afforestation/reforestation	44

reforestation and 7 Pg through the increased adoption of agroforestry practices (Table 8.11).

Studies of tropical regions indicate that an additional 11.5–28.7 Pg C may be sequestered through the regeneration of about 217 million hectares of degraded land. However, the present availability of land for forest management may be less when full account is taken of economic and social factors. In fact, only one third of ecologically suitable land may presently be available for reforestation/afforestation activities (Houghton et al., 1991). Considering this, afforestation/reforestation and agroforestry activities would only sequester about 0.25 Pg and the restoration of degraded land a further 0.13 Pg C $year^{-1}$.

8.4.4 Potential for C Sequestration by Agriculture

Terrestrial ecosystems can play an important role in mitigating CO_2 emissions through biotic processes of C sequestration in soils, biota and wetlands. Restoration of degraded ecosystems, land use and management, especially agriculture and forestry, can enhance terrestrial C sequestration. Degraded ecosystems have lost a large proportion of their native C pool and the present pool is much below the potential capacity. Such ecosystems include soils degraded by severe water and wind erosion, salinization, nutrient depletion, compaction, contamination and pollution, and drastic disturbance by mining activities. Restoring wetlands has a large potential of C sequestration, because erosion of top-soil and organic matter from upland catchment areas is deposited in wetlands and the decomposition rate is slow. It has been estimated that in temperate and cool climates annually 0.5–1 Mg C ha^{-1} can be sequestered by restoring wetlands, 0.2–0.8 Mg C ha^{-1} by restoring severely degraded soils and 0.2–0.5 Mg C ha^{-1} by mine soil reclamation (Lal, 2004a). Globally, restoration of degraded soils could increase C sequestration by 0.65–1.9 Pg C $year^{-1}$ (Batjes, 1999).

Because of historic losses of C from soils, estimated to be 41–55 Pg, the soils have significant capacity to mitigate atmospheric CO_2 through enhanced C sequestration. Improved management of existing agricultural lands can significantly enhance C sequestration in soils. Management practices or technologies that increase carbon input to the soil and decrease output/losses of carbon lead to carbon sequestration in soils (Fig. 8.4). Enhanced biomass production, humification of organic materials returned to the soil, aggregation by formation of organomineral complexes, deep placement of organic carbon beneath the plow zone, deep rooting, and calcification result in greater C sequestration. Management practice, which favor or facilitate these processes include return of above ground and below ground biomass to the soil, exogenous application of organic materials (e.g. animal manure, compost, sludge, etc.), adoption of agroforestry systems, intensification of agriculture adopting recommended management practices, reducing winter fallow or periods with no ground cover, changing from monoculture to rotation cropping, switching from annual crops to perennial vegetation, and increasing area under forests. Switching from annual crops to perennial vegetation increases residue production, plant roots and reduces soil disturbance, thus enhancing soil C sequestration (Paustian et al., 1997c). Average global C sequestration rates, when changing from agriculture to forest or grassland have been estimated to be 33.8 and 33.2 g C m^{-2} $year^{-1}$, respectively (Post & Kwon, 2000). But there is a large variation in the length of time for and the rate at which C may accumulate in the soil, related to the productivity of the recovering vegetation, physical and biological conditions in the soil, and the past history of soil organic carbon inputs and physical disturbance (Post & Kwon, 2000). For example, Silver et al. (2000) estimated that reforestation of abandoned tropical agricultural land and pasture can sequester 130 g C m^{-2}-$year^{-1}$ for the first 20 years, and then at an average rate of 41 g C m^{-2} $year^{-1}$ during the following 80 years. Forestry has been proposed as a means to sequester C and

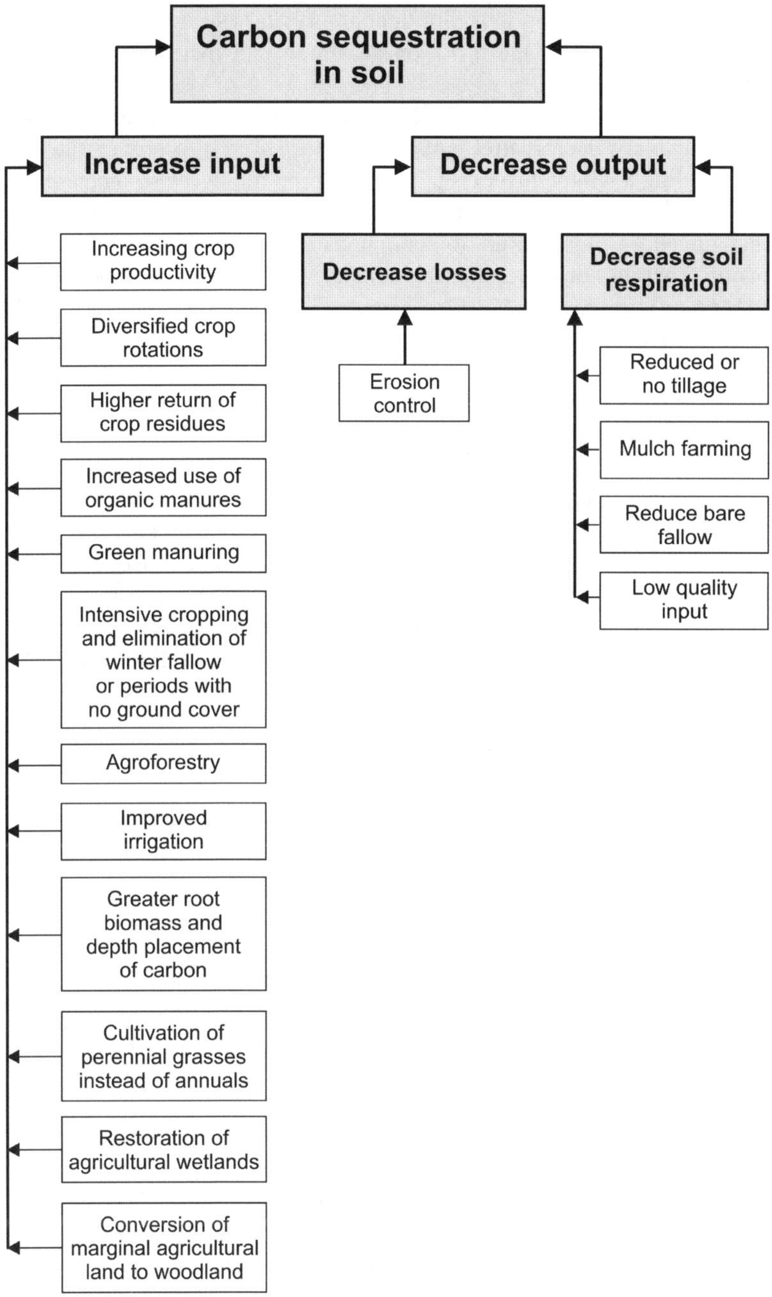

Fig. 8.4 Strategies for carbon sequestration in agricultural soils

reduce greenhouse gas emissions. Smith et al. (1997a) estimated that afforestation of 30% of present arable land in European Union will increase soil C stocks by about 8% over a century.

In some climatic regions, land dedicated to annual crops can be planted with a grass or legume cover crop after harvesting the cash crop to protect the soil over winter. Including a winter cover crop in annual crop rotation also increases residue inputs to the soil and hence soil C sequestration. Enhancing rotation complexity (i.e. changing from monoculture to continuous cropping, changing crop-fallow to continuous cropping, or increasing the number of crops in a rotation system) can sequester on an average 20 ± 12 g C m^{-2} $year^{-1}$ excluding a change from continuous corn to corn-soybean which may not result in significant accumulation of C (West & Post, 2002).

Management options that result in reduced output through decomposition or soil respiration include reduced or no-tillage practices, mulch farming, reduced bare fallow or increased cropping intensity (Fig. 8.4). Croplands under no-till systems have been shown to increase soil C compared to more intensive tillage operations. Analysis of results from a global database of 67 long-term experiments showed that a change from conventional tillage (CT) to no-till (NT) can sequester 57 ± 14 g C m^{-2} $year^{-1}$, excluding wheat-fallow system, which may not result in SOC accumulation with a change from CT to NT (West & Post, 2002). Carbon sequestration rates, with a change from CT to NT, can be expected to peak in 5–10 year with SOC reaching a new equilibrium in 15–20 year. No-till agriculture greatly reduces the degree of soil disturbance normally associated with annual cropping. Physical disturbance associated with intensive soil tillage increases the turnover of soil aggregates and accelerates the decomposition of aggregate associated SOM (Paustian et al., 2000). No-till increases aggregate stability and promote the formation of recalcitrant SOM fractions within stabilized micro- and macro-aggregate structures, and reduces soil erosion. Greater cropping intensity, i.e. by reducing the frequency of bare fallow in crop rotations and increasing the use of perennial vegetation, can increase water and nutrient use efficiency by plants, thereby increasing C inputs to soil and reducing organic matter decomposition rates (Paustian et al., 2000). It has been widely observed that soil C is lower in systems employing summer fallow than in continuous cropping systems (Campbell et al., 2000a; Janzen et al., 1998).

Sperow et al. (2003) analyzed the influence of several improved management strategies on potential soil C storage in the US cropland for a 15 year period. Their analysis showed that US cropland soils have the potential to increase sequestered soil C by an additional 60–70 Tg C $year^{-1}$, over present rates of 17 Tg C $year^{-1}$ with widespread adoption of soil C sequestering management practices. Adoption of no-till on all annually cropped area (129 Mha) would increase soil C sequestration by 47 Tg C $year^{-1}$. Elimination of summer fallow practices and conversion of highly erodible cropland to perennial grass cover could sequester around 20 and 28 Tg C $year^{-1}$, respectively. The soil C sequestration potential from including a winter cover crop on annual cropping system was estimated at 40 Tg C $year^{-1}$. The total sequestration potential estimated (Sperow et al., 2003) for the 15 year period (83 Tg

C year^{-1}) represents about 5% of 1999 total US CO_2 emissions or nearly double estimated CO_2 emissions from agricultural production (43 Tg C year^{-1}). Their analysis suggests that agricultural soil C sequestration could play a meaningful, but not predominant role in helping mitigate greenhouse gas increase.

Globally, potential CO_2 mitigation by agricultural has been estimated to be 49–126 Pg C over a 50 year period (0.9–2.5 Pg C year^{-1}) with dedicated biofuel crops and use of crop residues as biofuel accounting for 25–80 Pg C (0.5–1.6 Pg C year^{-1}), and enhanced C sequestration in soil contributing 24–43 Pg C (0.4–0.9 Pg C year-1) through improved management of existing agricultural soils, restoration of degraded lands, permanent set-asides of surplus agricultural lands in temperate developed countries and restoration of 10–20% of former wetlands now being used for agriculture (Table 8.12; Paustian et al., 1998). The exploratory scenarios developed by Batjes (1999) show that from 14 ± 7 Pg C (0.58–0.80 Pg C year^{-1}) may be sequestered over the next 25 years if the world's degraded and stable agricultural lands are restored and submitted to appropriate management. There is considerable uncertainty in the estimates, concerning both C flux rates and C storage capacity as well as in the level at which various mitigation options could be implemented. Since

Table 8.12 Estimates of CO_2 mitigation potential by agriculture (Adapted from Paustian et al., 1998)

Mitigation option	Assumption	Annual (Pg C)	Cumulative (Pg C)
Better management of existing agricultural soils	1/2 to 2/3 rd recovery of the estimated 43 Pg historical C loss	0.4–0.6	22–29
Set-aside of upland soils	15% of 640 Mha farm land at 10–15% implementation	0.003–0.03	0.15–1.5
Restoration of wetland soils	10–20% of former 8 Mha wetland now under cultivation	0.006–0.012	0.3–0.6
Restoration of soil C on degraded lands	10–50% of 1,200 Mha globally degraded land	0.024–0.24	1.2–12
Sub-total (soils)		0.43–0.88	24–43
Dedicated biofuel crops	10–15% of world cropland available for biofuels	0.3–1.3	15–65
Crop residues as biofuels	25% recovery of crop residues and assumptions on energy conversion and substitution	0.2–0.3	10–15
Reduction in fossil energy use	10–15% reduction in current use	0.01–0.05	0.5–2.5
Total		0.94–2.53	49–126

soils have a finite capacity to store additional C, the total amount of C sequestered and the estimates thereof depend on the time horizon considered. The C sequestration potential of a soil depends on the vegetation it supports, its mineralogical composition, the depth of solum, soil drainage, the edaphic environment, soil organic matter content and it ability to resist microbial decomposition (Swift, 2001). Further, the question of permanence that is how long the sequestered C will stay in the soil must also be addressed. Permanence of C sequestered in soil depends on the continuation of the recommended management practices (Lal, 2004a).

These estimates of C sequestration have been made assuming that best management practices and/or manipulation of a large portion of the global soils is possible. However, this may not be possible because of a variety of ecological, socioeconomic and policy reasons. The most appropriate management practices to increase soil C reserves are, therefore, site specific, which will require evaluation and adaptation with reference to soil type and land use system, preferably by agroecological region (Batjes, 1999). Since no single land-management strategy in isolation may be adequate to mitigate carbon emissions, it is important to evaluate the integrated combination of various land-management strategies, as done by Smith et al. (2000b) for European soils (Fig. 8.5). A realistic optimal combined land management scenario that have mitigation potential of 103 Tg C year^{-1} and can meet Europe's Kyoto Protocol reduction commitments includes level of bioenergy production and woodland growth, rates and areas of organic amendment, and an area for no-till farming. The realization of the optimal scenario would entail changes in European land management/agricultural policy such as using surplus arable land for alternative long-term land use, growing bioenergy crops and woodland regeneration on surplus land as per feasibility, greater adoption of conservation tillage and application of majority of organic amendments to arable land (Smith et al., 2000b).

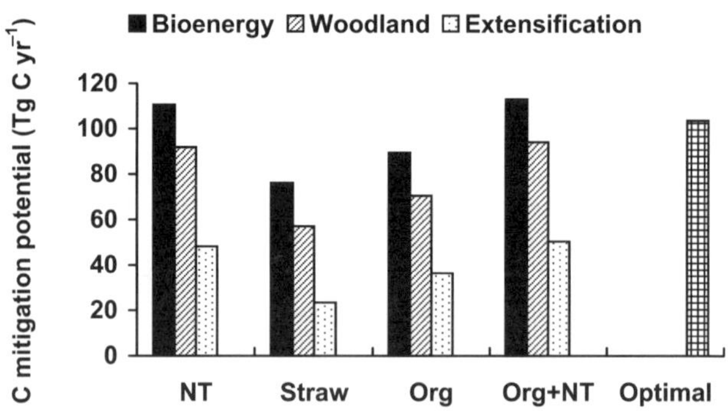

Fig. 8.5 Maximum yearly carbon mitigation potential in Europe through different combinations of land management scenario using 10% surplus arable either for bioenergy production or woodland regeneration or extensification and other management practices viz. no-tillage (NT), straw incorporation (Straw), application of organic amendments (Org). Optimal scenario uses 50% of surplus arable land for bioenergy production and the other 50% for woodland, application of organic amendments at the highest rates allowed and putting the remaining area into no-till (Adapted from Smith et al., 2000b)

8.5 Methane Emission

Methane (CH_4) is a potent greenhouse gas and is about 25 times more powerful at warming the atmosphere than CO_2 over a 100-year period (Forster et al., 2007). It has the second-largest radiative forcing (0.48 W m^{-2}) after CO_2 (1.66 W m^{-2}). Methane contributes some 16% of the global warming resulting from the increasing concentrations of greenhouse gases in the atmosphere. Methane has an atmospheric lifetime of about 12 years and plays an important role in atmospheric oxidation chemistry and affects stratospheric ozone and water vapor levels.

Since the preindustrial times, the atmospheric concentration of CH_4 has almost tripled and ice core records indicate that the abundance of CH_4 in atmosphere has varied from about 400 ppb during glacial periods to about 700 ppb during interglacials (Spahni et al., 2005). In 2005, the global average abundance of CH_4 was 1,774 ± 1.8 ppb (Forster et al., 2007). In recent years atmospheric growth rate of CH_4 seems to stagnate, or even decline (Fig. 8.6). The global growth rate of atmospheric methane decreased from nearly +12 ± 2 ppb $year^{-1}$ in the 1980s to +4 ± 4 ppb $year^{-1}$ in the last decade. However, there is a large interannual variability, with growth rates ranging from a high of 14 ppb $year^{-1}$ in 1998 to less than zero in 2001, 2004 and 2005. The reasons for the decrease in the atmospheric CH_4 growth rate and the implications for future changes in its atmospheric burden are not understood. Bouquet et al. (2006) attributed the interannual variability to wetland emissions and the long-term changes during the 1990s to a decrease in anthropogenic emissions.

Atmospheric CH_4 originates from both natural and anthropogenic sources. The natural sources of CH_4 include wetlands, oceans, forests, wildfires, termites, geological

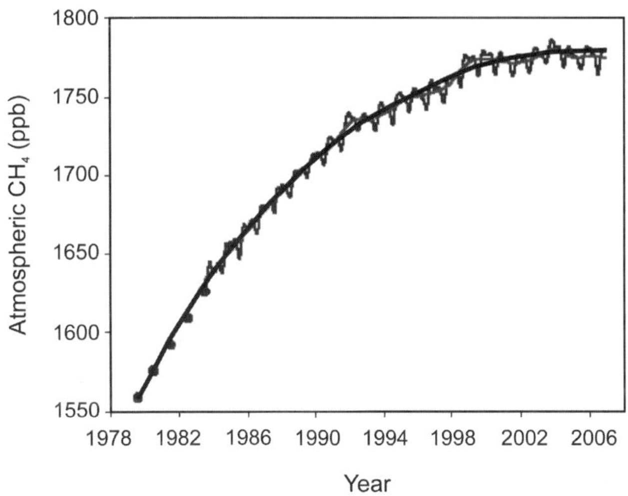

Fig. 8.6 Atmospheric CH_4 concentration from the NOAA global flask sampling network since 1978 (http://www.esrl.noaa.gov/gmd/aggi/. Reproduced with kind permission from the US National Oceanic and Atmospheric Administration)

sources and gas hydrates. The anthropogenic sources include rice agriculture, livestock, landfills and waste treatment, ruminants, biomass burning, and fossil fuel combustion (Denman et al., 2007). While emissions from natural sources dominated the preindustrial global budget of atmospheric methane, anthropogenic emissions dominate the current methane budget. Total global preindustrial emissions of CH_4 are estimated to be 200–250 Tg CH_4 year^{-1} (Denman et al., 2007) of which natural sources emitted between 190 and 220 Tg CH_4 year^{-1} and anthropogenic sources accounted for the rest. In contrast, anthropogenic emissions account for about 70% of the current global budget (Table 8.13).

The most important natural source for CH_4 emission is wetlands, which account for about 80% of the total natural emissions (Table 8.13). Several studies indicate a high sensitivity of wetland CH_4 emissions to temperature and water table.

Table 8.13 Global annual CH_4 emissions (Tg CH_4 year^{-1}) from natural and anthropogenic sources (Adapted from Denman et al., 2007)

	Wuebbles & Hayhoe (2002)	Wang et al. (2004)	Mikaloff Fletcher et al. (2004)	Olivier et al. (2005)	Chen & Prinn (2006)	Denman et al. (2007)
Natural sources	145	200	260		168	
Wetlands	100	176	231		145	
Termites	20	20	29		23	
Oceans	4					
Hydrates	5	4				
Geological sources	14					
Wild animal						
Wildfires	2					
Anthropogenic sources	358	307	350	320	428	
Energy		77				
Coal mining	46		30	34	48	
Gas, oil industry	60		52	64	36	
Landfills and waste	61	49	35	66		
Ruminants	81	83	91	80	189[a]	
Rice agriculture	60	57	54	39	112	
Biomass burning	50	41	88		43	
C_3 vegetation				27		
C_4 vegetation				9		
Global total	503	507	610		598	582
Sinks						
Soils	30				30	30
Tropospheric OH	445				506	511
Stratospheric loss	40				40	40
Total sinks	515				576	581
Imbalance (trend)					+22	+1

[a] Includes emissions from landfills and wastes

Observations indicate substantial increases in CH_4 released from northern peatlands that are experiencing permafrost melt (Christensen et al., 2004; Wickland et al., 2006). Termites, which produce methane as part of their normal digestive process, account for about 11% of the global natural emissions. The major anthropogenic emissions of CH_4 originate from agriculture (mainly from enteric fermentation by animals and animal waste, rice cultivation and savanna burning), energy production and transmission (mainly from coal and gas production and transmission) and from waste and landfills. Ruminant and rice agriculture together contribute ~120–145 Tg CH_4 year^{-1}. In 2004 agriculture accounted for 43% of the emissions, energy production and transmission for 36% and rest of the emissions originated from waste (18%), landfills wastewater and others (IEA, 2006).

Atmospheric CH_4 sources are both non-biogenic and biogenic. Non-biogenic sources include emission from fossil fuel mining and burning (natural gas, petroleum and coal), biomass burning, waste treatment and geological sources (fossil CH_4 from natural gas seepage in sedimentary basins and geothermal/volcanic CH_4). Biogenic emissions, which account for more than 70% of the global budget, originate from wetlands, rice agriculture, ruminants, landfill, forests, oceans and termites (Denman et al., 2007). As discussed later in this Chapter, these emissions result from the microbial breakdown of organic compounds in strictly anaerobic conditions. Rates of emission of methane from wetlands are affected by many factors: soil water status and temperature, soil type, pH, soil redox potential (Eh), nutrient inputs, and the presence of adapted vascular plants. These plants have a well-developed system of intracellular air spaces (aerenchyma) in stems, leaves and roots. This allows the transport of oxygen from the atmosphere to the root meristems and also serves as a pathway for the movement of methane from the soil into the atmosphere (Lloyd et al., 1998).

Though the major sources of CH_4 emissions have probably been identified, the individual source strengths are still uncertain because of the difficulty in assessing the global emission from biospheric sources, whose strengths are highly variable in space and time. Recently, terrestrial plants have been implicated as a global source of methane (Keppler et al., 2006). Using stable carbon isotopes Keppler et al. (2006) showed that methane is readily formed *in situ* in terrestrial plants under oxic conditions by a hitherto unrecognized process. Scaling up their data from incubation experiments on a global basis, they estimated a methane source strength of 62–236 Tg year^{-1} from living plants and 1–7 Tg year^{-1} for plant litter, which constitutes 10–30% of the present annual source strength. The detection of this additional source, though not confirmed by other studies lends some support to space-borne observations of CH_4 plumes above tropical rainforests reported by Frankenberg et al. (2005). These high methane emissions might also provide a link between the annual decline in growth rate of atmospheric methane and deforestation during the last decade (Dlugokencky et al., 1998). However, in a recent publication Dueck et al. (2007) has disputed the methane emission estimates presented by Keppler et al. (2006). Using stable isotope ^{13}C and a laser-based measuring technique, Dueck et al. (2007) indicated that the contribution of terrestrial plants to global methane emission is very small; maximally 0.3% of the values reported by Keppler et al. (2006).

The growth rate of atmospheric methane is determined by the balance between surface emissions and photochemical destruction by the hydroxyl radicals, the major atmospheric oxidant. Most CH_4 is removed from atmosphere by reaction with the hydroxyl free radical (OH), which is produced photochemically in the atmosphere (Table 8.13). Other major sinks include reaction with free chlorine (Platt et al., 2004) and destruction in the stratosphere (Born et al., 1990). The only known biological sink for atmospheric methane is its oxidation in aerobic soils by methanotrophic bacteria. This may account for ~30 Tg CH_4 $year^{-1}$ with an uncertainty range of 7 to >100 Tg CH_4 $year^{-1}$ (Smith et al., 2000a). Annual rates of CH_4 oxidation in northern Europe have been reported to vary from 0.1 to 9.1 kg CH_4 ha^{-1} $year^{-1}$ with a median value of 1–2 kg CH_4 ha^{-1} $year^{-1}$. Soil bulk density, water content and gas diffusivity have major impacts on CH_4 oxidation rates in soil. Conversion of natural soils to agriculture has been found to reduce the oxidation rates by two-thirds (Smith et al., 2000a). The grasslands may have an oxidation rate of 2.5 kg CH_4 ha^{-1} $year^{-1}$ compared to 1.5 kg CH_4 ha^{-1} $year^{-1}$ for arable land (Boeckx & van Cleemput, 2001).

8.5.1 Methane Emission from Rice Agriculture

Rice cultivation is one of the most important anthropogenic sources of atmospheric CH_4. Recent estimates of CH_4 emission from rice cultivation range between 39 and 112 Tg CH_4 $year^{-1}$ (Table 8.13). Using region-specific CH_4 emission factors, Yan et al. (2003c) estimated the global emission of 28.2 Tg CH_4 $year^{-1}$ from rice fields. Asian region accounts for 25.1 Tg CH_4 $year^{-1}$, of which 7.67 Tg is emitted from China and 5.88 Tg from India. But there is considerable uncertainty in the estimates and these differ from other published reports from India and China (Gupta et al., 2002; Mingxing & Jing, 2002; Jing et al., 2002). Using a process based model Matthews et al. (2000c) estimated annual methane emissions from rice fields in China and India to range from 3.73–7.22 and 2.1–4.99 Tg CH_4 $year^{-1}$ depending on the crop management scenario. Variable hydrological environments (irrigated, deep water, rainfed flood-prone and rainfed drought-prone) under which rice is grown, wide spectrum of agricultural practices, climatic conditions and complexity of the role of rice plants for regulating CH_4 fluxes to the atmosphere are the main reasons for the uncertainty in the global estimates of this CH_4 source. The emission factors for different rice ecosystems are postulated as irrigated = 1, drought-prone rainfed = 0.4, flood-prone rainfed and deepwater = 0.8 (IPCC, 1997). Based on the area under various rice ecosystems in different regions of Asia, it has been estimated that irrigated rice accounts for 97% of the CH_4 emissions from rice fields in East Asia and for 60% of the emissions from South and Southeast Asian rice fields (Fig. 8.7). While emissions from rainfed and deepwater are negligible for East Asia they contribute 24% and 16%, respectively to the source strength of South and Southeast

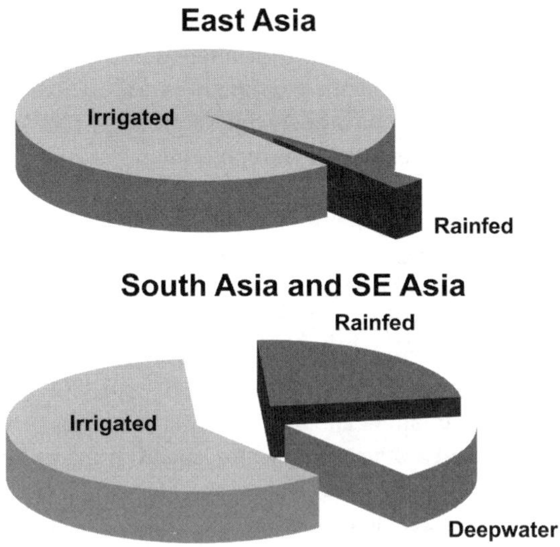

Fig. 8.7 Seasonal emission potential of different rice ecosystems in East, South and Southeast (SE) Asia (Adapted from Wassmann et al., 2000)

Asia (Wassmann et al., 2000). Based on the global distribution of the rice area, irrigated rice accounts for 70–80%, rainfed rice about 15% and deepwater rice about 10% of CH_4 emission from rice agriculture globally. But these regional and global estimates imply considerably uncertainty because of site-specific water management and other cultural practices. Sass et al. (2002) observed that in a single rice-growing region in Texas, there was 25% uncertainty in methane flux due to spatial variability and 49% uncertainty due to temporal variability. Denier van der Gon et al. (2000) proposed the use of combination of upscaling and downscaling methodologies as a potential method to reduce uncertainty in the regional CH_4 source strength of rice fields, but currently the approach is hampered due to the lack of regional-scale emission measurements.

8.5.2 *Methane Production in Rice Soils*

Strictly anaerobic condition and availability of readily decomposable organic substrates are essential for the process of CH_4 production in soil. Methane is produced in rice fields after the sequential reduction of O_2, nitrate, manganese, iron and sulfate, which serves as electron acceptors for oxidation of organic matter to CO_2. The decomposition of organic matter occurs through methanogenic fermentation, which produces CH_4 and CO_2 according to the reaction (Equation 8.5):

$$C_6H_{12}O_6 \rightarrow 3\ CO_2 + 3CH_4 \tag{8.5}$$

This transformation requires successive action of four populations of microorganisms that degrade complex molecules in simpler compounds through: (a) hydrolysis of polymers into monomers (glucides, fatty acids, amino acids) by hydrolytic organisms; (b) acidogenesis from monomeric compounds formed during fermentation (production of volatile fatty acids, organic acids, alocohols, H_2 and CO_2) by fermentative microflora; (c) acetogenesis from the previous metabolites of fermentation by homoacetogenetic or syntrophic microflora; and (d) methanogenesis from simple compounds that can be used by methanogens particularly $H_2 + CO_2$, acetate, simple methylated compounds or alcohols and CO_2 (Yao & Conrad, 2001). Methanogenesis, which requires low redox potentials ($Eh < -200\,mV$) is carried out by a specialized, strictly anaerobic microorganisms, called methanogenic archaea that can develop in synergy or in syntrophy with other anaerobic bacteria. In paddy soil, methanogens produce CH_4 from either the reduction of CO_2 with H_2 (hydrogenotrophic) or from the fermentation of acetate to CH_4 and CO_2 (acetoclastic) (Deppenmeir et al., 1996). The latter accounts for about two-third of the CH_4 emitted (Ferry, 1992).

Methane escapes to the atmosphere from soil via aerobic interfaces where CH_4 oxidation takes place. There are three pathways of CH_4-transport into the atmosphere- molecular diffusion, ebullition (gas transport via gas bubbles) and plant transport (Fig. 8.8). In the temperate rice fields more than 90% of the CH_4 is emitted through plant transport (Schütz et al., 1989) while in the tropical rice fields, significant amounts of CH_4 may evolve by ebullition in particular during the early period of the season and in the case of high organic input (Deniere van der Gon & Neue, 1995). Plant mediated transport is the primary mechanism for the CH_4 emission

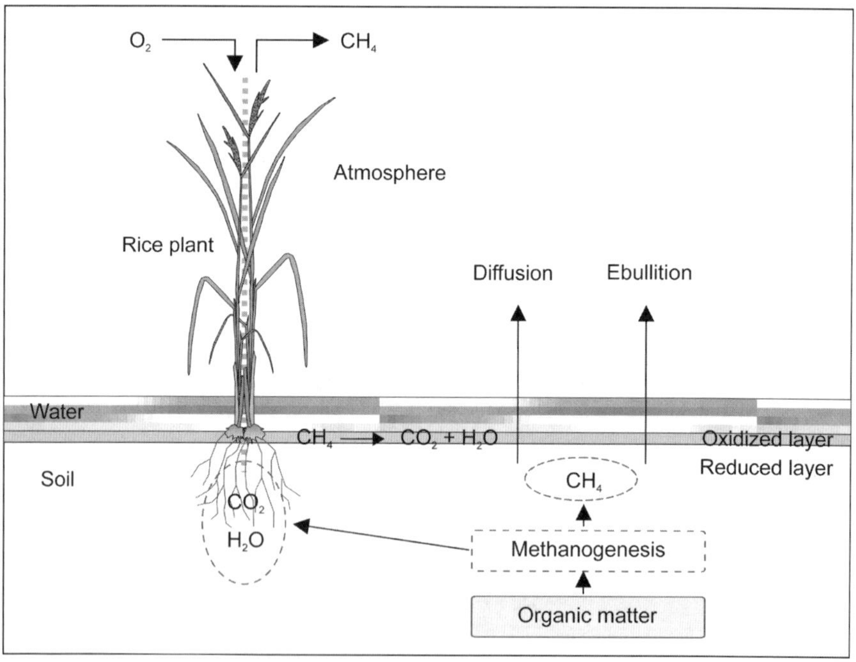

Fig. 8.8 Schematic representation of methane emission from rice fields

from paddy fields and contributes 50–90% of the total CH_4 flux (Wassmann & Aulakh, 2000). Methane is transported to the shoots via lysigenous intercellular spaces and aerenchyma and is released to the atmosphere through shoot nodes, the micropores in the leaf sheath of the lower leaf position and through the stomata in the leaf blade.

Although CH_4 flux rates are a function of the total amount of CH_4 in the soil, there is the possibility that the gas may be consumed in the thin oxidized layer close to the soil surface and the rhizosphere. Therefore, actual emissions to the atmosphere are less than the quantities of methane produced in flooded soils. The amount of CH_4 emitted range from 3% to 91% of total production in soils (Holzapfel-Pschorn et al., 1986; Nouchi et al., 1994; Yagi et al., 1994). It is known that soil methanotrophic bacteria can grow with CH_4 as their sole energy source, and that other soil bacteria, e.g. nitrosomonas species consume CH_4 (Seiler & Conrad, 1987). Methanotrophs oxidize CH_4 with the help of methanemonooxygenase enzyme. The quantity of CH_4 emitted from a rice field depends upon several important factors, including soil factors, nutrient management, water regime and cultivation practices. The gas transport resistance in the soil mainly controls methane oxidation rate and the oxidation rate increases with the increase of temperature from 5°C to 36°C (Mingxing & Jing, 2002). Temperature from 25°C to 35°C and pH from 6 to 8 is considered optimum for methane oxidation in paddy soils (Min et al., 2002).

8.5.3 Factors Regulating Methane Emission from Rice Fields

Rates of emission of methane from wetlands and rice fields are affected by a number of interacting soil, plant, management and climatic factors (Table 8.14). A statistical analysis of the CH_4 emission fluxes from rice fields in Asia (Yan et al., 2005) showed that the average CH_4 flux during the growing season is significantly affected by water management, organic matter application, content of soil organic carbon, soil pH, preseason water status and climate. Soil redox potential (Eh) is the most important factor that directly controls the production of CH_4 in soils and a negative relationship between the soil redox potential and methane emission has been reported (Yagi & Minami, 1990). Yagi & Minami (1990) observed that for the initiation of CH_4 production in paddy soils, the Eh values vary between −100 to −200

Table 8.14 Factors regulating CH_4 emission from rice agriculture

Main factor	Property
Soil	Redox potential, organic matter content, content of electron acceptors, pH, soil salinity, percolation rate, texture
Plant	Plant variety, root exudates, stage of crop growth, biomass production, CH_4 transport, oxidation of CH_4 in the rhizosphere
Management practices	Water management, mineral fertilizer application, organic matter application
Climate or environmental	Temperature, water regime

and methane is emitted to the atmosphere as Eh falls below −200 mV (Yamane & Sato, 1964). Results from laboratory studies show that Eh affects not only methanogenesis but also gas transfer through the plant as at lower Eh, aernchyma formation increases and the size of the roots decreases (Kludze & Delaune, 1995). A decrease in Eh from −200 to −300 mV induced a tenfold increase in CH_4 production and a 17-fold increase in its emission (Kludze et al., 1993). The intensity and capacity of soil reduction are controlled by degree of submergence, the nature and extent of organic substances (electron donors), temperature, and the nature and quantity of electron acceptors (Ponnamperuma, 1972). Soil submergence allows the development of the methanogenic activity and reduces methanotrophic activity by reducing the size of the oxidized zones. Soils containing high amounts of readily decomposable organic substrates (e.g. acetate, formate, methanol, methylated amines, etc.) and low amounts of electron acceptors (NO_3^-, Mn^{4+}, SO_4^{2-}) are likely to show a high production of CH_4 (Parashar et al., 1991). Several studies have shown that addition of organic matter markedly increases CH_4 emission (Merr & Roger, 2001) and the magnitude of increase depends on C:N ratio, biochemical composition and amount of the organic material added. Yan et al. (2003b) summarized data from a number of published studies in China and showed that input of organic material, such as green manure, animal waste, and straw increased CH_4 emission by a factor of 2.

Temperature influences CH_4 emission through its effect on the activity of soil microorganisms and decomposition of organic materials. Wassman et al. (1998) observed a faster CH_4 production rate and a higher maximum value with increasing temperatures between 25°C and 35°C. Methanogenesis is considered to be optimum between 30°C and 40°C. Low soil temperatures reduce CH_4 production by decreasing the activity of methanogens and other bacteria involved in methanogenic fermentation. Temperature also affects CH_4 transport through the rice plant (Nouchi et al., 1994). Diurnal variations in CH_4 emission, which generally increase rapidly after sunrise, reach a peak in the early afternoon then decline rapidly, have been related to temperature variations during the day (Schütz et al., 1990).

The other soil properties that influence CH_4 emission include soil pH and texture. It is generally recognized that the activity of methanogens is very sensitive to variations in soil pH and most CH_4 is formed in a very narrow pH range (6.4–7.8). The optimum pH for methane production ranges between 6.7 to 7.1 (Wang et al., 1993). Since the soil pH on flooding tends to be towards neutrality i.e. the pH of acid soils increases and that of alkaline soils decreases, the flooded soils provide favorable pH conditions for CH_4 production. In contrast, Yan et al. (2005) reported that soil pH of 5.0–5.5 yielded maximum CH_4 emissions. Methanotrophs are more tolerant to pH variations than methanogens (Dunfield et al., 1993). As texture determines various physicochemical properties of soils, it could influence CH_4 production indirectly. A negative correlation between CH_4 emission and clay content (Sass & Fisher, 1994) and a positive relationship with sand content (Huang et al., 2002) has been reported.

In addition to soil factors, plants exert a major influence on the magnitude and seasonality of emissions. The presence of rice plants increases CH_4 emission by providing C source (Dannenberg & Conrad, 1999) and by favoring CH_4 transfer to the atmosphere. In a Louisiana soil, CH_4 emission in 77 days was 50 kg ha^{-1} in unplanted control and 200 kg ha^{-1} in planted field (Lindau & Bollich, 1993). Methane emission correlates strongly with plant growth (Sass et al., 2002) as the plant growth determines how much substrate will be available for either methanogenesis or methanotrophy (Matthews & Wassmann, 2003). Since rice yield is usually higher during the dry season than during the rainy season, CH_4 emission is higher during the dry season. In the Philippines, a rice yield of 5.2–6.3 Mg ha^{-1} during the dry season corresponded to an average emission of 190 mg CH_4 m^{-2} day^{-1} and a yield of 2.4–3.3 Mg ha^{-1} during the wet season to 79 mg CH_4 m^{-2} day^{-1} (Wassmann et al., 1994). It has been argued that any climate change scenario that results in an increase in plant biomass in rice agriculture is likely to increase CH_4 emissions (Xu et al., 2004). However, the magnitude of increase will depend on other factors. Allen et al. (2003) observed fourfold higher total seasonal CH_4 emission under high CO_2, high temperature treatments as compared to under low CO_2, low temperature treatment. This was attributed to grater root exudation or root sloughing mediated by increased photosynthetic CO_2 uptake. The plant growth stage also influences methane fluxes from rice fields. Lower CH_4 fluxes are recorded in the early growth period of rice plant, which increases gradually during mid to late season and drops to very low level before or after harvest. Flowering period is generally considered as the peak period for methane emission. The peak emission value remains for a period of 10–15 days in the crop duration of 90–100 days. According to Holzapfel-Pschorn et al. (1986) this period emits 90% of the total methane during the whole crop season. High emission rates at the flowering stage have been attributed to recent plant-borne material, either root exudates or decaying tissue (Watanabe et al., 1997). Not only the rice plant and the growth stage but also the cultivar/variety strongly influences the magnitude of CH_4 emission. Rice cultivars have been reported to influence the magnitude of CH_4 emission due to variability in the content and composition of their exudates and the methane transport capacity of different rice cultivars (Wassmann & Aulakh, 2000).

The effect of fertilizers on CH_4 emission depends on rate, type and mode of application. Urea application enhances CH_4 fluxes by increasing soil pH following urea hydrolysis and the drop in redox potential, which stimulates methanogenic activities (Wang et al., 1993). The addition of sulfate or nitrate containing fertilizers can suppress the production of CH_4. Application of sulfate as chemical fertilizers results in production of H_2S, which is toxic to methanogens and its application, also enhances the activity of sulfate reducing bacteria, which outcompete methanogens for substrate. Application of fertilizer nitrate, which acts as a terminal electron acceptor in the absence of molecular oxygen, poises the soil redox potential at values such that the activity of strict anaerobes is prevented (Minami, 1995).

8.5.4 Mitigation Options for Agricultural Emission of Methane

Manipulation of the factors that regulate CH_4 emission, particularly appropriate water and nutrient management, cultural practices and choice of crop cultivar can help reduce CH_4 emission from rice fields. Since irrigated rice is considered to contribute about 70–80% of CH_4 emission from global rice fields it provides the most promising target for mitigation strategies. Several studies have shown that proper water management could reduce CH_4 emissions without affecting yields. Water management such as midseason drainage and intermittent irrigation is one of the most effective strategies for decreasing CH_4 emission, because it prevents the development of soil reductive conditions. One or multiple drainage systems have been reported to decrease CH_4 emission compared to continuous flooding. Numerous *in situ* studies report a significant decrease (60% to >90%) of CH_4 emission by rice fields that are drained once or several times during the crop cycle (Sass et al., 1992; Cai et al., 1994; Zheng et al., 2000). In Texas rice fields, average CH_4 emission (mg m^{-2} day^{-1}) were 106 for classical continuous irrigation, 56 when the field was drained in the middle of the crop cycle, 13 when the field was drained three times (Sass et al., 1992). Similarly, studies from India show that the mid season drainage may reduce methane emission by about 50%; compared to seasonal CH_4 flux of 15.3 ± 2.6 g m^{-2} with continuous flooding, the introduction of single and multiple mid-season aeration reduced methane flux to 6.9 ± 4.3 and 2.2 ± 1.5 g m^{-2}, respectively (Gupta et al., 2002). Yan et al. (2003b) summarized data from a number of published studies in China and found that average CH_4 flux from intermittently irrigated rice fields is 53% of that from continuously flooded rice fields. Compilation of the published CH_4 emission data from major rice growing areas in Asia shows that the average CH_4 flux with single and multiple drainages are 60% and 52% of that from continuously flooded rice fields (Yan et al., 2005). Suppression of CH_4 production due to field drainage usually persists for quite some time even when fields are flooded again (Yagi et al., 1996). Even short-term drainage is sufficient for rather long-term suppression of CH_4 emission. The reasons for this behavior are ascribed to the regeneration of oxidants during the short drainage and aeration period (Sigren et al., 1997; Ratering & Conrad, 1998). Short drainage induces the formation of sulfate and ferric iron, which allows the operation of sulfate-reducing and iron-reducing bacteria that utilize acetate and H_2 more efficiently than the methanogens. As a result, concentrations of H_2 and acetate decrease to values that are no longer thermodynamically permissive for CH_4 production (Sigren et al., 1997; Ratering & Conrad, 1998; Conrad, 2002).

Water management between crops is also an important factor. A dry fallow emitted less CH_4 during the next crop cycle than a wet fallow (Trolldenier, 1995). Cai et al. (2003) observed that compared to permanently flooded paddy soils in China, draining floodwater during the following upland winter crop not only prevented CH_4 emissions during the upland winter crop season but also reduced CH_4 emissions substantially during the following rice-growing period resulting in an annual reduction in methane emission by 68%. Methane flux from fields that were flooded in the previous season was 2.8 times that from fields previously drained for a long season and 1.9 times that from fields previously drained for a short season (Yan et al., 2005).

Water management as a mitigation practice is only feasible in areas that have the requisite physical characteristics (Sass et al., 1992). The strategy is best suited to areas with lowland and flatland rice fields that have highly secure and controllable water supplies. On rainfed areas, drainage may also be less feasible because farmers depend on the water stored in the bunded field. It is important that future research indicate how different water management methods, intended to reduce CH_4 emissions, affect emissions of N_2O.

Several studies confirm the advantage of ammonium sulfate fertilizer in reducing CH_4 emission (by 50–60%) as compared to urea. Fertilizing with ammonium sulfate supplies N and sulfate, which maintains soil Eh above that required to produce CH_4. Application of gypsum at 6.7 Mg ha^{-1} in saline and alkaline soils has been reported to reduce CH_4 emission by 50% and 70% in rice fields fertilized with urea or green manure, respectively (Denier van der Gon & Neue, 1994). However, sulfate addition might be detrimental to rice by favoring rhizosphere sulfate-reduction. Use of nitrification inhibitor, coated calcium carbide can reduce CH_4 production by producing small quantities of acetylene slowly over time (Banerjee & Mosier, 1989). Methane emission seems to be reduced when N-fertilizer is incorporated, as compared to surface application (Schütz et al., 1989). Combining organic and mineral fertilizers can mitigate the increased CH_4 emission due to organic manure. For example, Shao and Li (1997) observed that ammonium sulfate combined with organic manure reduced emission by 58% as compared with organic manure alone and increased yield by 32%. Emission peaks were suppressed at tillering and during the reproductive stages of rice. Certain tillage, seeding and weeding techniques used to minimize water use and mechanical soil disturbance may also offer some CH_4 mitigation potential (Neue, 1992). Another mitigation option is fertilization with iron. Increased soil iron contents in the rhizosphere helps suppress CH_4 formation (Jäckel & Schnell, 2000). Addition of iron containing revolving furnace slag has been reported to suppress CH_4 emission from paddy soils (Furukawa & Inubushi, 2002).

Yagi (2002) evaluated different mitigation strategies from the point of view of effectiveness, productivity and economics (Table 8.15). While mid-season drainage is an effective strategy but it is likely to increase labor costs and may promote N_2O production. Similarly, high percolation rates may promote NO_3^- leaching. Mosier et al. (1998b) estimated that adoption of a combination of water management, nutrient management, cultural practices and new cultivars have the potential to mitigate CH_4 emissions in flooded rice by about 20 (range 8–35) Tg CH_4 $year^{-1}$. However, much additional research is needed to establish and demonstrate that these practices will maintain or increase rice productivity while reducing methane emissions.

Global methane emissions due to burning of croplands, grasslands and forests may be reduced through sustained land management programs and land use policies (Mosier et al., 1998b) that aim at (i) increasing the productivity of existing agricultural lands and restoring the degraded lands, (ii) lengthening the rotation times and improving the productivity of shifting agriculture, (iii) improving grassland management to reduce frequency of fires, and (iv) returning crop residues to the field instead of burning. It has been estimated that using combinations of these techniques can potentially reduce CH_4 emissions due to biomass and other agricultural burning by 6 Tg CH_4 $year^{-1}$ (Mosier et al., 1998b)

For domesticated ruminants, the most appropriate strategy for reducing CH_4 emissions is to improve the nutrition and animal productivity for milk and growth

Table 8.15 Evaluation of some mitigation options for CH_4 emission from irrigated rice (Adapted from Yagi, 2002)

Mitigation option	Mitigation efficiency	Economy Cost	Economy Labor	Yield	Other effect
Water management					
Midseason drainage	H	~	↑	+	May promote N_2O
Short flooding	H	~	~	–	May promote N_2O
High percolation	H	↑	↑	+	May promote NO_3^- leaching
Soil amendments					
Sulfate fertilizer	H	↑	~	V	May cause H_2S injury
Oxidants	H	↑	↑	V	
Soil dressing	M	↑	↑	–	
Organic matter					
Composting	H	↑	↑	+	
Aerobic decomposition	H	~	↑	~	
Burning	M	~	↑	~	Atmospheric pollution

H: very effective; M: effective/applicable; ↑: increase; ~: about equal; V: variable case by case; +: positive; –: negative

through dietary supplementation (Mosier et al., 1998b). Supplementation of the diets of native cattle/buffalo in India has been shown to decrease CH_4 emission by a factor of 3 per liter of milk produced and by a factor of 6 per t of live weight gain (Leng, 1991). Most of the CH_4 produced in anaerobic digestion of livestock manure constitutes a wasted energy source that can be recovered by adopting manure management and treatment practices adapted to collect CH_4 (Hogan, 1993). With current technology, CH_4 emissions from manures can be reduced by 25–80%. The total potential for reducing methane emissions in agriculture is estimated to be 24–92 Tg year^{-1} depending on effectiveness of proposed options and degree of implementation (Cole et al., 1997)

8.6 Emission of Oxides of Nitrogen: N_2O and NO

8.6.1 Nitrous Oxide Emissions

Nitrous oxide, N_2O, is a greenhouse gas and is also one of the substances that destroy stratospheric ozone. It constitutes 6% of the anthropogenic greenhouse effect and its concentration in the atmosphere has been increasing by about 0.25% per year, from about 270 ppb in preindustrial times to 319 ppb in 2005 (Fig. 8.9). In the 1990s the concentration of N_2O in the atmosphere has increased by 0.8 ppb year^{-1}. The global warming potential (GWP) of N_2O is 296 times that of CO_2 and 13 times that of CH_4 over a 100-year time horizon (IPCC, 2001a). Nitrous oxide is

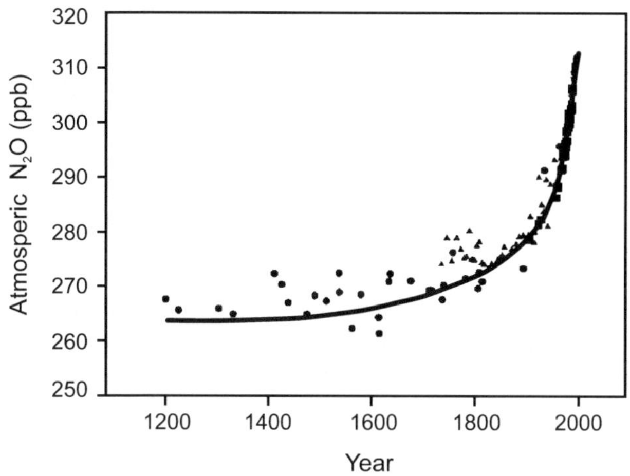

Fig. 8.9 Atmospheric N_2O abundance trend since the year 1200 (Adapted from IPCC, 2001a. Reproduced with kind permission from Cambridge University Press)

Table 8.16 Abundance, atmospheric life time and global warming potential (GWP) of N_2O and NO_x (IPCC, 2001a)

	N_2O	NO_x ($NO + NO_2$)
Preindustrial concentration (1750)	270 ppb	?
Concentration in 2005	319 ppb	5–999 ppt
Rate of concentration change (1990–1999)	0.8 ppb year^{-1}	?
Atmospheric life time (year)	114[a]	<0.01–0.03
100-year GWP	296	–
Radiative forcing (W m^{-2})	0.15	–

[a] This life time has been defined as an adjustment time that takes into account the indirect effect of the gas on its own residence time

fairly stable and has a global atmospheric lifetime of 120 years but because of its own feedback the lifetime decreases by about 0.5% for every 10% increase in N_2O and has thus adjusted lifetime of 114 years (Table 8.16).

8.6.1.1 Sources and Sinks of Nitrous Oxide

Nitrous oxide is emitted into the atmosphere both from natural and anthropogenic sources but there is considerable uncertainty about the contribution of different sources to the global N_2O emissions. Natural sources include soils, ocean and atmospheric NH_3 oxidation. Anthropogenic sources of N_2O can be both biogenic (biological nitrification and denitrification) and abiogenic (e.g. during biomass burning). During biomass burning the nitrogen in fuel in end groups, open chains and heterocyclic rings can be converted into gaseous forms such as ammonia, nitric oxide, nitrous oxide, dinitrogen and hydrogen cyanide (Galbally & Gillett, 1988) and emitted to the atmosphere. Biomass burning contributes about 0.5 Tg N year^{-1} to the global atmospheric N_2O budget. Most of the biomass burning (about 90%) takes place in the tropics as a result of forest clearing, savanna and sugarcane fires, and burning of agricultural wastes (Freney, 1997).

Biogenic emissions of N_2O and NO from soils result from microbial mediated nitrification and denitrification processes. Nitrification is the biological oxidation of NH_4^+ to NO_3^- through NO_2^- under aerobic conditions. However, under oxygen limited conditions nitrifiers can use NO_2^- as a terminal electron acceptor and result in the production of N_2O and NO. Denitrification is an anaerobic bacterial process by which nitrate is reduced to nitrite (NO_2) and further reduced to nitrous oxide (N_2O) or dinitrogen (N_2) which is lost to the atmosphere as a gas (Equation 8.6).

$$NO_3^- \rightarrow NO_2^- \rightarrow NO \rightarrow N_2O \rightarrow N_2 \tag{8.6}$$

There has been some doubt if nitric oxide (NO) is a true intermediate or a by-product (Amundson & Davidson, 1990) in the process. Depending on conditions, intermediate products can accumulate and eventually escape.

Nonbiological denitrification, called chemodenitrification occurs in subsoil, but is generally a less important source of emissions (Granli & Bøckman, 1994). Chemodenitrification, may also occur wherein NO_2^- can react with organic compounds (e.g. amines) to form N_2, NO_2 and N_2O (Bremner & Nelson, 1968). Nitrous oxide can also form in reactions between NO_3^-/NO_2^- and some inorganic compounds (e.g. Fe^{2+}, Cu^{2+}). These reactions may be important for slow denitrification of groundwater (Van Cleemput et al., 1987).

The only significant process that removes N_2O is its reaction in the stratosphere with excited oxygen atoms formed by photolysis of ozone (Crutzen, 1991). Microorganisms in soils can reduce N_2O into N_2 under anaerobic conditions (Ryden, 1981) but the significance of soil as a sink for N_2O remains uncertain and probably very small (Freney et al., 1978).

8.6.1.2 Nitrous Oxide Emission Estimates

Estimates of global N_2O emissions from natural and anthropogenic sources range from 14.7 to 17.7 Tg N year^{-1} (Table 8.17). The estimates of N_2O emissions made by Mosier et al. (1998a) and Kroeze et al. (1999) match the global loss rate. Natural sources contribute more than 50% (9–11 Tg N year^{-1}) to the global flux with natural soils being the largest emitter. Of the total anthropogenic emissions agricultural soils and biomass burning account for more than half of the emissions. However, there is large variation in the estimates for agricultural soils because of the diversity of agricultural systems and of the response of these systems to fertilizer application, according to the soil type, climate, and management practices. The estimates are further impacted by the sources considered and the emission factors used for computing the fluxes. Mosier et al. (1998a) distinguished three sources of N_2O emissions

Table 8.17 Estimates of the global nitrous oxide budget (in Tg N $year^{-1}$) from different sources

	IPCC (1996)	Olivier et al. (1998)	Mosier et al. (1998a) & Kroeze et al. (1999)	Denman et al. (2007)
Natural sources				
Ocean	3	3.6	3.0	3.8
Atmosphere (NH_3 oxidation)		0.6	0.6	0.6
Tropical soils				
Wet forest	3		3.0	
Dry savannas	1		1.0	
Temperate soils				
Forests	1		1.0	
Grasslands	1		1.0	
All soils		6.6		6.6
Anthropogenic sources				
Agricultural soils	3.5	1.9	6.3	2.8
Biomass burning	0.5	0.5	0.5	0.7
Industrial sources	1.3	0.7	1.3	0.7
Cattle and feedlots	0.4	1.0		
Huma excreta				0.2
Rivers, estuaries, coastal zones				1.7
Atmospheric deposition				0.6
Total sources	14.7	14.9	17.7	17.7

Table 8.18 Global N_2O emissions from agricultural soils (Tg N $year^{-1}$) (Adapted from Mosier et al., 1998a)

Source	N_2O (Tg N $year^{-1}$)
Direct soil emissions	2.1 (synthetic fertilizer: 0.9, animal waste: 0.6, crop residue: 0.4, biological nitrogen fixation: 0.1, and cultivated histosols: 0.1)
Animal production	2.1 (animal waste management systems)
Indirect emissions	2.1 (nitrogen leaching and runoff: 1.6, atmospheric deposition: 0.3, human sewage: 0.2)
Total	6.3

from agricultural soils viz. direct emissions, emissions from animal production systems, and emissions indirectly induced by agricultural activities; each source contributing 2.1 Tg N $year^{-1}$ (Table 8.18). This indicates that in earlier reports N_2O emissions from agricultural soils were probably underestimated and the fresh estimates can account for the missing N_2O sources presented in earlier IPCC reports (Mosier et al., 1998a). However, these estimates are considerably higher than those (Table 8.19) presented by Stehfest and Bouwman (2006). In the recent estimates (Denman et al., 2007), inclusion of several minor sources such as human excreta, atmospheric deposition and coastal, riverine and estuarine fluxes have closed the global N_2O budget (see Table 8.17)

Table 8.19 Nitrous oxide (N_2O) and nitric oxide (NO) emissions (Gg N year^{-1}) from arable land and grassland for nine regions of the world (Adapted from Stehfest & Bouwman, 2006)

	N_2O		NO	
Region	Cropland	Grassland	Cropland	Grassland
North America	459	240	116	86
Latin America	363	79	177	58
North Africa and Middle East	150	32	50	12
West, east and southern Africa	294	56	179	58
Europe	330	99	144	57
Former USSR	177	69	64	41
South Asia	617	22	265	14
East Asia	677	79	173	22
Southeast Asia, Oceania and Japan	278	134	220	70
World	3,345	809	1,388	417

Table 8.20 Anthropogenic N_2O emissions in different regions of the world (Recalculated from IAE, 2006)

			2000				
	1990	1995	Energy	Agriculture	Industrial processes	Other	Total
Region	Tg N year^{-1}			Gg N year^{-1}			Tg N year^{-1}
OECD N. America (Canada, Mexico and USA)	1.13	1.23	80.2	906.8	131	130	1.25
OECD Pacific (Australia, Japan, Korea and New Zealand)	0.37	0.34	18.5	308.6	20.3	11.5	0.36
OECD Europe[a]	1.08	1.01	33.1	721	191.3	58.1	1.00
Non-OECD Europe[b]	0.12	0.08	3.1	52.9	16.2	0.83	0.07
Former USSR	0.57	0.32	14.4	225.3	8.9	7.4	0.25
Africa	0.85	0.88	33.4	894.7	10.5	173.7	1.11
Middle East	0.17	0.18	5.5	186.5	3.9	0	0.19
Latin America	0.93	0.97	11.9	887.7	13.9	205	1.12
Asia (including India and China)	1.94	2.28	114.9	2191.6	22.4	73.7	2.41
India	0.47	0.53	32.5	532.3	3.9	4.9	0.57
China	0.92	1.01	48.1	1062.1	11.3	1.5	1.13
World	7.16	7.29	315	6375.1	418.4	660	7.77

[a] OECD(Organization for Economic Development) – Europe includes Austria, Belgium, Czech Republic, Denmark, Finland, France, Germany, Hungary, Iceland, Ireland, Italy, Luxemburg, the Netherlands, Norway, Poland, Portugal, Slovak Republic, Spain, Sweden, Switzerland, Turkey, the United Kingdom

[b] Non-OECD Europe includes Albania, Bulgaria, Cyprus, Gibraltar, Malta, Romania, Former Yugoslavia, Bosnia-Herzegovina, Croatia, FYR of Macedonia, Serbia/Montenegro, Slovenia

Regional distribution of anthropogenic emission indicates that Asia is the largest emitter accounting for about one third of the global emissions (Table 8.20; IAE, 2006). More than 90% of the emissions in Asian region are from agricultural activities. This is obviously related to fertilizer use as this region accounts for more than half of the global consumption of chemical N fertilizers. Emissions from agriculture related activities in South and East Asia increased by 2–3% annually during the 1980s but increased by 21% during 1990 and 2000. This increase was partially offset by decreasing emissions in the former USSR countries (55%). However, these estimates of N_2O emission are higher than those presented in other reports (see Table 8.19; Stehfest & Bouwman, 2006; Yan et al., 2003a; IPCC, 2007a). This clearly shows that there is great uncertainty in global inventory calculations, particularly from agricultural sources. Therefore, there is a need to improve inventorization and assessment methodologies. One approach could be the use of appropriate process based models (see Chapter 9) coupled to land use, crop and soil database for estimating N_2O emissions from different regions and components of agriculture.

8.6.2 Nitric Oxide Emissions

Atmospheric nitrogen oxides, NO_x, a substantial proportion of which comes from soils, are also environmentally important. Nitrogen oxides ($NO_x = NO + NO_2$) are produced in the troposphere, primarily in the form of nitric oxide (NO), which is readily oxidized into nitrogen dioxide (NO_2). Nitric oxide emissions originate from surface and troposheric sources. The surface sources include fossil fuel and biomass burning and biogenic emissions from soils. The troposheric sources include lightning, aircrafts and stratospheric injection. There is substantial spatial and temporal variability in the measured abundance of NO_x, which ranges from a few ppb near the surface over the remote tropical Pacific Ocean to >100 ppb in urban regions (IPCC, 2001a). The primary sink for NO_x and its reaction products is wet and dry deposition as described in the preceding section.

Nitric oxide concentration is directly linked to the proximity and magnitude of source because of its very short atmospheric lifetime (see Table 8.16). It reacts with CO and hydrocarbons in the atmosphere to form tropospheric ozone, and is a precursor of acid rain. Nitric oxide is a by-product of the nitrification pathway and the typical yield of NO in well-aerated soil ranges from 1% to 4% of the NH_4^+ oxidized (Hutchinson & Brams, 1992). Nitric oxide is also produced during denitrification of NO_3^- to N_2 but the release of NO from soil is greatly influenced by the gas phase diffusivity in soil and the rate of NO consumption by the denitrifiers. Soil pH appears to be an important factor determining the mechanism of NO formation, for example in an alkaline loamy clay soil (pH 7.8), nitrification was the main source of NO, whereas in an acid sandy clay loam (pH 4.7) denitrification dominated the NO production (Remde & Conrad, 1991). In agricultural soils of temperate climates, where high nitrification rates are sustained by maintaining the soil pH above 5, nitrification is the dominant source of NO in soils whereas in acid tropical soils denitrification may dominate

It is difficult to quantify the overall global importance of nitrification or denitrification as sources of atmospheric NO_x. The NO/N_2O emission ratio has been proposed as a useful indicator of the dominant underlying process. Laboratory studies indicate that for nitrifiers the NO/N_2O ratio is greater than unity while for denitrifiers this ratio is less than unity (Lipschultz et al., 1981; Anderson & Levine, 1986). Though it is difficult to extrapolate these results to field conditions as nitrification and denitrification occur simultaneously in soil, yet the NO/N_2O ratio may provide an indication of the dominant process responsible for NO emission.

8.6.2.1 Nitric Oxide Emission Estimates

The global estimates of NO_x flux range from 38.2 to 51.9 Tg N year^{-1} with most recent (Denman et al., 2007) estimates of 41.8–47.1 Tg N year^{-1} (Table 8.21). Because of short atmospheric lifetime of NO_x species it is difficult to measure their concentration and there is great temporal and spatial variability in their distribution. Fossil fuel combustion is the largest source of NO_x emission contributing about 50% of the total emissions followed by biomass burning (12–18%). The emission of NO_x has accelerated exponentially during the last few decades (Fig. 8.10), primarily due to the increase in fossil fuel combustion (Galloway et al., 1995; Holland & Lamarque, 1997). For example, in 1860 out of the total NO_x emissions of ~13.1 Tg N year^{-1} fossil fuel combustion contributed only 0.3 Tg N year^{-1} and majority of the emissions (10.5 Tg N year^{-1}) originated from natural sources (emissions from soil processes, lightning, wildfires and stratospheric injection). In comparison, in 1990s the fossil fuel combustion and biomass burning contributed 20.4 and 8.5 Tg N year^{-1}, respectively to the total NO_x emissions of ~46 Tg N year^{-1} (van Aardenne et al., 2001; Galloway et al., 2004).

Biogenic emissions from soil constitute the third important source and the emission from soil and biomass burning contribute the most closest to the surface and their concentration dissipate with height (Holland & Lamarque, 1997). The estimates of global NO_x flux from soils are highly variable and range from 4 to 21 Tg N year^{-1}. In the recent estimates (Denman et al., 2007), the soil NO_x emissions have been distributed between agriculture (1.6 Tg N year^{-1}) and natural vegetation (7.3 Tg N year^{-1}). Holland & Lamarque (1997), using 3-D chemistry transport models, estimated global NO_x flux from soils to range from 4–10 Tg N year^{-1}. However, Davidson & Kingerlee (1997), based on the data presented in 60 published papers estimated that 21 Tg N year^{-1} is emitted from soils. Inclusion of canopy reduction factor (adsorption of NO_x onto plant canopy surfaces), as used by Yienger & Levy (1995) reduces the NO_x flux to 13 Tg N year^{-1}, which is still considerably higher as compared to the other reports. The great variability in global estimates of NO_x flux from soils could be due to dynamic nature of the NO_x production and emission processes. These processes are controlled by a number of environmental and edaphic characteristics, all of which vary at short time and space scales. These sources of variation become more critical in understanding fluxes from agricultural systems due to the influence of management (Matson, 1997).

Table 8.21 Estimates of global NO_x emissions (Tg N year^{-1}) from different sources

NO_x source	Delmas et al. (1997)	Lee et al. (1997)	Holland & Lamarque[b] (1997)	Ehhalt (1999)	van Aardenne et al. (2001)	IPCC (2001a)	Denman et al. (2007)
Natural sources						3.3[c]	
Soils under natural vegetation	–	–	–	–	–		7.3
Lightning	2 (1–4)	5	3–10	7.0	5.4	5.0	1.1–6.4
Stratospheric injection	0.5 (0.4–0.6)	0.6	0.2–0.64	0.15	0.6	<0.5	–
Ammonia oxidation	1.0 (0.5–1.5)	0.9	–	3.0	–	–	
Anthropogenic sources							
Fossil fuel	22 (15–29)	22	20–22.4	21.0	25.5[a]	33.0	25.6
Aircrafts	0.55 (0.5–0.6)	0.85	0.23–0.6	0.45	0.5	0.7	–
Biomass and bio-fuel burning	6.7 (3–10.4)	7.9	4.4–10	7.5	8.5	7.1	5.9
Agriculture	–	–	–	–	–	2.3[c]	1.6
Biogenic soil emission	5.5 (3.3–7.7)	7.0	4–10	5.5	5.5	–	
Atmospheric deposition	–	–	–	–	–	–	0.3
Total	38.2 (23.7–53.8)	44.3	35–48.8	44.6	46.0	51.9	41.8–47.1

[a] Includes emissions from non-road transport (3.6 Tg N); and industrial processes (1.5 Tg N)
[b] Using five three-dimensional chemical transport models
[c] The total soil NO_x emission estimates of 5.6 shown in IPCC (2001a) has been distributed between Agriculture and natural vegetation in Denman et al. (2007)

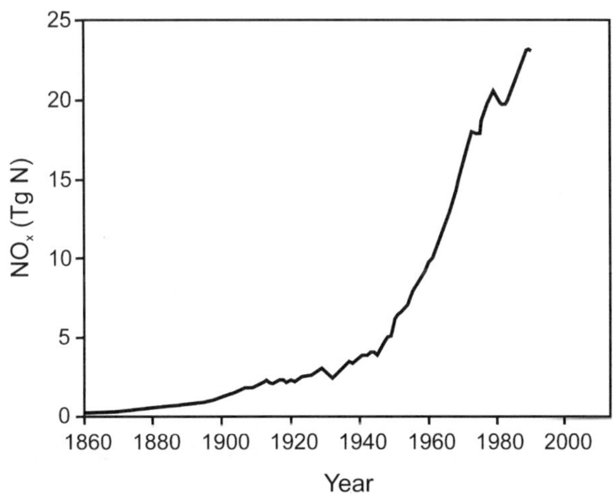

Fig. 8.10 Increase in global NO_x emission from fossil fuel combustion since 1860 (Holland & Lamarque, 1997, p. 8. Reproduced with kind permission from Springer)

Estimates of global NO_x flux from soils have been made for different ecosystems, land use or land cover categories. According to Davidson & Kingerlee (1997), largest annual emissions were from tropical savanna/woodland (7.4 Tg N), chaparral (4.7 Tg N), and cultivated agriculture (5.5 Tg N) but there was wide variation within the categories. These fluxes are much higher than those (1.4 Tg N for fertilized crops and 0.4 Tg N $year^{-1}$ from grasslands) reported recently by Stehfest & Bouwman (2006). Regional estimates of NO fluxes from cropland show that, similar to N_2O fluxes, Asia is the biggest emitter followed by North and Latin America with North Africa and Middle East being the least emitters. However, for grasslands North and Latin America together are highest emitters followed by Asia (see Table 8.19). But these regional estimates are based on the data that did not represent full variability of world agricultural systems.

8.6.3 Factors Regulating Emission of N_2O and NO_x

A number of soil, plant, environmental, and management factors interact to influence rate and extent of nitrogen oxide emissions. Availability of mineral nitrogen through fertilizer application or organic matter mineralization, the presence of NO_3^-, the presence of decomposable organic matter, soil moisture and aeration status, soil temperature, and soil and crop management determine the nitrogen oxide emissions.

8.6.3.1 Soil and Fertilizer Nitrogen

The concentration of mineral N in soil is the most important variable influencing the magnitude of NO emissions. The mineral N concentration in agricultural soils is raised by fertilizer application and mineralization of organic matter. The denitrification rate increases with increasing soil NO_3^- concentrations up to a certain level and then becomes constant. Production of N_2O by nitrification is also enhanced as the soil concentration of the substrate, NH_4^+, increases. Therefore, the magnitude of NH_3 volatilization also determines the availability of N for nitrification and denitrification. Studies have found denitrification to be lower when NH_3 losses are high in both upland soils and wetland rice systems.

Application of N fertilizers or manures is usually followed by an increase in N_2O emission, which is likely to be greater with urea than nitrate as the fertilizer source. Bouwman (1996) estimated the total N_2O emissions from fertilized crop from Equation 8.7:

$$N_2O\ (\text{kg N ha}^{-1}) = 1 + 0.0125 * \text{N application (kg ha}^{-1}) \qquad (8.7)$$

The value of 1 in the equation represents the background emission while the factor 0.0125 accounts for the contribution from fertilizer. Though the relationship is based on long-term data sets including a variety of mineral and organic N-fertilizers, yet several studies have shown that the background and fertilizer induced emission fluxes differ greatly depending on soil, crop, local climatic conditions and the type of fertilizer used. For example, the emission factor has been found to be 0.021 for clay soils, 0.008 for sandy soils and 0.012 for loam soils (Del Grosso et al., 2006). In agriculturally intensive regions in France, the emission factors from the fertilizers at the regional levels have been reported to range from 0.0007 to 0.0033 (Gabrielle et al., 2006). Yan et al. (2003a) estimated the background emission fluxes of N_2O and NO from croplands in Asia to be 1.22 kg N ha^{-1} $year^{-1}$, which account for 43% of the total emissions of 1.19 Tg N $year^{-1}$ and the fertilizer induced emissions account for another 30% (Fig. 8.11). The fertilizer induced emission factor for N_2O from paddy fields is one fifths (0.25%) that of upland fields (1.19%). However, many researchers have found that N_2O flux from fertilized paddy fields in the fallow season is much higher than N_2O flux in the cropping season (Bronson et al., 1997; Tsuruta et al., 1997). Moreover paddy fields cannot be completely differentiated from uplands, because mostly fields are rotated with rice and upland crops in the same year. This shows that the use of a single value for fertilizer induced emissions can lead to erroneous global budgets of nitrogen oxide emissions. The mean global fertilizer induced emissions are 0.91% of the N applied in cropland (excluding legumes) and grassland. The calculated fertilizer induced emission for NO from agriculture and grassland excluding legumes is 0.55%.

Similar to N_2O, emission of NO from agricultural soils is mainly governed by N application rate. Agricultural soils, immediately after the application of N fertilizer, are generally found to emit the largest rates of NO. The fertilizer effect usually lasts for periods of a few days to 2–3 weeks, depending on the rainfall and wetness of the soil (Skiba et al., 1993). In temperate regions, NO emission is linearly related to the amount of fertilizer applied and about 0.5% of the applied fertilizer N is emitted as NO during the crop growing season (Fig. 8.12; Veldkamp & Keller, 1997). Similar

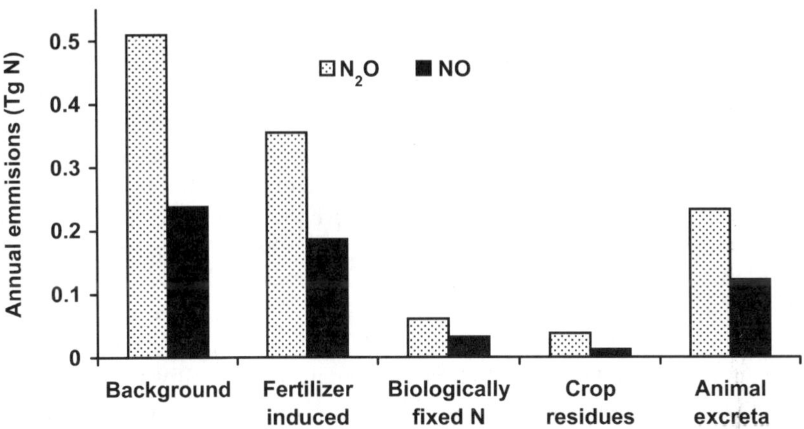

Fig. 8.11 Estimated emissions of N_2O and NO from croplands in East, Southeast and South Asia from different sources (Yan et al., 2003a. Reproduced with kind permission from Springer)

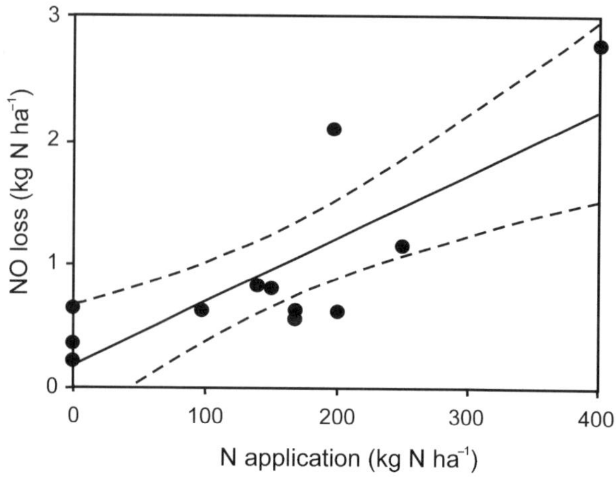

Fig. 8.12 Relationship between NO emission and fertilizer N application. Solid line represents regression ($R^2 = 0.64$) and dashed lines indicate 95% confidence intervals (Veldkamp & Keller, 1997, p. 74. Reproduced with kind permission from Springer)

fertilizer induced NO emission factor (0.48%) with background flux of 0.57 kg N ha^{-1} was estimated for the Asian region (Yan et al., 2003a). Fertilizer induced emissions account for 31% of the total (0.591 Tg N) from cropland in the region.

8.6.3.2 Soil Moisture and Aeration

Several studies have shown that N_2O emissions increase with increasing water filled pore space (WFPS) (Dobbie et al., 1999; Skiba & Ball, 2002; Bateman & Baggs, 2005) but the influence depends on the pathway responsible for the production of N_2O and NO and the diffusion properties of the soil. The rate of N_2O production from nitrification increases rapidly with increasing water content up to 55–65% water-filled pore space (WFPS). Above 60–70% WFPS, an increase in water content hinders aeration and promotes denitrification. Bateman and Baggs (2005) showed that between 35% to 60% WFPS, autotrophic nitrification was the dominant process contributing to N_2O emissions and above 60% WFPS denitrification was the main mechanism. The WFPS around 60% offers optimal conditions for nitrification because neither the diffusion of substrates nor the diffusion of O_2 is restricted (Parton et al., 1996). At 70% WFPS all N_2O was produced due to denitrification (Fig. 8.13) indicating the predominance of anaerobic microsites at this moisture content. There is an inverse relationship between the rate of denitrification and O_2 concentration (e.g. Focht, 1974; Arah et al., 1991). The inverse relationship is more pronounced at high (34.5°C), rather than at low (19.5°C), temperature (Focht & Verstraete, 1977). Reduction of N_2O to N_2 is more prone to inhibition by O_2 than reduction of NO_3^- to N_2O, thus the N_2O/N_2 ratio decreases with increasing O_2 concentration. The O_2 concentration in soil depends on soil water content, diffusion of O_2 into the soil, consumption of O_2 by soil microorganisms and plant roots (Smith, 1990). Diffusion of O_2 in soil is determined by texture, management (e.g. tillage) and water content. Sufficiently low oxygen diffusion rates that promote rapid denitrification are most likely under waterlogged conditions, as in rice paddies

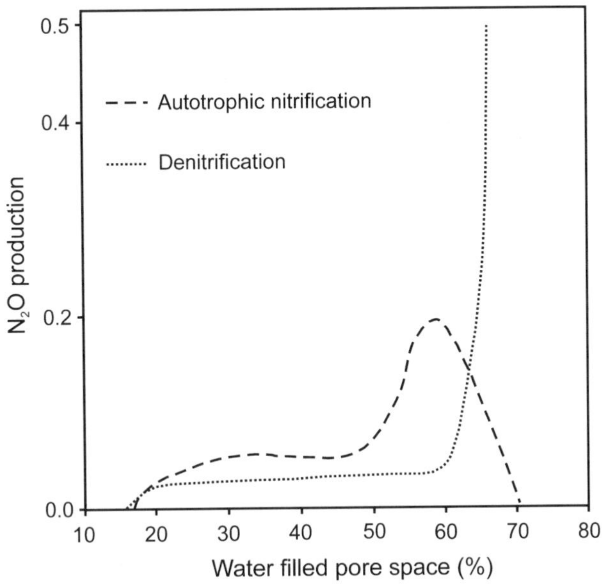

Fig. 8.13 Schematic representation of the influence of water filled pore space on the contribution of nitrification and denitrification towards N_2O emissions from soil (Bateman & Baggs, 2005. Reproduced with kind permission from Springer)

and in pasture systems with compacted soil. Other factors such as temperature, NO_3^- concentration, soil texture, and compaction influence the effect of soil water on denitrification rate.

The wetting of dry soils causes pulses in N mineralization, nitrification and NO and N_2O fluxes. The alternate drying and wetting of soils enhances the release of N_2O and NO from the soil due to stimulation of N mineralization and accumulation of NO_3^- during the dry periods.

8.6.3.3 Temperature

Some studies have shown that emission of N_2O increases with increasing soil temperature, at least up to 37°C, but N_2O/N_2 ratio declines with increasing temperatures above 37°C (Keeney et al., 1979; Castaldi, 2000). Skiba et al. (1994) observed that for a range of agricultural and seminatural soils, soil temperature and soil NO_3^- concentration accounted for 60% of the variability in the NO emission. However, the relationship between temperature and NO fluxes is subject to considerable uncertainty with numerous exceptions in temperate and tropical systems where no clear relationship was found. Generally, the response to temperature is positive within the intermediate range of soil moisture content.

8.6.3.4 Soluble and Readily Decomposable Carbon

Several studies have shown that soil denitrification capacity is positively correlated to organic C, water soluble C, and total C concentration in soil provided other factors are favorable. Although the ratio of N_2O/N_2 from denitrification decreases with increasing available C supply (Weier et al., 1993), the total amount of N_2O produced from denitrification may be enhanced by the addition of organic materials. The N_2O emission form crop residue incorporation has been reported to range from <0.1 to 8 kg N ha^{-1} depending on the type and management of the residue and the measurement period. The magnitude of N_2O emissions depends on the C:N ratio of the residue with comparatively large emissions after incorporation of material with low C:N ratio (Baggs et al., 2000) but the effect is generally short-lived and most of the emissions occur during the first 2 weeks. In general, addition of degradable organic materials (e.g. animal and green manures) increased N_2O production in soils containing NO_3^- or supplied with fertilizer NO_3^- (Murakami et al., 1987).

8.6.3.5 Soil pH

Šimek & Cooper (2002) reviewed the research work on the interaction between pH and denitrification in soils. They concluded that the gaseous emissions (N_2O, NO and N_2) are less in acidic than in neutral or slightly alkaline soils; and the ratio N_2O: N_2 is increased when the pH of soil is reduced. In a series of denitrification enzyme activity assays in arable soil in New Zealand, van der Weerden et al. (1999) found a strong relationship between soil pH and N_2O/N_2 ratio, and suggested that maintaining the soil pH at about 6.5 might help maintain a low N_2O mole fraction from denitrification.

8.6.3.6 Tillage

Greater rates of denitrification are usually observed with zero tillage compared to plowed soils (Nieder et al., 1989). The increase is related to increased soil organic matter and higher levels of available C in the topsoil, as well as to greater soil densities and decreased soil aeration (Myrold, 1988). On the contrary some studies report greater NO emissions in tilled soils due to its effect on mineralization of soil organic matter and exposure of greater surface area to the atmosphere (Sanhueza et al., 1994; Sanhueza, 1997). Skiba et al. (1997) estimated that if 50% of all fertilized cropland is tilled, then tillage accounts for 0.75 Tg NO-N of the total annual biogenic NO emission.

8.6.3.7 Crop Type

Legumes

Biological nitrogen fixation (BNF) by legume crops provides input of N in many agricultural soils and the emissions of N_2O from leguminous crops are supposed to be of the same level as those of fertilized nonleguminous crops. However, studies of Rochette & Janzen (2005) showed that average N_2O emissions from legumes are 1.0 kg N ha^{-1} for annual crops, 1.8 kg N ha^{-1} for pure forage crops and 0.4 kg N ha^{-1} for grass legume mixes, which are only slightly greater than background emissions from agricultural crops. They (Rochette & Janzen, 2005) argued that there is little justification in using an emission factor for BNF by legume crops equal to that for fertilizer N.

Flooded Rice

Results of several studies show that conditions in wetland rice systems are more prone to denitrification than those in upland cropping and grassland systems (Hofstra & Bouwman, 2005). In flooded soils, denitrification rate is regulated by the NO_3^- supplying capacity of the system and time of flooding relative to nitrate production. On the other hand, flux of NO_3^- is governed by denitrification rate in the reduced soil layer, floodwater depth and NH_3 concentration in the floodwater and oxidized soil layer. Results from temperate and tropical rice fields indicate that <0.1% of the applied nitrogen is emitted as nitrous oxide if the soils are flooded for a number of days before fertilizer application (Freney et., 1981; Mosier et al., 1989). However, this loss could be an underestimate as N_2O dissolved and retained in water is not measured (Heincke & Kaupenjohann 1999). Thus, total N_2O losses could be higher than measured from short-period observations. Water management in lowland rice can influence the magnitude of denitrification. Intermittent dry spells (aerobic conditions) could result in greater total N loss from the soil than would occur under continuous anaerobic conditions. When soil experiences wetting-drying cycles and if mineral nitrogen is present in the soil before flooding it could serve as a source of nitrous oxide. Thus dry seeded rice can be a source of considerable nitrous oxide.

Denitrification in the floodwater has been found to increase with increased availability of available C in the underlying anaerobic soil layer (Engler & Patrick, 1974). Denitrification losses are reduced in the presence of rice plants than in systems without plants (Reddy & Patrick 1980); mainly because of competition for NO_3^- uptake between denitrifying bacteria and rice roots. The growth of Azolla in rice fields has been reported to enhance N_2O emissions (Chen et al., 1997).

There could be secondary emissions of nitrous oxide due to ammonia volatilization from flooded rice (Freney, 1997). Since NH_3 has a short lifetime in the atmosphere, the volatilized NH_3 will provide a secondary source of N_2O when it is redeposited on the soil's surface.

8.6.3.8 Land Use

Based on the analysis of published estimates of N_2O emission rates for a range of land-uses across Europe, Machefert et al. (2002) showed that the emission rates were higher for agricultural lands (2.1–38.3 kg N_2O-N ha^{-1} $year^{-1}$) as compared to forests and grasslands (0.122–7.3 kg N_2O-N ha^{-1} $year^{-1}$). However, some forested sites in their study showed N_2O emissions within the same range as for the arable agricultural sites. Conversion of tropical forests to crop production and pasture has a significant effect on emission of nitrogen oxides in the atmosphere. An increase in NO emissions occurs immediately after deforestation (1–5 years) followed by a significant decrease (below forest levels) in old pastures and secondary successional forests, which is probably related to the availability of nitrogen and the bacterial population (Sanhueza, 1997). Similarly, conversion of natural grassland to cropland in Savanna soils resulted in ~7 times greater NO emission than the original grassland (Fig. 8.14) and the yearly variations were related to the availability of moisture (Pérez et al., 2007). Yienger & Levy (1995) suggested that in temperate regions agriculture dominates soil NO emissions and in tropical regions grassland dominates NO emissions.

Grazing has been shown to stimulate NO emissions from grasslands, by increasing the availability of inorganic N and N input to soils. During grazing the animals return large quantities of N in small concentrated urine and dung patches, which can produce high N_2O emissions. For both urine and dung patches, N_2O emissions generally range between 0.1% and 4% of the N returned (de Klein et al., 2001). The N_2O emission factors reported for grazed pastures range from 0.2% to 10% of N excreted with highest values from intensively managed dairy pastures in the UK and the Netherlands (Velthof et al., 1996). The emission factors for pastures grazed by sheep or beef cattle are generally lower than from pastures grazed by dairy cows (de Klein et al., 2001).

8.6.3.9 Biomass Burning

Nitric oxide and also N_2O emissions have been shown to increase significantly following burning. Measurements made in diverse ecosystems show that vegetation burning enhances NO soil emissions. However, it seems that different processes occur at various sites e.g., in the tropical savanna enhanced emission from dry soils are observed immediately after burning whereas in Californian chaparral burned

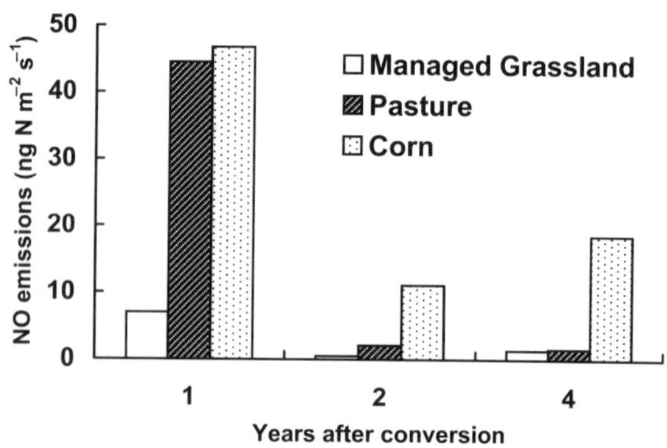

Fig. 8.14 Influence of conversion of natural grassland to cropland on average seasonal NO emission rates during 1st, 2nd and 4th year after conversion (Drawn from Pérez et al., 2007)

dry soil emit on average less than the unburned plots, and the fluxes only increase after soil wetting. Changes in the physical conditions of the soil surface and N availability are the most likely factors that explain the increased fluxes (Sanhueza, 1997). Yienger & Levy (1995) estimated that biomass burning is likely to account for 0.6 Tg N $year^{-1}$ of the global biogenic emission.

8.6.4 Nitrogen Oxide Emission Mitigation Options

One of the main approaches to mitigate emission of nitrogen oxides from the soil is to improve efficiency of N use. Higher the N recovery efficiency in plants, lesser is the amount of mineral N available for emission to the atmosphere. Approaches suggested for increasing fertilizer nitrogen use efficiency include the proper choice of fertilizer form to suit the conditions, the rate, mode and method of application, matching N supply with crop demand such as through use of optical sensor techniques, optimizing split application schemes, supplying fertilizer in the irrigation water, and using nitrification inhibitors and slow release fertilizers. Other practices that can mitigate emissions include adoption of minimum tillage, use of cover crops for canopy scavenging, replacement of the traditional slash and burn type agriculture with a more intensive agriculture and adoption of proper residue management practices to synchronize nutrient release and crop demand (e.g. by delaying incorporation of low C:N ratio crop residues until immediately prior to sowing of the following crop).

It has been estimated that if each of the suggested strategies are implemented, then the likely reduction in soil NO emissions from fertilizer application would be of the order of 0.4 Tg N $year^{-1}$, or 4% of the total biogenic NO emissions (Skiba et al., 1997). Since adoption of the suggested strategies will require substantial and expensive changes in current agricultural practices, these are unlikely to occur in the foreseeable future. Therefore, reducing the NO_x emissions from vehicles and fossil fuels burning could be a more realistic and straightforward option to reduce global atmospheric NO_x concentrations (Skiba et al., 1997).

8.7 Ammonia Emission

Ammonia (NH_3) is an important atmospheric pollutant with several impacts. It is involved in aerosol formation, plays a central role in the global N cycle, and is the most abundant atmospheric base with the ability to neutralize harmful acids. More than 80% of the NH_3 emissions to the atmosphere originate from anthropogenic sources, which include excreta from domestic animals, use of synthetic fertilizers, biomass burning and crops/crop decomposition. In animal excreta, nitrogen is in the form of NH_4^+, urea and organic nitrogen. Urea and much of the organic nitrogen in animal excreta is rapidly converted to NH_3, which can be directly volatilized from the animal production system or when the excreta is applied to the soil. It is estimated that 30% of the nitrogen excreted by farm animals is released to the atmosphere from animal houses, during storage, grazing and after application of animal waste to soil. When synthetic fertilizers such as urea and ammonium forms of fertilizers (e.g. ammonium bicarbonate, ammonium sulfate) are applied on moist soil surfaces, they undergo a series of chemical conversions to ammonia. The ammonia gas then escapes to the atmosphere rather than becoming a plant nutrient. During biomass and fossil fuel burning, a part of the N contained in these fuels is converted to NH_3 and emitted to the atmosphere.

Globally NH_3 emissions are estimated to be around ~58 Tg N year^{-1}, though the estimates vary considerably (Table 8.22). Compared to preindustrial estimates of about 20 Tg N year^{-1}, there is a threefold increase in the emissions of NH_3. Of the current global fluxes, ~43 to 47 Tg N year^{-1} originate from anthropogenic sources with excreta from domestic animals being the highest emitter (22.9 Tg N year^{-1}), followed by synthetic N fertilizers (9.7 Tg N year^{-1}), and biomass burning (4.6 Tg N year^{-1}). The emissions from industrial processes such as chemical and fertilizer manufacture are rather small. Among the natural sources the oceans contribute the

Table 8.22 Global estimates of NH_3 emissions (Tg N year^{-1}) to the atmosphere for 1980–1990s

	Bouwman et al. (1997)	Holland et al. (1999)	van Aardenne et al. (2001)	IPCC (2001a)	Denman et al. (2007)
Natural sources					
Soils under natural vegetation	2.4	2.4–10	4.6	2.4	2.4
Oceans	8.2	8.2–13	5.6	8.2	8.2
Wild animal excreta	0.1	0.1–6	–		
Natural burning at high altitudes	–	–	0.8		
Anthropogenic sources					
Fossil fuel combustion and industrial process	0.3	0.3–2.2	0.3	0.3	2.5
Agriculture	–	–	–	34.2	35.0
Domestic animal excreta	21.6	20–43	22.9	–	–
Synthetic fertilizer use	9.0	1.2–9.0	9.7	–	–
Biomass and biofuel burning	5.9	2.0–8.0	7.2	5.7	5.4
Crops	3.6	3.6	4.0	–	–
Humans and pets	2.6	2.6–4	3.1	2.6	2.6
Total	54	45–83	58.2	53.4	56.1

maximum followed by soils under natural vegetation. These estimates indicate that about two-thirds of the total global NH_3 is emitted from agricultural systems. The regions with highest emission rates are located in Europe, the Indian subcontinent and China, reflecting the patterns of animal densities, type and intensity of synthetic fertilizer use (Bouwman et al., 1997). In Western Europe, 74% of total NH_3 emissions (4.02 Gg N year^{-1}) originate from animal excreta, 12% from synthetic fertilizers and ~6% from crops (Ferm, 1998). In contrast, in east, southeast and south Asia ammonia emissions from croplands are estimated to be 11.8 Tg N year^{-1} with 6.5 Tg N originating from the use of chemical N fertilizer and 4.7 Tg N from the use of animal manures. The average NH_3 loss rate from chemical N fertilizer in the Asian region is 16.8%, which is much higher than the global average of 10%; mainly because of widespread use of urea and ammonium bicarbonate as fertilizers, occurrence of alkaline soils in some parts of the region (such as semiarid India and north China), and cultivation of rice on a large proportion of area. It is estimated that 22% of the urea fertilizer applied to rice fields is lost through ammonia volatilization as compared to 13.7% from upland crops. The emissions of NH_3 from ammonium bicarbonate are still higher by a factor of 1.5 (Yan et al., 2003a). In addition to type of fertilizers a number of soil, environmental and management factors influence the emission of NH_3 from soil to atmosphere. Bouwman et al. (2002a) developed a regression model to incorporate the effect of some of the factors related to agricultural management (crop type, fertilizer type and method of application), soil and environmental conditions (climate, soil pH and CEC) on NH_3 volatilization. The model, which is based on NH_3 loss measurements presented in 148 publications has been used to estimate NH_3 volatilization from global application of synthetic fertilizers (78 million tons N year^{-1}) and animal manure (33 million tons N year^{-1}) during 1995. The calculated NH_3 loss amounted to 14% (range 10–19%) from synthetic N fertilizers and 23% (range 19–29%) from application of animal manure. The estimated NH_3 loss from synthetic fertilizers is higher in developing countries (18%) as compared to industrialized countries (7%). This has been attributed to high temperatures and the widespread use of urea, ammonium sulfate and ammonium bicarbonate fertilizers. The differences are smaller for animal manure; 26% in developing and 21% in industrialized countries. However, there is considerable uncertainty in these estimates as the model incorporates the effect of only some selected environmental and management variables and a number of assumptions have been made in upscaling the emission estimates. In addition to the factors considered in the model a number of other soil factors such as buffering capacity, calcium carbonate content, soil texture, and organic matter content influence NH_3 volatilization. The dominant environmental and management factors include wind speed, temperature, soil moisture, source and method of N application. In submerged soils, losses of NH_3 are directly related to the concentration of aqueous NH_3 in the floodwater, which in turn is a function of total ammoniacal N, the pH, and the temperature of the floodwater. The process of NH_3 volatilization and the variables that influence NH_3 emission from agricultural soils have been discussed in detail by Benbi & Richter (2003).

8.7.1 Ammonia Emission Mitigation Options

Since the animal excreta and croplands constitute the main sources of anthropogenic emissions of NH_3, the mitigation options need to focus on better management of the two sources. The measures to reduce ammonia emissions from animal wastes include optimizing livestock densities, reducing the excretion of urea and urea like products by optimizing N intake and retention (Bussink & Oenema, 1998), dilution or acidification of slurry, injection or band placement of slurry in the soil instead of surface application, application of irrigation subsequent to application of slurry, covering of slurry storage tanks, keeping the temperature of animal wastes as low as possible, and storing solid and liquid wastes anaerobically. For example, addition of water to the slurry in the ratio of 3:1 slurry has been reported to reduce NH_3 losses by 44–91% as compared to undiluted slurry and injection of slurry in open slits in soil has been found to reduce ammonia loss by 80–90% (Huijsmans & Bussink, 1990). Practices to reduce NH_3 emissions in croplands are related to efficient fertilizer use as discussed in the preceding section

8.7.2 Ammonia Emission from Plants

It has long been known that plants can both emit and take up NH_3 from air. The presence of NH_4^+ in leaf tissues results in the existence of an NH_3 'compensation point' concentration (a concentration where NH_3 is neither emitted nor taken up) for sub-stomatal tissues, so that both emission and deposition are possible from stomata. Ammonia is emitted if the concentration of NH_3 in the atmosphere is lower than the compensation point, while in the opposite case NH_3 absorption takes place (Farquhar et al., 1980). The value of compensation point concentration has been reported to range from as low as 0.5 μg NH_3-N m^{-3} for an N-limited Colorado forest (Langford & Fehsenfeld, 1992) to 23 μg NH_3-N m^{-3} for wheat at late grain filling stage (Morgan & Parton, 1989). High compensation points are probably related to high tissue N status, rapid absorption of NH_4^+ from the root medium and/or low activity of glutamine synthetase (Schjoerring et al., 1998). The compensation point changes with stage of crop growth, diurnal cycle and temperature, which complicates evaluation of NH_3 emissions from plants. For each 5°C increase in temperature the compensation point doubles (Asman et al., 1998).

Estimating long-term bi-directional NH_3 fluxes between plant and atmosphere is still uncertain, though it is now possible to apply a single model concept to a range of ecosystem types and satisfactorily infer NH_3 fluxes over diurnal time scales (Sutton et al., 1995). On a seasonal basis emission amounting to 1–5 kg NH_3-N ha^{-1} was measured from differentially fertilized wheat, barley oilseeds and pea crops. The amount of NH_3 lost from plant foliage constituted 1–4% of the actual amount of nitrogen present in the shoots of crops (Schjoerring & Mattsson, 2001). Ammonia emission has been suggested to contribute to the decline in shoot nitrogen content which is often observed in agricultural crops during generative growth stages.

Another source of NH_3 emission from plants is their senescence and decomposition. When plants begin to senesce they typically lose N content through NH_3 volatilization (Bouwman et al., 1997). Globally NH_3 emissions from crops and crop decomposition add 3.6 Tg N $year^{-1}$.

8.8 Global Climate Change and Crop Yields

The growth in human population over the past century has been closely associated with increased production of food and forest products. In the last 30 years, grain production has increased slightly faster than population but this increase has not been uniformly distributed and the number of chronically undernourished has remained relatively stable at about 700–900 million (Dyson, 1996). Production of food and forest products will need to increase to meet the world's projected demand. Yield growth for 11 major annual crops in the USA has been rapid since 1939 ranging, across crops, from about 1% per year on average to over 3% per year (Reilly & Fuglie, 1998). Nevertheless, there is no evidence that yields plateaued for important crops in recent years in a strict sense that yield growth in absolute terms has slowed or stopped. The task of increasing production will be made more complex by the additional and interactive effects of changes in climate, atmospheric composition, land use and other global change drivers.

8.8.1 Projected Demand of Crop Yields

Overall, a population of about six billion today is projected to rise to about eight billion by about 2025 with most of the increase in the less developed nations of Asia and Africa (Fischer & Heilig, 1997). These estimates show regional differences in the expansion of population with large percentage increases in Africa (102%) and West Asia (77%) but with the largest increases in absolute numbers in south central Asia (737 million). Along with this change will be an increasing trend towards urbanization, which will reduce the area of prime agricultural land and increase the pressure toward the intensification of crop production (Gregory & Ingram, 2000). Given the close association of population and grain production, and allowing for changes in diet towards greater consumption of meat, it is possible to estimate the required grain production (wheat, rice and maize together supply about 60% of the human carbohydrate nutrition). Annual population growth of 80 million requires an annual increase in grain production of about 29 million tons (Crosson & Anderson, 1994). In fact, this will be a continuation of the past trend, which has seen the area harvested per capita decline from about 0.24 ha in 1950 to 0.125 ha in 1993, while global cereal yield increased from 1.2 to 2.8 Mg ha^{-1} over the same period (Dyson, 1996). Global average cereal yields will need to increase from the current 2.9–4.2 Mg ha^{-1} by 2025 (Gregory & Ingram, 2000).

At the same time, the production of forest products will also need to increase to keep up with the projected demands for lumber and paper. Calculations by Alexandratos (1995) show that world consumption of forest products will increase annually by 1.4% for fuel wood and charcoal, 3.1% for paper, and 4.6% for panels in the period 1990–2010.

8.8.2 Influence of Climate Change on Crop Yields

Global climate change will have influence on agricultural, forestry and other natural ecosystems. With respect to agriculture, changes in solar radiation, temperature, and precipitation will produce changes in crop yields, crop mix, cropping systems, scheduling of field operations, grain moisture content at harvest as well as on economics of agriculture including farm profitability. Individual crop growth processes are affected differently by climate change. Global production of annual crops will be affected by increases in mean temperature of 2–4°C expected towards the end of the 21st century. Within temperate regions, seasonal rise in temperature will increase the development rate of crop, resulting in an earlier harvest. Nevertheless, this negative effect of warmer temperatures should be countered by the increased rate of crop growth at elevated atmospheric CO_2 concentration, at least when there is enough water. Of more importance for the yield of annual seed crops may be changes in the frequency of hot temperatures which are associated with warmer mean climates. Seed yields are particularly sensitive to brief episodes of hot temperatures if these coincide with critical stages of crop development (Wheeler et al., 2000). Hot temperatures at the time of flowering can reduce the potential number of seeds or grains that subsequently contribute to crop yield. There are several research needs in order to provide a framework for predicting the impact of episodes of hot temperatures on the yields of annual crops, including reliable seasonal weather forecasts, robust predictions of crop development, and crop simulation models.

In contrast, increasing rainfall in drier areas may allow the photosynthetic rate of the crop to increase, resulting in higher yields. Across a long-term study in the midwestern USA, yields from long-season maize increased significantly in the northern part under predicted future climate conditions, i.e., under wetter conditions occurring during several years of the study (Southworth et al., 2000). Scenarios of increased climate variability produced diverse yields on a year-to-year basis and had increased risk of low yields particularly in drier years.

The atmospheric CO_2 concentration is projected to reach ~550 ppm by 2050. Plant responses to elevated CO_2 concentration have the potential to influence the global carbon cycle and climate in the future, but the complexity of scaling from the leaf to whole plant, canopy, ecosystem and biosphere scales make it unclear to what extent this will be realized. Elevated CO_2 concentration will probably offset some of the future losses in crop yield caused by increased temperature and drought stress. Plants with C_3 pathway respond directly to growth at elevated CO_2 concentration via stimulated photosynthesis and reduced stomatal conductance. The enhancement of photosynthesis is a result of increased velocity of carboxylation of CO_2 by the enzyme Rubisco (carboxylase-oxygenase) and inhibition of the competing oxygenation reaction. Long-term exposure of C_3 plants to elevated CO_2 concentration also leads to photosynthetic acclimation. The decrease in stomatal conductance at elevated CO_2 concentration can reduce canopy water use and indirectly enhance carbon gain by ameliorating drought stress. Reduced water use can indirectly enhance carbon gain by ameliorating stress in times and places of drought.

The mechanism of increased CO_2 concentration effects upon C_4 plants has received considerable research interest but still remains poorly understood. Leakey et al. (2006) conducted a rainfed-field experiment in 2002 and 2004, utilizing FACE (Free-Air Carbon Dioxide Enrichment) technology to determine the effect of elevated CO_2 concentration on *Zea mays*. Each year, crop performance was compared at ambient (~370 ppm CO_2) and the elevated (~550 ppm CO_2) CO_2 concentration predicted for 2050. The year 2004 was unusual climatically and provided a unique opportunity to test the effects of elevated CO_2 concentration in the absence of water stress. There was no CO_2-effect on photosynthesis, carbon metabolism, growth or yield. Nevertheless, elevated CO_2 concentration reduced stomatal conductance, transpiration and soil moisture depletion. The year 2002 was an "average" year in which plants experienced episodic water stress. During these dry periods, photosynthesis was greater under elevated CO_2 concentration. It was concluded that elevated CO_2 concentration can only indirectly enhance carbon gain during drought.

However, while relations described above can be used retrospectively to explain yields, their ability to predict is often poor and the current uncertainties in global climate models for predicting future climates at temporal and spatial scales appropriate to cropping systems mean that predicting the consequences for future production is imprecise. Reilly & Schimmelpfennig (1999) reviewed global climate change impacts on agriculture and concluded that the current uncertainties about future climate change mean that there still is limited ability to predict which regions will benefit or loose beyond the generalization that tropical regions are likely to suffer disproportionably.

8.8.3 *Potential to Increase Global Production*

There may be two major options whereby the projected increases in regional food production can be achieved. The first is the expansion of the area of cultivated land (extensification) and the second is by intensifying the production system either by increasing the number of crops sown on a particular area of land or by increasing the yield per unit area of individual crops, or both (intensification). Globally, no single means will be adopted and different regions will increase production in different ways. According to predictions by Alexandratos (1995), most (81%) of the increased production in developing countries is expected to come about through intensification via increased yield of individual crops (69%) and intensified use of existing land (12%). In South Asia there is little new land that can be brought into cultivation so that almost all of the increased production will be through increases in yield and crop intensification (e.g., more rice crops per year or rice grown in rotation with an upland crop such as wheat. In Central and South America, a large amount of new land is still available for cultivation and a substantial percentage (29%) of the projected increases in production might be met in this way. In Africa too, there is a potential for expansion of the cultivated area but the land is often far from centres and the infrastructure is poor. Therefore, yields will have to increase substantially on land currently under crops.

In summary, even under constant environmental conditions, it will be a challenging task to increase production for meeting the growing global demand. It will be complicated still further because global climate change means that optimum crop and forest management will be difficult to plan.

8.9 Economics of Carbon Sequestration

Atmospheric concentrations of greenhouse gases can be reduced by withdrawing carbon from the atmosphere and sequestering it in soils and biomass and by storing in underground geologic formations, and in the deep ocean. Geologic sequestration involves a direct injection of carbon dioxide into an underground geologic formation at high pressure and at great depth. Because geologic sequestration costs are site-specific, they make cost estimates difficult. The costs will depend on the option, available infrastructure, location, depth, and the individual characteristics of the storage reservoir formation. The ocean is the largest natural sink on Earth and is thought to have enormous potential for additional carbon storage. Two processes drive the natural activity of the ocean to take in carbon dioxide: a biological system that transports carbon to the ocean's interior, and the solubility of carbon dioxide in seawater that is further enhanced by ocean circulation. However, many of the ecological, chemical, and geological elements of the deep sea and, therefore, the effects of injecting carbon dioxide into the ocean, are widely unknown.

Because of the difficulties related to economic evaluation of C sequestration in geologic and ocean systems, we will focus only on the economics of C sequestration in terrestrial systems. Carbon sequestration in agricultural soils occurs only until a new SOM equilibrium is reached. In most cases, the new equilibrium is reached after a period of 20–30 years after a change in management. Carbon sequestration through afforestation spans more time period depending on a maturation period of a stand, which is estimated to be between 70 and 150 years (Stavins & Richards, 2005). From an economic perspective, soil carbon provides value in three dimensions: (i) as an essential component of soil that affects agricultural productivity, (ii) as a way to offset CO_2 emissions from other sources, and (iii) as an indirect source of benefits involving improved environmental quality.

A number of independent initiatives are presently coming up at national and international levels that use various approaches calculating carbon sequestration costs. Although this diversity can lead to innovation, it may also reflect conflicting interests that need to be reconciled. Ultimately, a global carbon market requires the support of an institutional infrastructure that can increase investor confidence and reduce transaction costs in international trading. This infrastructure may include national offices, regulatory agencies, and establishment of trust funds, trading platforms such as exchanges, brokers, certifiers and insurers. For example, establishing a national carbon registry can help to prevent double selling of carbon credits and also provide transparency for prices that are critical to fair negotiations. This section provides a brief overview of economic issues related to carbon sequestration.

8.9.1 Methods for Calculating Carbon Sequestration Costs

8.9.1.1 Monetary Unit

The cost (or value) of carbon sequestration is generally expressed in monetary units per Mg of carbon sequestered, and is based on the cost of the land, planting and management. The amount of carbon that can be sequestered varies with storage method, land management practices, the plant, crop, and vegetation species, and geographic location. Various studies in the literature differ in terms of the metric used for cost-effectiveness calculations, i.e., US Dollars ($) or Euros (€) per megagrams of carbon.

8.9.1.2 Type of Model

There are three broad categories of models in literature including (i) engineering or bottom-up, (ii) sectoral, and (iii) econometric models. The engineering models generally use reported land prices to estimate the social cost of converting land from one use or practice to another (e.g., Moulton & Richards, 1990; van Kooten et al., 1992). These models are generally limited in scope to the immediate on-site effects of a program or practice. On the other hand, sectoral models estimate the opportunity cost of land using supply and demand equations for each sector under consideration, say agriculture, forestry, or both (Alig et al., 1998; Adams et al., 1999). The costs are derived from estimates of changes in consumer and producer surplus as the amount of land dedicated to agriculture or forestry expands or contracts. The models incorporating both agriculture and forestry could include the effects of competing demands from the two sectors for land. Econometric studies use historic market responses to changes (e.g., in timber prices) that are analogous to a government carbon sequestration program to infer how landowners would respond to direct or indirect carbon prices (Stavins, 1999; Newell & Stavins 2000). These models inherently capture the interplay between agriculture and forestlands.

As always the case, the bottom-up models tend to be simple and transparent but fail to include dynamics of market interactions. The sectoral and econometric models on the other hand do capture the market interactions, but could be data demanding.

In general, econometric studies seem to provide higher estimates of carbon sequestration costs, while the bottom-up models provide lower estimates because they account for less of the overall counteracting results.

8.9.1.3 Lack of Permanence

Despite the potential of C sinks that have been identified, there is concern about how effective these sinks can be and whether they can contribute to decreases in the build-up of GHG in the long term. Even if carefully preserved, both forest biomass and soils have a finite capacity to sequester carbon. Relative to fossil carbon, they are volatile and subject to reemission into the atmosphere (Murray et al., 2007). This could possibly lead to a complete release of the sequestered C. For example, forests can be cleared and farming practices such as conservation tillage or residue return can be reversed. From a more dynamic perspective, temporary terrestrial sinks may be useful if technological progress reduces emissions of GHG by industry. However, this "buying time" argument is relevant only if new technologies will be invented and adopted by polluting industries.

To deal with lack of permanence of carbon credits a common approach has been either to acknowledge the same, assess the environmental and economic benefits of limited-term sequestration, and allot credits in proportion to the time period over which carbon is sequestered, or to provide reasonable assurance of indefinite sequestration. The first alternative has led to what has been called the ton-year approach, in which activities would accrue credits for each year that a ton of carbon is withheld from the atmosphere and some quantity of ton-years would be equated with a permanent ton. For the second alternative, three mechanisms have evolved for providing reasonable assurance of indefinite sequestration: (i) provide partial credits according to the perceived risk that they will be maintained for a specified time, (ii) link temporary sequestration projects with obligations for later action to assure permanence of the emissions reduction, and (iii) tax sequestration credits to finance research and development into emissions-saving technologies (Chomitz, 2000).

Several approaches have been suggested for defining the equivalency factor, i. e. the number of ton-years that is to be equated with permanence (IPCC, 2000a; Fearnside et al., 2000; Moura Costa & Wilson, 2000). Basically, one would integrate over time the number of tons sequestered and convert this to tons of carbon emissions offset by dividing the equivalency factor, i.e. Mg-years/f = permanent Mg, where f is the equivalency factor. There is no unique way to determine a conversion rate between ton-years and permanent tons and that the choice among a number of justifiable possibilities is thus a policy decision. A new approach that avoids many of the above mentioned problems is so called 'rental' approach suggested by Sedjo (2001). Just as a space can be rented to provide for the temporary parking of a car, space could be rented for 'parking' carbon.

8.9.2 Economics of Carbon Sequestration in Forestry

Early work on economics of carbon sequestration focused on examining whether expansion of forest sinks could play a major role in mitigating atmospheric CO_2. These studies typically used hypothetical government programs such as subsidies, or government purchases, to promote forestry management practices like afforestation of agricultural land, or conservation of forestland, and attempted to estimate cost of sequestration by assessing the costs of various inputs to production including land, labor and material.

Carbon is sequestered for long periods of time if the forest is never harvested. If the forest is harvested, long-term sequestration can only occur if slow-growing older trees are processed to lumber used in construction of buildings lasting for decades or centuries. Sequestration is short-lived if trees are burned for firewood, or processed into paper. Estimates by Heath et al. (1996) suggest that 35% of carbon removed from US forests is stored in the long term, 35% is burned for energy and 30% returns to the atmosphere through decay.

Numerous studies have estimated the cost of carbon sequestration in forests. However, analyses of apparently similar programs have led to disparate results. While most of the C sequestration cost studies contain the same components, they are not comparable due to inconsistent use of terms, geographic scope, assumptions and methods. Moreover, studies have not only used different ecosystem components and different data for CO_2 sequestration rates, but also different formats for carbon yield, including average yield, cumulative lifetime yield, and yield curves. Also, there are different methods for estimating land opportunity costs.

The data in Table 8.23 are adjusted for the described variation among the studies. The data vary according to geographic scope but even among studies that have focused on similar regions, there are vastly different estimates of costs for C sequestration. Globally, the costs are estimated to range between about \$10 and \$200 Mg^{-1} C. The data also suggest that there is a significant potential to sequester large quantities of carbon for less than \$50 Mg^{-1} C and that C sequestration in developing countries is more cost-effective than in developed countries.

Because forestry has the potential to sequester substantial carbon at prices consistent with current estimates of potential greenhouse gas abatement policies, it is useful to ask how forestry fits within the set of available policy tools. Carbon sequestration cost studies are of major importance for policy makers who consider the role of carbon sink programs in national and global strategies to mitigate greenhouse gas emissions. They will analyze these studies to compare cost-effectiveness between carbon sink enhancement programs and carbon source reduction programs. Understanding how forest C sequestration integrates with other climate change options is challenging. For the most part, climate policy is assessed with national or global economic models that capture important economic linkages in the world economy (e.g., Nordhaus & Boyer, 2000).

Table 8.23 Examples of costs for carbon sequestration in the forestry sector (Adapted from Richards & Stokes, 2004)

Authors	Region	Forest plantation ($ Mg^{-1} C)	Forest management ($ Mg^{-1} C)	Agroforestry ($ Mg^{-1} C)
Sedjo & Solomon (1989)	Global	3.5–7	–	–
Nordhaus (1991)	Global	42–114	–	–
IPCC (2000b)	Global	0.1–100	–	–
Dixon et al. (1991)	Boreal	5–8	7	–
	Temperate	2–6	1–13	23
	Tropical	7	1–9	5
Dixon et al. (1994)	South America	–	–	4–41
	Africa	–	–	4–69
	South Asia	–	–	2–66
	North America	–	–	1–6
Moulton & Richards (1990)	USA	9–41	6–47	–
Dudek & LeBlanc (1990)	USA	23.9–38.4	–	–
Adams et al. (1993)	USA	20–61	–	–
Richards et al. (1993)	USA	9–66	–	–
Parks & Hardie (1995)	USA	5–90	–	–
Alig et al. (1997)	USA	24–141	–	–
Stavins (1999)	USA	0–136	–	–
Stavins (1999)	Delta States/USA	0–66	–	–
Plantinga et al. (1999)	Maine/USA	0–250	–	–
Plantinga et al. (1999)	South Carolina/USA	0–40	–	–
Plantinga et al. (1999)	Wisconsin/USA	0–85	–	–
van Kooten et al. (1992)	Canada	6–18	8–23	–
van Kooten et al. (1992)	British Columbia and Alberta/Canada	0–50	–	–
Masera et al. (1995)	Mexico	5–11	0.3–3	–
Sedjo (1999)	Patagonia/Argentina	20	–	–
Kerr et al. (2001)	Costa Rica	–	10–30	–
Makundi & Okitingati (1995)	Tanzania	–	1.27–34.38	0.13–3.40
Xu (1995)[1]	China	(12)–2	(2)–1	(13)–(1)
Ravindranath & Somashekhar (1995)	India	0.13–1.06	0.09–1.22	0.95–2.78
Wangwacharakul & Bowonwiwat (1995)[a]	Thailand	(579)–0.92	0.9–12.5	(115)–(1.17)

[a] Numbers in parentheses indicate negative costs

–: Not available

Table 8.24 Examples of costs for carbon sequestration in the agricultural sector (Adapted from Paustian et al., 2006)

Authors	Region	Measure	Quantity (million megagrams year^{-1})	Average cost (\$ Mg^{-1} C)	Price or implicit (marginal) cost (\$ Mg^{-1} C)
Antle et al. (2003)	Montana/ USA	Reduced fallow or change from arable land to permanent grassland			
		– Per hectare payment	0.2–0.9	10–65	10–160
		– Per Mg C payment	0.4–0.8	10–40	10–100
Antle et al. (2005)	Central USA	Reduced fallow and conservation tillage			
		– Zero transaction cost	1.2–7.9	–	10–200
		– \$5 ha^{-1} transaction cost	0.9–7.4		10–200
Kurkalova et al. (2003)	Iowa	Conservation tillage with per Mg C payment	0.2–1.1	–	17–95
McCarl & Schneider (2001)	USA	Reduced tillage with emission tax, subsidy for CO_2, NO_2 and CH_4	44–70	–	10–500
Pautsch et al. (2001)	Iowa	Conservation tillage			
		– Per hectare payment	0–1.0	0–200	–
		– Per Mg C payment	0–1.0	0–300	–
Lewandrowski et al. (2004)	USA	– Conservation tillage with per Mg C payments	1–26.9	–	10–125
		– Change from arable land to permanent grassland	–	–	10–125

–: Not available

8.9.3 Economics of Carbon Sequestration in Agriculture

Increasing the potential of agricultural lands for carbon storage may also involve a change in land-use and/or management practices. For example, conservation tillage leaves a minimum of 30% crop residue on the soil after planting, decreases soil disturbance, and increases the amount of C that can accumulate in the soil. However, some practices may not be the most economical for farmers. One way to encourage carbon storage is for governments to subsidize new agricultural management practices that increase carbon storage, similar to those subsidies already provided for growing specific crops or for keeping land fallow. Recent model-based studies of economic potential for carbon sequestration on agricultural lands are listed in Table 8.24.

The studies indicate that agriculture can sequester carbon at a cost competitive with other forms of GHG reductions. The costs vary regionally and according to the cropland management change. They also depend on whether payments are offered per hectare or per megagrams of C sequestered. It is generally feasible that in US agricultural soils 70–220 million megagrams of C could be added to soils annually over 2–3 decades, offsetting about 4–11% of current US GHG emissions. The economic potential to store carbon varies by region, and current studies suggest that at prices of \$50 Mg^{-1} C, soil carbon increases in the US would be limited to 70 million megagrams of C.

8.9.4 Secondary Benefits from Carbon Sequestration Measures

Most studies on the economics of C sequestration focus on the costs of practices to protect and expand carbon sinks. Only a few have included the possible secondary benefits, which could be accounted as "negative costs" in the standard analysis of carbon sequestration effectiveness. The importance of SOM to maintain productive soils and key functions like acting as water and nutrient reservoir, promoting favorable soil structure for plant rooting, and acting as a filter for pesticides and other organic compounds, is well known. Soils rich in SOM are commonly well aggregated and tend to be less susceptible to soil erosion by water and wind. Reduction of sediment runoff leads to improved water quality and, therefore, to cost reduction of water treatment. Reduced wind erosion leads to better air quality through reduction of dust and a clearer viewscape. Sanders et al. (1995) reviewed different studies indicating that US on-farm costs caused by soil erosion range from \$500 million to \$1.2 billion per year. This contrasts with estimates of off-site damage costs ranging from \$2.2 billion (Clark et al., 1985) to \$7 billion (Ribaudo et al., 1989).

The buildup of SOM that accompanies C sequestration thus provides many environmental, social and economic co-benefits. These secondary benefits from C sequestration are sometimes believed to be as high as the cost of the conversion measure itself, implying that carbon sequestration might be nearly costless. However, besides their positive effects, some practices to sequester carbon in soils may have potential adverse effects. For example, enhanced leaching of soluble nutrients to the groundwater due to preferential flow under reduced tillage (Cole et al., 1993), and possible negative effects on the environment from application of manures, composts and sludges (Batjes, 1998). These effects have not been quantified, and there is almost no information about the extent to which these may offset ecological and economic benefits of C sequestration.

8.9.5 Leakage of Emissions Beyond Project Boundaries

In contrast to the secondary benefits, the effects of leakage may substantially raise the costs of C sequestration. The problem of leakage arises from two basic facts. The first is that land can be shifted back and forth among various forestry and agricultural uses. Second, the balance of activities on land will depend on the relative prices in the agricultural and forestry sectors, and individual projects and programs do little to change those prices or the resulting demand for land. For example, if forestland is preserved from harvest and conversion in one location, the unchanged demand for agricultural land and forest products could lead to increased forest clearing and conversion in another region. Thus the effects of the preservation may be partially or entirely undone by the 'leakage' (Richards & Anderson, 2001). Conversely, if agricultural land is converted to forest, the underlying demand for agricultural land may simply cause other forested land to be converted back to agriculture. The problem is not only of regional but also of global importance. For example, the timber market is global in scope, and forest protection and expansion programs in one part of the world can be counterbalanced by market responses in other regions.

Accounting for leakage would increase the cost of carbon sequestration (Stavins, 1999). However, leakage is not a problem concerned with carbon sequestration programs alone as similar concerns surround the other GHG mitigation options as well. Leakage can be avoided or minimized by proper project design, such as: (i) maintaining needed resources and providing socioeconomic benefits, including alternative economic opportunities to local populations; (ii) monitoring key products, such as timber extraction, to quantify and reduce carbon benefits if necessary; and (iii) monitoring deforestation rates during the project life and quantifying them to determine actual project carbon benefits. Leakage is an unresolved issue that may seriously affect the contribution of carbon sequestration to a greenhouse gas mitigation program. Since leakage is a serious issue at both the national and international levels, it may occur that governments will expend billions of dollars in subsidies or other forms of incentives, with no net gain in carbon, forest or secondary benefits (Richards & Stokes, 2004).

Chapter 9
Modeling Carbon and Nitrogen Dynamics in the Soil-Plant-Atmosphere System

It is generally agreed that soils play an important role in gaseous exchange and water and matter dynamics in terrestrial ecosystems. Much experimental work has been undertaken in the last few years in attempts to quantify the emission of greenhouse gases in terrestrial ecosystems and to identify the key factors that govern them. However, fluxes vary greatly between ecosystems, and are often subject to great spatial and temporal variations, within ecosystems. Consequently, there is an essential role for modeling, to scale up observations made on relatively small areas to larger regions and ultimately to global level, and also to be able to predict the effects of environmental changes and changes in land management on future emissions. This chapter examines current knowledge regarding modeling approaches employed to describe soil-atmosphere gaseous exchange and carbon and nitrogen dynamics in soils. The extent to which the models have been used for predictive purposes for different natural and managed systems and at different levels of spatial aggregation is discussed. Since the number of models available is far too great to cover comprehensively, only examples are given to illustrate different types of model, and the modeling approaches applicable to different scales.

9.1 Carbon Dioxide Exchange from Soils

Soils are the dominant terrestrial source of CO_2. Carbon dioxide is produced in soils primarily by heterotrophic organisms and by respiration of living roots and released to the atmosphere. This process, commonly called soil respiration results from decomposition of organic matter and is influenced by a number of soil and climatic factors. Several models of organic matter decomposition or carbon dynamics are available, which include soil respiration as a component or a sub-model and these could be used to estimate regional CO_2 efflux or exchange from soil. However, this section is restricted to the description of models specifically meant to predict fluxes of CO_2 between terrestrial ecosystems and the atmosphere. Modeling of soil organic matter or carbon and nitrogen dynamics is discussed in a subsequent section. Contrasting approaches have been used to predict CO_2 fluxes from different ecosystems and at different scales. For regional and global predictions relatively

R. Nieder, D.K. Benbi, *Carbon and Nitrogen in the Terrestrial Environment,*

simple empirical models, which establish relationship between CO_2 fluxes and climatic variables such as temperature and precipitation are generally employed. Raich & Schlesinger (1992) based on published data in the literature derived a model involving temperature and precipitation as variables. It was indicated that global variation of soil respiration is mostly accounted for by the variation of temperature. Raich & Potter (1995) presented a regression model (Equation 9.1) for estimating global soil CO_2 efflux using information on mean monthly air temperature (*Ta*, °C), and mean monthly precipitation (*P*, cm) for the period 1980–1994. The model contains three parameters viz. soil respiration rate at a mean monthly air temperature of 0°C (*F*, g C m^{-2} day^{-1}), rate of change of the soil respiration rate with respect to temperature (*Q*, $°C^{-1}$), and the half-saturation constant for a hyperbolic relationship between soil respiration and rainfall (*K*, mm $month^{-1}$)

$$R_s = F^* \, e^{(Q * Ta)} * [P/(K + P)] \tag{9.1}$$

where R_s refers to the mean monthly soil-CO_2 efflux in g C m^{-2} day^{-1}. Raich et al. (2002) recalculated the model parameters using a new data set of soil respiration measurements. The resulting model is:

$$R_s = 1.250 * e^{(0.05452 * Ta)} * [P/(4.259 + P)] \tag{9.2}$$

The model has been used to predict mean monthly soil respiration rates at 0.5° latitude × 0.5° longitude grid cells. The mean annual global soil-CO_2 flux over 1980 to 1994 was estimated to be 80.4 (range 79.3–81.8) Pg C. Monthly variations in global soil-CO_2 emissions followed closely the mean temperature cycle of the Northern Hemisphere. Tropical and subtropical evergreen broad-leaved forests contributed more (~30% of the total) soil-derived CO_2 to the atmosphere than did any other vegetation type. At the global scale, however, annual soil-CO_2 fluxes correlated with mean soil temperature (Fig. 9.1). The model results showed that

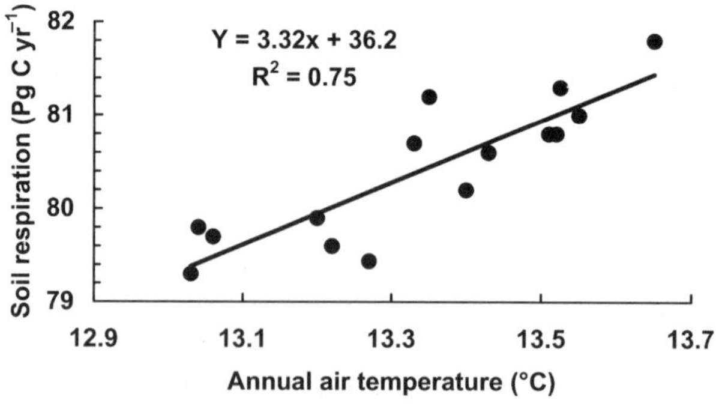

Fig. 9.1 Relationship between estimated global soil respiration and mean annual air temperature over land during 1980 to 1994 (Raich et al., 2002, p. 809. Reproduced with kind permission from Wiley-Blackwell)

soil-CO_2 emissions increase with increasing global temperature suggesting thereby that soils are losing organic C in response to global warming or that soil C cycles faster (i.e. more inputs and outputs) as temperatures increase.

Zamolodchikov & Karelin (2001) developed an empirical model based on GIS approach to estimate CO_2 fluxes over the entire Russian tundra zone. Eight geographical regions were divided into tundra landscape types (2–13 per region, totalling 69). Six years of field data on CO_2 fluxes and their environmental controls were collected from 423 sample plots between 65° and 74° N and 63° and 172° E, together with data on phytomass and diurnal temperature variations. Using stepwise multiple regression, the equations for gross primary production (*GPP*, g C m^{-2} day^{-1}) and gross ecosystem respiration (*GR*, g C m^{-2} day^{-1}) were obtained (Equations 9.3 and 9.4), which explained 67% and 75% of the variability, respectively.

$$GPP = 0.9 - 0.099 * PAR - 0.0867 * T_a - 0.0452 * GS - 0.0207 * GG - 0.00361 * WDS - 0.00179 * M \quad (9.3)$$

$$GR = -0.57 + 0.029 * PAR + 0.164 * T_a + 0.0205 * GS + 0.0049 * GDS + 0.0091 * GG + 0.00117 * WS + 0.00335 * WDS + 0.00146 * M - 0.00076 * L \quad (9.4)$$

where *PAR* is a diurnal sum of photosynthetically active radiation (MJ m^{-2} day^{-1}), *Ta* is mean diurnal air temperature (°C). The other parameters in the equations refer to oven dry mass (in grams) of: foliage of shrubs (*GS*), the foliage of dwarf shrubs (*GDS*), the above-ground living mass of grasses (*GG*), the aboveground mass of shrub woody components (*WS*), aboveground mass of dwarf shrub woody components (*WDS*), the living phytomass of mosses (*M*), and the living phytomass of lichens (*L*). Model estimates gave an annual *GPP* of Russian tundra zone (235 m ha area) as –485.8 ± 34.6 Tg C, *GR* as + 474.2 ± 35.0 Tg C, and a net flux of –11.6 ± 40.8 Tg C.

For forest soils, Hanson et al. (1993) described a relatively simple regression model, relating emissions to temperature, soil water content and a constant that was dependent on texture and density. The model explained between 50% and 74% of the variability in the flux data, and performed similarly for data from different topographical locations. Kicklighter et al. (1994) established that there is an exponential relationship between monthly mean CO_2 emissions from temperate forest soils and monthly mean air temperatures:

$$M_{CO_2} flux = 27.46 * \exp\,[0.06844\,(MT_{air})] \quad (9.5)$$

where M_{CO_2} *flux* is the monthly evolution of CO_2 from soils, in g C m^{-2} $month^{-1}$, and MT_{air} is the mean monthly air temperature in degree Celsius. They then used a georeferenced database of temperatures to provide estimates of monthly CO_2 fluxes for temperate forests around the world and found that the model provided good estimates of soil CO_2 effluxes for different sites regardless of forest type.

Since temperature and moisture has major influence on soil respiration, attempts have been made to better quantify their influence in model calculations. Many models use a constant Q_{10} temperature equation but Q_{10} values often vary depending on temperature range. Del Grosso et al. (2005b) proposed the use of an arctangent function, which allows for varying sensitivity of respiration to temperature and water.

$$F\ (Tsoil) = 056 + (1.46 * \arctan(\pi * 0.0309 * (Tsoil - 15.7)))/\pi \qquad (9.6)$$

$$F(RWC) = 5 * (0.287 + (\arctan(\pi * 0.009 * (RWC - 17.47)))/\pi) \qquad (9.7)$$

where *F* (*Tsoil*) is the temperature effect normalized to 1 at 30°C, *F (RWC)* is the water effect normalized to 1 at RWC = 100%, *RWC* is the measured soil relative water content. This temperature function predicts that Q_{10} values vary inversely with temperature and that CO_2 fluxes are significant below 0°C. The arctangent variable Q_{10} model agreed closely with observed Q_{10} values over a wide range of temperatures ($R^2 = 0.94$) and explained 16–85% of the observed intra-site variability in CO_2 flux rates.

The lack of a biological framework in regression models makes it difficult to explain the role of the environment on soil respiration or carbon cycle in ecosystem. Several factors such as root biomass, soil temperature and water content influence soil respiration. A few process-based models, which incorporate the effects of different factors, have been developed (e.g. Šimunek & Suarez, 1993; Fang & Moncrieff, 1999; Wang et al., 2002; Pumpanen et al., 2003). Šimunek & Suarez (1993) described soil respiration in terms of soil water potential, temperature, oxygen (O_2) concentration and depth in the soil. The model of Vleeshouwers & Verhagen (2002) includes the effects of crop (species, yield and rotation), climate (temperature, rainfall and evapotranspiration) and soil (carbon content and water retention capacity) on the carbon budget of agricultural land. The model calculates changes in SOC by quantifying the addition of organic matter to the soil and the decomposition of SOM. The model was used to evaluate the effects of different CO_2 mitigation measures on SOC in agricultural areas in Europe. Results showed that arable fields are carbon sources (–0.84 t C ha^{-1} $year^{-1}$), whereas the majority of grasslands are carbon sinks (0.52 Mg C ha^{-1} $year^{-1}$). A temperature rise of 1°C resulted in a –0.05 Mg C ha^{-1} $year^{-1}$ change whereas the rising CO_2 concentration gave a 0.01 t C ha^{-1} $year^{-1}$. Thus the higher decomposition rate as a result of a higher temperature is not compensated by greater crop yields owing to higher atmospheric CO_2 concentration.

Fang & Moncrieff (1999) presented a process-based model to simulate production and transport of CO_2 in soil (PATCIS). The model takes into account the production of CO_2 by the respiration of plant roots and microorganisms, and the transport of gases in the soil: CO_2 from soil to atmosphere and O_2 in the opposite direction. Gaseous diffusion and liquid phase dispersion are the major mechanisms governing the transport of CO_2. The major inputs of carbon are as leaf and root litter, and carbohydrates transported from leaves to the roots. The loss of soil carbon is assumed only to be as CO_2 emitted from the soil through respiration by roots and microbes. Soil temperature, moisture content, O_2 concentration in soil gas, and live and dead biomass are assumed to influence soil respiration. The response of soil respiration to soil temperature is described by the Arrhenius type relationship. The effect of soil moisture content on soil respiration is expressed by Equation 9.8.

$$df(W)/dW = a\,[f(W)_{max} - f(W)] \tag{9.8}$$

where W is soil moisture content, and a is a parameter defining the maximum increase in the rate of soil respiration with soil moisture, $f(W)_{max} = 1$ is the maximum value of $f(W)$ when soil moisture content does not limit respiration. The effect of oxygen concentration on soil respiration is described via the Michaelis-Menten equation and the oxygen concentrations in soil air at different depths are simulated using the method of Campbell (1985). The validation of the model by the authors (Moncrieff & Fang, 1999) with data from a slash pine plantation in Florida showed a close agreement between the simulated and measured hourly and daily CO_2 effluxes (Fig. 9.2).

The model of Wang et al. (2002) for simulating carbon dynamics of Boreal forests in addition to describing soil heterotrophic respiration includes plant photosynthesis, autotrophic respiration, root N uptake, litterfall and plant growth. The plant

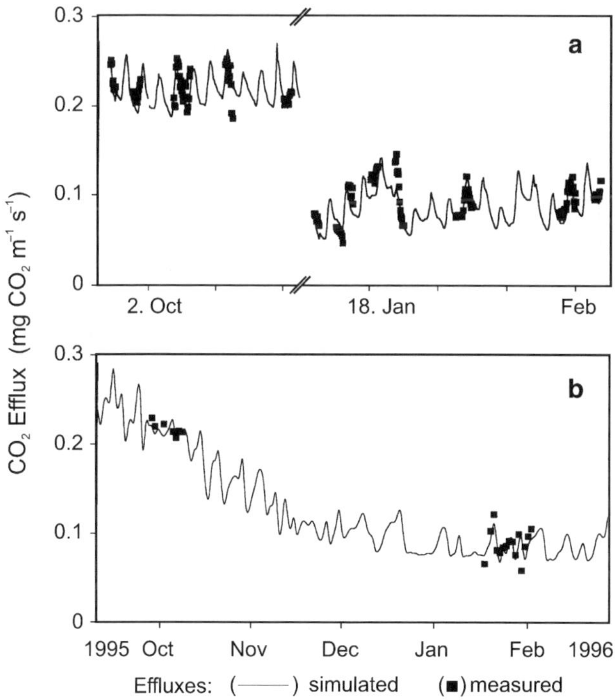

Fig. 9.2 Comparison between measured and simulated hourly (**a**) and daily average (**b**) CO_2 efflux in a slash pine plantation, Florida. Points represent measurements and lines represent simulations (Moncrieff & Fang, 1999, p. 241. Reproduced with kind permission from Elsevier)

C and N are divided into three physiologically active pools (foliage, stem and branches, and fine roots), and one inactive pool of heartwood. The physiological active pools are further differentiated into substrate and structural components. Soil organic matter is separated into one surface litter layer and three layers in the mineral soil. Litterfall from foliage and sapwood contributes to the surface litter layer and plant roots and exudates contribute to the layers in the soil. Decomposition of organic matter is simulated by dividing into different pools based on rate of decomposition using first-order kinetics. The evaluation of the model on two boreal forest ecosystems, deciduous (aspen) and conifers (black spruce) in Canada showed that modeled daily ecosystem CO_2 exchange explained 86% and 54% of the observed variance of eddy correlation flux at the two sites, respectively.

Another model, which needs mention here is the DeNitrification DeComposition (DNDC) model of Li et al. (1992), which has been developed for predicting emission of greenhouse gases (CO_2, CH_4 and N_2O) from soils in different ecosystems. The model is described further later in this chapter (section 9.3).

9.2 Methane Emissions from Rice Fields and Natural Wetlands

Several models have been developed to predict emissions of CH_4 from rice fields. These models range in complexity from empirical/semi-empirical to process-based models. Simple models use empirical relationships between rates of emission and grain yield or crop biomass (Anastasi et al., 1992; Bachelet & Neue, 1993; Kern et al. 1997). These models are based on the assumption that the higher the biomass production of the crop, the more substrate would be available for CH_4 production, either from increased crop residues or from higher rates of rhizodeposition. However, not only crop biomass but also a number of soil, plant, cultural and environmental factors influence the size of methane emissions from rice paddies (see Chapter 8). Yan et al. (2005) developed a statistical model (Equation 9.9) relating CH_4 flux in the rice-growing season to soil properties, water regime in the rice-growing seasons, water status in the previous seasons, organic amendment and climate. The statistical results showed that all these variables significantly affected CH_4 flux, and explained 68% of the variability.

$$\ln (flux) = constant + a * \ln (SOC) + pH_m + PW_i + WT_j + CL_k + OM_l * \ln (1 + AOM_l) \quad (9.9)$$

where $flux$ is the average CH_4 flux during the rice-growing season, SOC is the soil organic carbon content and a is the effect coefficient for SOC, pH_m is the effect of soil pH (m represents a pH class < 4.5, seven classes between 4.5 and 8.0 at intervals of 0.5 pH units each and ≥8.0), PW_i is the effect of preseason water status (i is flooded, long drainage, short drainage, double drainage, or unknown), WT_j is the effect of water regime in the growing seasons (j is continuous flooding, single drainage, multiple drainage, wet season rainfed, deepwater, or unknown), CL_k is the effect of climate (seven agroecological zones as per Food and Agricultural Organization), OM_l is the effect of added organic material (l is compost, farmyard manure, green manure, rice straw used on season, or rice straw used off-season); AOM_l is the amount of organic amendment in tons per hectare. Using the statistical model Yan et al. (2005) calculated the CH_4 emission factors for flooded rice fields without organic amendments to be 5.6 and 3.8 mg CH_4 m^{-2} h^{-1} for preseason short and long drainage, respectively as compared to 10.6 mg CH_4 m^{-2} h^{-1} for preseason flooded water status. While the statistical model is useful for estimating the effect of a number of variables on seasonal methane flux, it is oversimplified for site-specific simulation of CH_4 emission fluxes.

Nouchi et al. (1994) measured and modeled methane emissions from experimental plots with different fertilizer and straw treatments. Multiple linear regressions gave the following equation for the methane flux due to gas bubbles at 13:00 h on sunny summer days:

$$F = 19.34Ra + 8.13T - 197.85 \tag{9.10}$$

where F is the flux, Ra is the hourly global solar radiation (MJ m^{-2} h^{-1}) and T is the temperature (°C) at 1300 h. Daily flux was calculated from the observed diurnal cycle in emissions. Transport of methane through the plants was assumed to be governed by molecular diffusion:

$$F = (C_s - P_a/H)D \tag{9.11}$$

where C_s is the concentration in the soil solution, P_a is the concentration in the atmosphere, H is the Henry's Law constant for methane solubility in water, and D is the conductance (the reciprocal of the resistance) due to the biomass of the rice. P_a/H is negligible compared with C_s, so Equation 9.11 reduces to

$$F = C_s D \tag{9.12}$$

Equation 9.12 was parameterized using the measured values for the conductance and the flux was simulated.

Sass et al. (2000) developed a semiempirical model based on the hypothesis that rates of CH_4 production are determined by the availability of methanogenic substrates, which are primarily derived from rice plants and added organic matter. The rate of CH_4 production is influenced by temperature, texture and redox state of the soil (Equation 9.13).

$$P = 0.27 * F_{Eh} * (C_R + C_{OM}) \tag{9.13}$$

where P represents the production rate of CH_4 by methanogenic bacteria (g m^{-2} day^{-1}); C_R and C_{OM} represent the daily amount of carbohydrates (g m^{-2} day^{-1}) derived from rice plants and degradation of organic amendments, respectively; F_{Eh} (dimensionless) describes the time development of redox potential; and 0.27

assumes that three moles of CH_4 are derived from one carbohydrate unit and is the ratio of their molecular weights.

The amount of CH_4 transported from the soil to the atmosphere is determined by the rates of production and the emitted fraction (E_f) (Equation 9.14). The emitted fraction is estimated from seasonal maximum aboveground biomass (W_{max}; Equation 9.15).

$$E = P * E_f \tag{9.14}$$

$$E_f = 0.55 * (1-W/W_{max})^{0.25} \tag{9.15}$$

The constant 0.55 represents the initial fraction of produced CH_4, which is emitted. Model validation against measurements from irrigated rice in various regions of the world showed good agreement between the modeled (312 ± 138 mg C m^{-2} day^{-1}) and the measured (322 ± 144 mg C m^{-2} day^{-1}) average flux. A simplified version of the model has earlier been presented (Huang et al., 1998) in which seasonal emission values can be estimated using integrated or average values of the time-dependent parameters.

Some simple process based models that partition organic matter into two or three pools depending on the rate of decomposition have been presented. Cao et al. (1995) partitioned organic matter into three pools, which decompose according to first order kinetics. The carbon released from the decomposition of organic matter together with that released from the growing rice plants as root exudates and dead root tissue constitute substrate for methanogenesis. The methane production is calculated as a function of substrate available, adjusted for the influence of soil redox potential (Eh), pH, temperature, floodwater depth, and addition of mineral fertilizers. Methane emission rate is computed as the difference between rates of methane production and oxidation (estimated empirically). The model requires seasonal pattern of redox potential as an input. In contrast, the model of Lu et al. (2000a) is based on the assumption that only the active fraction of SOM is responsible for CH_4 production, which is further divided into two functional pools with a rapid and a slower rate of decomposition. The decomposition of both the pools is mediated by microorganisms rather than first order kinetics. The production of methanogenic substrate is assumed to be directly coupled to the anaerobic organic matter decomposition. The methanogenic substrate is converted into CH_4 and CO_2, with a factor of 0.5 to produce 0.5 CO_2 and 0.5 CH_4 for each carbon. Though the model pools are not measurable but it is hypothesized that the active pool includes acetate and glucose, which are immediately converted to CH_4 under anaerobic conditions whereas the slow pool includes cellulose and like. The decomposition of fast pool accounts for the initial phase and the peak of CH_4 production, while the slower pool contributes most during the late phase of CH_4 production. The test of the model on laboratory incubation data showed that the model simulated well the trends of CH_4 production in response to organic amendments (R^2 = 0.61 to 0.91), except for root exudates. However, the model represents only CH_4 production under controlled supply of substrate and anaerobic condition; for the model to be applicable to field conditions it needs to be coupled with other sub-models such as substrate production, soil aeration, and electron-acceptor reoxidation, etc.

Matthews et al. (2000a) developed a model for simulating *M*ethane *E*mission in *R*ice *E*co*S*ystems (MERES) by using an existing crop simulation model (CERES-Rice) as a basis and linking it to a model describing the steady-state concentrations of methane and oxygen in soils (Arah & Kirk, 2000). Subroutines are incorporated in the model to simulate the influence of the combined pool of alternative electron acceptors in the soil (i.e. NO_3^-, Mn^{4+}, Fe^{3+}, SO_4^{2-}) on CH_4 production. The MERES model explained well the seasonal patterns of methane emissions in an experiment involving mid- and end-season drainage and additions of organic material at International Rice Research Institute (IRRI) in the Philippines. Model predictions for a dry season rice at IRRI showed that of the total methanogenic substrate of 1,018 kg C ha^{-1} $season^{-1}$ when no organic amendment is added, 37% is contributed by rhizodeposition (root exudation and root depth), 22% by previous crop residues and the remaining 41% originate from the humic fraction of SOM. With the addition of organic matter, the absolute quantities from different sources remain the same, but their proportion decreases (Table 9.1).

Methane emissions from natural wetlands originate from anaerobic decomposition of organic matter through a mechanism mostly similar to that operative in rice paddies (see Chapter 8). Both empirical and process based models have been developed to predict methane emission from wetlands. The empirical models are based on direct relationship between CH_4 emission and the controlling variables such as water table, soil temperature, plant primary productivity or ecosystem productivity (Moore & Roulet 1993; Whiting & Chanton, 1993; Frolking & Crill, 1994; Bellisario et al., 1999). Since empirical models cannot be extrapolated to other sites, these have limited use. Several process-based models have been presented, which differ considerably with respect to the processes considered and the level of complexity. Cao et al. (1996) presented *W*etland *M*ethane *E*mission *M*odel (WMEM), which is based on the hypothesis that plant primary production and soil organic matter decomposition act to control the supply of substrate needed by methane-producing microorganisms. Net primary production, deposition of litter C into the soil, and soil organic matter decomposition are calculated with the Terrestrial Ecosystem Model (Raich et al., 1991). The wetland environment is assumed to be always saturated. The major environmental factors regulating the

Table 9.1 Predicted contribution from different sources to the total amount of methanogenic substrate in rice fields growing in the dry seasons at IRRI, Philippines with and without organic amendment (Adapted from Matthews et al., 2000b)

Methanogenic substrate	No organic amendment	Organic amendment[a]
Rhizodeposition (%)	37	21
Previous crop residue (%)	22	12
Humus (%)	41	24
Organic amendment (%)	0	43
Total (kg C ha^{-1} $season^{-1}$)	1,018	1,845

[a]Three tons dry matter ha^{-1} of green manure added 20 days before planting

production of methane are assumed to be water table position and temperature. Methane flux decreases exponentially with decreasing water table height, and an appropriate function for this is included in the model. A temperature function assumes a Q_{10} of 2.0 for methane production and an optimum temperature of 30°C. Methane production is assumed to occur only during thaw/wet seasons. Oxidation occurs above the water table and in the rhizosphere, where oxygen is likely to be available. The balance between methane production and oxidation determines the rate of emission into the atmosphere. The emission estimates obtained with WMEM (Cao et al., 1996) ranged from 3.6 g CH_4 m^{-2} $year^{-1}$ for non-forested bogs (mainly >50° N) to 38.2 g CH_4 m^{-2} $year^{-1}$ for forested swamps (mainly tropical). The model predicted a total emission of 92 Tg CH_4 $year^{-1}$, of which 56% came from tropical wetlands. This prediction is well within the range for natural wetlands (see Chapter 8).

Potter (1997) modified the CASA model of ecosystem production and C cycling in the terrestrial biosphere. Modification to represent flooded wetland soils and anaerobic decomposition include (i) layered soil temperature and water table depth as a function of daily climatic drivers, (ii) CH_4 production within the anoxic soil layer as a function of water table depth and CH_4 production under poorly drained conditions, and (iii) CH_4 gaseous transport pathways (molecular diffusion, ebullition, and plant vascular transport) as a function of water table depth and ecosystem type. The model has been applied and tested using climate data to characterize tundra wetland sites near Fairbanks, Alaska. Reasonably close agreement between the model's mean daily and seasonal estimates of CH_4 flux and observed emission rates for northern wetland ecosystems has been achieved.

Arah & Stephen (1998) and Walter & Heimann (2000) presented one-dimensional process-based models in which surface irregularities are ignored and only variations with depth are considered. In the model by Walter & Heimann (2000), the three different transport mechanisms by which methane can move from the zone of formation to the atmosphere, i.e. diffusion, plant-mediated transport, and ebullition, are modeled explicitly. Daily values of soil temperature, water table and net primary productivity (NPP) are used, and at permafrost sites the thaw depth is included. Testing of the model using experimental data from five wetland sites in North and Central America, and Europe, representing a wide range of environmental conditions, showed a satisfactory fit with experiment in most cases. Zhang et al. (2002) modified the DNDC model (Li et al., 2000) to predict CO_2 and CH_4 emission from wetlands. The modified model (Wetland-DNDC) includes functions and algorithms for simulating water table dynamics, the effects of soil properties and hydrological conditions on soil temperature, C fixation by mosses and herbaceous plants, and the effects of anaerobic conditions on decomposition, CH_4 production/oxidation, and other biological processes. For simulation of CH_4 production, oxidation and transport the model is linked to the model by Walter & Heimann (2000). The test of the model at three wetland sites in North America showed that the modeled values were generally in agreement with the observations. The model could capture general patterns of CH_4 emission at different sites with R^2 values ranging from 0.37 to 0.76 (sample size 47–214), except at one site where CH_4 emission was very low.

9.2.1 *Oxidation of Atmospheric Methane in Soils*

In addition to the oxidation in aerobic soil layers of methane diffusing upwards from underlying anaerobic environments, considered above, there is also the quite distinct process of oxidation of atmospheric methane. The model of Potter et al. (1996) predicts maximum methane oxidation rates at low soil water contents assuming that soil gas diffusivity is the major control. This is an over-simplification, in that there is good evidence that oxidation rates decline as soils become very dry and microbial activity is inhibited. The constraints have been taken account in the further development of the model (Del Grosso et al., 2000). The well-established decrease in oxidation rate as a result of soil disturbance is also included. The model is more empirically based and is parameterized using several data sets from the USA and one from the UK. The model predicted the variation in oxidation with soil water content very well in a German coniferous forest, but gave more mixed results with data for arable and former arable land in Scotland.

The more complex of the existing oxidation models simulate microbial dynamics (Grant, 1999) and methane oxidation gradients in the soil (Dörr et al., 1993). Ridgwell et al. (1999) developed a process-based model in which methane oxidation is controlled by gas diffusivity at high rates of microbial activity and by microbial activity at high diffusivities. The global soil sink calculated by their model is 20–51 Tg CH_4 $year^{-1}$, with a preferred value of 38 Tg. Dry tropical systems are predicted to account for almost a third of this total.

9.3 Nitrogen Trace Gas Emission

Both simple regression models and process-based models have been developed to compute regional estimates of N trace gas emission from soils. A very simple flux model for N_2O emission (Equation 9.16) currently used by the Intergovernmental Panel on Climate Change (IPCC) is based on a regression analysis by Bouwman (1996):

$$F = 1 + (1.25 \pm 1.0).N/100 \tag{9.16}$$

where F is the annual flux, in kg N_2O-N ha^{-1}, and N is the quantity of fertilizer N applied, in kg ha^{-1}. Thus agricultural land is regarded as having a background emission rate of 1 kg N_2O-N ha^{-1} $year^{-1}$, and to this is added an emission factor of 1.25 ± 1.0% of the N applied. However, new compilations of experimental data and results of several studies (see Chapter 8) show that the emission factor varies even more widely. In addition to fertilizer rate, a number of soil, crop and climatic variables influence N trace gas emissions. Based on nitrogen emission data from a large number of studies in the literature, Stehfest & Bouwman (2006) developed a model (Equation 9.17) that incorporates the effect values for different factors related to soil, crop/vegetation and climate.

$$\log(N_{emission}) = A + \sum_{i=1}^{n} E_i \tag{9.17}$$

where $N_{emission}$ is the emission of N_2O or NO expressed in kilograms per hectare of N over the time period covered by the measurements, A is a constant and E_i is the effect value for factor i. The factors that significantly influence agricultural N_2O emissions are N application rate, crop type, fertilizer type, soil organic C content, soil pH and texture. The factors that significantly influence NO emission from agricultural soils include N application rate, soil N content and climate. N_2O emissions from soils under natural vegetation are significantly influenced by vegetation type, soil organic C content, soil pH, bulk density and drainage, while vegetation type and soil C content are major factors for NO emissions. Statistical models of these factors were used to calculate global annual emissions from fertilized cropland (3.3 Tg N_2O-N and 1.4 Tg NO-N) and grassland (0.8 Tg N_2O-N and 0.4 Tg NO-N). However, as shown in Chapter 8, these emission estimates are much different from those reported in other studies and there is a great uncertainty in the calculated emission fluxes. The average 95% confidence interval for calculated N_2O and NO emissions was −51% to + 107% and −80% to + 406%, respectively.

Williams et al. (1992) related NO_x emission flux to soil temperature and lumped all other sources of variability into a single proportionality constant that differed only among major land use types. Using the empirical relationships and the land use classification based on a coarse-scale segregation of land into natural and agricultural types with further subdivisions based on biome for natural areas and fertilizer usage for agricultural areas, the authors (Williams et al., 1992) published an inventory of NO emissions from soils in the USA. The approach was used by Müller (1992) to estimate global soil NO_x emissions (6.6 Tg N year^{-1}). Yienger & Levy (1995) extended the temperature dependence approach and added empirically derived functions to account for several other influences on the NO_x exchange rate. The additions included separate exponential temperature dependence for wet soils and linear dependence for dry soils, an optimal temperature above which the flux becomes temperature independent, scalar adjustments to account for pulsing (flush of NO_x emissions following wetting of very dry soil) and canopy reduction (uptake by the canopy of soil emitted NO_x), synoptic-scale temperature and precipitation forcing, explicit linear dependence on the N fertilizer application rate, and a simple scheme to account for the stimulation of NO_x emissions by biomass burning. Developed on a 1° × 1° grid with a global data set, these relationships provided a reasonable estimate of soil NO_x exchange in regions with few or no data. The Yinger and Levy model's estimates of global NO_x emission from the soil surface (10.2 Tg N year^{-1}) was reduced nearly by 50% by canopy capture (5.5 Tg N year^{-1}). Pulsing accounted for 1.3 Tg N year^{-1}, or 24% of the global soil NO_x source. This suggests that pulsing and canopy reduction are important determinants of global N trace gas budgets.

The empirical models for N_2O and NO emissions may be useful for getting a broad estimate of regional emissions but these are too simplistic and generalized ignoring all site-specific controls. Modeling at the regional or global scale requires capturing the overall dynamics of nitrification and denitrification processes at a more aggregated level through linkages to major biogeochemical nutrient cycling processes (Schimel & Potter, 1995).

Several process based models of intermediate complexity, such as DNDC (Li et al., 1992), CASA (Potter et al., 1993) and the NASA-Ames version of the CASA model (NASA CASA; Potter et al., 1997), Century-NGAS (Parton et al., 1996) now further developed in the DAYCENT model of Del Grosso et al. (2001a), ExpetN (Engel & Priesack, 1993), Hole-in-the-Pipe (Firestone & Davidson. 1989), NLOSS (Riley & Matson, 2000), and others have been developed for scaling up gas emission estimates. The models differ with regard to the level of complexity, number of parameters and empirical relationships used to capture basic patterns of gas fluxes. However, most of these models have several features in common. Generally each model simulates (1) soil climate dynamics; (2) plant growth, nutrient uptake, and litterfall; (3) decomposition of soil organic matter; and (4) nitrogen mineralization and transformations. A brief description of the processes simulated by some of the commonly used models to provide site-specific and regional estimate of N_2O emissions from soils is presented here.

The NGAS model is a hybrid between the detailed process oriented models and the simplistic nutrient cycling models. The model uses a daily time step. The model was developed using laboratory denitrification gas flux data and field observed N_2O gas fluxes from different sites The factors which have been considered to influence N_2O fluxes from nitrification include soil pH, soil water content, soil temperature, and soil NH_4^+ level. The implicit assumptions are that N_2O formation during nitrification is directly proportional to soil nitrification rates, that the soil temperature and soil water controls on nitrification are similar to those used to control soil microbial activity, and that nitrification rates are proportional to soil N turnover rate. Total denitrification (N_2 plus N_2O) gas fluxes are a function of soil heterotrophic respiration rates, soil NO_3^-, soil water content, and soil texture. The N_2:N_2O ratio is a function of soil water content, soil NO_3^-, and soil heterotrophic respiration rates. Comparison of simulated model results with field N_2O data for several sites (Parton et al., 1996) showed that the model results compared well with the observed data ($r^2 = 0.62$). However, winter denitrification events were poorly simulated by the model.

The NGAS model has been modified and linked to the Century (Parton et al., 1987, 1994) model. Input from Century is used to derive the NGAS model. Inputs from the Century model (Parton et al., 1994) used to drive the NGAS model include soil water-filled pore space (WFPS), soil temperature, soil CO_2 fluxes, N availability, and nitrification rates. The site specific parameters include the sand, silt and clay content, the field capacity and wilting point, the saturated hydrologic conductivity, and the soil water potential versus water content curve. The hydrologic conductivity and water potential curves are estimated from the soil texture using standard equations, while the wilting point and field capacity are estimated from observed soil water data. The only site-specific curve for the denitrification model is the water-filled pore space curve for total denitrification gas flux (Parton et al., 1996).

The DAYCENT is the daily time-step version of the Century model (Parton et al., 1994) and simulates the fluxes of carbon and nitrogen among the atmosphere, vegetation and soil (Del Grosso et al., 2001a; Parton et al., 1998). Key sub-models include soil water content and temperature by layer, plant production and allocation of net primary production (NPP), decomposition of litter and soil organic matter, mineralization of nutrients, N gas emissions from nitrification and denitrification, and CH_4 oxidation in non-saturated soils. Flows of C and N between different pools are controlled by the size of the pools, C:N ratio and lignin content of the material, and abiotic water and temperature controls. Plant production is a function of genetic potential, phenology, nutrient availability, water and temperature stress, and solar radiation. Nitrogen gas fluxes from nitrification and denitrification are driven by soil NH_4^+ and NO_3^- concentrations, water content, temperature, texture, and labile C availability (Parton et al., 2001). The input required for the model include daily maximum and minimum air temperature and precipitation, surface sol texture class, and land cover-land use data (e.g. vegetation type, cultivation and planting schedules, amount and timing of nutrient amendments). Outputs include daily N-gas flux (N_2O, NO_x, N_2), CH_4 uptake, CO_2 flux from heterotrophic soil respiration, soil NO_3^-, water content, and temperature by horizon, soil NH_4^+ in top 15 cm, water and nitrate leaching and other ecosystem parameters. The ability of DAYCENT to simulate NPP, SOC, N_2O emissions, and NO_3^- leaching has been tested with data from various native and managed systems (Del Grosso et al., 2001b, 2002, 2005a). Simulated and observed nitrous oxide emission data from eight cropped sites in North America and NO_3^- leaching data from three cropped sites showed reasonable agreement with r^2 values of 0.74 and 0.96 for N_2O and NO_3^-, respectively (Del Grosso et al., 2005a).

Building on the Carnegie-Ames-Stanford (CASA) Biosphere model (Potter et al., 1993), Potter et al. (1996) described a process-based approach for simulating soil-atmosphere NO_x, N_2O and N_2 exchange on 1° global grid. The model runs on a monthly time interval and explicitly includes interactions among substrate availability, soil water content, temperature, soil texture and microbial turnover. The model is based on the hypothesis that N mineralization and soil aeration are the dominant factors controlling both the amount and composition of N gases exchanged between soils and the atmosphere. The modified NASA-Ames daily version of CASA (NASA CASA; Potter et al., 1997) simulates seasonal patterns in carbon fixation, nutrient allocation, litterfall, and soil nitrogen mineralization, net CO_2 exchange, and soil N_2O and NO production. In the soil component of the CASA model, first-order equations simulate exchanges of decomposing plant residue (metabolic and structural fractions) at the soil surface; and the surface soil organic matter. The soil organic matter is compartmentalized into active (microbial biomass and labile substrates), slow (chemically protected), and passive (physically protected) fractions. NASA CASA includes a multilayer model of soil heat and moisture content calculations. The soil profile is treated as three layers: surface organic matter, topsoil, and subsoil to rooting depth. These layers can differ in soil texture, moisture holding capacity, and carbon-nitrogen dynamics. Daily water balance in the soil and soil water-filled pore space (WFPS) in each layer are calculated, which are coupled to N mineralization fluxes and N trace gas emissions. Processes of ammonification and nitrification are lumped into combined mineralization fluxes from litter, microbial and soil organic matter pools to a common mineral N pool. Gaseous nitrogen losses are modeled as a 'leaky pipe' (Firestone & Davidson, 1989) where emissions are a function of the N mineralization rate (pipe flow rate), and soil water content (size of 'holes' through which gas can 'leak'). Production and emission of both NO and N_2O occur at intermediate levels of WFPS. At higher moisture levels, N_2O emissions increase exponentially with WFPS, while NO emission declines. Under very wet soil conditions, production of only N_2 occurs. The potential loss of either N_2O-N or NO-N as a percentage of total mineralized nitrogen has a default setting of 2%.

The DNDC model (Li et al., 1992) was developed to predict N_2O fluxes produced by nitrification and denitrification, and CO_2 fluxes produced by decomposition and root respiration. The DNDC model consists of four interacting sub-models: soil climate, crop growth, decomposition, and denitrification. The soil climate sub-model uses daily meteorological data to predict soil moisture conditions and to capture anaerobic microsite formation and sequential substrate reduction. The crop growth/vegetation sub-model simulates growth of various crops from planting to harvest. Above-ground biomass is accumulated based on daily N and water uptake. Yield and N content of above and below-ground plant components are modeled. The decomposition sub-model has four soil carbon pools: litter, microbial biomass, labile humus and passive humus. Each pool has a fixed C:N ratio and decomposition rates are influenced by soil texture (clay content), soil moisture and temperature, and potentially by nitrogen limitation. The decomposition sub-model provides initial NO_3^- and soluble C pools for the initiation of denitrification, which is also activated by rainfall, flooding (as in irrigated rice) and soil freezing. An increase in the WFPS caused by rain or irrigation events decreases soil oxygen availability. The WFPS, soluble C, soil temperature, soil pH, available N and denitrifier biomass control the rate of denitrification (Frolking et al., 1998). DNDC has been designed so that soil moisture has a large influence on nitrous oxide fluxes through its impact on the volume of soil in which denitrification occurs and the duration of denitrifying conditions. As the soil dries following rain, the denitrifying portion of each model layer decreases with soil water content.

The model has been validated against a number of field data sets observed in China, the USA, Japan, and Thailand and widely used over the last 10 years by many researchers. Simulated results show a close match between the modeled and observed N_2O fluxes from agricultural sites in the US, Canada, the UK, Germany, China, New Zealand, Costa Rica and Zimbabwe (Fig. 9.3; Saggar et al., 2004b). DNDC has been used to produce regional estimates of N_2O production for the US (Li et al., 1996), China (Li et al., 2001), Canada (Smith et al., 2002), Germany (Butterbach-Bahl et al., 2001), and the UK (Brown et al., 2002).

The predictive capacities of four models viz. DNDC, Century-NGAS, ExpertN, NASA–Ames version of the CASA model have been explored through a simulation exercise involving three data sets from contrasting environments: the semiarid prairie land of Colorado, USA, the cool temperate and moist conditions of SE Scotland,

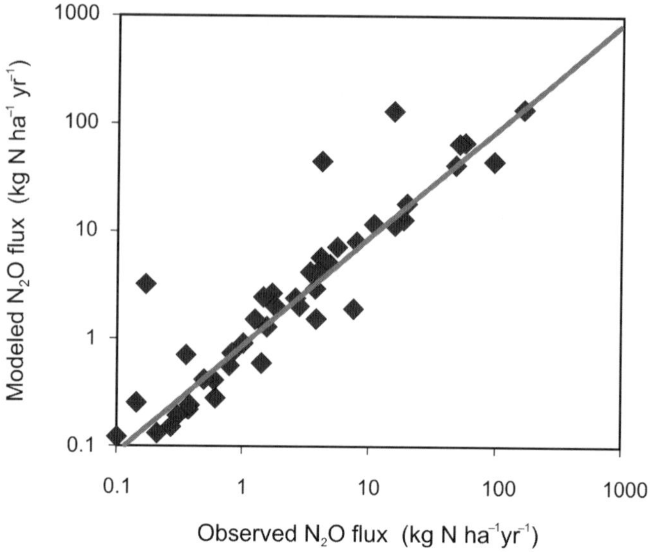

Fig. 9.3 Comparison of modeled and observed N_2O fluxes from agricultural sites in the US, Canada, the UK, Germany, China, New Zealand, Costa Rica and Zimbabwe (Saggar et al., 2004b, p. 2. Reproduced with kind permission from The Regional Institute Ltd.)

and the more continental climate of Germany (Frolking et al., 1998). In most cases the simulated N_2O emissions were within a factor of about 2 of the observed annual emissions, but even when the models produced similar N_2O fluxes they often produced very different estimates of gaseous loss as N_2 and NO. Accurate simulation of soil water content was identified as a key requirement for predicting N_2O fluxes. All the models simulated the general pattern of low background fluxes interspersed by high fluxes following fertilization, at the Scottish site, but they were not able to simulate the large pulses of N_2O emitted during winter thaws at the German site. All the models except DNDC simulated the very low fluxes at the dry site in Colorado.

The DNDC model has been modified for predicting emission fluxes from paddy rice ecosystems (Li et al., 2004). The modifications mainly focus on simulations of anaerobic biogeochemistry and rice growth as well as the parameterization of paddy rice management. When estimating emission fluxes under specific management conditions at regional scale, the spatial heterogeneity of soil properties (e.g. texture, soil organic carbon content, pH) are the major sources of uncertainty. An approach, the most sensitive factor method, has been developed for DNDC to reduce the magnitude of the uncertainty. The modified DNDC has been used for estimating emissions of CO_2, CH_4, and N_2O from rice paddies in China with two different management practices, i.e. continuous flooding and midseason drainage.

The DNDC model has also been modified to represent grazed pasture systems prevalent in New Zealand (Saggar et al., 2004a). The modifications include incorporation of a multiplicative day-length factor to reflect the seasonal pattern of growth and N uptake generally observed in pastures; changes in water infiltration and drainage component to exhibit saturated soil condition during winter; changing the WFPS-denitrification threshold to field capacity instead of 35% originally used in the model, and accounting for excretal N input from grazing animals. Validation of the model on two sites for grazed and ungrazed pastures showed that the modified model was able to predict the annual N_2O emissions within 10% of the measured values (Fig. 9.4). These values were within the uncertainty range of the measured values. In contrast, anthropogenic emission estimates based on the New Zealand refined IPCC methodology were about 25–60% lower than the measured values for the grazed pastures. The NZ-DNDC model had limited success in predicting the size and timing of very high emissions observed immediately after a summer rainfall event probably because it under-predicted the temperature effect.

In addition to the modeling of fluxes from agricultural soils, there has been a considerable development in the modeling of fluxes from forest soils. Li et al. (2000) have coupled the DNDC model to a forest physiology model, PnET (Aber & Federer, 1992), which predicts photosynthesis, respiration, organic carbon production and allocation, and litter production, and also to a model for predicting nitrification-induced NO and N_2O production. In the combined model viz. "PnET-N-DNDC" (Photosynthesis and EvapoTranspitaion-Nitrification-Denitrification and Decomposition), five submodels predict forest growth, soil climate, decomposition, nitrification and denitrification. While the soil climate submodel calculates soil oxygen diffusion in the forest soil profile, the forest growth model simulates forest growth and litter production for use in the decomposition submodel. The decomposition submodel simulates turnover of the litter and organic matter in the soil and passes ammonium, nitrate and, dissolved organic carbon (DOC) to the nitrification or denitrification submodels. The production of NO and N_2O from nitrification is calculated based on soil temperature, moisture, ammonium and DOC concentration. The emission of NO and N_2O from denitrification is simulated based on the dynamics of soil aeration status, substrate limitation and gas diffusion. A so-called "anaerobic balloon" concept is employed to calculate the anaerobic status of the soil and divide it into aerobic and anaerobic fractions. Nitrification is only allowed to occur in the aerobic fraction, while denitrification only occurs in the anaerobic fraction. The size of the balloon is defined by the simulated oxygen partial pressure, calculated from oxygen diffusion and consumption in the soil.

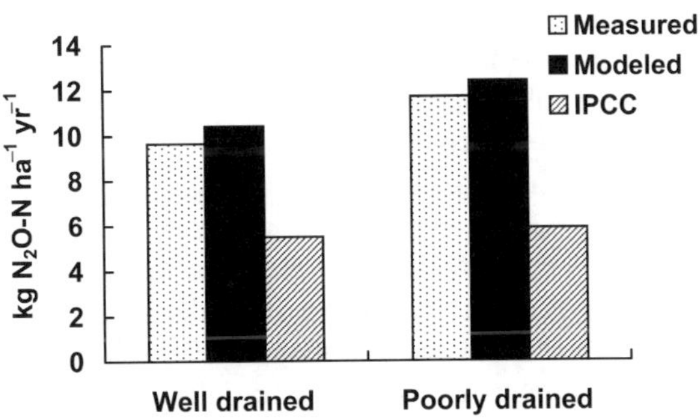

Fig. 9.4 Comparison of measured, modeled and IPCC-calculated annual nitrous oxide emissions from two dairy grazed pasture sites in New Zealand (Adapted from Saggar et al., 2004a)

Although the allocation of processes entirely to one or other volume fraction in PnET-N-DNDC may be a simplification of the true situation, but it is useful for simulating soil environments in which aerobic and anaerobic microsites coexist, and where nitrification and denitrification can go on simultaneously. Sensitivity analysis of the model showed that soil texture, soil carbon content, site fertility and soil pH are most important parameters influencing NO and N_2O emissions from forest soils in south-east Germany (Butterbach-Bahl et al., 2001). Validation of the PnET-N-DNDC model on seven temperate forest ecosystems in the USA and Europe showed that the differences between predicted and measured N_2O fluxes were <27%, except for one site, and that for NO fluxes (for three sites) were <13% (Stange et al., 2000).

9.4 Modeling Nitrogen Dynamics in Soils

Nitrogen exists in many different forms in soils, plants, animals and the atmosphere. The major forms of N in soil are organic, ammonium (NH_4^+) and nitrate (NO_3^-). The amount and distribution of nitrogen in different forms in the soil depends upon a number of interlinked input processes, transformations and loss processes. The N input pathways include fertilizer application, atmospheric wet and dry deposition and biological nitrogen fixation by legumes. Mineralization and immobilization of nitrogen, nitrification, fixation and release of NH_4^+ comprise the nitrogen transformation processes. Nitrogen can be lost from the soil-plant system through erosion and runoff, leaching, ammonia volatilization and denitrification. While the magnitude of different processes in terrestrial ecosystems have been discussed in earlier chapters, here we describe modeling of some of the N cycling processes influencing fluxes of N between soils and the environment. A comprehensive review of the N cycling processes and their modeling has been presented by Benbi & Richter (2003).

9.4.1 Denitrification

A variety of approaches have been used to model denitrification losses in soil. Depending on the description of the underlying processes and the level of complexity the models for soil denitrification have been grouped into three distinct classes (Parton et al., 1996) viz.: (i) microbial growth models that include an explicit description of microbial dynamics responsible for nitrification, denitrification, and other nutrient cycling processes (e.g. McGill et al., 1981; Li et al., 1992; Riley & Matson, 2000); (ii) simplified process models that represent different N cycling processes as a function of soil water, temperature, and pH controls on microbial activity without representing the microbial dynamics (e.g. Parton et al., 1988; Bradbury et al., 1993; Bergstrom & Beauchamp, 1993; Potter et al.,1996); and (iii) soil structural models that simulate denitrification by describing soil physical processes like diffusion of gases and solutes into soil aggregates and explicitly representing the distribution of soil aggregates. Models of this type assume that denitrification primarily occurs in anaerobic soil aggregates and that diffusion of O_2, N_2O, and NO_3^- into and out of soil aggregates needs to be represented (Arah & Smith, 1989; Smith, 1990; Arah, 1990). The most complex models, mechanistically, are those which calculate the soil anaerobic fraction in which denitrification can occur (e.g. Leffelaar, 1979; Arah & Smith, 1989; Renault & Stengel, 1994). These diffusion-based mechanistic models require too much detailed parameterization to be applicable at sufficiently large scales to be useful for flux prediction purposes. These models are useful for understanding the processes responsible for emissions, and the sensitivity of the fluxes to changes in particular variables. Similarly, the microbial growth models are too complex and require specific information on microbial growth rates, reaction rate constants, solubility coefficients, diffusivities, and numerous other terms. While it may be possible to parameterize such models for short periods at individual sites, these are too complex to be used for estimating larger-scale fluxes.

The simplified process models consider denitrification to be a first order decay process that is influenced by soil and environmental variables. Generally, a maximum or potential denitrification rate (D_p) is defined, which is adjusted for substrate availability (such as NO_3^- concentration), soil temperature and moisture (Equation 9.18) to calculate actual denitrification rate (D_a).

$$D_a = D_p f_N f_W f_T \tag{9.18}$$

Where f_N, f_w and f_T are the empirical reduction functions for NO_3^- concentration, soil water content, and temperature, respectively. Though the coefficients for reduction functions differ considerably among studies but mostly similar functional forms, particularly for NO_3^- concentration and temperature are used. The reduction function for NO_3^- concentration is generally described by a Michaelis-Menten function and the effect of soil temperature is described by a Q_{10} or Arrhenius function. The formulation for water content reduction functions vary from power, exponential sigmoidal, segmented and Michaelis-Menten type. Some models have also used reduction function for pH (Engel & Priesack, 1993; Parton et al., 1996), which is particularly important when the predictions are to be scaled up for different soils and regions. These simple models for denitrification are generally meant to describe field data and some of these, as discussed in the preceding section, have been customized to predict N trace gas emissions at the regional level.

9.4.2 Ammonia Volatilization

When urea and ammonium forms of fertilizers are applied on moist soil surfaces, they undergo a series of chemical conversions to ammonia that escapes to the atmosphere. This loss process, termed volatilization, is governed by a number of equilibria that may be represented as (Equation 9.19)

$$NH^{+}{}_{4(exch)} \longleftrightarrow NH^{+}{}_{4(soil\ solution)} \xleftrightarrow{K} NH_{3(soil\ solution)} \xleftrightarrow{K_H} NH_{3(gas\ soil)} \xleftrightarrow{K_a} NH_{3(atmosphere)} \tag{9.19}$$

Factors that promote these reactions toward the right result in increased ammonia (NH_3) volatilization. The values of equilibrium constant (K) and Henry's law constant (K_H) depend on temperature, whereas the value of transfer coefficient K_a depends on many factors, of which wind speed, temperature, surface roughness and surface area are the most important. In submerged soils, losses of NH_3 are directly related to the concentration of aqueous NH_3 in the floodwater, which in turn is a function of total ammoniacal N, the pH, and the temperature of the floodwater (Vlek & Stumpe, 1978; Vlek & Craswell, 1979).

Empirical models have been used to describe cumulative NH_3 volatilization as a function of time after urea application. These models generally use different formulation of the exponential function such as Gompertz, logistic or sigmoid. Hengnirun et al. (1999) used a first-order kinetic equation (Equation 9.20) to simulate NH_3 volatilization from the soil and the surface applied manure.

$$A_t = A_0 \exp(-K_v t) \tag{9.20}$$

where A_t is the total ammoniacal N remaining in soil at time, t; A_0 is the amount of ammoniacal N in soil at $t = 0$; K_v is the first-order rate constant adjusted to incorporate the effect of temperature, cation exchange capacity, and air flow rate. Although the empirical models can be used to smoothen data and compare results from different treatments under a given set of soil, environmental and management conditions, but these do not express the individual and interactive effects of a large number of factors. While soil factors determine the NH_3 volatilization potential, the actual volatilization in the field is of stochastic nature varying with time and space depending on the environmental and management conditions.

A few process based models have been developed to predict NH_3 losses in the soil-plant system from application of fertilizers to upland (Rachhpal-Singh & Nye, 1986; Sadeghi et al., 1988) and submerged conditions (Bouwmeester & Vlek, 1981; Moeller & Vlek, 1982; Jayaweera & Mikkelsen, 1990). The model developed by Rachhpal-Singh & Nye (1986) for predicting NH_3 volatilization from urea applied to non-calcareous soils includes simultaneous numerical solution of three continuity or flow equations to describe the diffusive movement of urea, ammoniacal-N and soil alkalinity, involving both solution and gas phase diffusion in the soil. The model has been modified by including pH-buffering action of the soil carbonates in calcareous soils (Roelcke et al., 1996) and to account for the effects of water drainage and evaporation on NH_3 volatilization (Kirk & Nye, 1991a, b). Génermont & Cellier (1997) presented a phenomenological model for estimating NH_3 volatilization from slurry applied to bare soil under field conditions using simple meteorological data. The model includes sub-models to simulate ammoniacal N transfers and equilibria between ammoniacal N species, as well as heat and water transfer in the soil. The aqueous ammoniacal N transfers between the soil layers are described by the classical convection-diffusion equation and the gaseous NH_3 transfer by the analogous diffusive part of the equation.

The models for NH_3 volatilization from flooded soils predict NH_3 loss as a function of floodwater chemistry and atmospheric conditions (Bouwmeester & Vlek, 1981; Moeller & Vlek, 1982; Jayaweera & Mikkelsen, 1990). The model presented by Bouwmeester & Vlek (1981) considers the mass transfer of the air-water interface and is similar to the penetration theories of Higbie and Danckwerts (e.g. Danckwerts, 1970). The model was used to study some of the factors that control the rate of NH_3 volatilization such as floodwater chemistry and meteorological conditions. In the development of this model, it was assumed that pH, and therefore the NH_3/NH_4^+ ratio, was constant throughout the diffusion layer. The model developed by Moeller & Vlek (1982) predicts a gradient in the hydrogen ion concentration through the liquid-phase diffusion barrier.

Jayaweera & Mikkelsen (1990) based their model on the two-film theory of mass transfer. They included depth of floodwater as one of the primary variables in addition to the other four (floodwater NH_4^+-N concentration, pH, temperature, and wind speed) considered by earlier workers. They derived the following relationship (Equation 9.21) to determine the rate of NH_3 volatilization from a flooded system.

$$\frac{d[NH^+{}_4]}{dt} = k_a \left\{ \frac{k_d\,(A-[NH_3]_{aq})}{k_a[H^+]+k_{vN}} \right\} - k_d\,(A-[NH_3]_{aq}) \tag{9.21}$$

where A is ammoniacal N concentration, $[NH_3]_{aq}$, is aqueous ammonia concentration, $[H^+]$ is hydrogen ion concentration in floodwater at equilibrium, k_d and k_a are dissociation and association rate constants for NH_4^+/NH_{3aq} equilibrium, and k_{vN} is the first-order volatilization constant for NH_3. The volatilization rate constant is calculated as a ratio of the overall mass transfer coefficient for NH_3 and mean depth of floodwater.

Though most of the models for NH_3 volatilization explicitly incorporate underlying process of volatilization, these are difficult to be parameterized under field conditions. While such models are valuable for understanding the processes involved in NH_3 volatilization, they may have limited value in the field.

9.4.3 Nitrate Leaching

The movement of NO_3^- through soil is governed by three mechanisms viz. convection or mass flow with the moving water, diffusion, and hydrodynamic dispersion. The mathematical modeling of N leaching in soils has proceeded on two paths – the first kind of models consider transport under steady state water flow conditions. The main advantage of these models is their simplicity but these do not represent real field situation. The second kind of models consider transient flow of soil N using finite difference solution of one dimensional solute flow equation. Using this equation, a close match between the measured NO_3^--N profile and that computed from the model for wheat grown in Indian Punjab was observed (Benbi et al. 1991b). The model-predicted residual NO_3^--N in the soil profile at wheat harvest matched well with the measured ones (Benbi, 1992).

Heterogeneous nature of soil and spatial variability are the major problems associated with the modeling of NO_3^- leaching. There is considerable uncertainty with respect to real system output. To overcome this problem use of Transfer Function Models (TFM) in a stochastic framework has been suggested. In this approach, the volume of soil between the soil surface and the outflow surface where solute is monitored is characterized with regard to its solute transport properties by the solute travel time probability density functions (pdf). The transfer function approach has the advantage that it is general enough to encompass all linear transport models, and it does not require the direct measurement of soil water transport properties. These models attempt to simulate spatially variable field processes with minimum input of data.

Addiscott & Wagenet (1985) classified solute leaching models as deterministic mechanistic, deterministic functional, stochastic mechanistic and stochastic non-mechanistic or transfer function models. The spatial variability of soil properties caused problems for deterministic models using rate parameters, but stochastic elements can be incorporated in these models. Soil variability appears to be the major problem with modeling and measurement of nitrate leaching.

9.4.4 Nitrogen Mineralization Kinetics

Nitrogen mineralization is the process of conversion of organic forms of nitrogen to inorganic forms viz. NH_3 or NH_4^+ and NO_3^-. The first step in the process, called ammonification is an enzymatic process carried out exclusively by heterotrophic organisms. The subsequent conversion of NH_4^+ to NO_3^-, termed nitrification, is mediated primarily by two groups of autotrophic bacteria (*Nitrosomonas* and *Nitrobacter*). Mineralization is always coupled with immobilization, which operates in the reverse direction, with the soil microbial biomass (SMB) assimilating inorganic forms of N and transforming them into organic N constituents in their cells and tissues during the oxidation of suitable C substrates.

$$\text{Organic N} \underset{\text{Immobilzation}}{\overset{\text{Mineralization}}{\rightleftarrows}} NH_4 \text{ and } NO_3 \tag{9.22}$$

The immobilized N is likely to be available subsequently for mineralization as the microbial population turns over. Total release of NH_4^+ prior to any immobilization back into the organic forms is called gross mineralization. The difference between gross mineralization and immobilization constitutes net mineralization or net immobilization. A decrease in mineral N level with time indicates net immobilization and an increase suggests net mineralization. Since gross mineralization is difficult to measure, most often it is the net mineralization that is measured or modeled. For a further description of the N mineralization-immobilization processes in soils see Chapter 5.

The models for simulating N mineralization kinetics in soils vary considerably in complexity and conceptual treatment of the processes involved. Benbi & Richter (2002) categorized these models into simple functional models to predict net N mineralization and mechanistic models that include a description of microbial biomass processes to predict long-term C and N turnover in soils. Empirically determined equations such as polynomials and parabolic functions have also been used to describe net N mineralization in soils. Broadbent (1986) found that parabolic function of the form $Y = aX^b$, provided a good fit to the net N mineralization data obtained by various investigators. Although empirically determined equations have been found to provide a better fit to the data, but no physical meaning can be attached to the regression coefficients.

Simple functional models quantify one or more active fractions of organic matter with associated rate constants to predict net N mineralization in soils (Table 9.2). The simplest way of modeling these processes is to consider net mineralization as a zero-order process. Stanford & Smith (1972) proposed the use of first-order single compartment (FOSC) model. The model is based on the assumption that the rate of N mineralization is proportional to the size of the mineralizable pool termed as N mineralization potential (N_o). Several authors (Molina et al., 1980; Nuske & Richter, 1981; Deans et al., 1986) have proposed that more than one fraction of soil organic N may be directly mineralized each with its specific rate of decomposition. In the simplest form of the multi-fraction approach, two main fractions of organic N are assumed to mineralize at different rates, following first-order kinetics (FODC model). One fraction consists of N compounds of easily decomposable plant material (dpm, N_d), while the second fraction represents the more resistant or recalcitrant

Table 9.2 The integral form of kinetic models used to predict net N mineralization in soils

Sr. No.	Model	Equation
I	Zero-order	$N_t = Kt$
II	First order single compartment (FOSC)	$N_t = N_o(1 - e^{-kt})$
III	First order double compartment (FODC)	$N_t = N_d(1 - e^{-k_d t}) + N_r(1 - e^{-k_r t})$
IV	First-order multicompartment	$N_t = \sum_{i-1}^{n} N_{0i}(1 - e^{-k_i t})$
V	First order plus zero order (FOZO)	$N_t = N_d(1 - e^{-k_d t}) + Kt$
VI	Modified first order (MFO)	$N_t = N_o - (N_o - N_e)e^{-kt}$
VII	Three-half (3/2) order	$N_t = N_0(1 - e^{-k_1 t - k_2 t^2/2}) + Kt$

N_t = cumulative N mineralized (mg kg^{-1}) at time 't' (days); N_d = readily decomposable organic N fraction (mg kg^{-1}); N_r = recalcitrant organic N fraction (mg kg^{-1}); N_e = intercept on the Y-axis (mg kg^{-1}) amount of N mineralized in first 7 days, N_0 = N mineralization potential (mg kg^{-1}); k_d, k_r = first-order rate constants (day^{-1}); K = zero-order rate constant (day^{-1}); *i* represents a specific N fraction and n is the total number of fractions

plant material (rpm, N_r). The dpm fraction mineralizes faster than the rpm fraction. A number of studies have confirmed that the first order double compartment (FODC) model is superior to the single compartment (FOSC) model in describing N mineralization. For dried soil samples, Richter et al. (1982) and Nordmeyer & Richter (1985) adopted a three-fraction approach that accounted for dead biomass N, dpm and rpm. Beauchamp et al. (1986) argued that the relatively fast mineralizable fraction may be a consequence of drying and subsequently rewetting the soil rather than due to N mineralization from a separate soil organic N fraction. They (Beauchamp et al., 1986) modified the FOSC model to account for this 'artifact mineralization'.

The single and multi-fraction first-order kinetic models are based on the assumption that the N mineralization potential of soil is a definable quantity and may be quantified from short-term or long-term incubations. However, many researchers have suggested that, with the incubation methods, it may not be possible to define the N mineralization potential of soils because the slowly mineralizable N fraction may, in fact, follow zero-order kinetics (Bonde & Rosswall, 1987; Lindemann et al., 1988; Seyfried & Rao, 1988). These authors advocated the use of mixed first-order and zero-order (FOZO) kinetic model. Since the first-order term of the model was interpreted as accounting for pretreatment effects (e.g. air-drying), it has been argued that in the absence of air-drying, i.e. in field-moist soils, net N mineralization may be described by zero-order only, rather than by combined first- plus zero-order kinetics (Addiscott, 1983). The implication of this approach is that there may be any number of N pools and each is sufficiently large not to be significantly depleted by mineralization. However, other authors have documented the applicability of first-order kinetic models to net N mineralization in field moist samples as well. Haer & Benbi (2003) evaluated different models on a number of arable soils cultivated mainly to rice-wheat (Fig. 9.5). Both the FODC and FOZO models provided good fit to the laboratory incubation data, however in soils with OC content > 0.55%, the FODC model invariably provided the best fit. The recalcitrant fraction (N_r) of the FODC model has been found to be positively related to SOC whereas the decomposable fraction (N_d) was not related to SOC (Fig. 9.6). This is obviously because N_d originates due to residues from the last crop and due to drying and rewetting of soils prior to incubation. Drying and rewetting the soils causes changes in soluble organic matter and some of the solubilized organic compounds may come from the microbial biomass, which is killed by drying of soil samples. On rewetting, the dead biomass becomes mineralized rapidly. Conversely, the recalcitrant fraction represents the plant material within the soil and is thus better related to SOC.

In situations where crop residues or organic manures of wider C:N ratios are added to soils, initially N immobilization may occur followed by net mineralization with the narrowing down of the C:N ratio. The description of such a data requires the inclusion of a term for initial lag phase in the model. Bonde & Lindberg (1988) modified the FOSC and FOZO models to accommodate the initial lag phase and described a model similar to the one presented by Brunner & Focht (1984) for microbial decomposition of C substrates, referred to as three-half-order (3/2 order), as it is suitable for substrate metabolism with (pseudo-second order) or without (pseudo-first order) growth.

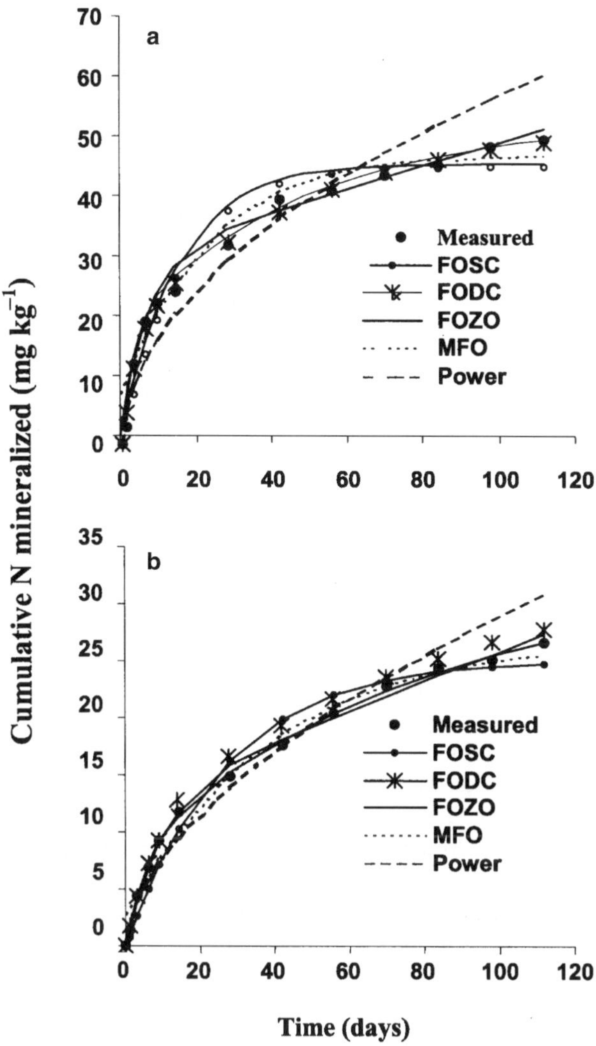

Fig. 9.5 Typical fit of different models to cumulative N mineralization in a sandy loam (a) and a clay loam (b) soil from northwestern India (Haer & Benbi, 2003, p. 163. Reproduced with kind permission from Taylor & Francis, Copyright Clearance Center)

Benbi & Richter (2002) made a critical assessment of the above discussed kinetic models on previously published data and concluded that (i) the single pool first-order kinetic model is inadequate to describe net N mineralization in soils as it provides the worst fit to the data and systematically deviates from the measured cumulative N mineralization; (ii) in disturbed soil samples, a minimum of two pools of mineralizable N are considered essential to significantly contribute towards

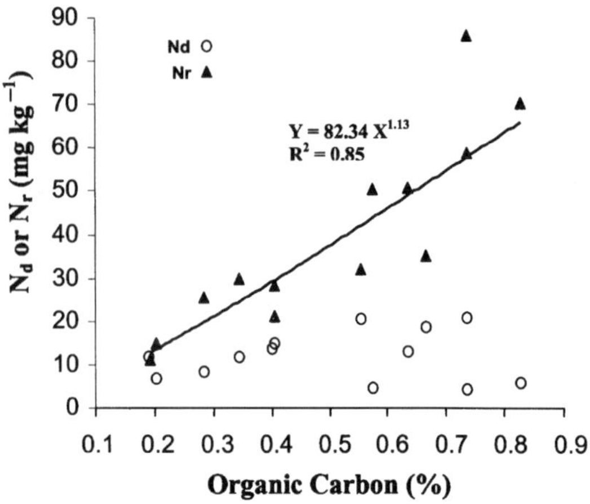

Fig. 9.6 Relationship of nitrogen mineralization potential of easily decomposable (N_d) and recalcitrant fraction (N_r) estimated from first-order two compartment model (Haer & Benbi, 2003, p. 161. Reproduced with kind permission from Taylor & Francis, Copyright Clearance Center)

N mineralization, irrespective of the fact whether the soil samples were dried and subsequently rewetted or kept field moist prior to incubation; and (iii) for precisely estimating potentially mineralizable N, the incubation needs to be continued until the mineralization rate approaches a small constant value.

While simple functional models predict net N mineralization, mechanistic models attempt to simulate mineralization-immobilization turnover. These models are designed to simulate C and N cycling during organic matter decomposition using microbial biomass kinetics. The models range in complexity from non-compartment models considering the decomposition of soil organic matter as a continuum (Bosatta & Ågren, 1985) to multicompartment models assuming different fractions of organic matter as functional pools (Jenkinson & Rayner, 1977; van Veen & Frissel, 1981; Molina et al., 1983; Parton et al. 1987; Jenkinson 1990; Hansen et al. 1991). In each of these pools the turnover rates are simulated by first-order kinetics. The pools in different models vary greatly in number, C and N content and turnover rates. The underlying concepts in different process-based models are described further in a subsequent section on modeling of organic matter dynamics.

Hunt et al. (1987) and De Ruiter et al. (1993) presented food-web models in which C and N fluxes in the soil are related to the abundance and activity of soil organisms, constituting the soil food web. In these models, organisms are classified as functional groups according to food choice and life-history parameters (Moore et al., 1988). Consumption rates among the groups of organisms are calculated based on biomass and turnover rates. Nitrogen mineralization rates are computed from consumption rates using information on energy conversion efficiencies and C:N ratios of the organisms. The food web models require information on physiological parameters such as assimilation efficiency, production efficiency, death rate, etc. for different functional groups of soil biota. As these values are not constant throughout the year/season, such models can calculate N mineralization kinetics only during a period under observation and are thus of descriptive value.

9.4.5 Nitrification

Models for nitrification range from simple empirical relationships through first-order kinetics to a set of mathematical equtions based on enzyme kinetics (e.g. Paul & Domsch, 1972). The empirical formulations that have been used to describe nitrification include exponential or logistic and Gompertz function. For example, Hadas et al. (1986) described nitrate production and ammonia disappearance by using a modified logistic equation (Equation 9.23)

$$NO_3^- = \frac{a}{1+(a/NO_{3\ 0}^- - 1)\exp[-ak(t-t_0)]} \tag{9.23}$$

where a and $NO_{3\,0}^-$ are the initial and final nitrate concentrations, t_0 is the initial time, and k is a constant. For incubation studies, Seifert (1980) proposed the use of Gompertz function

$$NO_3^- = NO_{3\max}^- \exp[-m\exp(-kt)] \tag{9.24}$$

where m and k are constants and $NO_{3\max}^-$ is the final nitrate concentration. In the N regime models that simulate nitrogen dynamics in soil-plant system, nitrification is usually described by zero- or first-order kinetics and the rate coefficient is generally adjusted to inorporate the effect of temperature and moisture. Mechanistic models use Michaelis-Menten kinetics and describe the rate of nitrification as a function of substrate (NH_4^+) concentration and several biological parameters such as number of bacteria and their specific growth rate (van Veen & Frissel, 1981).

9.5 Modeling Organic Matter Dynamics in Soils

Soils play an important role in global C cycling. The availability and turnover of organic matter in soils is central to most of the cycling processes. The turnover of organic matter is a complex biological process influenced by numerous physical, chemical and biological factors controlling the activity of microorganisms and soil fauna. In efforts to better understand SOM turnover processes and to predict and quantify the effect of C management strategies or land use changes on carbon dynamics in soil, a multitude of models have been presented in literature. Though the models vary considerably in conceptual details and the manner in which the effect of different variables have been incorporated, but essentially these could be categorised into two main classes viz. continuous or non-compartment models and the compartment models. The non-compartment models (e.g. Bosatta & Ågren, 1985) have been developed on the presumption that SOM is represented by a changing continuum. The decomposition is described by a continuous quality equation (CQE), beginning with the input of fresh organic matter and leading to the formation of refractory humic substances. The CQE, which is an integro-differential equation, is defined by three organism-related functions: growth rate of decomposers, efficiency of carbon utilization by decomposers, and a transition probability between different states (Ågren & Bosatta, 1998; Bosatta & Ågren, 2003). A continuous structure has merit in that it more closely represents the continuum of decomposability of SOM than a pool structure does, but its use has been impeded by complicated mathematics. However, Bosatta & Ågren (2003) derived exact solutions to the CQE and showed that efficient algorithms for numerical solutions are possible. A model (SOMKO) based on cohort approach similar to QSOIL (Bosatta & Ågren, 1991) but without imposing any type of distribution in cohorts or quality types has been presented by Gignoux et al. (2001). At each time step in the model, a new cohort of fresh organic matter is incorporated into SOM, and its fate is followed until it reaches a very low contribution to total SOM. Each SOM cohort of a given age comprises: q carbon pools, q nitrogen pools, and two microbial biomass pools. The q pools reflect SOM chemical heterogeneity and are similar to the variable of QSOIL. The model has not yet been validated on independent data.

In the compartmentalization approach, the organic matter is divided into conceptual compartments varying from 1 (Jenny, 1941; Woodruff, 1949) to several, sometimes 10 (McGill et al., 1981). The breakdown of organic matter in each compartment is assumed to follow first-order kinetics, which may be generalized as:

$$\frac{dX}{dt} = -kX \tag{9.25}$$

where X is soil C content at a given time; k is first order decomposition rate constant, and t is time.

Jenny (1941) presented the simplest formulation incorporating single compartment to represent the rate of change of organic C or N (dX/dt) with time in cultivated soils.

$$\frac{dX}{dt} = A - kX \tag{9.26}$$

where A is rate of addition of C or N. The solution of the differential equation is

$$X = X_E + (X_0 - X_E)e^{-kt} \tag{9.27}$$

where X_0 is the initial C or N content and X_E the content at equilibrium (when A and kX are equal and $dX/dt = 0$). The main problem with the use of this equation is that it does not allow for changes in the decomposition rate (k) that would result from changes in the composition of soil organic matter with time. Soil organic matter is intrinsically heterogeneous and consists of a complex of different components, varying in chemical composition and C:N ratios. These compounds have different decomposition rates and their proportion changes during the decomposition process leading to change in the overall decomposition rate. Modelers have attempted to describe the changing values of k by developing time-dependent functions to modify

k, using decay series in which residual portion becomes more stable or by adding more components.

Kolenbrander (1969) while considering organic matter as a single pool, assumed that the decomposition rate decreases with time. Janssen (1984) defined the relative rate of decomposition as a function of 'apparent initial age' of the organic material. This age was related to the humification coefficient and varied from 1 year for green matter to 14 years for some peats. Earlier, Hénin & Dupuis (1945) suggested a simple model (Equation 9.28) in which the annual input of plant carbon, A_c, decomposes very rapidly, forming a quantity of humus carbon fA_c, identical in nature to that already in the soil. The constant f is described as the isohumic-coefficient. It is usually about 0.3 for material from agricultural crops but much greater for peat and similar materials.

$$C = fA_c / k + (C_0 - fA_c / k)e^{-kt} \tag{9.28}$$

Here C is the quantity of organic carbon at time t and C_0 the initial quantity; k is the fraction of C that decomposes each year.

Yang & Janssen (2000) proposed a mono-component model with varying rate constant. The logarithm of the average relative mineralization rate, K of a substrate was linearly related to the logarithm of time, provided prevailing soil conditions remain unchanged

$$K = R\,t^{-S} \tag{9.29}$$

where R represents K at $t = 1$, and S (dimensionless, $1 \geq S \geq 0$) called the speed of ageing of the substrate is a measure of the rate at which K decreases with time. The quantity of the remaining substrate, Y_t is calculated as:

$$Y_t = Y_0 \exp(-Rt^{1-S}) \tag{9.30}$$

where Y_0 is the initial quantity of the substrate. The actual relative mineralization rate, k, at time t is proportional to K (Equation 9.31).

$$k = (1 - S)K \tag{9.31}$$

The test of the model on several data sets from different countries including a range of organic materials showed that the model could describe well the changes in organic matter from months to tens of years provided major environmental conditions remain unchanged.

In the multicompartment models, small homogenous pools with a high turnover rate and pools of greater size with slower turnover rate governed by first-order kinetics are distinguished. The decomposition rates are generally modified by climatic and edaphic reduction factors. In the simplest formulation of multicompartment approach, the SOM is partitioned into two compartments: labile and resistant or stable. The labile pool decomposes faster as compared to the other pool.

Jenkinson (1977) observed that a two-compartment model (Equation 9.32) fitted well to the decomposition of ^{14}C labeled ryegrass leaves in soil in the field.

$$C = 71e^{-2.83t} + 29e^{-0.087t} \tag{9.32}$$

Most of the multicompartment models that simulate long-term SOM dynamics in soils follow a similar general structure. These models partition organic matter into two main pools viz. recently added organic material such as plant residue or litter and native soil organic matter (SOM). Each of the main pool is further subdivided into different pools or compartments. Some of the SOM models and the pools considered therein are listed in Table 9.3. A generalized scheme of organic matter

Table 9.3 Pools of soil and added organic matter in different models

Model	Pools	Reference
ANIMO	Humus, fraction 2, fraction 3, fraction 4, fraction 5, fraction 6, fraction 7	Rijtema & Kroes (1991)
APSIM	Biomass, Humus, inert C	McCown et al. (1996)
CANDY	Active organic matter, stable organic matter, inert organic matter	Franko et al. (1995)
Century	AOM metabolic, AOM structural, active SOM, slow SOM, passive SOM	Parton et al. (1987)
CN-SIM	AOM1, AOM2, SMB1, SMB2, soil microbial residue, native humified OM, inert OM	Petersen et al. (2005)
DAISY	AOM 1, AOM 2, biomass pool 1, biomass pool 2, SOM pool 1, SOM pool 2	Hansen et al. (1991)
DNDC	Labile microbial biomass, resistant microbial biomass, labile humads, resistant humads, passive humads	Li et al. (1994)
Ecosys	Soluble SOM, adsorbed SOM, microbial SOM, microbial residues, active SOM, particulate SOM, acetate, methane	Grant (1995)
EPIC	Active SOM, stable SOM	Williams & Renard (1985)
ICBM	Young SOM, old SOM	Andrén & Kätterer (1997)
NCSOIL	Plant residues, labile microbial biomass, resistant microbial biomass, humads, stable SOM	Molina et al. (1983) and Nicolardot et al. (1994)
NICCE	Microbial biomass, metabolic, structural, humic, stable, resistant	Van Dam & Van Breemen (1995)
Rothamsted	DPM, RPM, microbial biomass, humified OM, inert OM	Jenkinson (1990)
SOMM	Undecomposed litter, partially decomposed litter, humus	Chertov & Komarov (1997)
SUNDIAL	Microbial biomass, humus	Smith et al. (1995)
SWATNIT	Litter, manure, humus	Vereecken et al. (1991)
Verberne	DPM, SPM, RPM, microbial biomass, protected biomass, non-protected biomass, protected SOM, non-protected SOM, stabilized OM	Verberne et al. (1990)

OM = organic matter; SOM = soil organic matter; AOM = added organic matter; SMB = soil microbial biomass; DPM = decomposable plant material; RPM = resistant plant material; SPM = structural plant material partitioning into different pools has been given by Nieder et al. (2003a). Some models operate with one litter input pool only (Molina et al., 1983; Nicolardot et al., 1994), while others consider two- (metabolic or labile and structural or resistant e.g. Hunt, 1977; Jenkinson et al., 1987; Parton et al., 1987; Hansen et al., 1991; Sallih & Pansu, 1993) or three pools (readily decomposable, structural and resistant lignified e.g. van Veen & Paul, 1981; Verberne et al., 1990). The litter pools are distinguished based on either lignin to nitrogen ratio (Parton et al., 1987), C:N ratio (Verberne et al., 1990) or determined from curve fitting of short-term decomposition data (Jenkinson et al., 1987; Coleman & Jenkinson, 1996). Paustian et al. (1997c) proposed a conceptual frame-work in which litter quality can be broken down into chemical, physical and inhibitory factors.

The SOM is commonly distinguished into the microbial biomass, and one or more pools of nonliving SOM. The soil microbial biomass (SMB) is further subdivided into two or more pools such as labile (non-protected or dynamic) and resistant (physically protected or stable) biomass (Molina et al., 1983; van Veen et al., 1984; Verberne et al., 1990; Hansen et al., 1991); cell walls and cytoplasm (Paustian & Schnürer, 1987); or labile cell carbon and assimilated live biomass (Knapp et al., 1983). Some models distinguish SMB based on their activity state such as active and inactive biomass (Hunt, 1977; Blagodatsky et al., 1994) or zymogenous growing very fast on readily available carbon and autochthonous biomass that react slowly and consumes the most resistant fraction of carbon (Jenkinson et al., 1987; Kersebaum & Richter, 1994). The Century model (Parton et al., 1987) includes the microbial biomass together with microbial metabolites and other SOM with a short turnover time (1–5 years) into a pool termed active SOM. The current version of the Rothamsted model (Jenkinson, 1990) operates with one biomass pool only.

The nonliving soil organic matter is differentiated into 'slow' or 'physically stabilized' pools with turnover times of a few decades and 'passive' or 'chemically stabilized pools' that remain in soil for hundreds or thousands of years (Jenkinson & Rayner, 1977; Parton et al., 1987). The physically stabilized pool is assumed to consist of compounds protected against biological attack by adsorption to soil colloids or entrapment within soil aggregates whereas the chemically stabilized pool includes compounds with a chemical structure resistant to biological attack (Hansen et al., 1991). A wide range of compounds (e.g. acid-hydrolysis residues, humin, humic acids, and interlayer organic complexes have been identified as highly resistant components of SOM, but model passive or stable pools are not defined with respect to chemical composition (Falloon & Smith, 2000).

9.5.1 Measured Versus Functional Soil Organic Matter Pools

The distribution of SOM within conceptual pools is an important consideration in developing a better understanding of SOM dynamics. As compared to continuous quality structure, the pool approach offers the advantage in ease of use and transferability. However, a major limitation is that most of the conceptual pools, except microbial biomass, do not correspond to experimentally verifiable fractions. The reasons for making widely different assumptions with regard to pool numbers, size, and decomposition behavior are not explicitly clear (Richter & Benbi, 1996). Further, it is difficult to comprehend how the authors or the users have parameterized these arbitrary pools. Most model pools are empirically fitted to total C values, and pool sizes estimated from other parameters. For example, parameterization of the Century model based on measured microbial plus soluble C for the active pool and light fraction for the slow pool resulted in underestimation of the C mineralization in the laboratory experiments (Motavalli et al., 1994). Similarly, Magid et al. (1997) found that there is no firm relationship between the measurable quality parameters of the added plant residues and an adequate parameterization of the DAISY model. Attempts have been made to establish linkages between the modeled and measured organic matter pools either by devising advanced laboratory fractionation procedures to match measurable organic matter fractions with model pool definitions or by revising model pool definitions to coincide with measurable quantities. New methods of fractionation are being developed to base models on measurable fractions. Christensen (1996) and Elliott et al. (1996) have described in detail attempts to match pools in models with measured fractions.

As discussed in Chapter 3, physical fractionation methods such as wet sieving, density flotation or chemical dispersal have been used to separate SOM into pools of different sizes and stability classes. Based on their differential turnover rates, some of the physical fractions have been related to conceptual pools considered in the existing SOM turnover models (Cambardella & Elliott, 1992; Buyanovsky et al., 1994). For example, it has been suggested that fractions that are 53–2,000 μm may provide an accurate estimate of the slow pool, while those finer than 53 μm may provide an accurate estimate of the passive pool (Cambardella & Elliott, 1992). Introduction of physically isolated SOM fractions as experimental equivalents to pools in models will require redefinition of the structure of current models and the factors that regulate the material fluxes between compartments. A revised model structure presented by Christensen (1996) includes two dynamic pools of particulate organic matter (free and intra-aggregate light fraction OM), one pool of inert light fraction organic matter, and two pools of organomineral associated SOM (silt- and clay- SOM). Gaunt et al. (2001) presented an alternative approach using analytically defined pools and measurement of ^{13}C and ^{15}N stable isotope tracers to derive model parameters. The approach considers all possible transformations between measured C and N pools and devises a system of equations using observed changes in total C and N and ^{13}C and ^{15}N for each fraction to solve all model unknowns. Sohi et al. (2001) proposed the use of free light, intra-aggregate and organomineral fractions isolated by two-stage density separation as measurable pools in SOM turnover models. But introduction of these proposed pools in model structures has either not been accomplished or the proposed model structures have not been tested to ensure that the measured fraction can be equated to a pool in the model. A measured fraction is equivalent to a model pool only if it is unique as well as non-composite (Smith et al., 2002). Most of the currently employed physical or chemical fractionation methods do not yield homogenous pools, particularly the conceptual passive pool, which is stabilized by various mechanisms. All efforts to isolate passive conceptual pool so far have yielded heterogeneous SOM fractions in terms of turnover times and do not correspond to specific stabilization mechanism and hence do not describe functional SOM pool (von Lützow et al., 2007). Smith et al. (2002) developed methods to examine if a fraction is both unique and non-composite. Their results suggest that the debris, biomass and humus pools are unique, but biomass and humus are composed of two or more sub-pools. The availability of new and advanced analytical techniques will help in better characterization of SOM and differentiation of functionally meaningful fractions. Therefore, the major challenge in SOM research in future will be establishing a close linkage between modeled and measured fractions through the use of advanced analytical techniques coupled with isotopic tracers and molecular markers, which could give functionally meaningful results.

9.5.2 Classification of Models

Though most of the soil organic matter models share a common structure yet these vary considerably with respect to the number of pools considered (Table 9.3), treatment of flux partitioning and microbial biomass kinetics. For the sake of comparison and organizing information it is important to classify models based on their distinct attributes. Jenkinson (1990) classified SOM turnover models into: (i) *non-compartmental decay models* which assume that decomposition is a continuum, organic matter moving down a quality scale as it decays, fresh decomposable organic matter having a high quality (set at 1), with the most resistant material present in the system being of zero quality (e.g. Bosatta & Ågren, 1985); (ii) *single homogenous compartment* models that assume a single compartment and have been applied mainly to organic N (e.g. Jenny, 1941; Woodruff, 1949); (iii) *two compartment models* in which incoming plant carbon is split into two compartments, each decomposing by a first-order kinetics, but one much faster than the other (e.g. Jenkinson, 1977); and (iv) *multicompartmental models* in which material in a compartment is assumed to decay by first order kinetics so that the rate of decomposition in a particular compartment is deemed to be a feature of organic matter itself. These may be short-term models (Smith 1979; McGill et al. 1981; van Veen et al., 1984) or longer term models (van Veen & Paul, 1981; Parton et al., 1987; Jenkinson et al. 1987). Van Keulen (2001) classified the multicompartment models into two broad groups: (i) models that consider only decomposition, i.e. mineralization (e.g. Ladd et al., 1981; Gregorich et al., 1989); and (ii) models that simulate both transformation and decomposition processes (e.g. Nicolardot et al., 1994; Jenkinson et al., 1987; Parton et al., 1987; van Veen & Paul, 1981).

The classification scheme proposed by Paustian (1994) categorises the multicompartment models into *organism-oriented*, and *process-oriented*. While organism-oriented models focus on flow of energy and matter through food-webs; the process-oriented models focus on processes controlling energy and matter transformations. Organisms are represented in process models as a form of genetic biomass or active organic matter. The process-oriented models share some common elements such as dominance of first order kinetics, discrete SOM components with characteristic rates of decomposition, and interconnected dynamics of C and N. McGill (1996) proposed a classification scheme consisting of a partially catenary sequence of attributes to organize information about ten SOM models. He grouped information about models under the following eight headings: (i) model type (dynamic or static), (ii) scale over which a model will function, (iii) soil horizons, (iv) regulation by soil properties, (v) biotic component, (vi) litter/ SOM distinction, (vii) litter and SOM compartmentalization, and (viii) control (process level or biological activity of organisms). This scheme showed convergence with respect to kinetic compartmentalization, use of clay content and inclusion of an inert organic matter component.

Currently, SOM models are being increasingly used for regional and global analysis of soil C dynamics. The models used for this purpose has been classified as ecosystem-level models and macro-scale models (Paustian et al., 1997d). The ecosystem models (e.g. Parton et al., 1987; Jenkinson & Rayner, 1977; Jenkinson, 1990; Pastor & Post, 1986) were originally developed to simulate ecosystems processes at local scales but these are increasingly being applied at the regional scales. The macro-scale models (e.g. TEM, Osnabrück; Raich et al., 1991; Esser, 1987, 1989) refer to a class of models which have been developed primarily for global modeling in which the spatial scale is defined as a latitudinal-longitudinal grid cells. The primary objective of most of these models has been to model vegetation patterns and net C fluxes to and from terrestrial ecosystems.

9.5.3 Evaluation and Use of Soil Organic Matter Models

One of the first steps in model evaluation and usage is its validation vis-à-vis real system output. Since SOM levels change relatively slowly compared to a large and variable background level, measurements over the long-term with well-documented management histories are required. Therefore, long-term experiments provide an important avenue for the development and evaluation of SOM models (Powlson, 1996). The compilation of many of the long-term experiment data sets (e.g. Paul et al., 1997b; Powlson et al., 1998) have helped to address this need by providing data, in standardized formats, which can be used to test and validate models for broad geographic regions and multiple types of land use, management, soil and climate conditions. The establishment of a formal network of global metadata (SOMNET) describing both long-term experimental data and SOM models has facilitated collaboration between modelers and data holders (Smith et al., 1997c).

In a modeling workshop held at IACR-Rothamsted, UK, nine long-term, process-oriented multicompartment SOM models were evaluated and compared using 12 datasets from long-term experiments representing a variety of climate, soil and land use conditions (Powlson et al., 1996; Smith et al., 1997b). In terms of overall performance of models across all datasets in simulating SOC dynamics, models fell into two groups with one group (SOMM, ITE and Verberne) performing significantly less well than the other (RothC, CANDY, DNDC, Century, DAISY and NCSOIL). This was attributed to the differences in the level of site specific calibration used by the two group of models (Smith et al., 1997d). Since some of the model pools are arbitrarily defined, these have to be site-specifically calibrated and the initialization procedure, even for short-term simulations influences the simulation results to a great extent (Puhlmann et al., 2006). This considerably minimizes the predictive utility of the models. Therefore, in future a major challenge in SOM modeling will be to free simulations from the calibration process and to devise experimental methods that will provide initial values relevant to the dynamic requirements of the model (Molina et al., 1997).

Several concepts used in the different SOM models have been incorporated into nutrient cycling models that are linked to crop plant production models such as Erosion-Productivity Impact Calculator (EPIC; Williams et al., 1985) Nitrogen-Tillage Residue Management (NTRM; Shaffer et al., 1992) DeNitrification and DeComposition Model (DNDC; Li et al., 1992). The models for SOM turnover in soils are being used to explore atmospheric CO_2 mitigation options through increased carbon sequestration in soil and vegetation. Grogan & Matthews (2002) used a process-based model similar in concept to Century and Rothamsted models to evaluate the potential for soil carbon sequestration in short rotation coppice willow plantations in the UK. The model estimates show that the potential for C sequestration in these plantations is comparable to, or even greater than that of naturally regenerating woodland. Soil organic matter models have been applied at the global scale to explore the potential role of soils in C uptake or release. SOM models have also been used to study the impact of different crop and land management practices on soil C stabilization and nutrient dynamics in different regions of the world (Jenkinson et al., 1987; Paustian et al., 1992; Parton & Rasmussen, 1994; Metherell et al., 1995), to estimate net primary production of a soil and to assess the impact of soil texture on SOM dynamics (Parton et al., 1994).

Models such as Century, DNDC have also been used to simulate trace gas fluxes. Using Rothamsted model, a few studies have examined the potential impacts of climate change on the organic carbon stocks of mineral soils in Europe (Smith et al., 2005, 2006). The carbon stocks of mineral soils in European croplands and grasslands are projected to decrease due to enhanced decomposition but increased net primary productivity is likely to slow the loss (Smith et al., 2005). Similar projections have been made for changes in mineral soil C in European forests for the period 1990–2100 (Smith et al., 2006). Whilst climate change will be a key driver of change in forest soil carbon, changes in ageclass structure and land-use change are estimated to have greater effects. Niklaus & Falloon (2006) used RothC model (Coleman & Jenkinson, 1996) along with C isotope measurements to estimate the effect of elevated CO_2 on soil C sequestration in grassland. The model could accurately reproduce soil C pools and fluxes under elevated CO_2 both in terms of total soil organic C as well as microbial biomass C and new C sequestration. Simulations with RothC model has shown that the increase in global temperature will result in enhanced soil respiration rates and hence decreased soil carbon contents, estimated at 54 Pg C by the year 2100. Globally, this effect dominates over increases in organic carbon input to the soil as a result of increased vegetation growth in many parts of the world (Jones et al., 2005). Smith et al. (2007) used the RothC model to estimate the effect of future climate and agricultural management on soil carbon stocks in European Russia and Ukraine. The results showed that SOC stocks will be lost, irrespective of the scenario considered but optimal management is able to reduce this loss of SOC by up to 44% compared with business as usual management. Results of above studies show that SOM models can help in estimating the impact of future climate change and identify the relative merits of different management practices in choosing mitigation option.

References

Abbt-Braun, G., Lankes, U., & Frimmel, F. H. (2004). Structural characterization of aquatic humic substances- the need for a multiple method approach. *Aquatic Sciences, 66*, 151–170.

Aber, J. D., & Federer, C. A. (1992). A generalized lumped-parameter model of photosynthesis, evaporation and net primary production in temperate and boreal forest ecosystems. *Oecologia, 92*, 463–474.

Aber, J. D., McDowell, H. W., Nadelhoffer, K. J., Magill, A., Bernston, G., Kamakea, M., et al. (1998). Nitrogen saturation in temperate forest ecosystems. *BioScience, 48*, 921–934.

Adams, D., Alig, R., McCarl, B., Calloway, J., & Winnett, S. (1999). Minimum cost strategies for sequestering carbon in forests. *Land Economics, 75*(3), 360–374.

Adams, R., Adams, D., Callaway, J., Chang, C., and McCarl, B. (1993). Sequestering carbon on agricultural land: Social cost and impacts on timber markets. *Contemporary Policy Issues, 11*(1), 76–87.

Addiscott, T. M. (1983). Kinetics and temperature relationships of mineralization and nitrification in Rothamsted soils with differing histories. *Journal of Soil Science, 34*, 343–353.

Addiscott, T. M., & Cox, D. (1976). Winter leaching of nitrate from autumn-applied calcium nitrate, ammonium sulphate, urea and sulphur-coated urea in bare soil. *Journal of Agricultural Science (Cambridge), 87*, 381–389.

Addiscott, T. M., & Wagenet, R. J. (1985). Concepts of solute leaching in soils:A review of modelling approaches. *Journal of Soil Science, 36*, 411–424.

Adger, W. N, & Brown, K. (1995). *Land use and the causes of global warming* (p. 271). Chichester, UK: Wiley.

Aerts, R. (1995). The advantages of being evergreen. *Trends in Ecology and Evolution, 10*, 402–407.

AF (Agriculture and Food). (2000). Greenhouse gas emissions and Alberta's cropping industry – things you need to know. http://www1.agric.gov.ab.ca/$department/deptdocs.nsf/all/cl3010

AG Boden (2005). *Bodenkundliche Kartieranleitung* (438 pp.). Stuttgart, Germany: Schweizerbart.

Agboola, A. A. (1981). The effects of different soil tillage and management practices on the physical and chemical properties of soils and maize yield in a rainforest zone of western Nigeria. *Agronomy Journal, 73*, 247–251.

Agostini, B. B., & Khan, D. (1986). Trends, situations and outlook for the world pulse economy. In R. J. Summerfield (Ed.), *World crops: Cool season food legumes* (pp. 461–482). Boston, MA: Kluwer.

Ågren, G. I., & Bosatta, E. (1998). *Theoretical ecosystem ecology. Understanding element cycles.* Cambridge: Cambridge University Press.

Aiken, G. R., & Gillam, A. H. (1989). Determination of molecular weights of humic substances by colligative property measurements. In M. H. B. Hayes, P. McCarthy, R. L. Malcolm, & R. S. Swift (Eds.), *Humic substances II. In search of structure* (pp. 516–543). New York: Wiley.

Aiken, G. R., McKnight, D. M., Wershaw, R. L., & MacCarthy, P. (1985). An introduction to humic substances in soil, sediment, and water. In G. R. Aiken, D. M. McKnight, R. L. Wershaw, & P. MacCarthy (Eds.), *Humic substances in soil, sediment and water: Geochemistry, isolation and characterization* (pp. 1–9). New York: Wiley-Interscience.

Ajtay, G. l., Ketner, P., & Duvigneaud, P. (1979). Terrestrial primary production and phytomass. In B. Bolin, E. T. Degens, S. Kemps, & P. Ketner (Eds.), *The global carbon cycle* (pp. 129–198). SCOPE 13, New York: Wiley.

Alexander, M. (1977). *Introduction to soil microbiology* (pp. 251–271). New York: Wiley.

Alexandratos, N. (1995). *World agriculture: Towards 2010, an FAO study*. Chichester, UK: Wiley.

Alig, R., Adams, D., McCarl, B., Callaway, J. M., & Winntt, S. (1997). Assessing effects of mitigation strategies for global climate change with an intertemporal model of the US forest and agricultural sectors. *Environmental Resources Economy, 9*, 259–274.

Alig, R., Adams, D., & McCarl, B. (1998). Ecological and economic impacts of forest policies: Interactions across forestry and agriculture. *Ecological Economics, 27*, 63–78.

Allen, J. C. (1985). Soil response to forest clearing in the United States and the tropics: Geological and biological factors. *Biotropica, 17*, 15–27.

Allen, L. H. J., Albrecht, S. L., Colón-Guasp, W., Covell, S. A., Baker, J. T., Pan, D., et al. (2003). Methane emission of rice increased by elevated carbon dioxide and temperature. *Journal of Environmental Quality, 32*, 1978–1991.

Allison, F. E., Roller, E. M., & Doetsch, J. H. (1953). Ammonium fixation and availability in vermiculite. *Soil Science, 75*, 173–180.

Almquist-Jacobson, H., & Foster, D. R. (1995). Toward an integrated model for raised-bog development: Theory and field evidence. *Ecology, 76*, 2503–2516.

Alvarez, C. R., Alvarez, R., Grigera, S., & Lavado, R. S. (1998). Association between organic matter fractions and the active soil microbial biomass. *Soil Biology and Biochemistry, 30*, 767–773.

Amador, J. A., & Jones, R. D. (1995). Carbon mineralization in the pristine and phosphorus-enriched peat soil of the Florida Everglades. *Soil Science, 159*, 129–141.

Ambus, P., & Robertson, G. P. (2006). The effect of increased N deposition on nitrous oxide, methane and carbon dioxide fluxes from unmanaged forest and grassland communities in Michigan. *Biogeochemistry, 79*, 315–337.

Amelung, W., Flach, K. W., & Zech, W. (1997). Climatic effects on soil organic matter composition in the Great Plains. *Soil Science Society of America Journal, 61*, 115–123.

Amon, B., Amon, T., Boxberger, J., & Wagner-Alt, C. (2001). Emissions of NH_3, N_2O and CH_4 from dairy cows housed in a farmyard manure tying stall (housing, manure storage, manure spreading). *Nutrient Cycling in Agro Ecosystems, 60*, 103–113.

Amundson, R. (2001). The carbon budget in soils. *Annual Review of Earth and Planetary Sciences, 29*, 535–562.

Amundson, R. G., & Davidson, E. A. (1990). Carbon dioxide and nitrogenous gases in the soil atmosphere. *Journal of Geochemical Exploration, 38*, 13–41.

Anastasi, C., Dowding, M., & Simpson, V. J. (1992). Future CH_4 emission from rice production. *Journal of Geophysical Research, 97*, 7521–7525.

Anderson, D. W. (1977). Early stages of soil formation of glacial till mine spoils in a semiarid climate. *Geoderma, 19*, 11–19.

Anderson, D. W. (1995). The role of nonliving organic matter in soils. In R. G. Zepp, & C. H. Sonntag (Eds.), *The role of nonliving organic matter in Earth's carbon cycle*. Chichester, UK: Wiley.

Anderson, H. A., Berrow, M. L., Farmer, V. C., Hepburn, A., Russell, J. D., & Walker, A. D. (1982). A reassessment of podzol formation processes. *Journal of Soil Science, 33*, 125–136.

Anderson, H. A., Stewart, M., Miller, J. D., & Hepburn, A. (1991), Organic nitrogen in soils and associated surface waters. In W. S. Wilson (Ed.), *Advances in soil organic matter research: The impact on agriculture and the environment* (pp. 97–106). Cambridge: The Royal Society of Chemistry.

Anderson, I. C., & Levine, J. S. (1986). Relative rates of nitric oxide and nitrous oxide production by nitrifiers, denitrifiers and nitrate respirers. *Applied Environmental Microbiology, 51*, 938–945.

Anderson, J. P. E., & Domsch, K. H. (1980). Quantities of plant nutrients in the microbial biomass of selected soils. *Soil Science, 130*, 211–216.

Anderson, K. A., & Downing, A. (2006). Dry and wet atmospheric deposition of nitrogen, phosphorus and silicon in an agricultural region. *Water, Air, and Soil Pollution, 176*, 351–374.

Anderson, T. H., & Domsch, K. H. (1985a). Determination of ecophysiological maintenance requirements of soil microorganisms in a dominant state. *Biology and Fertility of Soils, 1*, 81–89.

Anderson, T. H., & Domsch, K. H. (1985b). Maintainance carbon requirements of actively metabolising microbial populations under *in situ* conditions. *Soil Biology and Biochemistry, 17*, 197–203.

Anderson, T. H., & Domsch, K. H. (1989). Ratio of microbial biomass carbon to total organic carbon in arable soils. *Soil Biology and Biochemistry, 21*, 471–479.

Andreae, M. O., & Merlet, P. (2001). Emission of trace gases and aerosols from biomass burning. *Global Biogeochemical Cycles, 15*, 955–966.

Andrén, O., & Kätterer, T. (1997). ICBM- the Introductory Carbon Balance Model for exploration of soil carbon balances. *Ecological Modelling, 7*, 1226–1236.

Andrén, O., Rajkai, K., & Kaetterer, T. (1993). Water and temperature dynamics in a clay soil under winter wheat-influence on straw decomposition and N immobilization. *Biology and Fertility of Soils, 15*, 1–8.

Andrén, O., Zhao, X., & Liu, X. (1994). Climate and litter decomposition in Naiman, Inner Mongolia, China. *Ambio, 23*, 222–224.

Antia, N. J., Harrison, P. J., & Oliveira, L. (1991). The role of dissolved organic nitrogen in phytoplankton nutrition, cell biology and ecology. *Phycologia, 30*, 1–89.

Antle, J. M., Capalbo, S. M., Mooney, S., Elliott, E. T., & Paustian, K. H. (2003). Spatial heterogeneity, contract design, and the efficiency of carbon sequestration policies for agriculture. *Journal of Environmental Economics and Management, 46*, 231–250.

Antle, J. M., Capalbo S. M., Paustian, K. H., & Ali, K. (2005) *Estimating the economic potential for agricultural soil carbon sequestration in the central United States using an aggregate econometric-process simulation model.* Working paper, Program on Greenhouse Gas Mitigation and Climate Change, Montana State University.

Aoyama, M., Anger, D. A., & Ndayegamiye, A. (1999). Particulate and mineral-associated organic matter in water-stable aggregates as affected by mineral fertilizer and manure applications. *Canadian Journal of Soil Science, 79*, 295–302.

Apps, M., Cerri, C., Fujimori, T., Janzen, H., Krankin, O., Makundi, W., et al. (2001). Technological and economic potential of options to enhance, maintain, and manage biological carbon reservoirs and geo-engineering. In *Climate Change 2001:Working Group III: Mitigation*. http://www.grida.no/climate/ipcc_tar/wg3/index.htm

Aquaye, D. K., & MacLean, A. J. (1966). Influence of form and mode of nitrogen fertilizer application on the availability of soil and fertilizer potassium. *Canadian Journal of Soil Science, 46*, 23–28.

Arah, J. R. M. (1990). Diffusion-reaction models of denitrification in soil microsites. In N. P. Revsbech, & J. Sorensen (Eds.), *Denitrification in soil and sediment* (pp. 245–258). New York: Plenum.

Arah, J. R. M., & Kirk, G. J. D. (2000). Modeling rice plant-mediated methane emission. *Nutrient Cycling in Agroecosystems, 58*, 221–230.

Arah, J. R. M., & Smith, K. A. (1989). Steady state denitrification in aggregated soils: a mathematical model. *Journal of Soil Science, 40*, 139–149.

Arah, J. R. M., & Stephen, K. D. (1998). A model of the processes leading to methane emission from peatland. *Atmospheric Environment, 32*, 3257–3264.

Arah, J. R. M., Smith, K. A., Crichton, I. J., & Li, H. S. (1991). Nitrous oxide production and denitrification in Scottish arable soils. *Journal of Soil Science, 42*, 351–367.

Arbeitskreis Standortskartierung. (1996). *Forstliche Standortsaufnahme* (5th ed.). Eching, Germany: IHW-Verlag.

Archer, S., Schimel, D. S., & Holland, E. A. (1995). Mechanisms of shrubland expansion: Land use, climate or CO_2? *Climatic Change, 29*, 91–99.

Armentado, T. V., & Menges, E. S. (1986). Patterns of change in the carbon balance of organic soil-wetlands of the temperate zone. *Ecology, 74*, 755–774.

Armstrong, J. L. (1990). Runoff and soil loss from bare fallow plots at Inverell, New-South-Wales. *Australian Journal of Soil Research, 28*, 659–675.

Arrhenius, S. (1896). On the influence of carbonic acid in the air upon the temperature of the ground. *Philosophical Magazine and Journal of Science Series, 5*(41), 237–276.
Arshad, M. A., & Martin, S. (2002). Identifying critical limits for soil quality indicators in agro-ecosystems. *Agriculture, Ecosystems and Environment, 88*, 153–160.
Arshad, M. A., Ripmeester, J. A., & Schnitzer, M. (1988). Attempts to improve solid-state ^{13}C NMR spectra of the whole mineral soils. *Canadian Journal of Soil Science, 68*, 593–602.
Asman, W. A. H., Sutton, M. A., & Schjørring, J. K. (1998). Ammonia: Emission, atmospheric transport and deposition. *New Phytologist, 139*, 27–48.
Atanasiu, N., Westphal, A., & Banerjee, A. K. (1967). Studien über die Wirkung gedüngten Stickstoffs auf Ertrag und N-Aufnahme der Pflanze bei Böden mit verschiedenem Ammoniu mfixierungsvermögen. *Agrochimica, 11*, 257–274.
Augustin, J., Blasinski, F., Steffens, L., & Merbach, W. (1995a). Einfluß von Pflanzenbewuchs und N-Angebot auf die Spurengasemissionen aus dem Boden. In W. Merbach, & R. Bork (Eds.), *Kohlenstoff und Stickstoffumsatz im System Pflanze-Boden* (pp. 81–85). ZALF Bericht Nr. 23, Selbstverlag des Zentrums für Agrarlandschafts- und Landnutzungsforschung (ZALF) e.V. Müncheberg.
Augustin, J., Käding, H., Merbach, W., & Schmidt, W. (1995b). Lachgasemissionen aus degradierten Niedermoorstandorten Norddeutschlands. In W. Merbach, & R. Bork (Eds.), *Kohlenstoff und Stickstoffumsatz im System Pflanze-Boden* (pp. 125–128). ZALF Bericht Nr. 23, Selbstverlag des Zentrums für Agrarlandschafts- und Landnutzungsforschung (ZALF) e.V. Müncheberg.
Bachelet, D., & Neue, H. U. (1993). Methane emissions from wetland rice areas of Asia. *Chemosphere, 26*, 219–237.
Baes, C. F., Goeller, H. E., Olson, J. S., & Rotty, R. M. (1977). Carbon dioxide and climate: The uncontrolled experiment. *American Science, 65*, 310–320.
Baggs, E. M., Rees, R. M., Smith, K. A., & Vinten, A. J. A. (2000). Nitrous oxide emission from soils after incorporating crop residues. *Soil Use and Management, 16*, 82–87.
Bajwa, M. I. (1985). Ammonium and potassium fixation by wetland rice soils. *Pakistan Journal of Agricultural Research, 6*, 23–25.
Baldock, J. A., & Ballard, R. A. (2004). Editorial: Fixed nitrogen in sustainable farming systems: A symposium examining factors influencing the extent of biological nitrogen fixation and its role in southern Australian agricultural systems. Setting the scene. *Soil Biology and Biochemistry, 36*, 1191–1193.
Baldock, J. A., & Nelson, P. N. (2000). Soil organic matter. In M. E. Sumner (Ed.), *Handbook of soil science* (pp. B-25–B-84). Boca Raton, FL: CRC.
Baldock, J. A., Oades, J. M., Waters, A. G., Peng, X., Vassallo, A. M., & Wilson, M. A. (1992). Aspects of the chemical structure of soil organic materials as revealed by solid-state ^{13}C NMR spectroscopy. *Biogeochemistry, 16*, 1–42.
Balesdent, J. (1996). The significance of organic separates to carbon dynamics and its modelling in some cultivated soils. *European Journal of Soil Science, 47*, 485–493.
Balesdent, J., Chenu, C., & Balabane, M. (2000). Relationship of soil organic matter dynamics to physical protection and tillage. *Soil and Tillage Research, 53*, 215–230.
Balesdent, J., Mariotti, A., & Guillet, B. (1987). Natural ^{13}C abundance as tracer for soil organic matter dynamics studies. *Soil Biology and Biochemistry, 19*, 25–30.
Balesdent, J., Wagner, G. H., & Mariotti, A. (1988). Soil organic matter turnover in long-term field experiments as revealed by the ^{13}C natural abundance in maize field. *Soil Science Society of America Journal, 52*, 118–124.
Bandick, A. K., & Dick, R. P. (1999). Field management effects on soil enzyme activities. *Soil Biology and Biochemistry, 31*, 1471–1479.
Banerjee, M. R., Burton, D. L., McCaughey, W. P., & Grant, C. A. (2000). Influence of pasture management on soil biological quality. *Journal of Range Management, 53*, 127–133.
Banerjee, N. K., & Mosier, A. R. (1989). Coated calcium carbide as a nitrification inhibitor in upland and flooded soils. *Journal Indian Society of Soil Science, 37*, 306–313.
Barber, S. A. (1979). Corn residue management and soil organic matter. *Agronomy Journal, 71*, 625–627.

Barnard, R., Barthes, L., Le Roux, X., Harmens, H., Raschis, A., Soussana, J. F., et al. (2004). Atmospheric CO_2 elevation has little effect on nitrifying and denitrifying enzyme activity in four European grasslands. *Global Change Biology, 10*, 488–497.

Barraclough, D., & Puri, G. (1995). The use of ^{15}N pool dilution and enrichment to separate the heterotrophic and autotrophic pathways of nitrification. *Soil Biology and Biochemistry, 27*, 17–22.

Barson, M., & Gifford, R. (1990). Carbon dioxide sinks: The potential role of tree planting in Australia. In D. Swaine (Ed.), *Greenhouse and energy* (pp. 433–443). Australia: CSIRO.

Bartlett, R. J., & Simpson, T. J. (1967). Interaction of ammonium and potassium in a potassium-fixing soil. *Soil Science Society of America Proceedings, 31*, 219–222.

Barunke, A. (2002), *Die Stickstoffproblematik in der Landwirtschaft* (227 p.) (P. Weingarten Ed.). Germany: Wissenschaftsverlag Vauk Kiel AG, IAMO/Halle.

Bass, S., Dubois, O., Moura Costa, P., Pinard, M., Tipper, R., & Wilson, C. (2000). *Rural livelihood and carbon management*. IIED Natural Resources Issues Paper No. 1, London.

Bateman, E. J., & Baggs, E. M. (2005). Contribution of nitrification and denitrification to N_2O emissions from soils at different water-filled pore space. *Biology and Fertility of Soils, 41*, 379–388.

Batjes, N. H. (1996). Total carbon and nitrogen in the soils of the world. *European Journal of Soil Science, 47*, 151–163.

Batjes, N. H. (1997). World soil carbon stocks and global change. In V. R. Squires, E. P. Glenn, & A. T. Ayoub (Eds.), *Proceedings of the Workshop (held in Nairobi, Kenya, 4–8 September 1995) Combating Global Climate Change by Combating Land Degradation* (pp. 51–78). Wageningen, The Netherlands: ISRIC (International Soil Reference and Information Centre).

Batjes, N. H. (1998). Mitigation of atmosphere CO_2 concentrations by increased carbon sequestration in the soil. *Biology and Fertility of Soils, 27*, 230–235.

Batjes, N. H. (1999). *Management options for reducing CO_2-concentration in the atmosphere by increasing carbon sequestration in the soil, Report 410-200-031*. Dutch National Research Programme on Global Air Pollution and Climate Change & Technical Paper 30, International Soil Reference and Information Centre, Wageningen.

Batjes, N. H., & Sombroek, W. G. (1997). Possibilities for carbon sequestration in tropical and subtropical soils. *Global Change Biology, 3*, 161–173.

Baum, C., Leinweber, P., & Schlichting, A. (2003). Effects of chemical conditions in re-wetted peats on temporal variation in microbial biomass and acid phosphatase activity within the growing season. *Applied Soil Ecology, 22*, 167–174.

Bayer, C., Mielniczuk, J., Martin-Neto, L., & Ernani, P. R. (2002). Stocks and humification degree of organic matter fractions as affected by no tillage on a subtropical soil. *Plant and Soil, 238*, 133–140.

Bazilevich, N. L. (1974). Energy flow and biological regularities of the world ecosystem. *Proceedings of the 1st International Congress of Ecology* (pp. 468–470). Wageningen, The Netherlands: Pudoc.

Beare, M. H., Cabrera, M. L., Hendrix, P. F., & Coleman, D. C. (1994a). Aggregate-protected and unprotected organic matter pools in conventional- and no-tillage soils. *Soil Science Society of America Journal, 58*, 787–795.

Beare, M. H., Hendrix, P. F., & Coleman, D. C. (1994b). Water-stable aggregates and organic matter fractions in conventional and no-tillage soils. *Soil Science Society of America Journal, 58*, 777–786.

Beauchamp, E. G., Reynolds, W. D., Brasche-Villeneuve, D., & Kirkby, K. (1986). Nitrogen mineralization kinetics with different soil pretreatments and cropping histories. *Soil Science Society of America Journal, 50*, 1478–1483.

Becher, H. H. (1996). Einfluss organischer Düngungen auf bodenphysikalische Parameter. *Zeitschrift für Planzenernährung und Bodenkunde, 159*, 121–127.

Beck, T. (1984). Methods and applications of soil microbiological analyses at the Landesanstalt für Bodenkultur and Pflanzenbau (LBB) in Munich for the determination of some aspects of soil fertility. In M. P. Nemes, S. Kiss, P. Papacostea, G. Stefanic, & M. Rusan (Eds.), *Fifth symposium on soil biology* (pp. 13–20). Bucharest, Romania: Roman National Society of Soil Science.

Becker, K. W. (1990). Rates of denitrification as indicated by ^{15}N labelled soil nitrogen balance experiments in Germany – A critical review. *Mitteilungen der Deutschen Bodenkundlichen Gesellschaft, 60*, 31–36.

Beese, F. (1986). Parameter des Stickstoffumsatzes in Ökosystemen mit Böden unterschiedlicher Acidität. *Göttinger Bodenkundliche Berichte, 90*, 1–344.

Bellini, G., Sumner, M. E., Radcliffe, D. E., & Qafoku, N. P. (1996). Anion transport through columns of highly weathered acid soils: Adsorption and retardation. *Soil Science Society of America Journal, 60*, 132–137.

Bellisario, L. M., Bubier, J. L., Moore, T. R., & Chanton, J. P. (1999). Controls on CH_4 emissions from a northern peatland. *Global Biogeochemical Cycles, 13*, 81–91.

Belotti, E. (1989). Untersuchungen zur Variabilität und Stabilität von Humusprofilen in Wäldern. *Ph.D. thesis*. Hohenheim University, Germany, 183 p.

Belotti, E., & Babel, U. (1993). Variability in space and time and redundancy as stabilizing principles of forest humus profiles. *European Journal of Soil Biology, 29*, 17–27.

Benbi, D. K. (1990). Efficiency of nitrogen use by dryland wheat in a subhumid region in relation to optimizing the amount of available water. *Journal of Agricultural Science (Cambridge), 115*, 7–10.

Benbi, D. K. (1992). Development and validation of a soil water nitrogen model for wheat. In M. S. Bajwa, N. S. Pasricha, P. S. Sidhu, M. R. Chaudhary, D. K. Benbi, & V. Beri, (Eds.), *Proceedings of the International Symposium on Nutrient Management for Sustained Productivity* (Vol. 2, pp. 173–175). Ludhiana: Punjab Agricultural University.

Benbi, D. K., & Biswas, C. R. (1997). Nitrogen balance and N recovery after 22 years of maize-wheat-cowpea cropping in a long-term experiment. *Nutrient Cycling in Agroecosystems, 47*, 107–114.

Benbi, D. K., & Chand, M. (2007). Quantifying the effect of soil organic matter on indigenous soil N supply and wheat productivity in semiarid sub-tropical India. *Nutrient Cycling in Agroecosystems, 79*, 103–11.

Benbi, D. K., & Richter, J. (1996). Nitrogen mineralization kinetics in sewage water irrigated and heavy metal treated sandy soils. In O. Van Cleemput, G. Hofman, & A. Vermoesen (Eds.), *Progress in nitrogen cycling studies* (pp. 17–22). Dordrecht, The Netherlands: Kluwer.

Benbi, D. K., & Richter, J. (2002). A critical review of some approaches to modelling nitrogen mineralization. *Biology and Fertility of Soils, 35*, 168–183.

Benbi, D. K., & Richter, J. (2003). Nitrogen dynamics. In D. K. Benbi, & R. Nieder (Eds.), *Handbook of processes and modeling in the soil-plant system* (pp. 409–481). New York: Haworth.

Benbi, D. K., Biswas, C. R., & Kalkat, J. S. (1991a). Nitrate distribution and accumulation in an Ustochrept soil profile in a long-term fertilizer experiment. *Fertilizer Research, 28*, 173–178.

Benbi, D. K., Prihar, S. S., & Cheema, H. S. (1991b). A model to predict changes in soil moisture, NO_3-N content and N uptake by wheat. *Fertilizer Research, 28*, 73–84.

Benbi, D. K., Biswas, C. R., Bawa, S. S., & Kumar, K. (1998). Influence of farmyard manure, inorganic fertilizers and weed control practices on some soil physical properties in a long-term experiment. *Soil Use and Management, 14*, 52–54.

Benedetti, A., Antisari, L. V., Canali, S., Gioacchini, P., & Sequi, P. (1996). Relationship between the fixed ammonium and the mineralization of the organic nitrogen in soil. In O. Van Cleemput, G. Hofman, & A. Vermoesen (Eds.), *Progress in nitrogen cycling studies* (pp. 23–26). Dordrecht, The Netherlands: Kluwer.

Benner, R., Maccubbin, A. E., & Hodson, R. E. (1984). Anaerobic biodegradation of the lignin and polysaccharide components of lignocellulose and synthetic lignin by sediment microflora. *Applied Environmental Microbiology, 47*, 998–1004.

Bentham, H., Harris, J. A., Birch, P., & Short, K. C. (1992). Habitat classification and soil restoration assessment using analyses of soil microbiological and physiochemical characteristics. *Journal of Applied Ecology, 29*, 711–718.

Berg, M. P., & Verhoef, H. A. (1998). Ecological characteristics of a nitrogen-saturated coniferous forest in the Netherlands. *Biology and Fertility of Soils, 26*, 258–267.

Bergstrom, D. W., & Beauchamp, E. G. (1993). An empirical model of denitrification. *Canadian Journal of Soil Science, 73*, 421–431.

Bergstrom, D. W., Monreal, C. M., Tomlin, A. D., & Miller, J. J. (2000). Interpretation of soil enzyme activities in a comparison of tillage practices along a topographic and textural gradient. *Canadian Journal of Soil Science, 80*, 71–79.

Bergström, L. F., & Kirchmann, H. (1999). Leaching of total nitrogen from nitrogen-15 labelled poultry manure and inorganic nitrogen fertilizer. *Journal of Environmental Quality, 28*, 1283–1290.

Berner, R. A. (1992). Weathering, plants, and the long-term carbon cycle. *Geochimica et Cosmochimica Acta, 56*, 3225–3231.

Bernier, N. (1998). Earthworm feeding activity and development of the humus profile. *Biology and Fertility of Soils, 26*, 215–223.

Bernier, N., & Ponge, J. F. (1994). Humus form dynamics during the sylvogenetic cycle in a mountain spruce forest. *Soil Biology and Biochemistry, 26*, 183–220.

Besnard, E., Chenu, C., Balesdent, J., Puget, P., & Arrouays, D. (1996). Fate of particulate organic matter in soil aggregates during cultivation. *European Journal of Soil Science, 47*, 495–503.

Bettany, J. R., Steward, J. B. W., & Saggar, S. (1979). The nature and forms of sulphur in organic matter fractions of soils selected along an environmental gradient. *Soil Science Society of America Journal, 43*, 981–985.

Beyer, L. (1996a). The chemical composition of soil organic matter in classical humic compound fractions and in bulk samples- a review. *Zeitschrift für Pflanzenernährung und Bodenkunde, 159*, 527–539.

Beyer, L. (1996b). Humusformen und -typen. Chapter 2.2.1. In H. P. Blume, P. Felix-Henningsen, W. R. Fischer, H. G. Frede, R. Horn, & K. Stahr (Eds.), *Handbuch der Bodenkunde*. Landsberg, Germany: Ecomed.

Beyer, L., Pingpank, K., Boelter, M., & Seppelt, R. D. (1998). Small-distance variation of carbon and nitrogen storage in mineral Antarctic Cryosols near Casey Station (Wilkes Land). *Journal of Plant Nutrition and Soil Science, 161*, 211–220.

Beyer, L., Blume, H. P., & Koebbemann, C. (1999). Colluvisols under cultivation in Schleswig-Holstein. 3. Soil organic matter transformation after translocation. *Journal of Plant Nutrition and Soil Science, 162*, 61–69.

Bhatti, J. S., Apps, M. J., & Jiang, H. (2001). Examining the carbon stocks of Boreal forest ecosystems at stand and regional scales. In R. Lal, M. Kimble, R. F. Follett, & B. A. Steward (Eds.), *Assessment methods for soil C pools* (pp. 513–532). Advances in Soil Science. Boca Raton, FL: Lewis.

Billings, W. D. (1987). Carbon balance of Alaskan tundra and taiga ecosystems: Past, present and future. *Quaternary Science Reviews, 6*, 165–177.

Binkley, D. (1981). Nodule biomass and acetylene reduction rates of red alder and Sitka alder on Vancouver Island, B.C. *Canadian Journal of Forest Research, 11*, 281–286.

Birch, H. F. (1958). The effect of soil drying on humus decomposition and nitrogen availability. *Plant and Soil, 12*, 9–13.

Bird, M., Santruckova, H., Lloyd, J., & Veenendahl, E. (2001). The soil carbon pool and global change. In E. D. Schulze, M. Heimann, S. Harrison, E. Holland, J. Lloyd, I. C. Prentice et al. (Eds.), *Global biogeochemical cycles in the climate system*. San Diego, CA: Academic.

Bird, M. I. (1995). Fire, prehistoric humanity, and the environment. *Interdisciplinary Science Review, 20*, 141–154.

Bird, M. I., Veenendahl, E. M., Moyo, C., Lloyd, J., & Frost, P. (2000). Effect of fire and soil texture on soil carbon in a sub-humid savannah (Matopos, Zimbabwe). *Geoderma, 94*, 71–90.

Black, A. S., & Waring, S. A. (1972). Ammonium fixation and availability in some cereal producing soils of Queensland. *Australian Journal of Soil Research, 10*, 197–207.

Black, K. E., Harbron, C. G., Franklin, M., Atkinson, D., & Hooker, J. E. (1998). Differences in root longevity in some tree species. *Tree Physiology, 18*, 259–264.

Blagodatsky, S. A., Blagodatsky, E. V., & Rozanova, L. N. (1994). Kinetics and strategy of microbial growth in chernozemic soil as affected by different long-term fertilization. *Microbiology, 63*, 298–307.

Blake, D. R., & Rowland, F. S. (1988). Continuing worldwide increase in tropospheric methane 1978 to 1987. *Science, 239*, 1129–1131.

Bleuten, W., Vasiliev, S. V., & Yefremov, S. P. (2000). *The West Siberian peat accumulating ecosystems, the missing sink for atmospheric carbon depletion*. In 11th International Peat Congress, Quebec 2000. http://www.aswm.org/science/carbon/quebec/cont1.html

Bloom, P. R. (2001), Soil pH and pH Buffering. In M. E. Sumner (Ed.), *Handbook of soil science* (pp. B-333–B-352). Boca Raton, FL: CRC.

Blume, H. P., Beyer, L., Boelter, M., Erlenkeuser, H., Kalk, E., Kneesch, S., et al. (1997). Pedogenic zonation in the southern circumpolar region. *Advances in GeoEcology, 30*, 69–90.

Bobbink, R., Ashmore, M., Braun, S., Flückiger, W., & van den Wyngaert, I. J. J. (2002). Empirical nitrogen critical loads for natural and semi-natural ecosystems: 2002 update. Available online http://www.iap.ch/publikationen/nworkshop-background.pdf, pp. 128.

Boeckx, P., & van Cleemput, O. (2001). Estimates of N_2O and CH_4 fluxes from agricultural lands in various regions in Europe. *Nutrient Cycling in Agroecosystems, 60*, 35–47.

Bognonkpe, J. P. I. (2004). Effect of land use on soil N dynamics at watershed scale in Africa. *Bonner Agrikulturchemische Reihe, Band* 19. Available online http://www.ipe.uni-bonn.de/BAR/Band_19.htm

Bohn, H. L. (1978). Organic soil carbon and CO_2. *Tellus, 30*, 472–475.

Bol, R., Kandeler, E., Amelung, W., Glaser, B., Marx, M. C., Preedy, N., et al. (2003). Short-term effects of dairy slurry amendment on carbon sequestration and enzyme activities in a temperate grassland. *Soil Biology and Biochemistry, 35*, 1411–1421.

Bolan, N. S., Hedley, M. J., & White, R. E. (1991). Processes of soil acidification during nitrogen cycling with emphasis on legume-based pastures. *Plant and Soil, 134*, 53–63.

Bolin (1970). The carbon cycle. *Science America, 223*, 124–130.

Bolin, B., & Keeling, C. D. (1963). Large-scale atmospheric mixing as deduced from the seasonal and meridional variations of carbon dioxide. *Journal of Geophysical Research, 68*, 3899–3920.

Bolin, B., & Sukumar, R. (2000). Global perspective. In R. T. Watson, I. R. Noble, B. Bolin, N. H. Ravindranath, D. J. Verardo, & D. J. Doken (Eds.), *Land use, land-use change, and forestry* (pp. 23–51). Cambridge: Cambridge University Press.

Bonan, G. B., & Shugart, H. H. (1989). Environmental factors and ecological processes in boreal forests. *Annual Review of Ecology and Systematics,* 20, 1–28.

Bonde, T. A., & Lindberg, T. (1988). Nitrogen mineralization kinetics in soil during long-term aerobic laboratory incubations:a case study. *Journal of Environmental Quality, 17*, 414–417.

Bonde, T. A., & Rosswall, T. (1987). Seasonal variation of potentially mineralizable nitrogen in four cropping systems. *Soil Science Society of America Journal, 15*, 1508–1514.

Bork, R., Dreibrodt S., & Mieth A. (2006), Klimaänderungen und Maßnahmen in der Fläche – eine Betrachtung im gesellschaftspolitischen Kontext. In J. Mathews, St. Marahrens, & S. Schmidt (Eds.), *UBA-Workshop:Klimaänderungen – Herausforderungen für den Bodenschutz, 28. and 29. September 2005* (pp. 1–11). Dessau, Germany: Umweltbundesamt.

Borken, W., & Matzner, E. (2004). Nitrate leaching in forest soils:an analysis of long-term monitoring sites in Germany. *Journal of Plant Nutrition and Soil Science, 167*, 277–283.

Borken, W., Xu, Y. J., & Beese, F. (2004). Leaching of dissolved organic carbon and carbon dioxide emission after compost application to six nutrient-depleted forest soils. *Journal of Environmental Quality, 33*, 89–98.

Born, M., Dorr, H., & Levin, I. (1990). Methane consumption in aerated soils of the temperate zone. *Tellus, 42B*, 2–8.

Bornebusch, C. H. (1930). The fauna of forest soil. *Det Forstlige Forsogsvaesen, 11*, 1–158.

Bosatta, E., & Ågren, G. I. (1985). Theoretical analysis of decomposition of heterogeneous substrates. *Soil Biology and Biochemistry, 17*, 601–610.

Bosatta, E., & Ågren, G. I. (1991). Dynamics of carbon and nitrogen in the organic matter of the soil: A generic theory. *American Naturalist, 138*, 227–245.

Bosatta, E., & Ågren, G. I. (2003). Exact solutions to the continuous-quality equation for soil organic matter turnover. *Journal of Theoretical Biology, 224*, 97–105.

Botch, M. S., Kobak, K. I., Vinson, T. S., & Kolchugina, T. P. (1995). Carbon pools and accumulation in peatlands of the former Soviet Union. *Global Biogeochemical Cycles, 9*, 37–46.

Böttcher, J., & Springob, G. (2001). A carbon balance model for organic layers of acid forest soils. *Journal of Plant Nutrition and Soil Science, 164*, 399–405.

Bottner, P. (1985). Response of microbial biomass to alterate moist and dry conditions in a soil incubated with ^{14}C and ^{15}N-labelled plant material. *Soil Biology and Biochemistry, 17*, 329–337.

Bouma, J. (2002). Land quality indicators of sustainable land management across scales. *Agriculture, Ecosystems and Environment, 88*, 129–136.

Bouman, O. T., Curtin, D., Campbell, C. A., Biederbeck, V. O., & Ukrainnetz, H. (1995). Soil acidification from long-term use of anhydrous ammonia and urea. *Soil Science Society of America Journal, 59*, 1488–1494.

Bouquet, E., Bononat, A., & Ramos-Cormenzana, A. (1973). A production of calcite crystals (calcium carbonate) by soil bacteria is a general phenomena. *Nature, 246*, 527–528.

Bousquet, P., Ciais, P., Miller, J. B., Dlugokencky, E. J., Hauglustaine, D. A., Prigent, C., et al. (2006). Contribution of anthropogenic and natural sources to atmospheric methane variability. *Nature, 443*, 439–443.

Bouwman, A. F. (1990). Exchange of greenhouse gases between terrestrial ecosystems and the atmosphere. In A. F. Bouwman (Ed.), *Soils and the Greenhouse Effect* (pp. 61–129). Chichester, UK: Wiley.

Bouwman, A. F. (1996). Direct emissions of nitrous oxide from agricultural soils. *Nutrient Cycling in Agroecosystems, 46*, 53–70.

Bouwman, A. F., van der Hoek, K. W., & Batjes, N. H. (1995). Uncertainty in the global source distribution of nitrous oxide. *Journal of Geophysical Research, 100*, 2785–2800.

Bouwman, A. F., Lee, D. S., Asman, W. A. H., Dentener, F. J., Van Der Hoek, K. W., & Olivier, J. G. J. (1997). A global high-resolution emission inventory for ammonia. *Global Biogeochemical Cycles, 11*, 561–587.

Bouwman, A. F., Boumans, L. J. M., & Batjes, N. H. (2002a). Estimation of global NH_3 volatilization loss from synthetic fertilizers and animal manure applied to arable lands and grasslands. *Global Biogeochemical Cycles, 16*, 10.1029/2000GB001389.

Bouwman, A. F., van Vuuren, D. P., Derwent, R. G., & Posch, M. (2002b). A global analysis of acidification and eutrophication of terrestrial ecosystems. *Water, Air, and Soil Pollution, 141*, 349–382.

Bouwman, A. F., Van Drecht, G., Knoop, J. M., Breusen, A. H. W., & Meinardi, C. R. (2005). Exploring changes in river nitrogen export the world's oceans. *Global Biogeochemical Cycles, 19*, GB1002, doi:10.1029/2004GB002314.

Bouwmeester, R. J. B., & Vlek, P. L. G. (1981). Rate control of ammonia volatilization from rice paddies. *Atmospheric Environment, 15*, 131–140.

Bowden, R. D., Rullo, G., Stevens, G. R., & Steudler, P. A. (2000). Soil fluxes of carbon dioxide, nitrous oxide, and methane at a productive temperate deciduous forest. *Journal of Environmental Quality, 29*, 268–276.

Bowden, W. B. (1986). Gaseous nitrogen emissions from undisturbed terrestrial ecosystems: An assessment on their impacts on local and global nitrogen budgets. *Biogeochemistry, 2*, 249–279.

Box, E. O. (1988). Estimating the seasonal carbon source-sink geography of a natural, steady-state terrestrial biosphere. *Journal of Applied Meteorology, 27*, 1109–1124.

Boyer, E. W., Goodale, C. L., Jaworski, N. A., & Howarth, R. W. (2002). Anthropogenic nitrogen sources and relationships to riverine nitrogen export in the northwestern USA. *Biogeochemistry, 57/58*, 137–169.

Boyer, E. W., Howarth, R. W., Galloway, J. N., Dentener, F. J., Cleveland, C., Asner, P., et al. (2004). Current nitrogen inputs to world regions. In A. R. Mosier, J. K. Sycrs, & J. R. Freney (Eds.), *Agriculture and the nitrogen cycle: Assessing the impacts of fertilizer use on food production and the environment* (pp. 221–230). Washington, DC: Island.

Bracewell, J. M., Robertson, G. W., & Tate, K. R. (1976). Pyrolysis-gas chromatography studies on a climosequence of soils in tussock grassland, New Zealand. *Geoderma, 15*, 209–215.

Bradbury, N. J., Whitmore, A. P., Hart, P. B. S., & Jenkinson, D. S. (1993). Modelling the fate of nitrogen in crop and soil in the years following application of ^{15}N labelled fertilizer to winter wheat. *Journal of Agricultural Sciences, 121*, 363–379.

Brady, N. C., & Weil, R. R. (1999). *The nature and properties of soils*. Upper Saddle River, NJ: Prentice Hall.

Brandes, J. A., & Devol, A. H. (2002). A global marine-fixed nitrogen isotopic budget: Implications for Holocene nitrogen cycling. *Global Biogeochemical Cycles, 16*(4), Article No. 1120.

Bremner, J. M., & Nelson, D. W. (1968). Chemical decomposition of nitrite in soils. *Transactions 9th International Congress of Soil Science, 2*, 495–503.

Bridges, E. M., & Mukhopadhyay, S. (2003). Processes of soil formation. In D. K. Benbi, & R. Nieder (Eds.), *Handbook of processes and modeling in the soil-plant system* (pp. 121–148). New York: Haworth.

Bridgham, S. D. J., Megonigal, P., Keller, J. K., Bliss, N. B., & Trettin, C. (2006). The carbon balance of North American Wetlands. *Wetlands, 26*, 889–916.

Broadbent, F. E. (1986). Empirical modeling of soil nitrogen mineralization. *Soil Science, 141*, 208–210.

Bronson, K. F., Neue, H. U., Singh, U., et al. (1997). Automated chamber measurements of methane and nitrous oxide flux in a flooded rice soil. II. Fallow period emissions. *Soil Science Society of America Journal, 61*, 988–993.

Brookes, P. C. (1995). The use of microbial parameters in monitoring soil pollution by heavy metals. *Biology and Fertility of Soils, 19*, 269–279.

Brookes, P. C., Landman, A., Pruden, G., & Jenkinson, D. S. (1985). Chloroform fumigation and the release of soil nitrogen:a rapid direct extraction method to measure microbial biomass nitrogen in soil. *Soil Biology and Biochemistry, 17*, 837–842.

Brovkin, V., Stich, S., von Bloh, W., Claussen, M., Bauer, E., & Cramer, W. (2004). Role of land cover changes for atmospheric CO_2 increase and climate change during the last 150 years. *Global Change Biology, 10*, 1253–1266.

Brown, L., Syed, B., Jarvis, S. C., Sneath, R. W., Phillips, V. R., Goulding, K. W. T., et al. (2002). Development and application of a mechanistic model to estimate emissions of nitrous oxide from UK Agriculture. *Atmospheric Environment, 36*, 917–928.

Brown, S., Sathaye, J., Cannel, M., & Kauppi, P. (1996). Management of forests for mitigation of greenhouse gas emissions. In R. T. Watson, M. C. Zinyowera, & R. H. Moss (Eds.), *Climate change 1995, impacts, adaptations and mitigation of climate change: Scientific-technical analyses* (pp. 773–797). Report of Working Group II, Assessment Report, IPCC—. Cambridge: Cambridge University Press.

Bruinsma, J. E. (2003). *World agriculture: Towards 2015/2030. An FAO Perspective* (p. 432). London: Earthscan.

Brunner, W., & Focht, D. D. (1984). Deterministic three half order kinetic model for microbial degradation of added C substrates in soil. *Applied and Environmental Microbiology, 47*, 167–172.

Buffle, J. A. E. (1977). Les substances humiques et leurs interactions avec les ions minereux. In *Conference Proceedings de la Commission d' Hydrologie Appliquee de l' A.G* (pp. 3–10). H.T.M., I' Universite d'Orsay.

Burgess, D., Baldock, J. A., Wetzel, S., & Brand, D. G. (1995), Scarification, fertilization and herbicide treatment effects on planted conifers and soil fertility. *Plant and Soil, 168–169*, 513–522.

Burns, R. C., & Hardy, R. W. F. (1975). *Nitrogen fixation in bacteria and higher plants*. New York: Springer-Verlag.

Bussink, D. W., & Oenema, O. (1998). Ammonia volatilization from dairy farming systems in temperate areas: A review. *Nutrient Cycling in Agroecosystems, 51*, 19–33.

Butterbach-Bahl, K., Stange, F., Papen, H., & Li, C. (2001). Regional inventory of nitric oxide and nitrous oxide emissions for forest soils of southeast Germany using the biogeochemical model PnET-N-DNDC. *Journal of Geophysical Research, 106*, 34155–34166.

Buyanovsky, G. A., Aslam, M., & Wagner, G. H. (1994). Carbon turnover in soil physical fractions. *Soil Science Society of America Journal, 58*, 1167–1173.

Cadish, G., & Giller, K. E. (2001). Soil organic matter management: The roles of residue quality in C sequestration and N supply. In R. M. Rees, B. C. Ball, C. D. Campbell, & C. A. Watson (Eds.), *Sustainable management of soil organic matter* (pp. 97–111). Wallingford, UK: CABI.

Cadisch, G., Imhof, H., Urquiaga, S., Boddey, R. M., & Giller, K. E. (1996). Carbon turnover ($\delta^{13}C$) and nitrogen mineralization potential of particulate light soil organic matter after rainforest clearing. *Soil Biology and Biochemistry, 28*, 1555–1567.

Cai, Z., Xu, H., Zhang, H., & Jin, J. (1994). Estimate of methane emission from rice paddy fields in Taihu region. *Pedosphere, 4*, 297–306.

Cai, Z., Tsuruta, H., Gao, M., Xu, H., & Wei, C. (2003). Options for mitigating methane emission from a permanently flooded rice field. *Global Change Biology, 9*, 37–45.

Cambardella, C. A., & Elliott, E. T. (1992). Particulate soil organic matter changes across a grassland cultivation sequence. *Soil Science Society of America Journal, 56*, 777–783.

Campbell, C. A. (1978). Soil organic carbon, nitrogen and fertility. In M. Schnitzer, & S. U. Khan (Eds.), *Soil organic matter* (pp. 173–271). New York: Elesvier.

Campbell, C. A., McConkey, B. G., Zentner, P. R., Dyck, F. B., Selles, F., & Curtin, D. (1995). Carbon sequestration in a Brown Chernozem as affected by tillage and rotation. *Canadian Journal of Soil Science, 75*, 449–458.

Campbell, C. A., McConkey, B. G., Zentner, P. R., Selles, F., & Curtin, D. (1996). Tillage and crop rotation on soil organic C and N in a coarse-textured typic Haploboroll in southwestern Saskatchewan. *Soil Research, 37*, 3–14.

Campbell, C. A., Bienderbeck, V. G., Zentner, R. P., & Iafond, G. P. (2000a). Organic C accumulation in soil over 30 years in semiarid southwestern Saskatchewan: Effects of crop rotations and fertilizers. *Canadian Journal of Soil Science, 80*, 179–192.

Campbell, G. S. (1985). *Soil physics with basic: Transport models for soil-plant systems*. New York: Elsevier.

Campbell, I. B., & Claridge, G. G. C. (1987). *Antarctica: Soils, weathering processes and environment*. Amsterdam: Elsevier.

Campbell, J. L., Hornbeck, J. W., McDowell, W. H., Buso, D. C., Shanley, J. B., & Likens, G. E. (2000b). Dissolved organic nitrogen budgets for upland, forested ecosystems in New England. *Biogeochemistry, 49*, 123–142.

Cannell, R. Q. (1985). Reduced tillage in Northwest Europe – a review. *Soil and Tillage Research, 5*, 129–177.

Cannell, R. Q., & Hawes, J. D. (1994). Trends in tillage practices in relation to sustainable crop production with special reference to temperate climates. *Soil and Tillage Research, 30*, 245–282.

Cao, M., Dent, J. B., & Heal, O. W. (1995). Modelling methane emission from rice paddies. *Global Biogeochemical Cycles, 9*, 183–195.

Cao, M., Marshall, S., & Gregson, K. (1996). Global carbon exchange and methane emissions from natural wetlands: Application of a process-based model. *Journal of Geophysical Research, 101*, 14399–14414.

Cardon, Z. G. (1996). Influence of rhizodeposition under elevated CO_2 on plant nutrition and soil organic matter. *Plant and Soil, 187*, 277–288.

Carter, M. R. (1991). The influence of tillage on the proportion of organic carbon and nitrogen in the microbial biomass of medium-textured soils in a humid climate. *Biology and Fertility of Soils, 11*, 135–139.

Carter, M. R., Gregorich, E. G., Anderson, D. W., Doran, J. W., Janzen, H. H., & Pierce, F. J. (1997). Concepts of soil quality and their significance. In E. G. Gregorich, & M. R. Carter (Eds.), *Soil quality for crop production and ecosystem health*. Amsterdam: Elsevier.

Castaldi, S. (2000). Responses of nitrous oxide, dinitrogen and carbon dioxide production and oxygen consumption to temperature in forest and agricultural light-textured soils determined by model experiment. *Biology and Fertility of Soils, 32*, 67–72.

Cerri, C. C., Volkoff, B., & Andreuy, F. (1991). Nature and behaviour of organic matter in soils under natural forest, and after deforestation burning and cultivation, near Manaus. *Forest Ecology and Management, 38*, 247–257.

CGR, California Government report. (2006). http://www.energy.ca.gov/2006publications/CEC-500–2006–108/CEC-500–2006–108.PDF

Chakrabarti, K., Sarkar, B., Chakraborti, A., Banik, P., & Bakchi, D. K. (2000). Organic recycling for soil quality conservation in a sub-tropical plateau region. *Journal of Agronomy and Crop Science*, 184, 137–142.

Chander, K., & Joergensen, R. G. (2001). Decomposition of ^{14}C glucose in two soils with different levels of heavy metal contamination. *Soil Biology and Biochemistry, 33*, 1811–1816.

Chang, F. H., & Alexander, M. (1984). Effects of simulated acid precipitation on decomposition and leaching of organic carbon in forest soils. *Soil Science, 138*, 226–234.

Chantigny, M. H. (2003). Dissolved and water-extractable organic matter in soils: A review on the influence of land use and management practices. *Geoderma, 113*, 357–380.

Chantigny, M. H., Angers, D. A., Prevost, D., Vezina, L. P., & Chalifour, F. P. (1997). Soil aggregation and fungal and bacterial biomass under annual and perennial cropping systems. *Soil Science Society of America Journal, 61*, 262–267.

Chapin, F. S., Schulze, E. D., & Mooney, H. A. (1990). The ecology and economics of storage in plants. *Annual Reviews of Ecological Systems, 21*, 423–447.

Chen, C. C., Turner, F. T., & Dixon, J. B. (1989). Ammonium fixation by high-charge smectite in selected Texas Gulf Coast soils. *Soil Science Society of America Journal, 53*, 1035–1040.

Chen, C. T. A. (2004). Exchanges of carbon in the coastal seas. In C. B. Field, & M. R. Raupach (Eds.), *The global carbon cycle* (pp. 341–351). Washington, DC: Island.

Chen, G. X., Huang, G. H., Huang, B., Yu, K. W., Wu, J., & Xu, H. (1997). Nitrous oxide and methane emissions from soil-plant systems. *Nutrient Cycling in Agroecosystems, 49*, 41–45.

Chen, Y., & Aviad, T. (1990). Effects of humic substances on plant growth. In P. McCarthy, C. E. Clapp, R. L. Malcolm, & P. R. Bloom (Eds.), *Humic substances in soil and crop sciences: Selected readings* (pp. 161–186). Madison, WI: Soil Science Society of America.

Chen, Y. H., & Prinn, R. G. (2006). Estimation of atmospheric methane emission between 1996–2001 using a 3-D global chemical transport model. *Journal of Geophysical Research*, 111, D10307, doi:10.1029/2005JD006058.

Cheng, H. H., & Kimble, J. M. (2001). Characterization of soil organic carbon pools. In R. Lal, J. M. Kimble, R. F. Follett, & B. A. Stewart (Eds.), *Assessment methods for soil carbon* (pp 117–129). Boca Raton, FL: Lewis.

Chertov, O. G., & Komarov, A. S. (1997). SOMM: A model of soil organic matter dynamics. *Ecological Modelling, 94*, 177–189.

Cheshire, M. V. (1979). *Nature and origin of carbohydrates in soils*. London: Academic.

Cheshire, M. V., & Hayes, M. H. B. (1990). Composition, origins, structure, and reactivities of soil polysaccharides. In M. F. De Boodt, M. H. B. Hayes, & A. Herbillon (Eds.), *Soil colloids and their associations in aggregates* (pp. 307–336). New York: Plenum.

Cheshire, M. V., Dumat, C., Fraser, A. R., Hillier, S., & Staunton, S. (2000). The interaction between soil organic matter and soil clay minerals by selective removal and controlled addition of organic matter. *European Journal of Soil Science, 51*, 497–509.

Chomitz, K. M. (2000). *Evaluating carbon offsets from forestry and energy projects: How do they compare?* World Bank Policy Research Working Paper *2357*, New York, 25 pp.

Choudhury, A. T. M. A., & Kennedy, I. R. (2004). Prospects and potentials for systems of biological nitrogen fixation in sustainable rice production. *Biology and Fertility of Soils, 39*, 219–227.

Christensen, B. T. (1992). Physical fractionation of soil and organic matter in primary particle size and density separates. *Advances in Soil Science, 20*, 1–90.

Christensen, B. T. (1996). Matching measurable soil organic matter fractions with conceptual pools in simulation models of carbon turnover: Revision of model structure. In D. S. Powlson,

P. Smith, & J. U. Smith, (Eds.), *Evaluation of soil organic matter models using existing long-term datasets* (pp. 143–159). Heidelberg, Germany: Springer.
Christensen, B. T. (2001). Physical fractionation of soil and structural and functional complexity in organic matter turnover. *European Journal of Soil Science, 52*, 345–353.
Christensen, J. P., Murray, J. W., Devol, A. H. & Codispoti, C. A. (1987). Denitrification is continental shelf sedimens has a major impact on the oceanic nitrogen budget. Global Biogeochemical Cycles 1, 97–116.
Christensen, T. R., Jonasson, S., Callaghan, T. V., & Havaström, M. (1995). Spatial variation in high latitude methane flux – a transect across tundra environments in Siberiaand the European Arctic. *Journal of Geophysical Research, 100*, 21035–21045.
Christensen, T. R., Johansson, T., Åkerman, H. J., Mastepanov, M., Malmer, N., Friborg, T., et al. (2004). Thawing sub-arctic permafrost: Effects on vegetation and methane emissions. *Geophysical Research Letters, 31*, L04501, doi:10.1029/2003GL018680.
Christou, M., Avramides, E. J., Roberts, J. P., & Jones, D. L. (2005). Dissolved organic nitrogen in contrasting agricultural ecosystems. *Soil Biology and Biochemistry, 37*, 1560–1563.
Churchman, G. J., & Foster, R. C. (1994). The role of clay minerals in the maintenance of soil structure. In *Transaction of the 15th Congress of the International Soil Science Society* (pp. 17–34). Acapulco.
Ciais, P., Tans, P. P., Trolier, M., White, J. W. C., & Francey, R. J. (1995). A large Northern Hemisphere terrestrial CO_2 sink indicated by the C-13/C-12 ratio of atmospheric CO_2. *Science, 269*, 1098–1102.
Clapp, C. E., Stark, S. A., Clay, D. E., & Larson, W. E. (1986). Sewage sludge organic matter and soil properties. In Y. Chen, & Y. Avnimelech (Eds.), *The role of organic matter in modern agriculture* (pp. 209–253). Jerusalem, Israel: The Hebrew University of Jerusalem.
Clark, E., Haverkamp, J., & Chapman, W., (1985). *Eroding soils: The off-farm impact*. Washington, DC: Conservation Foundation.
Clemens, J., & Ahlgrimm, H. J. (2001). Greenhouse gases from animal husbandry: Mitigation options. *Nutrient Cycling in Agroecosystems, 60*, 287–300.
Clemens, J., Trimborn, M., Weiland, P., & Amon, B. (2006). Mitigation of greenhouse gas emissions by anaerobic digestion of cattle slurry. *Agriculture, Ecosystems and Environment, 112*, 171–177.
Cleveland, C. C., Townsend, A. R., Schimel, D. S., Fisher, H., Howarth, R. W., Hedin, L. O., et al. (1999). Global patterns of terrestrial biological nitrogen (N_2) fixation in natural ecosystems. *Global Biogeochemical Cycles, 13*, 623–645.
Cocking, E. C. (2003). Endophytic colonization of plant roots by nitrogen-fixing bacteria. *Plant and Soil, 252*, 169–175.
Codispoti, L., Brandes, J., Christensen, J., Devol, A., Naqvi, S., Paerl, H., et al. (2001). The oceanic fixed nitrogenand nitrous oxide budgets:moving targets as we enter the anthropocene? *Scientia Marina, 65*, 85–105.
Cole, C. V., Steward, J. W. B., Ojima, D. S., Parton, W. J., & Schimel, D. S. (1989), Modeling land use effects of soil organic matter dynamics in the North American Great Plains. In *Challenges in dryland agriculture – a global perspective. Proceedings of the International Conference on Dryland farming*. Bushland, TX: Amarillo.
Cole, C. V., Flach, K., Lee, J., Sauerbeck, D., & Steward, B. (1993). Agricultural sources and sinks of carbon. *Water, Air and Soil Pollution, 70*, 111–122.
Cole, C. V., Cerri, C., Minami, K., Mosier, A., Rosenberg, N., & Sauerbeck, D. (1996). Agricultural options for mitigation of greenhouse gas emissions. In R. T. Watson, M. C. Zinyovera, & R. H. Moss (Eds.), *Climate change 1995 – impacts, adaptions and mitigation of climate change: Scientific-technical analyses. Contribution of Working Group II to the Second Assessment Report of the Intergovernmental Panel of Climate Change* (pp. 745–771). Cambridge: Cambridge University Press.
Cole, C. V., Duxbury, J. M., Freney, J. R., Heinemeyer, O., Minami, K., Mosier, A., et al. (1997). Global estimates of potential mitigation of greenhouse gas emissions by agriculture. *Nutrient Cycling in Agroecosystems, 49*, 221–228.

Cole, D. W., & Rapp, M. (1981). Elemental cycling in forests. In D. E. Reichle (Ed.), *Dynamic properties of forest ecosystems, International Biological Programme 23* (pp. 341–409). London: Cambridge University Press.

Coleman, K., & Jenkinson, D. S. (1996). RothC 26.3- a model for the turnover of carbon in soil. In D. S. Powlson, P. Smith, & J. U. Smith (Eds.), *Evaluation of soil organic matter models using existing long-term datasets* (pp. 237–246). Berlin: Springer.

Coleman, M. D., Dickson, R. E., & Isebrands, J. G. (2000). Contrasting fine-root production, survival and soil CO_2 efflux in pine and poplar plantations. *Plant and Soil*, 225, 129–139.

Collatz, G. J., Berry, J. A., & Clark, J. S. (1998). Effects of climate and atmospheric CO_2 partial pressure on the global distribution of C_4 grasses: Present, past, and future. *Oecologica, 114*, 441–454.

Conant, R. T., Paustian, K., & Elliott, E. T. (2001). Grassland management and conversion into grassland: Effects on soil carbon. *Ecological Applications, 11*, 343–355.

Conrad, R. (2002). Control of microbial methane production in wetland rice fields. *Nutrient Cycling in Agroecosystems, 64*, 59–69.

CTIC (Conservation Technology Information Center). (2000). 2000 National crop residue management survey, West Lafayette, IN.

Cooper, A. D. (1990). Nitrate depletion in the riparian zone and stream channel of a small headwater catchment. *Hydrobiology, 202*, 13–26.

Cornell, S., Rendell, A., & Jickels, T. (1995). Atmospheric inputs of dissolved organic nitrogen to the oceans. *Nature, 376*, 243–246.

Cortez, J., & Schnitzer, M. (1979). Purines and pyrimidines in soils and humic substances. *Soil Science Society of America Journal, 43*, 958–961.

Cotrufo, M. F., Ineson, P., & Rowland, A. P. (1994). Decomposition of tree leaf litters grown under elevated CO_2: Effect of litter quality. *Plant and Soil, 163*, 121–130.

Couteaux, M. M., Mousseau, M., Celerier, M. L., & Bottne, P. (1991). Increased atmospheric CO_2 and litter quality:decomposition of sweet chesnut leaf litter with animal food webs of different complexities. *Oikos, 61*, 54–64.

Craft, C. B., & Richardson, C. J. (1993). Peat accretion and phosphorus accumulation along an eutrophication gradient in the Northern Everglades. *Biogeochemistry, 22*, 133–156.

Crews, T. E. (1999). The presence of nitrogen fixing legumes in terrestrial communities: Evolutionary vs. ecological considerations. *Biogeochemistry, 46*, 233–246.

Cromarty, P. (1996). New Zealand country report. In R. Jaensch (Ed.), *Wetland conservation in the Pacific Islands region* (pp. 79–92). *Proceedings of the Regional Workshop on Wetland Protection and Sustainable Use in Oceania, Port Moresby, Papua New Guinea, 6–10 June 1994. Wetlands International – Asia, Pacific, Canberra, Publication No. 118.*

Cronan, C. S. (1985). Comparative effects of precipitation acidity on three forest soils: Carbon cycling responses. *Plant and Soil, 88*, 101–112.

Crosson, P., & Anderson, J. R. (1994). Demand and supply: Trends in global agriculture. *Food Policy, 19*, 105–119.

Crutzen, P. J. (1991). Methane's sinks and sources. *Nature, 350*, 380–381.

Crutzen, P. J. (1981). Atmospheric chemical processes of the oxides of nitrogen, including nitrous oxide. In C. C. Delwiche (Ed.), *Denitrification, nitrification and atmospheric N_2O* (pp. 17–44). New York: Wiley.

Crutzen, P. J., & Goldhammer J. D. (1993). Fire in the environment. In *The ecological, atmospheric and climatic importance of vegetation fires. Dahlem Konferenz* (pp. 245–266), 15–20 March, 1992, Berlin, ES 13. Chichester, UK: Wiley.

Curtin, D., Campbell, C. A., & Messer, D. (1996). Prediction of titratable acidity and soil sensitivity to pH change. *Journal of Environmental Quality, 25*, 1280–1284.

Cuttle, S. P., & Scholefield, D. (1995). Management options to limit nitrate leaching from grassland. *Journal of Contaminant Hydrology, 20*, 299–312.

D'Angelo, E. M., & Reddy, K. R. (1994). Diagenesis of organic matter in a wetland receiving hypereutrophic lake water, II. Role of inorganic electron acceptors in nutrient release. *Journal of Environmental Quality, 23*, 928–936.

Dalal, R. C. (1998). Soil microbial biomass– what do the numbers really mean? *Australian Journal of Experimental Agriculture, 38*, 649–665.

Dalal, R. C., & Mayer, R. J. (1986). Long-term trends in fertility of soils under continuous cultivation and cereal cropping in southern Queensland. III Distribution and kinetics of soil organic carbon in particle-size fractions. *Australian Journal of Soil Research, 24*, 293–300.

Dalal, R. C., Wang, W., Robertson, G. P., & Parton, W. J. (2003). Nitrous oxide emissions from Australian agricultural lands and mitigation options: A review. *Australian Journal of Soil Research, 41*, 165–195.

Danckwerts, P. V. (1970). *Gas-liquid reactions*. London: McGraw-Hill.

Dannenberg, S., & Conrad, R. (1999). Effect of rice plants on methane production and rhizospheric metabolism in paddy soils. *Biogeochemistry, 45*, 53–71.

Darrah, P. R. (1996). Rhizodeposition under ambient and elevated CO_2 levels. *Plant and Soil, 187*, 265–275.

da Silva, A. P., & Kay, B. D. (1997). Estimating the least limiting water range of soils from properties and management. *Soil Science Society of America Journal, 61*, 877–883.

Davidson, E. A. (1994). Climate change and soil microbial processes: Secondary effects are hypothesised from better known interacting effects. In M. D. A. Rounsevell, & P. J. Lovelend (Eds.), *Soil responses to climate change*. NATO ASI Series, Vol. 23, Heidelberg, Germany: Springer.

Davidson, E. A., & Ackerman, I. L. (1993). Changes in soil carbon inventories following cultivation of previously untilled soils. *Biogeochemistry, 20*, 161–164.

Davidson, E. A., & Kingerlee, W. (1997). A global inventory of nitric oxide emissions from soils. *Nutrient Cycling in Agroecosystems, 48*, 37–50.

Davidson, N. C. D., Laffoley, d'A., Doody, D., Way, J. P., Gordon, L. S., Key, R., Drake, C. M., Pienkowski, M. W., Mitchell, R., & Duff, K. L. (1991). *Nature conservation and estuaries in Great Britain*. Peterborough, UK: Nature Conservation Council.

Davis, R. C. (1981). Structure and function of two Antarctic terrestrial moss communities. *Ecological Monographs, 51*, 125–143.

Deaker, R., Roughley, R. I., & Kennedy, I. R. (2004). Legume seed inoculation technology – a review. *Soil Biology and Biochemistry, 36*, 1275–1288.

Deans, J. R., Molina, J. A. E., & Clapp, C. E. (1986). Models for predicting potentially mineralizable nitrogen and decomposition rate constants. *Soil Science Society of America Journal, 50*, 323–326.

De Boer, W., Tietema, A., Klein Goldewiek, P. J. A., & Laanbroek, H. J. (1992). The autotrophic ammonium-oxidizing community in a nitrogen-saturated acid forest soil in relation to pH-dependent nitrifying activity. *Soil Biology and Biochemistry, 24*, 229–234.

DeBusk, W. F., & Reddy, K. R. (1998). Turnover of detrital organic carbon in a nutrient-impacted Everglades marsh. *Soil Science Society of America Journal, 62*, 1460–1468.

DeBusk, W. F., White, J. R., & Reddy, K. R. (2001). Carbin and nitrogen dynamics in wetland soils. In M. J. Shaffer, M. Liwang, & S. Hansen, (Eds.), *Modeling carbon and nitrogen dynamics for soil management*. Boca Raton, FL: Lewis.

Decau, M. L., Simon, J. C., & Jacquet, A. (2004). Nitrate leaching under grassland as affected by mineral fertilization and cattle urine. *Journal of Environmental Quality, 33*, 637–644.

Deckers, J. A., Nachtergaele, F. O., & Spaargaren, O. C. (Eds.). (1998). *World reference base for soil resources – introduction*. Leuven, Belgium: Acco.

de Klein, C. A. M., Sherlock, R. R., Cameron, K. C., & van der Weerden, T. J. (2001). Nitrous oxide emissions from agricultural soils in New Zeland- a review of current knowledge and directions for future research. *Journal of the Royal Society of New Zealand, 31*, 543–574.

de Nobili, M., Contin, N., & Bookes, P. C. (2006). Microbial biomass dynamics in recently air-dried and rewetted soils compared to others stored air-dry for up to 103 years. *Soil Biology and Biochemistry, 38*, 2871–2881.

Degens, B. P., & Harris, J. A. (1997). Development of a physiological approach to measuring the catabolic diversity of soil microbial communities. *Soil Biology and Biochemistry, 29*, 1309–1320.

Degryze, S., Six, J., Paustian, K., Morris, S., Paul, E. A., & Merckx, R. (2004). Soil organic carbon pool changes following land-use conversions. *Global Change Biology, 10*, 1120–1132.
Del Grosso, S. J., Parton, W. J., Mosier, A. R., Ojima, D. S., Potter, C. S., Borken, W., et al. (2000). General CH_4 oxidation model and comparisons of CH_4 oxidation in natural and managed systems. *Global Biogeochemical Cycles, 14*, 999–1019.
Del Grosso, S. J., Parton, W. J., Mosier, A. R., Hartman, M. D., Brenner, J., Ojima, D. S., et al. (2001a). Simulated interaction of carbon dynamics and nitrogen trace gas fluxes using the DAYCENT model. In M. J. Schaffer, L. Ma, & S. Hansen (Eds.), *Modeling carbon and nitrogen dynamics for soil management* (pp. 303–332). Boca Raton, FL: CRC.
Del Grosso, S. J., Parton, W. J., Mosier, A., Hartman, M. D., Keough, C. A., Peterson, G. A., et al. (2001b). Simulated effects of land use, soil texture, and precipitation on N gas emissions using DAYCENT. In R. F. Follett, & J. L. Hatfeld (Eds.), *Nitrogen in the environment: Sources, problems, and management* (pp. 413–431). Amsterdam: Elsevier Science.
Del Grosso, S. J., Ojima, D. S., Parton, W. J., Mosier, A. R., Peterson, G. A., & Schimel, D. S. (2002). Simulated effects of dryland cropping intensification on soil organic matter and greenhouse gas exchanges using the DAYCENT ecosystem model. *Environmental Pollution, 116*, S75–S83.
Del Grosso, S. J., Mosier, A. R., Parton, W. J., & Ojima, D. S. (2005a). DAYCENT model analysis of past and contemporary soil N_2O and net greenhouse gas flux for major crops in the USA. *Soil and Tillage Research, 83*, 9–24.
Del Grosso, S. J., Parton, W. J., Mosier, A. R., Holland, E. A., Pendall, E., Schimel, D. S., et al. (2005b). Modeling soil CO_2 emissions from ecosystems. *Biogeochemistry, 73*, 71–91.
Del Grosso, S. J., Parton.W. J., Mosier, A. R., Walsh, M. K., Ojima, D. S., & Thornton, P. E. (2006). DAYCENT national-scale simulations of nitrous oxide emisions from cropped soils in the United States. *Journal of Environmental Quality, 35*, 1451–1460.
DeLaune, R. D., Feijtel, T. C., & Patrick, W. H. (1989). Nitrogen flows in Louisiana Gulf Coast salt marsh: Spatial considerations. *Biogeochemistry, 8*, 25–37.
Deleporte, S., & Tillier, P. (1999). Long-term effects of mineral amendments on soil fauna and humus in an acid beech forest floor. *Forest Ecology and Managenment, 118*, 245–252.
Delgado, J. A., Mosier, A. R., Valentine, D. W., Schimel, D. S., & Parton, W. J. (1996). Long-term N^{15} studies in a catena of the shortgrass steppe. *Biogeochemistry, 32*, 41–52.
Delmas, R., Serca, D., & Jambert, C. (1997). Global inventory of NO_x sources. *Nutrient Cycling in Agroecosystems, 48*, 51–60.
Delschen, T. (1999). Impacts of long-term application of organic fertilizers on soil quality parameters in reclaimed loess soils of the Rhineland lignite mining area. *Plant and Soil, 213*, 43–54.
deMontigny, L. E., Preston, C. M., Hatcher, P. G., & Kögel-Knabner, I. (1993). Comparison of humus fractions from two ecosystem phases on Northern Vancouver Island using ^{13}C CPMAS NMR spectroscopy and CuO oxidation. *Canadian Journal of Soil Science, 73*, 9–25.
Denf, K., Six, J., Bossuyt, H., Frey, S. D., Elliott, E. T., Merckx, R., et al. (2001). Influence of wet-dry cycles on the interrelationship between aggregate, particulate organic matter, and microbial community dynamics. *Soil Biology and Biochemistry, 33*, 1599–1611.
Denier van der Gon, H. A. C., & Neue, H. U. (1994). Impact of gypsum application on the methane emission from a wetland ricefield. *Global Biogeochemical Cycles, 8*, 127–134.
Denier van der Gon, H. A. C., & Neue, H. U. (1995). Influence of organic matter incorporation on the methane emission from a wetland rice field. *Global Biogeochemical Cycles, 11*, 11–22.
Denier van der Gon, H. A. C., van Bodegom, P. M., Houweling, S., Verburg, P. H., & van Breemen, N. (2000). Combining upscaling and downscaling of methane emissions from rice fields: Methodologies and preliminary results. *Nutrient Cycling in Agroecosystems, 58*, 285–301.
Denman, K. L., Brasseur, G., Chidthaisong, A., Ciais, P., Cox, P. M., Dickinson, R. E., et al. (2007). Couplings between changes in the climate system and biogeochemisry. In S. Solomon, D. Qin, M. Manning, Z. Chen, M. Marquis, K. B. Averyt et al. (Eds.), *Climate change 2007: The physical science basis* (pp. 499–587). *Contribution of Working Group I to the Fourth Assessment Report of the Intergovernmental Panel on Climate Change*. Cambridge, UK/ New York: Cambridge University Press.

Dentener, F., Drevet, J., Lamarque, J. F., Bey, I., Eickhout, B., Fiore, A. M., et al. (2006). Nitrogen and sulphur deposition on regional and global scales: A multi-model evaluation. *Global Biogeochemical Cycles, 20*, doi:1029/2005GB002672.

Dentener, F. J., & Crutzen, P. J. (1994). A three-dimensional model of the global ammonia cycle. *Journal of Atmospheric Chemistry, 19*, 331–369.

Deppenmeir, U., Muller, V., & Gottschalk, G. (1996). Pathways of energy conservation in methanogenic archae. *Archieves MIcrobiology, 165*, 149–163.

De Ruiter, P. C., Moore, J. C., Zwart, K. W., Bouwman, L. A., Hassink, J., Bloem, J., et al. (1993). Simulation of nitrogen mineralization in the below ground food webs of two winter-wheat fields. *Journal of Applied Ecology, 30*, 95–106.

Des Marais, D. J., Strauss, H., Summons, R. E., & Hayes, J. M. (1992). Carbon isotope evidence for the stepwise oxidation of the Proterozoic environment. *Nature, 359*, 605–609.

Detwiler, R. P. (1986). Land use change and the global carbon cycle: The role of the tropical soils. *Biogeochemistry, 2*, 67–93.

Devol, A. H. (1991). Direct Measurement of Nitrogen gas fluxes from continental shelf sediments. Nature 349, 319–322.

Di, H. J., & Cameron, K. C. (2000). Calculating nitrogen leaching losses and critical nitrogen application rates in dairy pasture systems using a semi-empirical model. *New Zealand Journal of Agricultural Research, 43*, 139–147.

Di, H. J., & Cameron, K. C. (2002). Nitrate leaching in temperate agroecosystems: Sources, factors and mitigating strategies. *Nutrient Cycling in Agroecosystems, 46*, 237–256.

Di, H. J., Cameron, K. C., Moore, S., & Smith, N. P. (1998). Nitrate leaching and pasture yields following the application of dairy shed effluent or ammonium fertilizer under spray or flood irrigation:results of a lysimeter study. *Soil Use and Management, 14*, 209–214.

Dick, W. A. (1983). Organic carbon, nitrogen and phosphorus concentrations and pH in soil profiles as affected by tillage intensity. *Soil Science Society of America Journal, 47*, 102–107.

Diem, H. G., & Dommergues, Y. R. (1990). Current and potential uses and management of casuarinaceae in the tropics and subtropics. In C. R. Schnitzer, & J. D. Tjepkema (Eds.), *The biology of Frankia and Actinorhizal plants*. New York: Academic.

Di-Giovanni, C., Disnar, J. R. Campy, M. and Macaire, J. J. (1999). Variability of the ancient organic supply in modern humus. *Analysis* 27, 398–402.

Dilly, O., & Blume, H. P. (1998). Indicators to assess sustainable land use with reference to soil microbiology. *Advances in GeoEcology, 31*, 29–39.

Dilly, O., & Munch, J. C. (1998). Ratios between estimates of microbial biomass content and microbial activity in soils. *Biology and Fertility of Soils, 27*, 374–379.

Dilly, O., Blume, H. P., & Munch, J. C. (2003). Soil microbial activities in Luvisols and Anthrosols during 9 years of region-typical tillage and fertilization practices in northern Germany. *Biogeochemistry, 65*, 319–339.

Dise, N. B., Matzner, E., & Gundersen, P. (1998). Synthesis of nitrogen pools and fluxes from European forest ecosystems. *Water, Air, and Soil Pollution, 105*, 143–154.

Duiker, S. (2005). Carbon sequestration and farming. http://fcn.agronomy.psu.edu

Dixon, R., Schroeder, P., & Winjum, J. (Eds.) (1991). *Assessment of promising forest management practices and technologies for enhancing the conservation and sequestration of atmospheric carbon and their costs at the site level.* Report of the US Environmental Protection Agency, EPA/600/3-91/067, Environmental Research Laboratory, Corvallis, OR.

Dixon, R., Winjum, J., Andrasko, K., Lee, J., & Schroeder, P. (1994). Integrated land-use systems assessment of promising agroforestry and alternative land-use practices to enhance carbon conservation and sequestration. *Climatic Change, 30*, 1–23.

Dixon, R. K., Andrasko, K. J., Sussman, F. G., Lavinson, M. A., Trexler, M. C., & Vinson, T. S. (1993). Forest sector carbon offset projects: Near-term opportunities to mitigate greenhouse gas emissions. *Water, Air, and Soil Pollution, 70*, 561–577.

Dixon, R. K., Brown, S., Houghton, R. A., Solomon, A. M., Trexler, M. C., & Wisniewski, J. (1994). Carbon pools and flux in global forest ecosystems. *Science, 263*, 185–190.

Dlugokencky, E. J., Masa rie, K. A., Lang, P. M., & Tans, P. P. (1998). Continuing decline in the growth rate of the atmospheric methane burden. *Nature, 393*, 447–450.

Dobbie, K. E., McTaggart, I. P., & Smith, K. A. (1999). Nitrous oxide emission from intensive agricultural systems: Variations between crops and seasons; key driving variables; and mean emission factors. *Journal of Geophysical Research, 104*, 26891–26899.

Dobermann, A., & Witt, C. (2000). The potential impact of crop intensification on carbon and nitrogen cycling in intensive rice systems. In G. J. D. Kirk, & D. C. Olk (Eds.), *Carbon and nitrogen dynamics in flooded soils* (pp. 1–26). Makati City, Philippines: International Rice Research Institute.

Doram, D. R., & Evans, L. J. (1983). Native fixed ammonium and fixation of added ammonium in relation to clay mineralogy in some Ontario soils. *Canadian Journal of Soil Science, 63*, 631–639.

Doran, J. W. (1987). Microbial biomass and mineralizable nitrogen distribution in non-tillage and ploughed soils. *Biology and Fertility of Soils, 5*, 68–75.

Doran, J. W., & Parkin, T. B. (1996). Quantitative indicators of soil quality: A minimum data set. In J. W. Doran, & A. J. Jones (Eds.), *Methods for assessing soil quality*. Madison, WI: Soil Science Society of America, Special Publication 49.

Doran, J. W., Mielke, L. N., & Stamatiadis, S. (1988). Microbial activity and N cycling as regulated by soil water-filled pore space. In *Proceedings of the 11th International Soil Tillage Research Organization*, Paper No. 132, ISTRO, July 11–15, 1988, Edinburgh, UK.

Dörr, H., Katruff, L., & Levin, I. (1993). Soil texture parameterization of the methane uptake in aerated soils. *Chemosphere, 26*, 697–713.

Dou, H., & Steffens, D. (1995). Recovery of ^{15}N labelled urea as affected by fixation of ammonium by clay minerals. *Zeitschrift für Pflanzenernährung und Bodenkunde, 158*, 351–354.

Drake, B. G. (1992). A field study of the effects of elevated CO_2 on ecosystem processes in a Chesapeake Bay wetland. *Australian Journal of Botany, 40*, 579–595.

Drees, L. R., Wilding, L. P., Smeck, N. E., & Senkayi, A. L. (1989). Silica in soils: Quartz, and disordered silica polymorphs. In J. B. Dixon, & S. B. Weeds (Eds.), *Minerals in soil environments* (pp. 471–552). Madison, WI: Soil Science Society of America.

Driessen, P. M., & Dudal, R. (Eds.) (1991). *The major soils of the world*. Zutphen, The Netherlands: Koninklijke Wöhrmann B.V.

Driessen, P. M., Deckers, J., Spaargaren, O., & Nachtergaele, F. (Eds.). (2001). *Lecture notes of the major soils of the world*. World Soil Sources Report 94. Rome, Italy: Food and Agriculture Organization of the United Nations.

Drury, C. F., & Beauchamp, E. G. (1991). Ammonium fixation, release, nitrification and immobilization in high- and low-fixing soils. *Soil Science Society of America Journal, 55*, 980–985.

Drury, C. F., Beauchamp, E. G., & Evans, L. J. (1989). Fixation and immobilization of recently applied $^{15}NH_4$ in selected Ontario and Quebec soils. *Canadian Journal of Soil Science, 69*, 391–400.

Dšbereiner, J., Reis, V. M., Paula, M. A., & Olivares, F. (1993). Endophytic diazotrophs in sugarcane, cereals and tuber plants. In R. Palacios, J. Mora, & W. E. Newton (Eds.), *New horizons in nitrogen fixation* (pp. 671–676). Boston, MA: Kluwer.

Duan, S., Zhang, S., & Huang, H. (2000). Transport of dissolved inorganic nitrogen from the major rivers to estuaries in China. *Nutrient Cycling in Agroecosystems, 57*, 13–22.

Dudek, D., & LeBlanc, A. (1990). Offsetting new CO_2 emissions: A rational first greenhouse policy step. *Contemporary Policy Issues, 8*, 29–42.

Dueck, T. A., de Visser, R., Poorter, H., Presijn, S., Gorissen, A., de Visser, W., et al. (2007). No evidence for substantial aerobic methane emission by terrestrial plants: A ^{13}C -labelling approach. *New Phytologist, 175*, 29–35.

Dunfield, P., Knowles, R., Dumont, R., & Moore, T. R. (1993). Methane production and consumption in temperate and subarctic peat soils- response to temperature and pH. *Soil Biology and Biochemistry, 25*, 321–326.

Dunsmore, H. E. (1992). A geological perspective on global warming and the possibility of carbon dioxide removal as calcium carbonate mineral. *Energy Conservation Management, 33*, 565–572.

Duxbury, J. M., Smith, M. S., & Doran, J. W. (1989). Soil organic matter as a source and a sink of plant nutrients. In D. C. Coleman, J. M. Oades, & G. Uehara (Eds.), *Dynamics of soil organic matter in tropical ecosystems* (pp. 33–67). Honolulu, HI: University of Hawaii Press.
Dyson, T. (1996). *Population and food*. London: Routledge.
Eden, M. J., Furley, P. A., McGregor, D. F. M., Milliken, W., & Ratter, J. A. (1991). Effect of forest clearance and burning on soil properties in northern Roraima, Brazil. *Forest Ecology and Management, 38*, 283–290.
Ehhalt, D. H. (1974). The atmospheric cycle of methane. *Tellus, 26*, 58–70.
Ehhalt, D. H. (1999). Gas phase chemistry of the troposhere. In *Global aspects of atmospheric chemistry* (pp. 21–109). Deutsche Bunsen-Gesellschaft für Physicalische Chemie e.V. New York: Springer.
Ehleringer, J. R., & Cerling, T. E. (2001). Photosynthetic pathways and climate. In E. D. Schulze, M. Heimann, S. Harrison, E. Holland, J. Lloyd, I. C. Prentice, et al. (Eds.), *Global biogeochemical cycles in the climate system* (pp. 267–277). San Diego, CA: Academic.
Ehleringer, J. R., & Monson, R. K. (1993). Evolutionary and ecological aspects of photosynthetic pathway variation. *Annual Reviews of Ecological Systematics, 24*, 411–439.
Ehleringer, J. R., Sage, R. F., Flanagan, L. B., & Pearcy, R. W. (1991). Climate change and the evolution of C4 photosynthesis. *Trends in Ecological Evolution, 6*, 95–99.
EIA (Environmental Information Administration). (1998). Emissions of Greenhouse Gases in the United States. http://www.eia.doe.gov/oiaf/1605/gg98rpt/preface.html
EIA (Environmental Information Administration). (2000). Voluntary reporting of greenhouse gases 2000. http://www.eia.doe.gov/oiaf/1605/vrrpt/chapter4.html
Eichhorn, J., & Huettermann, A. (1999). Mechanisms of humus dynamics and nitrogen mineralization. In N. Rastin, & J. Bauhus (Eds.), *Going underground – ecological studies in forest soils* (pp. 239–277). Trivandrum, India: Research Signpost.
Elliott, E. T., Anderson R. V., Coleman D. C. & Cole (1980). Habitable pore space and microbial trophic interactions. Oikos 35, 327–335.
Elliott, E. T. (1986). Aggregate structure and carbon, nitrogen, and phosphorus in native and cultivated soils. *Soil Science Society of America Journal, 50*, 627–633.
Elliott, E. T., Paustian, K., & Frey, S. D. (1996). Modeling the measurable or measuring the modelable: A hierarchical approach to isolating meaningful soil organic matter fractionations. In D. S. Powlson, P. Smith, & J. U. Smith (Eds.), *Evaluation of soil organic matter models using existing long-term datasets* (pp. 161–179). Heidelberg, Germany: Springer.
Emerson, W. W., Foster, R. C., & Oades, J. M. (1986). Organo-mineral complexes in relation to soil aggregation and structure. In P. M. Huang, & M. Schnitzer, (Eds.), *Interactions of soil minerals with natural organics and microbes*. Soil Science Society of America Special Publication 17, MA, WI: Soil Science Society of America.
Engel, T., & Priesack, E. (1993). Expert-N, a building block system of nitrogen models as a resource for advice, research, water management and policy. In H. J. P. Eijsackers, & T. Hamers (Eds.), *Integrated soil and sediment research: A basis for proper protection* (pp. 503–507). Dordrecht, The Netherlands: Kluwer.
Engler, R. M., & Patrick, W. H. J. (1974). Nitrate removal from floodwater overlying flooded soils and sediments. *Journal of Environmental Quality, 3*, 409–412.
Enquete Commission. (1995). *Protecting our green earth. How to manage global warming through environmentally sound farming and preservation of the world's forests* (p. 683). Bonn, Germany: Eco-nomia Verlag.
EPA (Environmental Protection Agency) (2006). Land use, land use change and forestry. http://epa.gov./climatechange/emissions/downloads06/06LULUCF.pdf.
Epstein, H. E., Lauenroth, W. K., Burke, I. C., & Coffin, D. P. (1997). Productivity patterns of C_3 and C_4 functional types in the US Great Plains. *Ecology, 78*, 722–731.
Eriksson, K. E., & Johnsrud, S. C. (1982). Mineralization of carbon. In R. G. Burns, & J. H. Slater (Eds.), *Experimental microbial ecology* (pp. 134–153). Oxford: Blackwell Scientific.
Erisman, J. W., de Vries, W., Kros, H., Oenema, O., van der Eerden, L., van Zeijts, H., et al. (2001). An outlook for a national integrated nitrogen policy. *Environmental Science Policy, 4*, 87–95.

Esser, G. (1987). Sensitivity of global carbon pools and fluxes to human and potential climatic impacts. *Tellus, 39B*, 245–260.
Esser, G. (1989). Global land-use changes from 1860 to 1980 and future projections to 2500. *Ecological Modelling, 44*, 307–316.
Eswaran, H., Van den Berg, E., Reich, P., & Kimble, J. (1995). Global soil carbon reserves. In R. Lal, J. Kimble, E. Levine, & E. Steward (Eds.), *Soils and global change* (pp. 27–43). Boca Raton, FL: Lewis.
Eswaran, H. E., Berg, V. D., & Reich, P. (1993). Organic carbon in soils of the world. *Soil Science Society of America Journal, 57*, 192–194.
Evans, C. D., Reynolds, B., Jenkins, A., Helliwell, R. C., Curtis, C. J., Goodale, C. L., et al. (2006). Evidence that soil carbon pool determines susceptibility of semi-natural ecosystems to elevated nitrogen leaching. *Ecosystems, 9*, 453–462.
Evers, G. W. (2001). Legume nitrogen fixation and transfer. Available online http://overton.tamu.edu/clover/cool/nfix.htm
Falloon, P. D., & Smith, P. (2000). Modelling refractory soil organic matter. *Biology and Fertility of Soils, 30*, 388–398.
Fang, C., & Moncrieff, B. (1999). A model for soil CO_2 production and transport 1: Model development. *Agriculture and Forest Meteorology, 95*, 225–236.
FAO (Food and Agriculture Organisation). (1971–1981). *Soil map of the world 1:5,000,000*, Vols. 1–10. Rome: FAO-UNESCO.
FAO (Food and Agriculture Organisation). (1988). *Soil map of the world*. Revised Legend. Rome: FAO-UNESCO.
FAO (Food and Agriculture Organisation). (1990). *Guidelines for soil profile description* (3rd ed.). Rome: Soil Resources, Management and Conservation Service, Land and Water Development Division.
FAO (Food and Agriculture Organisation). (1991). *Digitized soil map of the world*. World Soil Resources Report 67, Rome: FAO-UNESCO.
FAO (Food and Agriculture Organisation). (1995). Planning for sustainable use of land resources: Towards a new approach. In W. G. Sombroek, & D. Sims (Eds.), *Land and water bulletin No. 2*. Rome: FAO.
FAO (Food and Agriculture Organization). (1996). *Forest resources assessment 1990. Survey of tropical forest cover and study of change processes*. FAO Forestry paper 130, Rome: FAO.
FAO (Food and Agriculture Organisation). (1998a). *World reference base for soil resources, ISSS, ISRIC, FAO*. World Soil Resources Report 84, Rome: FAO-UNESCO, FAO, 88 p.
FAO (Food and Agriculture Organisation). (1998b). FAOSTAT database result, http://www.fao.org/servlet ff9712-e.htm
FAO (Food and Agriculture Organization). (2001). Report of the 21st session of the International Poplar Commission and 40th session of its Executive Committee, Portland, OR 24–28, September 2000.
FAO (Food and Agriculture Organization). (2006). *Global forest resources assessment 2005. Progress towards sustainable forest management*. FAO Forestry Paper 147, 320 p.
FAO (Food and Agriculture Organisation). (2007). *State of the world's forests 2007* (p. 144). Rome: Food and Agricultural Organization of the United Nations.
FAOSTAT. (2006). FAOSTAT Agricultural Data. Available at:http://faostat.fao.org
Farmer, V. C., Russell, J. D., & Berrow, M. L. (1980). Imogolite and proto-imogolite allophane in spodic horizons: Evidence for a mobile aluminium silicate complex in podzol formation. *Journal of Soil Science, 31*, 673–684.
Farmer, V. C., McHardy, W. J., Robertson, L., Walker, A., & Wilson, M. J. (1985). Micromorphology and sub-microscopy of allophane and imogolite in a podzol Bs horizon: Evidence for translocation and origin. *Journal of Soil Science, 36*, 87–95.
Farquhar, G. D., Firth, P. M., Weselaar, R., & Weir, B. (1980). On the gaseous exchange of ammonia between leaves and the environment: Determination of the ammonia compensation point. *Plant Physiology, 66*, 710–714.

Fearnside, P. M., & Barbosa, R. I. (1998). Soil carbon changes from conversion of forest to pasture in Brazilian Amazonia. *Forest Ecology and Management, 108*, 147–166.

Fearnside, P. M., Lashof, D. A., & Moura-Costa, P. (2000). Accounting for time in mitigating global warming through land-use change and forestry. *Mitigation and Adaptation Strategies for Global Change, 5*, 239–270.

Feigin, A., & Yaalon, D. H. (1974). Non-exchangeable ammonium in soils of Israel and its relation to clay and parent material. *Journal of Soil Science, 25*, 384–397.

Feller, C., Balesdent, J., Nicolardot, B., & Cerri, C. (2001). Approaching "functional" soil organic matter pools through particle-size fractionation: Examples for tropical soils. In R. Lal, J. M. Kimble, R. F. Follett, & B. A. Stewart (Eds.), *Assessment methods for soil carbon* (pp. 53–67). Boca Raton, FL: Lewis.

Ferm, M. (1998). Atmospheric ammonia and ammonium transport in Europe and critical loads: A review. *Nutrient Cycling in Agroecosystems, 51*, 5–17.

Ferry, J. G. (1992). Methane from acetate. *Journal of Bacteriology, 174*, 5489–5495.

Field, C. B., & Raupach, M. R. (2004). *The global carbon cycle – integrating humans, climate, and the natural world* (p. 526). Washington, DC: Island.

Filip, Z. (2002). International approach to assessing soil quality by ecologically-related biological prrameters. *Agriculture, Ecosystems and Environment, 88*, 169–174.

Finck, A. (1979). *Dünger und Düngung*. Weinheim, Germany: Verlag Chemie.

Fioretto, A., Papa, S., Sorrentino, G., & Fuggi, A. (2001). Decomposition of *Cistus incanus* leaf bitter in a Mediterranean maquis ecosystem: Mass loss, microbial enzyme activities and nutrient changes. *Soil Biology and Biochemistry, 33*, 311–321.

Firestone, M. K., & Davidson, E. A. (1989). Microbiological basis of NO and N_2O production and consumption in soil. In M. O. Andreae, & D. S. Schimel (Eds.), *Exchange of trace gases between terrestrial ecosystems and the atmosphere* (pp. 7–21). Chichester, UK: Wiley.

Fischer, G., & Heilig, G. K. (1997). Population momentum and the demand on land and water resources. *Philosophical Transactions of the Royal Society of London*, B *352*, 869–889.

Fischer, H., Bens, O., & Hüttl, R. (2002). Changes in humus form, humus stock and soil organic matter distribution caused by forest transformation in the North-Eastern lowlands of Germany. *Forstwirtschaftliches Centralblatt, 121*, 322–334.

Fischer, W. R., Pfanneberg, T., Niederbudde, E. A., & Medina, R. (1981). Transformation of ^{15}N-labelled ammonium in two soils differing in NH_4^+-fixing capacity. *Journal of Soil Science, 32*, 409–418.

Fisher, M. J., Rao, I. M., Ayarza, M. A., Lascano, C. E., Sanz, J. L., Thomas, R. J., et al. (1994). Carbon storage by introduced deep rooted grasses in the South American Savannas. *Nature, 371*, 236–238.

Fisher, M. J., Rao, I. M., Lascano, C. E., Sanz, I. J., Thomas, R. J., Vera, R. R., et al. (1995). Pasture soils as carbon sink. *Nature, 376*, 472–473.

yFlaig, W. (1958). Die Chemie der organischer Stoffe in Boden und deren physiologische Wirkung. Verhandlungen, II Urteil IV. *Bodenkunde Gesellschaft, 2*, 23–45.

Flaig, W. (1964). Chemische Untersuchungen an Huminstoffen. *Zeitschrift für Chemie, 4*, 253–265.

Flaig, W. (1988). Generation of model chemical precursors. In F. H. Frimmel, & R. F. Christman (Eds.), *Humic substances and their role in environment* (pp. 75–92). New York: Wiley.

Fleige, H., & Meyer, B. (1975). Mineralisch fixiertes Ammonium in jungpleistozänen Sedimenten Norddeutschlands und ihren fossilen und holozänen Böden: Ein Indikator für litho- oder pedogenetische Prozesse und Herkünfte. *Göttinger Bodenkundliche Berichte, 34*, 315–328.

Fließbach, A., & Mäder, P. (2000). Microbial biomass and size-density fractions differ between soils of organic and conventional agricultural systems. *Soil Biology and Biochemistry, 32*, 757–768.

Flinn, J. C., & Marciano, V. P. (1984). Rice straw and stubble management. In *organic matter and rice* (pp. 593–611). Los Baños, Philippines: International Rice Research Institute.

Focht, D. D. (1974). The effect of temperature, pH and aeration on the production of nitrous oxide and gaseous nitrogen- a zero order kinetic model. *Soil Science, 118*, 173–179.

Focht, D. D., & Verstraete, W. (Eds.) (1977). *Biochemical ecology of nitrification and denitrification* (Vol. 1, pp. 135–214). New York: Plenum.

Fog, K. (1988). The effect of added nitrogen on the rate of decomposition of organic matter. *Biological Reviews, 63*, 433–462.

Forster, P., Ramaswamy, V., Artaxo, P., Berntsen, T., Betts, R., Fahey, D. W., et al. (2007). Changes in atmospheric constituents and in radiative forcing. In S. Solomon, D. Qin, M. Manning, Z. Chen, M. Marquis, K.B. Averyt, et al. (Eds.), *Climate change 2007: The physical science basis* (pp. 129–234). Contribution of Working Group I to the Fourth Assessment Report of the Intergovernmental Panel on Climate Change. Cambridge, UK/New York: Cambridge University Press.

Fortin, M. C. (1993). Soil temperature, soil water, and no-till corn development following in row residue removal. *Agronomy Journal, 85*, 571–576.

Fowler, D. (2002). Pollutant deposition and uptake by vegetation. In J. N. B. Bell, & M. Treshow (Eds.), *Air pollution and plant life* (pp. 42–67). Chichester, UK: Wiley.

Fox, R. L. (1980). Soils with variable charge: Agronomic and fertility aspects. In B. K. G. Theng (Ed.), *Soils with variable charge* (pp. 195–224). Lower Hutt, New Zealand: New Zealand Society of Soil Science.

Francez, A. J., & Vasander, H. (1995). Peat accumulation and peat decomposition after human disturbance in French and Finnish mires. *Oecologia Plantarum, 16*, 599–608.

Francis, G. S. (1995). Management practices for minimising nitrate leaching after ploughing temporary pastures in Canterbury, New Zealand. *Journal of Contaminant Hydrology, 20*, 313–327.

Frankenberg, C., Meirink, J. F., van Weele, M., Platt, U., & Wagner, T. (2005). Assessing methane emissions from global space-borne observations. *Science, 308*, 1010–1014.

Franko, U., Oelschlägel, B., & Schenk, S. (1995). Simulation of temperature, water and nitrogen dynamics using the model CANDY. *Ecological Modelling, 81*, 213–222.

Frede, H. G., Beisecker, R., & Gäth, S. (1994). Long-term impacts of tillage on the soil ecosystem. *Journal of Plant Nutrition and Soil Science, 157*, 197–203.

Freibauer, A., & Schrumpf, M. (2006). Ergebnisse aus dem EU-Projekt CarboEurope zu Fragen der standortgerechten Bodennutzung und des Klimaschutzes. In J. Mathews, St. Marahrens, & S. Schmidt (Eds.), *UBA-Workshop: Klimaänderungen – Herausforderungen für den Bodenschutz* (pp. 137–147) 28 and 29 September 2005. Umweltbundesamt Dessau.

Freney, J. R. (1997). Emission of nitrous oxide from soils used for agriculture. *Nutrient Cycling in Agroecosystems, 49*, 1–6.

Freney, J. R., Denmead, O. T., Watanabe, I., & Simpson, J. R. (1978). Soil as a source or sink for atmospheric nitrous oxide. *Nature, 273*, 530–532.

Freney, J. R., Denmead, O. T., Watanabe, I., & Craswell, E. T. (1981). Ammonia and nitrous oxide losses following applications of ammonium sulfate to flooded rice. *Australian Journal of Agricultural Research, 32*, 37–45.

Frolking, S., & Crill, P. (1994). Climate controls on temporal variability of methane flux from a poor fen in southeastern New Hampshire: Measurements and modeling. *Global Biogeochemical Cycles, 8*, 385–397.

Frolking, S., Bubier, J. L., Moore, T. R., Ball, T., Bellisario, L. M., Bhardwaj, A., et al. (1998). The relationship between ecosystem productivity and photosynthetically active radiation for northern peatlands. *Global Biogeochemical Cycles, 12*, 115–126.

Frolking, S. E., Mosier, A. R., Ojima, D. S., Li, C., Parton, W. J., Potter, C. S., et al. (1998). Comparison of N_2O emissions from soils at three temperate agricultural sites: Simulations of year-round measurements by four models. *Nutrient Cycling in Agroecosystems, 52*, 77–105.

Fuchs, W. (1931). *Die Chemie der Kohle*. Berlin: Springer.

Furrer, O. J., & Stauffer, W. (1983). Influence of sewage sludge application on physical properties and its contribution to the humus balance. In G. Catroux, P. L'Hermite, & E. Suess (Eds.), *The influence of sewage sludge application on physical and biological properties of soils* (pp. 65–74). Dordrecht, The Netherlands/Holland: D. Reidel Publication Company.

Furukawa, Y., & Inubushi, K. (2002). Feasible supression technique of methane emission from paddy soil by iron amendment. *Nutrient Cycling in Agroecosystems, 64*, 193–201.

Gabrielle, B., Laville, P., Duval, O., Nicoullaud, B., Germon, J. C., & Hénault, C. (2006). Process-based modeling of nitrous oxide emissions from wheat-cropped soils at the subregional level. *Global Biogeochemical Cycles, 20*, GB4018, doi 10:1029/2006GB002686.

Galbally, I. E., & Gillett, R. W. (1988). Processes regulating nitrogen compounds in tropical atmosphere. In H. Rodhe, & R. Herrera, (Eds.), *Acidification in tropical countries* (pp. 73–115). Chichester, UK: Wiley.

Gale, P. M., Devai, I., Reddy, K. R., & Graetz, D. A. (1993). Denitrification potential of soils from constructed and natural wetlands. *Ecological Engineering, 2*, 119–130.

Galloway, J. N. (1995). Acid deposition: Perspctive in time and space. *Water, Air, and Soil Pollution, 85*, 15–24.

Galloway, J. N. (2000). Nitrogen mobilization in Asia. *Nutrient Cycling in Agroecosystems, 57*, 1–12.

Galloway, J. N., & Cowling, E. B. (2002). Reactive nitrogen and the world: Two hundred years of change. *Ambio, 31*, 64–77.

Galloway, J. N., Schlesinger, W. H., Levy, H. I., Michaels, A., & Schnoor, J. L. (1995). Nitrogen fixation: Anthropogenic enhancement-environment response. *Global Biogeochemical Cycles, 9*, 235–252.

Galloway, J. N., Aber, J. D., Erisman, J. W., Seitzinger, S. P., Howarth, R. W., Cowling, E. B., et al. (2003). The nitrogen cascade. *Bioscience, 53*, 341–356.

Galloway, J. N., Dentener, F. J., Capone, D. G., Boyer, E. W., Howarth, S. P., Seitzinger, S. P., et al. (2004). Nitrogen cycles: Past, present, and future. *Biogeochemistry, 70*, 153–226.

Gansen, R. (1972). *Bodengeographie* (p. 325), Stuttgart, Germany: K.F. Koehler Verlag.

Garcia, F. O., & Rice, C. W. (1994). Microbial biomass dynamics in tallgrass prairie. *Soil Science Society of America Journal, 58*, 816–823.

Garland, J. L. (1996). Patterns of potential C source utilization by rhizosphere communities. *Soil Biology and Biochemistry, 28*, 55–63.

Garten, C. T., & Wullschleger, S. D. (2000). Soil carbon dynamics beneath switchgrass as indicated by stable isotope analysis. *Journal of Environmental Quality, 29*, 645–653.

Garz, J. (1995) C- und N-Umsetzungen in Dauerversuchen auf Sandlöss-Braunschwarzerde in Halle. In M. Körschens, & E. G. Mahn (Eds.), *Strategien zur Regeneration belasteter Agrarökosysteme des mitteldeutschen Schwarzerdegebietes* (pp. 463–498). Stuttgart/Leipzig, Germany: Teubner.

Gäth, S., Anthony, F., Becker, K. W., Geries, H., Höper, H., Kersebaum, K. C., et al. (1997). Bewertung des standörtlichen Denitrifikations- und Immobilisationspotenzials von Böden. *Mitteilungen der Deutschen Bodenkundlichen Gesellschaft, 85*, 1373–1376.

Gaunt, J. L., Sohi, S. P., Yang, H., Mahieu, N., & Arah, J. R. M. (2001). A procedure for isolating soil organic matter fractions suitable for modelling. In R. M. Rees, B. C. Ball, C. D. Campbell, & C. A. Watson (Eds.), *Sustainable management of soil organic matter* (pp. 83–90). Wallingford, UK: CABI.

Génermont, S., & Cellier, P. (1997). A mechanistic model for estimating ammonia volatilization from slurry applied to bare soil. *Agricultural and Forest Meteorology, 88*, 145–167.

Gerzabek, M. H., Haberhauer, G., & Kirchmann, H. (2001). Soil organic matter pools and carbon-13 natural abundance in particle-size fractions of a long-term agricultural field experiment receiving organic amendments. *Soil Science Society of America Journal, 65*, 352–358.

Ghani, A., Dexter, M., Kear, M., Lindsey, S., & Ledgard, S. (2006). *Influence of plant litter and animal excreta on leaching of dissolved organic nitrogen and carbon in pastoral soil.* In Abstracts 18th World Congress of Soil Science, Philadelphia.

Giddens, J. (1957). Rate of loss of carbon from Georgia soils. *Proceedings of the Soil Science Society of America, 21*, 513–515.

Gignoux, J., House, J., Hall, D., Masse, D., Nacro, H. B., & Abbadie, L. (2001). Design and test of a generic cohort model of soil organic matter decomposition: The SOMKO model. *Global Ecology and Biogeography, 10*, 639–660.

Gijsman, A. J. (1996). Soil aggregate stability and soil organic matter fractions under agropastoral systems established in native savanna. *Australian Journal of Soil Research, 34*, 891–907.

Gilliom, R. J., Alley, W. M., & Gurtz, M. E. (1995). Design of the national water quality assessment program: Occurrence and distribution of water quality condition. *US Geological Survey Circular, 1112*, 33.

Gil-Sotres, F., Trasar-Cepeda, C., Ciardi, C., Ceccanti, B., & Leiros, M. C. (1992). Biochemical measures of biological activity in very young mine soils. *Biology and Fertility of Soils, 13*, 25–30.

Gil-Sotres, F., Trasar-Cepeda, C., Leiros, M. C., & Seoane, S. (2005). Different approaches to evaluating soil quality using biochemical properties. *Soil Biology and Biochemistry, 37*, 877–887.

GLCCD (Global Land Cover Characteristics Database), 1998, Version 1.2. (1998). T. R. Loveland, B. C. Reed, J. F. Brown, D. O. Ohlen, Z. Zhu, L. Yang, et al. Development of a global land cover characteristics database and IGBP dISCover from 1 km AVHRR data. *International Journal of Remote Sensing, 21*, 1303–1330.

Glenn, E. P., Squires, V. R., Olsen, M. W., & Frye, R. J. (1993). Potential for carbon sequestration in the drylands. *Water, Air and Soil Pollution, 70*, 341–355.

Goh, K. M., Rafter, T. A., Stout, J. D., & Walker, T. W. (1976). The accumulation of soil organic matter and its carbon isotope composition in a chronosequence of soils developed on Aeolian sand in New Zealand. *Soil Science, 27*, 89–100.

Golchin, A., Baldock, J. A. and Oades, J. M. (1997). A model linking organic matter decomposition, chemistry, and aggregate dynamics, in: Soil Processes and the Carbon Cycle, R. Lal et al., ed., CRC Press, Boca Raton, pp 245–266.

Golchin, A., Baldock, J. A., & Oades, J. M. (1998). A model linking organic matter decomposition, chemistry and aggregate dynamics. In R. Lal, J. M. Kimble, R. F. Follett, & B. A. Steward (Eds.), *Soil processes and the carbon cycle* (pp. 245–266). Boca Raton, FL: CRC.

Gordon, A. A., & Millero, F. J. (1985). Adsorption mediated decrease in the biodegradation rate of organic compounds. *Microbial Ecology, 11*, 289–298.

Gorham, E. (1987). The ecology and biogeochemistry of *Sphagnum* bogs in central and eastern North America. In A. D. Laderman (Ed.), *Atlantic white cedar wetlands*. Boulder, CO: Westview.

Gorham, E. (1991). Northern peatlands: Role in the carbon cycle and probable responses to climatic warming. *Ecological Applications, 1*, 185–192.

Gorlach, E., & Grywnowicz, P. (1988). Distribution of various nitrogen forms in the soil profile and their relationship with nitrogen taken up by plants. *Polish Journal of Soil Science, 21*, 43–49.

Gosz, J. R., Likens, G. E., & Bormann, F. H. (1973). Nutrient release from decomposing leaf and branch litter in the Hubbard Brook forest. *New Hampshire Ecology Monograph, 43*, 173–191.

Goulding, K. W. T., Bailey, N. J., Bradbury, N. J., Hargreaves, P., Howe, M., Murphy, D. V., et al. (1998). Nitrogen deposition and its contribution to nitrogen cycling and associated soil processes. *New Phytologist, 139*, 49–58.

Gouyon, A., De Foresta, H., & Levang, P. (1993). Does jungle rubber deserve its name? An analysis of rubber agroforestry systems in southeast Sumatra. *Agroforestry Systems, 22*, 181–206

Grace, J., Malhi, Y., Lloyd, J., McIntyre, J., Miranda, A. C., Meir, P., et al. (1996). The use of eddy covariance to infer the net carbon dioxide uptake of Brazilian rain forest. *Global Change Biology, 2*, 209–217.

Graham, R. L., Turner, M. G., & Dale, V. H. (1990). How increasing CO_2 and climate change affect forests. *BioScience, 40*, 575–587.

Granli, T., & Bøckman, O. C. (1994). Nitrous oxide from agriculture. *Norwegian Journal of Agricultural Sciences. Supplement, 12*, 7–128.

Grant, R. F. (1995). Dynamics of water, carbon, and nitrogen in agricultural ecosystems:simulation and experimental validation. *Ecological Modelling, 81*, 169–181.

Grant, R. F. (1997). Changes in soil organic matter under different tillage and rotations:mathematical modeling in ecosystems. *Soil Science Society of America Journal, 61*, 1159–1175.

Grant, R. F. (1999). Simulation of methanotrophy in the mathematical model ecosys. *Soil Biology and Biochemistry, 31*, 287–297.

Green, P., Vörösmarty, C. J., Meybeck, M., Galloway, J. N., Petersen, B. J., & Boyer, E. W. (2004). Pre-industrial and contemporary fluxes of nitrogen through rivers:A global assessment based on typology. *Biogeochemistry, 68*, 71–105.

Green, R. E., Cornell, S. J., Scharlemann, J. P. W., & Balmford, A. (2005). Farming and the fate of wild nature. *Science, 307*, 550–555.

Green, R. N., Trowbridge, R. L., & Klinka, K. (1993). Towards a taxonomic classification of humus forms. *Forest Science Monographs, 29*, 1–49.

Greenblatt, J. B., & Sarmiento, J. L. (2004). Variability and climate feedback mechanisms in ocean uptake of CO_2. In C. B. Field, & M. R. Raupach (Eds.), *The global carbon cycle* (pp. 257–275). Washington, DC: Island.

Greenfield, L. G. (1991). Fixed ammonium in Antarctic rocks and soil and a possible cause of underestimation. *Soil Biology and Biochemistry, 23*, 397–399.

Greenland, D. J. (1971). Interactions between humic and fulvic acids and clays. *Soil Science, 3*, 34–41.

Greenland, D. J., & Ford, G. W. (1964). Separation of partially humified organic materials from soils by ultrasonic dispersion. *8th International Congress of Social Science* 3, 137–148. (pp. 137–148). Budapest, Hungary.

Greenland, D. J., & Oades, J. M. (1975). Saccharides. In J. E. Gieseking (Ed.), *Soil components: Vol. I. Organic components*. New York: Springer.

Gregorich, E. G., & Janzen, H. H. (1996). Storage of soil carbon in the light fraction and macroorganic matter. In M. R. Carter, & B. A. Stewart (Eds.), *Structure and organic matter storage in agricultural soils* (pp 167–190). Boca Raton, FL: CRC/Lewis.

Greogorich, E. G., Kachanoski, R. G., & Voroney, R. P. (1989). Carbon mineralization in soil size fractions after various amounts of aggregate disruption. *Journal of Soil Science, 40*, 649–659.

Gregorich, E. G., Carter, M. R., Angers, D. A., Monreal, C. M., & Ellert, B. H. (1994). Towards a minimum data set to assess soil organic matter quality in agricultural soils. *Canadian Journal of Soil Science, 74*, 367–385.

Gregorich, E. G., Ellert, B. H., & Monreal, C. M. (1995). Turnover and storage of corn residue carbon in a Gleysolic soil in eastern Canada. *Canadian Journal of Soil Science, 75*, 161–167.

Gregory, P. J., & Ingram, J. S. I. (2000). Global change and food and forest production: Future scientific challenges. *Agriculture, Ecosystems and Environment, 82*, 3–14.

Griffin, T. M., & Rabenhorst, M. C. (1989). Processes and rates of pedogenesis in some Maryland tidal marsh soils. *Science Society of America Journal, 53*, 862–870.

Griffith, S. M., & Schnitzer, M. (1989). Oxidative degradation of soil humic substances. In M. H. B. Hayes, P. MacCarthy, R. L. Malcolm, & R. S. Swift (Eds.), *Humic substances II. In search of structure* (pp. 69–98). New York: Wiley.

Grime, J. P., & Hodgson, J. G. (1987). Botanical contributions to contemporary ecological theory. *The New Phytologist, 106*, 283–295.

Grisi, B., Grace, C., Brookes, P. C., Benedetti, A., & Dell'Abate, M. T. (1998). Temperature effects on organic matter and microbial biomass dynamics in temperate and tropical soils. *Soil Biology and Biochemistry, 30*, 1309–1315.

Groffman, P. M., McDowell, W. H., Myers, J. C., & Merriam, J. L. (2001). Soil microbial biomass and activity in tropical riparian forests. *Soil Biology and Biochemistry, 33*, 1339–1348.

Grogan, P., & Matthews, R. (2002). A modelling analysis of the potential for soil carbon sequestration under short rotation coppice willow bioenergy plantations. *Soil Use and Management, 18*, 175–183.

Grubb, P. J. (1971). Interpretation of the "Massenerhebung" effect on tropical mountains. *Nature, 229*, 44–45.

Grubb, P. J., & Edwards, P. J. (1982). Studies of mineral cycling in a montane rain forest in New Guinea, III, The distribution of mineral elements in the above-ground material. *Journal of Ecology, 70*, 623–648.

Guggenberger, G., & Beudert, G. (1989). Zur Dynamik des gelösten Kohlenstoffs (DOC) in unterschiedlich immissionsbelasteten Waldstandorten. *Mitteilungen der Deutschan Bodenkundlichen Gesellschaft, 59*, 367–372.

Guggenberger, G., & Kaiser, K. (2003). Dissolved organic matter in soil: Challenging the paradigmof sorptive preservation. *Geoderma, 113*, 293–310.

Guggenberger, G., & Zech, W. (1993). Dissolved organic carbon control in acid forest soils of the Fichtelgebirge (FGR) as revealed by distribution patternsand structural composition analyses. *Geoderma, 59*, 109–129.

Guggenberger, G., Christensen, B. T., & Zech, W. (1994). Land-use effects on the composition of organic matter in particle-size separates of soil: I. Lignin and carbohydrate signature. *European Journal of Soil Science, 45*, 449–458.

Guggenberger, G., Frey, S., Six, J., Paustian, K., & Elliott, E. T. (1999). Bacterial and fungal cell-wall residues in conventional and no-tillage ecosystems. *Soil Science Society of America Journal, 63*, 1188–1198.

Gundersen, P., Callesen, I., & de Vries, W. (1998). Nitrate leaching in forest ecosystems is related to forest floor C/N ratios. *Environmental Pollution, 102*, 403–407.

Gundersen, P., Schmidt, I. K., & Raullund-Rasmussen, K. (2006). Leaching of nitrate from temperate forests- effects of air pollution and forest management. *Environmental Reviews, 14*, 1–57.

Guo, L. B., & Gifford, R. M. (2002). Soil carbon stocks and land use change: A meta analysis. *Global Change Biology, 8*, 345–360.

Gupta, P. K., Sharma, C., Bhattacharya, S., & Mitra, A. P. (2002). Scientific basis for establishing country greenhouse gas estimates for rice-based agriculture: An Indian case study. *Nutrient Cycling in Agroecosystems, 64*, 19–31.

Gupta, R. K., & Rao, D. L. N. (1994). Potential of wastelands for sequestering carbon by reforestation. *Curriculum Science, 66*, 378–380.

Haas, E., Nieder, R., & Richter, J. (1993). Temporal variability of non-exchangeable ammonium in a loess-derived luvisol. *Modeling Geobioshere Processes, 2*, 163–171.

Haas, H. J., Evans, C. E., & Miles, E. F. (1957). Nitrogen and carbon changes in Great Plains soils as influenced by cropping and soil treatments. *USDA Technical Bulletin 1164*. Washington, DC: US Government Printing Office.

Hadas, A., Feigenbaum, S., Feigin, A., & Portoy, R. (1986). Nitrification rates in profiles of differently managed soil types. *Soil Science Society of America Journal, 50*, 633–639.

Haer, H. S., & Benbi, D. K. (2003). Modeling nitrogen mineralization kinetics in arable soils of semiarid India. *Arid Land Research and Management, 17*, 153–168.

Haider, K. (1992). Problems related to the humification process in soils of temperate climates. In E. A. Paul, & A. D. Hammel (Eds.), *Soil biochemistry*, Vol. 7. New York: Marcel Dekker.

Halsted, R. L., & McKercher, R. B. (1975). Biochemistry and cycling of phosphorus. In E. A. Paul, & A. D. McLaren (Eds.), *Soil biochemistry* (pp. 31–63). New York: Marcel Dekker.

Hanawald, R. B., & Whittaker, R. H. (1976). Altitudinally coordinated patterns of soils and vegetation in the San Jakinto Mountains, California. *Soil Science, 121*, 114–124.

Hansen, E. M., Djurhuus, J., & Kristensen, K. (2000). Nitrate leaching as affected by introduction or discontinuation of cover crop use. *Journal of Environmental Quality, 29*, 1110–1116.

Hansen, K., Rosenqvist, L., Vesterdal, L., & Gundersen, P. (2007). Nitrate leaching from three afforestation chronosequences on former arable land in Denmark. *Global Change Biology, 13*, 1250–1264.

Hansen, S., Jensen, H. E., Nielsen, N. E., & Svendsen, H. (1991). Simulation of nitrogen dynamics and biomass production in winter wheat using the Danish simulation model DAISY. *Fertilizer Research, 27*, 245–259.

Hanson, P. J., Wullschleger, S. D., Bohlman, S. A., & Todd, D. E. (1993). Seasonal and topographic patterns of forest floor CO_2 efflux from an upland oak forest. *Tree Physiology, 13*, 1–15.

Hao, W. M., & Liu, M. H. (1995). Spatial and temporal distribution of tropical biomass burning. *Global Biogeochemical Cycles, 8*, 495–503.

Happell, J. D., & Chanton, J. P. (1993). Carbon remineralization in a North Florida swamp forest: Effects of water level on the pathways and rates of soil organic matter decomposition. *Global Biogeochemical Cycles, 7*, 475–490.

Harden, T. (1998). Erfassung und Modellierung mikrobieller C- und N-Umsetzungsprozesse in südostniedersächsischen Löss-Ackerböden, *Ph.D. thesis,* Braunschweig Technical University, Germany.

Harden, T., Joergensen, R. G., Meyer, B., & Wolters, V. (1993). Soil microbial biomass estimated by fumigation-extraction and substrate induced respiration in two pesticide treated soils. *Soil Biology and Biochemistry, 25*, 679–683.

Hare, F. K., & Ritchie, J. C. (1972). The boreal bioclimates. *Geographical Reviews, 62*, 333–365.

Harms, B. P., Dalal, R. C., & Cramp, A. P. (2004). Changes in soil carbon and soil nitrogen after tree clearing in the semi-arid rangelands of Queensland. *Australian Journal of Botany, 53*, 639–650.

Harrison, A. F., Harkness, D. D., & Bacon, P. J. (1990). The use of bomb-^{14}C for studying organic matter and N and P dynamics in a woodland soil. In A. F. Harrison, P. Ineson, & O. W. Heal (Eds.), *Nutrient cycling in terrestrial ecosystems: Field methods, application and interpretation*. Amsterdam: Elsevier Applied Science.

Harrison, J. A. (2003). The nitrogen cycle: Of microbes and men. Visionlearning Vol. EAS-2(4), 2003. http://www.visionlearning.com/library/module_viewer.php?mid=98

Hassink, J. (1995). Density fractions of soil macroorganic matter and microbial biomass as predictors of C and N mineralization. *Soil Biology and Biochemistry, 27*, 1099–1108.

Hassink, J. (1997). The capacity of soils to preserve organic C and N by their association with clay and silt particles. *Plant and Soil, 191*, 77–87.

Hassink, J., & Whitmore, A. P. (1997). A model of the physical protection of organic matter in soils. *Soil Science Society of America Journal, 61*, 131–139.

Hatch, D. J., Jarvis, S. C., & Reynolds, S. E. (1991). An assessment of the contribution of net mineralization to N cycling in grass swards using a field incubation method. *Plant and Soil, 138*, 23–32.

Hatcher, P. G. (1987). Chemical structural studies of natural lignin by dipolar dephasing solid-state ^{13}C nuclear magnetic resonance. *Organic Geochemistry, 11*, 31–39.

Hatcher, P. G., van der Hart, D. L., & Earl, W. L. (1980). Use of solid-state ^{13}C NMR in structural studies of humic acids and humin from holocene sediments. *Organic Geochemistry, 2*, 87–92.

Hatton, R. S., DeLaune, R. D., & Patrick, W. H. (1983). Sedimentation, accretion, abd subsidence in marshes of Barataria Basin, Louisiana. *Limnology and Oceanography, 28*, 494–502.

Haubold, W., Katzur, J., & Oehme, W. (1998). Standortkundliche Grundlagen. In W. Pflug (Ed.), *Braunkohle-tagebau und Rekultivierung: Landschaftsökologie– Folgenutzung– Naturschutz* (pp. 536–558). Berlin: Springer.

Hawkes, G. E., Powlson, D. S., Randall, E. W., & Tate, K. R. (1984). A 31_P nuclear magnetic resonance study of the phosphorus species in alkali extracts of soils from long-term field experiments. *Journal of Soil Science, 35*, 35–45.

Hawkins, J. M. B., Scholefield, D., & Jarvis, S. C. (1997). Organic N losses from a poorly drained grassland soil. In A. M. Petchey, B. J. D'Arcy, & C. A. Frost (Eds.), *Proceedings of the Scottish Agricultural College Diffuse Pollution and Agriculture Conference II* (pp. 246–248). Edinburgh, UK: Edinburgh University—.

Hayes, M. H. B. (1991). Concepts of the origin, composition and structures of humic substances. In W. S. Wilson (Ed.), *Advances in soil organic matter research: The impact of agriculture and the environment* (pp. 3–22). Cambridge: The Royal Society of Chemistry.

Haynes, R. J. (1999). Size and activity of the soil microbial biomass under grass and arable management. *Biology and Fertility of Soils, 30*, 210–216.

Heath, L., Birdsey, R., Row, C., & Platinga, A. (1996). Carbon pools and fluxes in US Forest Products. *NATO ASI Series, 1*(40), 271–278.

Hedin, L. O., Armesto, J. J., & Johnson, A. H. (1995). Patterns of nutrient loss from unpolluted, old-growth temperate forests:Evaluation of biogeochemical theory. *Ecology, 76*, 493–509.

Heffer, P. (2004). Medium-term outlook for global agricultural situation and fertilizer demand: 2003/04-2008/09. *IFA Annual Conference*, May 2004.

Hegarty, E. E. (1991). Leaf litter production by lianes and trees in a subtropical Australian rain forest. *Journal of Tropical Ecology, 7*, 201–214.

Heilig, G. K. (1999). China food. Can China feed itself? Availble online http://www.iiasa.ac.at/Research/LUC/ ChinaFood/argu/ trends/trend_50.htm

Heincke, M., & Kaupenjohann, M. (1999). Effects of soil solution on the dynamics of N_2O emissions: A review. *Nutrient Cycling in Agroecosystems, 55*, 133–157.

Heinsdorf, D., & Krauss, H. H. (1991). Massentierhaltung und Waldschäden auf dem Gebiet der ehemaligen DDR. *Forst- und Holzwirt, 46*, 356–361.

Helal, H. M., & Sauerbeck, D. S. (1991). Short-term determination of the actual respiration rate of intact plant roots. In B. L. McMichael, & H. Persson (Eds.), *Plant roots and their environments* (pp. 88–92). Amsterdam: Elsevier.

Helyar, K. R., & Porter, W. M. (1989). Soil acidification: Its measurement and the processes involved. In A. D. Robson (Ed.), *Soil acidity and plant growth* (pp. 61–101). Sydney, Australia: Academic.

Hempfling, R., & Schulten, H. R. (1989). Selective preservation of biomolecules during humification of forest litter studied by pyrolysis-field ionization mass spectrometry. *Science of Total Environment, 81/82*, 31–40.

Henderson, S., Hattersley, P., von Caemerer, S. & Osmond, C. B. (1994). Are C_4 pathway plants threatened by global climate change? In E. D. Schulze (Ed.), Ecophysiology of Photosynthesis (pp. 529–549). Heidelberg, Germany: Springer.

Hengnirun, S. S., Barrington, S., Prasher, S. O., & Lyew, D. (1999). Development and verification of a model simulating ammonia volatilization from soil and manure. *Journal of Environmental Quality, 28*, 108–114.

Hénin, S., & Dupuis, M. (1945). Essai de bilan de la matière organique du sol. *Annals of Agronomy, 15*, 17–29.

Herbert, B. E., & Bertsch, P. M. (1995). Characterization of dissolved and colloidal organic matter in soil solution: A review. In J. Kelly, & W. W. McFee (Eds.), *Carbon forms and functions in forest soils* (pp. 63–68). Madison, WI: Soil Science Society of America.

Hernández, I., Gallardo, J. F., Santa Regina, I., & Quilchano, C. (1992). Decomposition rate of Quercus rotundifolia leaves in an evergreen oak forest of the Duero Basin (Province of Zamora, Spain). *Vegetatio, 99–100*, 259–262.

Hinman, W. C. (1964). Fixed ammonium in some Saskatchewan soils. *Canadian Journal of Soil Science, 44*, 151–157.

Hinman, W. C. (1966). Ammonium fixation in relation to exchangeable K and organic matter content in two Saskatchewan soils. *Canadian Journal of Soil Science, 46*, 223–225.

Hoeft, R. G. (1984). Current status of nitrification inhibitor use in US agriculture. In R. D. Hauck (Ed.), *Nitrogen in crop production* (pp. 561–570). Madison, WI: ASA, CSSA, SSSA.

Hofmann, G. (1997). Mitteleuropäische Wald- und Forstökosysteme in Wort und Bild. *AFZ/Der Wald, B 1089*, 91.

Hofstra, N., & Bouwman, A. F. (2005). Denitrification in agricultural soils:summarizing published data and estimating global annual rates. *Nutrient Cycling in Agroecosystems, 72*, 267–278.

Hogan, K. B. (1993). Methane reductions are a cost-effective approach for reducing emissions of greenhouse gases. In A. R. van Amstel (Ed.), *Methane and nitrous oxide: Methods in national emissions inventories and options for control* (pp. 187–201). RIVM Report No. 481507003. The Netherlands: Bithooven.

Holdridge, L. R. (1947). Determination of world plant formations from simple climate data. *Science, 105*, 367–368.

Holland, E. A., & Lamarque, J. F. (1997). Modeling bio-atmospheric coupling of the nitrogen cycles through NO_x emissions and NO_y deposition. *Nutrient Cycling in Agroecosystems, 48*, 7–24.

Holland, E. A., Dentener, F. J., Braswell, B. H., & Sulzman, J. M. (1999). Contemporarory and pre-industrial global reactive nitrogen budgets. *Biogeochemistry, 46*, 7–43.

Hollis, G. E. (1992). The causes of wetland loss and degradation in the Mediterranean. In C. M. Finlayson, G. E. Hollis, & T. J. Davis (Eds.), *Managing Mediterranean wetlands and their birds* (pp. 285). *Symposium Proceedings*, Grado, Italy, 1991. Slimbridge, UK: IWRB Special Publication No. 20.

Holmen, K. (2000) The global carbon cycle. In M. C. Jacobson, R. J. Charlston, H. Rodhe, & G. H. Orians (Eds.), *Earth system science* (pp. 282–321). Amsterdam: Academic.

Holt, J. A., & Coventry, R. J. (1990). Nutrient cycling in Australian savannas. *Journal of Biogeography, 17*, 427–432.

Holzapfel-Pschorn, A., Conrad, R., & Seiler, W. (1986). Production, oxidation and emission of methane in rice paddies. *FEMS Microbial Ecology, 31*, 149–158.

Hood, E. W., Williams, M. W., & Caine, M. (2003). Landscape controls on organic and inorganic nitrogen leaching across an alpine/subalpine ecotone. Green Lakes Valley, Colorado Front Range. *Ecosystems, 6*, 31–45.

Hook, P. B., & Burk, I. C. (2000). Biogeochemistry in a shortgrass landscape:control by topography, soil texture, and microclimate. *Ecology, 81*, 2686–2703.

Hope, D., Billett, M. F., & Cresser, M. S. (1994). A review of the export of carbon in river water: Fluxes and processes. *Environmental Pollution, 84*, 301–324.

Houghton, J. T., Meira Filho, L. G., Callander, A., Harris, N., & Kattenburg, A. (1995). *The science of climate change: Contribution of Working Group I to the second assessment report of the intergovernmental panel on climate change (IPCC) 1994* (400 pp.). Cambridge: Cambridge University Press.

Houghton, J. T., Meira Filho, L.G., Lim, B., Tréanton, K., Mamaty, I., Bonduki, Y., et al. (1997). *Revised 1996 IPCC guidelines for National Greenhouse Gas Inventories*. UK: Hadley Centre Metrological Office.

Houghton, R. A. (2003). Revised estimates of the annual net flux of carbon to the atmosphere from changes in land use and land management 1850–2000. *Tellus, 55B*, 378–390.

Houghton, R. A., & Hackler, J. L. (2002). Carbon flux to the atmosphere from land-use changes. In *Trends: A compendium of data on global change*. Carbon Dioxide Information Analysis Centre, Oak Ridge national Laboratory, Oak Ridge, TN: US Department of Energy. Available online http://cdiac.ornl.gov/trends/landuse/houghton/houghton. html

Houghton, R. A., & Skole, D. L. (1990). Carbon. In B. L. Turner, W. C. Clark, R. W. Kates, J. F. Richards, J. T. Matthews, & W. B. Meyer (Eds.), *The earth as transformed by human action* (pp. 393–408). Cambridge, UK: Cambridge University Press.

Houghton, R. A., Hobbie, J. E., Melillo, J. M., Moore, B., Peterson, B. J., Shaver, G. R., et al. (1983). Changes in the carbon content of terrestrial biota and soils between 1860 and 1980: A net release of CO_2 to the atmosphere. *Ecological Monographs, 53*, 235–262.

Houghton, R. A., Unruh, J., & Lebebvre, P. A. (1991). Current land use in the tropics and its potential for sequestering carbon. *Proceedings of the Technical Workshop to explore Options for Global Forest Management*, April 1991. Bangkok/London: IIED.

Houot, S., Chaussod, R., Hounemenou, C., Barriusos, E., & Bourgeois, S. (1991). Differences induced in the soil organic matter characteristics and microbial activity by various management practices in long-term field experiments. In J. Berthelin (Ed.), *Diversity of environmental biogeochemistry* (pp. 435–443). Amsterdam: Elsevier.

Howard, P. J. A., & Howard, D. M. (1990). Titratable acids and bases in tree and shrub leaf litter. *Forestry, 63*, 177–196.

Howard-Williams, C., & Downes, M. T. (1994). Nitrogen cycling in wetlands. In T. P. Burt, A. L. Heatherwaite, & S. T. Trudgill (Eds.), *Nitrate: Processes, patterns and management* (pp. 141–167). New York: Wiley.

Howarth, R. W. (1993). Microbial processes in salt-marsh sediments. In T. E. Ford (Ed.), *Aquatic microbiology* (pp. 239–259). Oxford: Blackwell Scientific.

Howarth, R. W., Marino, R., Lane, J., & Cole, J. J. (1988). Nitrogen fixation in freshwater, estuarine, and marine ecosystems, 1. Rates and importance. *Limnologic Oceanography, 33*, 669–687.

Howieson, J., & Ballard, R. (2004). Optimising the legume symbiosis in stressful and competitive environments within southern Australia – some contemporary thoughts. *Soil Biology and Biochemistry, 36*, 1261–1273.

Huang, Y., Sass, R. L., & Fisher, F. M. (1998). A semi-empirical model of methane emission from flooded rice paddy soils. *Global Change Biology, 4*, 809–822.

Huang, Y., Jiao, Y., Zong, L., Zheng, X., Sass, R. L., & Fisher, F. M. (2002). Quantitative dependence of methane emission on soil properties. *Nutrient Cycling in Agroecosystems, 64*, 157–167.

Huebert, B. J., & Robert, C. H. (1985). The dry deposition of nitric acid to grass. *Journal of Geophysical Research-Atmosphere, 90*, 2085–2090.

Huijsmans, J., & Bussink, D. W. (1990). Mesttoediening op grassland. *Landbouwmechanisatie, 41*, 38–39.

Hunt, H. W. (1977). A simulation model for decomposition in grassland. *Ecology, 58*, 451–455.

Hunt, H. W., Coleman, D. C., Ingham, E. R., Ingham, R. E., Elliott, E. T., Moore, J. C., et al. (1987). The detrital food web in a shortgrass prairie. *Biology and Fertility of Soils, 3*, 57–68.

Hutchinson, G. L., & Brams, E. A. (1992). NO versus N_2O emissions from an NH_4^+ amended Bermuda pasture. *Journal of Geophysical Research, 97*, 9889–9896.

Hutchinson, G. L., & Davidson, E. A. (1992). Processes for production and consumption of gaseous nitrogen oxides in soil. In L. A. Harper, A. R. Mosier, J. M. Duxbury, & D. E. Rolston (Eds.), *Agricultural ecosystems effects on trace gases and global climate change* (pp. 79–93). Madison, WI: American Society of Agronomy.

IAEA (International International Atomic Energy Agency) (1968). *Isotopes and radiation in soil organic matter studies*. Vienna: FAO/IAEA.

IEA (2006). *CO_2 emissions from fuel combustion, Part III: Greenhouse gas emissions*. Paris: International Energy Agency.

IFA/FAO (International Fertilizer Industry Association/Food and Agriculture Organization of the United Nations) (2001). *Global estimates of gaseous emissions of NH_3, NO and N_2O from agricultural land* (106 pp.). Rome: IFA/FAO.

IIASA (International Institute for Applied Systems Analysis) (2006). *Land use change and agriculture program*. Available online http://www.iiasa.ac.at/Research/LUC/

Immirzi, C. P., Maltby, E., & Clymo, R. S. (1992). *The global status of peatlands and their role in carbon cycling*. A Report for Friends of the Earth by the Wetland Ecosystems Research Group, Department of Geography, University of Exeter. London: Friends of the Earth.

Inglett, P. W. (2000). Spatial and temporal patterns of periphyton N_2 fixation in a nutrient impacted Everglades ecosystem. *M.S. thesis*, University of Florida, Gainesville, 67 pp.

Inubishi, K., Shibahara, F., Hasegawa, K., & Yamamuro, S. (1997). Effects of added organic matter on microbial biomass nitrogen dynamics and plant uptake in paddy soils. In Ando et al. (Eds.), *Plant nutrition for sustainable food production and environment* (pp. 777–778). Dordrecht, The Netherlands: Kluwer.

IPCC (Intergovernmental Panel on Climate Change) (1995). Climate change 1995. impacts, adaptations and mitigation of climate change. In R. T. Watson, M. C. Zinyovera, & R. H. Moss (Eds.), *Contribution of working group II to the second assessment report of the intergovernmental panel on climate change* (878 p.). Cambridge: Cambridge University Press.

IPCC (Intergovernmental Panel on Climate Change) (1996). Climate change 1995: The science of climate change. In J. T. Houghton, L. G. Meira Filho, B. A. Callender, N. Harris, A. Kattenberg, & K. Maskell (Eds.), *Contribution of working group I to the second assessment report of the intergovernmental panel on climate ahange* (572 p.). Cambridge: Cambridge University Press.

IPCC (Intergovernmental Panel on Climate Change) (1997). *Guidelines for national greenhouse gas inventories*. Cambridge: Cambridge University Press.

IPCC (Intergovernmental Panel on Climate Change) (2000a). *Special report on emissions scenarios* (570 pp). Cambridge: Cambridge University Press.

IPCC (Intergovernmental Panel on Climate Change) (2000b). *Technological and economic potential of options to enhance, maintain, and manage biological carbon reservoirs and geo-engineering, climate change 2001: Mitigation*. Port Chester, NY: Cambridge University Press.

IPCC (Intergovernmental Panel on Climate Change) (2000c). Land-use, land-use change, and forestry. In R. T. Watson, I. R. Noble, B. Bolin, N. H. Ravindranath, D. J. Verado, & D. J. Dokken, et al. (Eds.), *A special report of the IPCC published for IPCC by the World Meteorological Society and UNEO*. Cambridge: Cambridge University Press.

IPCC (Intergovernmental Panel on Climate Change) (2001a). Climate change 2001: The scientific basis. In J. T. Hoghton, Y. Ding, D. J. Griggs, M. Noguer, P. J. van der Linden, X. Dai, K. Maskell,

& C. A. Johnson, et al. (Eds.), *Contribution of working group I to the third assessment report of the intergovernmental panel on climate change* (pp. 881). Cambridge, UK/New York: Cambridge University Press.

IPCC (Intergovernmental Panel on Climate Change) (2001b). Climate change 2001-mitigation. *The third assessment report of the intergovernmental panel on climate change*. Cambridge: Cambridge University Press.

IPCC (Intergovernmental Panel on Climate Change) (2005). IPCC special report on carbon dioxide capture and storage. In B. Metz, O. Davidson, H. C. de Coninck, M. Loos, L. A. Meyer (Eds.), *Prepared by working group III of the intergovernmental panel on climate change* (442 pp.). Cambridge, UK/New York: Cambridge University Press.

IPCC (Intergovernmental Panel on Climate Change) (2007a). Climate change 2007: The physical science basis. In S. Solomon, D. Qin, M. Manning, Z. Chen, M. Marquis, K. B. Averyt, et al. (Eds.), *Contribution of working group I to the fourth assessment report of the intergovernmental panel on climate change*. Cambridge, UK/New York: Cambridge University Press.

IPCC (Intergovernmental Panel on Climate Change) (2007b). Climate change 2007: Mitigation of climate change. *Working group III contribution to the intergovrenmental panel on climate change. Summary for policymakers*.

Isermann, K., & Isermann, R. (1995). Die Landwirtschaft als einer der Hauptverursacher der neuartigen Waldschäden. *AFZ/Der Wald, 5*, 1–6.

Isermann, K., & Isermann, R. (1997). Ausgangslage, Lösungsansätze und Lösungsaussichten zur nachhaltigen Landnutzung des deutschen Donaueinzugsgebietes auf der Grundlage seiner Stickstoff- und Phosphorbilanz. *VDLUFA-Schriftenreihe, 45*, 623–626.

Isermann, K., & Isermann, R. (1999). Bodenkundliche Anforderungen an die fachliche Praxis einer nachhaltigen Landwirtschaft/Landnutzung aus der Sicht ihrer Nährstoffhaushalte. *Mitteilungen der Deutschen Bodenkundlichen Gesellschaft, 91*, 59–62.

Iversen, T. M., Grant, R., Nielsen, K., & van der Hoek, K. W. (1998). Nitrogen enrichment of European inland and marine waters with special attention to Danish policy measures. *Environmental Pollution, 102*, 771–780.

Izaurralde, R. C., Williams, J. R., Post, W. M., Thomson, A. M., McGill, W. B., Owens, L. B., et al. (2007). Long-term modeling of soil C erosion and sequestration at the smaller watershed scale. *Climatic Change, 80*, 73–90.

Jäckel, U., & Schnell, S. (2000). Supression of methane emission from rice paddies by ferric iron fertilization. *Soil Biology and Biochemistry, 32*, 1811–1814.

Jaeger, C. H., Monson, R. K., Fisk, M. C., & Schmidt, S. K. (1999). Seasonal partitioning of nitrogen by plants and soil microorganisms in an alpine ecosystem. *Ecology, 80*, 1883–1892.

Jaffe, D. A. (2000). The nitrogen cycle. In M. C. Jacobson, R. J. Charlson, H. Rohde, & G. H. Orians (Eds.), *Earth system science – from biogeochemmical cycles to global change* (pp. 322–342). Amsterdam: Academic.

James, B. R., & Riha, S. J. (1986). pH buffering in forest soil organic horizons: Relevance to acid precipitation. *Journal of Environmental Quality, 15*, 229–234.

Jansen, S. A., Malaty, M., Nwabara, S., Johnson, E., Ghabbour, G. D., & Varnum, J. M. (1996). Structural modeling in humic acids. *Materials Science and Engineering: C, 4*, 175–179.

Janssen, B. H. (1984). A simple method for calculating decomposition and accumulation of 'young' soil organic matter. *Plant and Soil, 76*, 297–304.

Janzen, H. H. (2004). Carbon cycling in earth systems- a soil science perspective. *Agriculture, Ecosystems and Environment, 104*, 377–417.

Janzen, H. H., Campbell, C. A., Brandt, S. A., Lafond, G. P., & Townley-Smith, L. (1992). Light-fraction organic matter in soils from long-term crop rotations. *Soil Science Society of America Journal, 56*, 1799–1806.

Janzen, H. H., Campbell, C. A., Izaurralde, R. C., Ellert, B. H., Juma, N., McGill, W. B., et al. (1998). Management effects on soil C storage on the Canadian Prairies. *Soil and Tillage Research, 47*, 181–195.

Jarvis, S. C., & Barraclough, D. (1991). Variation in mineral nitrogen content under grazed grassland swards. *Plant and Soil, 138*, 177–188.

Jastrow, J. D. (1996). Soil aggregate formation and the accrual of particulate and mineral-associated organic matter. *Soil Biology and Biochemistry, 28*, 665–676.

Jastrow, J. D., Boutton, T. W., & Miller, R. M. (1996). Carbon dynamics of aggregate-associated organic matter estimated by carbon-13 natural abundance. *Soil Science Society of America Journal, 60*, 801–807.

Jayaweera, G. R., & Mikkelsen, D. S. (1990). Assessment of ammonia volatilization from flooded soil systems: A computer model. I. Theoretical aspects. *Soil Science Society of America Journal, 54*, 1447–1455.

Jenkinson, D. (1990). The turnover of organic carbon and nitrogen in soil. *Philosophical Transactions of Royal Society of London B, 329*, 361–369.

Jenkinson, D., & Rayner, J. H. (1977). The turnover of soil organic matter in some of the Rothamsted classical experiments. *Soil Science, 123*, 298–305.

Jenkinson, D. S. (1971). Studies on the decomposition of ^{14}C-labelled organic matter in soil. *Soil Science, 111*, 64–70.

Jenkinson, D. S. (1977). Studies on the decomposition of plant material in soil. V. *Journal of Soil Science, 28*, 424–434.

Jenkinson, D. S. (1987). Determination of microbial biomass carbon and nitrogen in soil. In J. R. Wilson (Ed.), *Advances in nitrogen cycling in agricultural systems* (pp. 368–386). Wallingford, UK: CABI.

Jenkinson, D. S. (1990). The turnover of organic carbon and nitrogen in soil. *Philosophical Transactions of the Royal Society of London, B 329*, 361–368.

Jenkinson, D. S., & Ayanaba, A. (1977). Decomposition of carbon-14 labelled plant material under tropical conditions. *Soil Science Society of America Journal, 41*, 912–915.

Jenkinson, D. S., & Johnston, A. E. (1977). Soil organic matter in the Hoosfield continuous barley experiment. *Report, Rothamsted Experimental Station 1976, Part 2*, 87–101.

Jenkinson, D. S., & Ladd, J. N. (1981). Microbial biomass in soils: Measurement and turnover. *Soil Biochemistry, 5*, 415–471.

Jenkinson, D. S., & Powlson, D. S. (1976). The effect of biocidal treatments on metabolism in soil. V. A method for measuring soil biomass. *Soil Biology and Biochemistry, 8*, 209–213.

Jenkinson, D. S., Hart, P. B. S., Rayner, J. H., & Parry, L. C. (1987). Modeling the turnover of organic matter in long-term experiments at Rothamsted. *INTECOL Bulletin, 15*, 1–8.

Jenkinson, D. S., Bradbury, N. J., & Coleman K. (1994). How the Rothamsted classical experiments have been used to develop and test models of the turnover of carbon and nitrogen in soil. In R. A. Leigh, & A. E. Johnston (Eds.), *Long-term experiments in agricultural and ecological sciences* (pp. 117–138). Wallingford, UK: CABI.

Jenny, H. (1941). *Factors of soil formation*. New York: McGraw Hill.

Jensen, B. (1993). Rhizodeposition by $^{14}CO_2$-pulse-labelled spring barley grown in small field plots on sandy loam. *Soil Biology and Biochemistry, 25*, 1553–1559.

Jensen, B. (1994). Rhizodeposition by field-grown winter barley exposed to $^{14}CO_2$ pulse-labelling. *Applied Soil Ecology, 1*, 65–74.

Jensen, E. S., Christensen, B. T., & Sörensen, L. H. (1989). Mineral fixed ammonium in clay- and silt-size fractions of soils incubated with ^{15}N-ammonium sulphate for five years. *Biology and Fertility of Soils, 8*, 298–302.

Jing, L., Mingxing, W., Yao, H., & Yuesi, W. (2002). New estimates of methane emissions from Chinese rice paddies. *Nutrient Cycling in Agroecosystems, 64*, 33–42.

Jobaggy, E., & Jackson, R. (2000). The vertical distribution of soil organic carbon and its relation to climate and vegetation. *Ecological Applications, 10*, 423–436.

Johnson, M. G. (1995). The role of soil management in sequestering soil carbon. In R. Lal, J. Kimble, E. Levine, & B. A. Steward (Eds.), *Soil management and greenhouse effect* (pp. 351–363). Boca Raton, FL: CRC/Lewis.

Jones, C., Mcconnell, C., Coleman, K., Cox, P., Falloon, P., Jenkinson, D., et al. (2005). Global climate change and soil carbon stocks; predictions from two contrasting models for the turnover of organic carbon in soil. *Global Change Biology, 11*, 154–166.

Jones, D. L., & Kieland, K. (2002). Soil amino acid turnover dominates the nitrogen flux in permafrost-dominated taiga forest soils. *Soil Biology and Biochemistry, 34*, 209–219.

Jones, D. L., Shannon, D., Murphy, D., & Farrar, J. F. (2004). Role of dissolved organic nitrogen (DON) in soil N cycling in grassland soils. *Soil Biology and Biochemistry, 36*, 749–756.

Jones, D. L., Healey, J. R., Willett, V. B., Farrar, J. F., & Hodge, A. (2005). Dissolved organic nitrogen uptake by plants- an important N uptake pathway. *Soil Biology and Biochemistry, 37*, 413–428.

Jones, T. A., & Hughes, J. M. R. (1993). Wetland inventories and wetland loss studies: A European perspective. In M. Moser, R. C. Prentice, & J. van Vessem (Eds.), *Water flow and wetland conservation in the 1990s*. IWRB special publication No. 26. Slimbridge, UK: IWRB.

Jörgensen, R. G. (2000). Ergosterol and microbial biomass in the rhizosphere of grassland soils. *Soil Biology and Biochemistry, 32*, 647–652.

Jugnia, L. B., Roy, R., Planas, D., Lucotte, M., Greer, C. W. (2005). Potential activity of methanogenesis in the soils, peats, lacustrine sediments and in the Robert-Bourassa hydroelectric reservoir in central North-Canada. *Canadian Journal of Microbiology, 51*, 79–84.

Kaiser, E. A., Mueller, T., Jörgensen, R. G., Insam, H., & Heinemeyer, O. (1992). Evaluation of methods to estimate the soil microbial biomass and the relationship with soil texture and organic matter. *Soil Biology and Biochemistry, 224*, 675–683.

Kaiser, K., Guggenberger, G., & Zech, W. (1997). Natural dissolved organic matter in forest soils: 2. Biotic immobilization in the mineral soil. In *Proceedings Symposium on Refractory Organic Substances in the Environment* (pp. 232–235), Karlsruhe.

Kaiser, K., Guggenberger, G., Haumaier, L., & Zech, W. (2001). Seasonal variations in the chemical composition of dissolved organic matter in organic forest floor layer leachates of old-growth Scots pine (*Pinus sylvestris* L.) and European beech (*Fagus sylvatica L.*) stands in northeastern Bavaria, Germany. *Biogeochemistry, 55*, 103–143.

Kalbitz, K., Solinger, S., Park, J. H., Michalzik, B., & Matzner, E. (2000). Controls on the dynamics of dissolved organic matter in soils: A review. *Soil Science, 165*, 277–304.

Kalbitz, K., Schwesig, D., Rethemeyer, J., & Matzner, E. (2005). Stabilization of dissolved organic matter by sorption to the mineral soil. *Soil Biology and Biochemistry, 37*, 1319–1331.

Kandeler, E., & Böhm, K. E. (1996). Temporal dynamics of microbial biomass, xylanase activity, N mineralization and potential nitrification in different tillage systems. *Applied Soil Ecology, 4*, 181–191.

Kandeler, E., & Eder, G. (1993). Effect of cattle slurry in grassland on microbial biomass and on activities of various enzymes. *Biology and Fertility of Soils, 16*, 249–254.

Kandeler, E., Kampichler, C., & Horak, O. (1996). Influence of heavy metals on the functional diversity of soil microbial communities. *Biology and Fertility of Soils, 23*, 299–306.

Kandeler, E., Stemmer, M., & Klimanek, E. M. (1999a). Response of soil microbial biomass, urease and xylanase within particle size fractions to long-term soil management. *Soil Biology and Biochemistry, 31*, 261–273.

Kandeler, E., Tscherko, D., & Spiegel, H. (1999b). Long-term monitoring of microbial biomass, N mineralization and enzyme activities of a Chernozem under different tillage management. *Biology and Fertility of Soils, 28*, 343–351.

Karlen, D. L., Wollenhaupt, N. C., Erbach, D. C., Berry, E. C., Swan, J. B., Eash, N. S., et al. (1994). Long-term tillage effects on soil quality. *Soil and Tillage Research, 32*, 313–327.

Karlen, D. L., Mausbach, M. J., Doran, J. W., Kline, R. G., Harris, R. F., & Schuman, G. E. (1997). Soil quality: A concept, definition, and framework for evaluation. *Soil Science Society of America Journal, 61*, 4–10.

Kasang, D. (2004). *Climate change and the greenhouse effect*. Available online http://lbs.hh.schule.de/welcome.phtml?unten=/klima/greenhouse/

Katou, H., Clothier, B. E., & Green, S. R. (1996). Anion transport involving competitive adsorption during transient water flow in an Andisol. *Soil Science Society of America Journal, 60*, 1368–1375.

Kauppi, P. E., Posch, M., Hänninen, P., Henttonen, H. M., Ihalainen, A., Lappalainen, E., et al. (1997). Cabon reservoirs in peatlands and forests in the boreal regions of Finland. *Silva Fennica, 31*, 13–25.

Kay, B., da Silva, A. P., & Baldock, J. A. (1997). Sensitivity of the structure of different soils to changes in organic carbon content: Predictions using pedotransfer functions. *Canadian Journal of Soil Science, 77*, 655–667.

Kearney, M. S., & Stevenson, J. C. (1991). Island land loss and marsh vertical accretion rate evidence for histosol sea level changes in Chesapeake Bay. *Journal of Coastal Research, 7*, 403–415.

Keeling, C. D., & Whorf, C. D. (Eds.) (2005). *Atmospheric CO_2 records from sites in the SIO air sampling network.* Carbon Dioxide Information Analysis Center, Oak Ridge National Laboratory. Oak Ridge, TN: US Department of Energy. Available online http://cdiac.esd.ornl.gov/trends/co2/sio-keel-flask/sio-keel-flask.html

Keeling, C. D., Adams, J. A., Ekdahl, A. A., & Guenther, P. R. (1976). Atmospheric carbon dioxide variations at the South Pole. *Tellus, 28*, 552–564.

Keeling, C. D., Carter, A. F., & Mook, W. G. (1984). Seasonal, latitudinal, and secular variations in the abundance and isotopic ratios of atmospheric CO_2, 2. Results from oceanographic cruises in the tropical Pacific ocean. *Journal of Geophysical Research, 89*, 4615–4628.

Keeling, C. D., Whorf, T. P., Wahlen, M., & van der Pflicht, J. (1995). Interannual extremes in the rate of rise of atmospheric carbon dioxide since 1980. *Nature, 375*, 666–670.

Keeling, C. D., Chin, J. F. S., & Whorf, T. P. (1996). Increased activity of northern vegetation inferred from atmospheric CO_2 measurements. *Nature, 382*, 146–149.

Keeney, D. R., Filery, I. R., & Marx, G. P. (1979). Effect of temperature on the gaseous nitrogen products of denitrification in a silt loam soil. *Soil Science Society of America Journal, 43*, 1124–1128.

Keerthisinghe, G., Mengel, K., & De Datta, S. K. (1984). The release of nonexchangeable ammonium (^{15}N labelled) in wetland rice soils. *Soil Science Society of America Journal, 48*, 291–294.

Kelly, K. R., & Stevenson, F. J. (1987). Effects of carbon source on immobilization and chemical distribution on fertilizer nitrogen in soil. *Soil Science Society of America Journal, 51*, 946–951.

Kempe, S. (1979). Carbon in the rock cycle. In B. Bolin, E. T. Degens, S. Kempe, & P. Ketner (Eds.), *The global carbon cycle* (pp. 343–377). New York: Wiley.

Keppler, F., Hamilton, J. T., Braß, M., & Röckmann, T. (2006). Methane emissions from terrestrial plants under aerobic conditions. *Nature, 439*, 187–191.

Kern, J. S., & Johnson, M. G. (1993). Conservation tillage impacts on national soil and atmospheric carbon levels. *Soil Science Society of America Jourmal, 57*, 200–210.

Kern, J. S., Gong, J., Zhang, G., Zhuo, H., & Luo, G. (1997). Spatial analysis of methane emissions from paddy soils in China and the potential for emission reduction. *Nutrient Cycling in Agroecosystems, 49*, 181–195.

Kerr, S., Pfaff, A., & Sanchez, A. (2001). The dynamics of deforestation and the supply of carbon sequestration: Illustrative results from Costa Rica. In T. Panayoutou (Ed.), *Environment: Conservation and competitiveness*. Central America Project, Harvard Institute for International Development.

Kersebaum, K. C., & Richter, O. (1994). A model approach to simulate C and N transformations through microbial biomass. In J. J. Neetson, & J. Hassink (Eds.), *Nitrogen mineralization in agricultural soils* (pp. 221–229). Wageningen, The Netherlands: AB-DLO.

Kersebaum, K. C., Nieder, R., & Köster, W. (2007). Neue Chancen für eine effizientere N-Düngung. *WasserWirtschaft*, 1–2, 58–63.

Khaleel, R., Reddy, K. R., & Overcash, M. R. (1981). Changes in soil physical properties due to organic waste applications: A review. *Journal of Environmental Quality, 10*, 133–141.

Khanna, M., & Stotzky, G. (1992). Transformation of *Bacillus subtilis* by DNA bound on montmorillonite and effect of Dnase on the transforming ability of bound DNA. *Applied Environmental Microbiology, 58*, 1930–1939.

Khanna, P. K., Ludwig, B., Bahus, J., & O'Hara, C. (2001). Assessment and significance of labile organic C pools in forest soils. In R. Lal, J. M. Kimble, R. F. Follett, & B. A. Stewart (Eds.), *Assessment methods for soil carbon* (pp. 167–182). Boca Raton, FL: Lewis.

Khomutova, T. E., Shirshova, T., Tinz, S., Rolland, W., & Richter, J. (2000). Mobilization of DOC from sandy loamy soils under different land use (Lower Saxony, Germany). *Plant and Soil, 219*, 13–19.

Kiala, A. (1963). Organic phosphorus in Finnish soils. *Soil Science, 95*, 38–44.

Kicklighter, D. W., Melillo, J. M., Peterjohn, W. T., Rastetter, E. B., McGuire, A. D., & Steudler, P. A. (1994). Aspects of spatial and temporal aggregation in estimating regional carbon dioxide fluxes from temperate forest soils. *Journal of Geophysical Research, 99*, 1303–1315.

Kieft, T. L., Soroker, E., & Firestone, M. K. (1987). Microbial biomass response to a rapid increase in water potential when dry soil is wetted. *Soil Biology and Biochemistry, 19*, 119–126.

Killham, K., Amato, M., & Ladd, J. N. (1993). Effect of substrate location in soil and soil pore-water regime on carbon turnover. *Soil Biology and Biochemistry, 25*, 57–62.

Kim, J., & Rees, D. C. (1992). Structural models for the metal centers in the nitrogenase molybdenum-iron protein. *Science, 257*, 1677–1682.

Kinchesh, P., Powlson, D. S., & Randall, E. W. (1995). ^{13}C NMR studies of organic matter in whole soils: I. Quantitation possibilities. *European Journal of Soil Science, 46*, 125–138.

King, J. A. (1997). The development of methods for the measurement of carbon fluxes from soils and vegetation, final report to MAFF on seedcorn project No. VLAFD/A/2. Boxworth, Cambridge: ADAS Boxworth.

Kira, T. (1978). Community architecture and organic matter dynamics in tropical lowland rain forests of Southeast-Asia with special reference to Pasoh Forest, West Malaysia. In P. B. Tomlinson, & M. H. Zimmermann (Eds.), *Tropical trees as living systems* (pp. 561–590). Cambridge: Cambridge University Press.

Kira, T., Ono, Y., & Hosokawa, T. (1978). Biological production in a warm-temperate evergreen oak forest of Japan. *JIBP Synthesis 18* (288 pp.). Tokyo: Tokio.

Kirk, G. J. D., & Nye, P. H. (1991a). A model of ammonia volatilization from applied urea. V. The effect of steady-state drainage and evaporation. *Journal of Soil Science, 42*, 103–113.

Kirk, G. J. D., & Nye, P. H. (1991b). A model of ammonia volatilization from applied urea VI. The effects of transient-state water evaporation. *Journal of Soil Science, 42*, 115–125.

Kirschbaum, M. U. F. (1994). The sensitivity of C_3 photosynthesis to increasing CO_2 concentration: A theoretical analysis of its dependence on temperature and background CO_2 concentration. *Plant, Cell and Environment, 17*, 747–754.

Kirschbaum, M. U. F. (1995). The temperature dependence of soil organic matter decomposition, and the effect of global warming on soil organic C storage. *Soil Biology and Biochemistry, 27*, 753–760.

Kirschbaum, M. U. F. (2004). Modelled soil changes after reforestation. In *Super Soil 2004: 3rd Australian New Zealand Soils Conference*, 5–9 December 2004. Australia: University of Sydney.

Kladivko, E. J., & Nelson, D. W. (1979). Changes in soil properties from application of an-aerobic sludge. *Journal of Water Pollution Control, 51*, 325–332.

Klinge, H., Rodrigues, E., Brunig, E., & Fittkau, E. J. (1975). Biomass and structure in a Central Amazonian rain forest. In F. B. Golley, & E. Medina (Eds.), *Tropical ecological systems* (pp. 115–122). Berlin, Germany: Springer.

Klose, S., Wernecke, K. D., & Makeschin, F. (2004). Microbial activities in forest soils exposed to chronic depositions from a lignite power plant. *Soil Biology and Biochemistry, 36*, 1913–1923.

Kludze, H. K., & Delaune, R. D. (1995). Gaseous exchange and wetland plant-response to soil redox intensity and capacity. *Soil Science Society of America Journal, 59*, 939–945.

Kludze, H. K., Delaune, R. D., & Patric, W. H. (1993). Aerenchyma formation and methane and oxygen exchange in rice. *Soil Science Society of America Journal, 57*, 386–391.

Knapp, E. B., Elliott, L. F., & Campbell, G. S. (1983). Carbon, nitrogen and microbial biomass interrelationships during the decomposition of wheat straw: A mechanistic simulation model. *Soil Biology and Biochemistry, 15*, 455–461.

Knicker, H., Schmidt, M. W. I., & Kögel-Knabner, I. (2000). Nature of organic nitrogen in fine particle separates of sandy soils of highly industrialized areas as revealed by NMR spectroscopy. *Soil Biology and Biochemistry, 32*, 241–252.

Knof, G. (1985). Eine Gerätekombination zur Bestimmung des Kohlenstoffs und seines ^{14}C-Anteils in Wurzeln und anderen pflanzlichen Substanzen. *Archiv für Acker- und Pflanzenbau und Bodenkunde, 29*, 23–30.

Knowles, R. (1986). Aspects of nitrification and denitrification. In F. Megusar, & M. Gantar (Eds.), *Perspectives in microbial ecology* (pp. 319–324). Ljubljana, Slovenia: Slovene Society of Microbiology.

Koch, G. W., & Mooney, H. A. (1996). *Carbon dioxide and terrestrial ecosystems*. San Diego, CA: Academic.

Kögel-Knabner, I. (1993). Biodegradation and humification processes in forest soils. In J. M. Bollag, & G. Stotzky (Eds.), *Soil biochemistry* (pp. 101–135). New York: Marcel Dekker.

Kögel-Knabner, I., Zech, W., Hatcher, P. G., & de Leeuw, J. W. (1991). Fate of plant components during biodegradation and humification in forest soils: Evidence from structural characterization of individual biomacromolecules. *Advances in Soil Organic Matter Research: The Impact of Agriculture and the Environment*, 61–70.

Kögel-Knabner, I., de Leeuw, J. W., & Hatcher, P. G. (1992a). Nature and distribution of alkyl carbon in forest soil profiles: implications for the origin and humification of aliphatic biomacromolecules. *Science of the Total Environment, 117/118*, 175–185.

Kögel-Knabner, I., Hatcher, P. G., Tegelaar, E. W., & de Leeuw, J. W. (1992b). Aliphatic components of forest soil organic matter as determined by solid-state ^{13}C NMR and analytical pyrolysis. *Science of the Total Environment, 113*, 89–106.

Kölbl, A., & Kögel-Knabner, I. (2004). Content and composition of free and occluded particulate organic matter in a differently textured arable Cambisol as revealed by solid-state ^{13}C NMR spectroscopy. *Journal of Plant Nutrition and Soil Science, 167*, 45–53.

Kolenbrander, G. J. (1969). Nitrate content and nitrogen loss in drainwater. *Netherlands Journal of Agricultural Science, 17*, 246–255.

Kolenbrander, G. J. (1981). Leaching of nitrogen in agriculture. In J. C. Brogan (Ed.), *Nitrogen losses and surface run-off* (pp. 199–216). Den Haag, The Netherlands: Nijhoff-Junk.

Kononova, M. M. (1966). *Soil organic matter: Its nature, its role in soil formation and in soil fertility*. Oxford: Pergamon.

Körschens, M. (1995). Zur Frage optimaler Humusgehalte in Ackerböden. *VDLUFA-Schriftenreihe, 40*, 157–160.

Körschens, M. (1997). Dependence of soil organic matter (SOM) on location and management, and its influence on yield and soil properties (in German). *Archives of Agronomy and Soil Science, 41*, 435–463.

Körschens, M., & Müller, A. (1996). The static experiment bad Lauchstädt, Germany. In D. S. Powlson, P. Smith, & J. U. Smith (Eds.), *Evaluation of soil organic matter models* (pp. 369–376). Berlin: Springer.

Körschens, M., & Schulz, E. (1999). Die organische Bodensubstanz. *UFZ-Bericht, 13*, 1–46.

Koskinen, W. C., & Harper, S. S. (1990). The retention process: Mechanisms. In H. H. Cheng (Ed.), *Pesticides in the soil environment: Processes, impacts, and modeling* (pp. 51–77). Madison, WI: Soil Science Society of America.

Kowalenko, C. G. (1978). Nitrogen transformations and transport over 17 months in field fallow microplots using ^{15}N. *Canadian Journal of Soil Science, 58*, 69–76.

Kramer, D. A. (1999). *Minerals Yearbook, Nitrogen, US Geological Survey Minerals Information*. Available online http://minerals.usgs.gov/minerals/pubs/commodity/nitrogen

Kreutzer, K. (1995). Effects of forest liming on soil processes. *Plant and Soil, 168–169*, 447–470.

Kreuzer, M., & Hindrichsen, I. K. (2006). Methane mitigation in ruminants by dietary means: The role of their methane emission from manure. In C. R. Soliva, J. Takahashi, & M. Kreuzer (Eds.), *Greenhouse gases and animal agriculture: An update* (pp. 199–208). International Congress Series No. 1293, The Netherlands: Elsevier.

Kroeze, C., Mosier, A., & Bouwman, L. (1999). Closing the global N_2O budget: a retrospective analysis 1500–1994. *Global Biogeochemical Cycles, 13*, 1–8.

Kuhry, P., & Vitt, D. H. (1996). Fossil carbon/nitrogen ratios as a measure of peat decomposition. *Ecology, 77*, 271–275.

Kurkalova, L., Kling, C., & Zhao, J. (2003). *Institutions and the value of nonpoint source measurement technology: Carbon sequestration in agricultural soils*. Working Paper 03-WP-338, Center for Agricultural and Rural Development, Iowa State University, USA.

Kurpick, P., Kurpick, U., & Huth, A. (1997). The influence of logging on a Malaysian dipterocarp rain forest: A study using a forest gap model. *Journal of Theoretical Biology, 185*, 47–54.

Kurz, W. A., & Apps, M. J. (1999). A 70-year retrospective analysis of carbon fluxes in the Canadian forest sector. *Ecological Applications, 9*, 526–547.

Kuylenstierna, J. C. I., Cambridge, H., Cinderby, S., & Chadwick, M. J. (1995). Terrestrial ecosystem sensitivity to acidic deposition in developing countries. *Water, Air and Soil Pollution, 85*, 2319–2324.

Kuzyakov, Y. (2002). Review: Factors affecting rhizosphere priming effects. *Journal of Plant Nutrition and Soil Science, 165*, 382–396.

Kuzyakov, Y., & Domanski, G. (2000). Carbon input by plants into soil, Review. *Journal of Plant Nutrition and Soil Science, 163*, 421–431.

Kuzyakov, Y., & Schneckenberger, H. (2004). Review of estimation of plant rhizodeposition and their contribution to soil organic matter formation. *Archives of Agronomy and Soil Science, 50*, 115–132.

Kuzyakov, Y., Kretzschmar, A., & Stahr, K. (1999). Contribution of *Lolium perenne* rhizode-position to carbon turnover of pasture soil. *Plant and Soil, 213*, 127–136.

Kuzyakov, Y., Ehrensberger, H., & Stahr, K. (2001). Carbon partitioning and below-ground translocation by *Lolium perenne*. *Soil Biology and Biochemistry, 33*, 61–74.

Lackner, K. S., Wendt, C. H., Butt, D. P., Joyce, E. L., & Sharp, D. H. (1995). Carbon dioxide disposal in carbonate minerals. *Energy, 20*, 1153–1170.

Ladd, J. N., & Amato, M. (1980). Studies of nitrogen immobilization and mineralization in calcareous soils. IV. Changes in the organic nitrogen of light and heavy subfractions of silt- and fine-clay sized particles during nitrogen turnover. *Soil Biology and Biochemistry, 12*, 185–189.

Ladd, J. N., & Butler, J. H. A. (1975). Humus-enzyme systems and synthetic organic polymer-enzyme analogs. In E. A. Paul, & A. D. McLaren (Eds.), *Soil biochemistry* (pp. 143–194). New York: Marcel Dekker.

Ladd, J. N., Oades, J. M., & Amato, M. (1981). Microbial biomass from ^{14}C, ^{15}N-labelled plant material decomposing in soils in the field. *Soil Biology and Biochemistry, 13*, 119–126.

Ladd, J. N., Amato, M., & Oades, J. M. (1985). Decomposition of plant materials in Australian soils. III. Residual organic and microbial biomass C and N from isotope-labelled legume materials and soil organic matter, decomposing under field conditions. *Australian Journal of Soil Research, 23*, 603–611.

Ladd, J. N., Amato, M., Li-Kai, Z., & Schultz, J. E. (1994). Differential effects of rotation, plant residue and nitrogen fertilizer on microbial biomass and organic matter in an Australian Alfisol. *Soil Biology and Biochemistry, 26*, 821–831.

Ladd, J. N., Foster, R. C. Nannipieri, P., & Oades, J. M. (1996). Soil structure and biological activity. In G. Stotzky, & J. M. Bollag (Eds.), *Soil biochemistry* (pp. 23–78). New York: Marcel Dekker.

Lal, R. (1995a). *Sustainable management of soil resources in the humid tropics* (146p). The United Nations University, Tokyo: United Nations Press.

Lal, R. (1995b). Global soil erosion by water and carbon dynamics. In R. Lal, J. Kimble, E. Levine, & B. A. Steward (Eds.), *Soils and global change* (pp. 131–142). Boca Raton, FL: CRC/Lewis

Lal, R. (2000). Soil carbon and the accelerated greenhouse effect. In A. Sapek (Ed.), *Poland agriculture and water quality protection. Proceedings of the conference "Scientific basis to mitigate the nutrient dispersion into the environment"* (pp. 106–118). Wydanie publikacji dofinaso-wane przez KBN.

Lal, R. (2004a). Agricultural activities and the global carbon cycle. *Nutrient Cycling in Agroecosystems, 70*, 103–116.

Lal, R. (2004b). Soil carbon sequestration to mitigate climate change. *Geoderma, 123*, 1–22.

Lal, R. (2005). Enhancing crop yields in the developing countries through restoration of the soil organic carbon pool in agricultural lands. *Land Degradation and Development, 17*, 197–209.

Lal, R., & Logan, T. J. (1995). Agricultural activities and greenhouse gas emissions from soils of the tropics. In R. Lal, J. Kimble, E. Levine, & B. A. Steward (Eds.), *Soil management and greenhouse effect* (pp. 293–307). Boca Raton, FL: CRC/Lewis.

Lal, R., Kimble, J. M., Levine, E., & Steward, B. A. (Eds.). (1995). *Soils and global change* (440 pp). Boca Raton, FL: Lewis.

Lamarque, J. F., et al. (2005). Assessing future nitrogen deposition and carbon cycle feedback using a multimodel approach: analysis of nitrogen deposition. *Journal of Geophysical Research, 110*, D19303:doi:1029/2005JD005825.

Landi, L., Renella, G., Moreno, J. L., Falchini, L., & Nannipieri, P. (2000). Influence of cadmium on the metabolic quotient, L-:D-glutamatic acid respiration ratio and enzyme activity: Microbial biomass ratio under laboratory conditions. *Biology and Fertility of Soils, 32*, 8–16.

Langer, U., & Gunther, T. (2001). Effects of alkaline dust deposits from phosphate fertilizer production on microbial biomass and enzyme activities in grassland soils. *Environmental Pollution, 112*, 321–327.

Langford, A. O., & Fehsenfeld, F. C. (1992). Natural vegetation as a source or sink for atmospheric ammonia: A case study. *Science, 255*, 581–583.

Larcher, W. (Ed.) (2003). *Physiological plant ecology*. Heidelberg, Germany: Springer.

Larionova, A. A., Rozanova, L. N., Yevdokimov, I. V., & Blagodatsky, S. A. (2003). Carbon balance in arable grey forest soils. *Annales UMCS, Sec E, 58*, 193–203.

Larson, W. E., & Pierce, F. J. (1991). Conservation and enhancement of soil quality. In *Evaluation for sustainable land management in the developing world*. Proc. 12 (Vol. 2, pp. 175–203). Bangkog, Thailand: International Board of Soil Research and Management (IBSRAM),

Lathja, K., & Bloomer, S. H. (1988). Factors affecting phosphorus adsorption and phosphorus retention in an arid ecosystem. *Soil Science, 146*, 16–167.

Lauer, W. (1975). *Vom Wesen der Tropen, Klimaökologische Studien zum Inhalt und zur Abgrenzung eines irdischen Landschaftsgürtels* (pp. 1–70), Abhandlungen der Mathematisch-Naturwissenschaftlichen Klasse 3, Akademie der Wissenschaften und der Literatur, Mainz, Germany.

Lavahun, M. F. E., Jörgensen, R. G., & Meyer, B. (1996). Activity and biomass of soil microorganisms at different depths. *Biology and Fertility of Soils, 23*, 38–42.

Le Houerou, H. N. (1993). Salt tolerant plants for the arid regions of the Mediterranean isoclimatic zone. In H. Leith, & A. Al Masoom (Eds.), *Towards the rational use of high salinity tolerant plants* (pp. 403–422). Dordrecht, The Netherlands: Kluwer.

Leakey, A. B., Ainsworth, E. A., Bernacchi, C. J., Dohleman, F. G., Naidu, S., Rogers, A., et al. (2006). Elevated CO_2 does not stimulate C_4 photosynthesis directly, but impacts water relations and indirectly enhances carbon gain during drought stress in maize (zea mays) grown under free-air CO_2 enrichment (FACE), ASA-CSSA-SSSA Annual Meeting. Available online http://crops/confec.com/crops/2006am/techprogram/P22427.htm

Lee, D. S., Köhler, I., Grobler, E., Rohrer, F., Sauen, R., Gallardo-Klenner, L., et al. (1997). Estimates of global NO_x emissions and their uncertainties. *Atmospheric Environment, 31*, 1735–1749.

Lee, J. A. (1999). The calcicole-calcifuge problem revisited. *Advances in Botanical Research, 29*, 1–30.

Lee, J. J., Phillipy, D. L., & Liu, R. (1993). The effects of trends in tillage practicees on erosion and carbon content of soils in the US Corn Belt. *Water, Air, and Soil Pollution, 70*, 389–401.

Leffelaar, P. A. (1979). Simulation of partial anaerobiosis in a model soil in respect to denitrification. *Soil Science, 128*, 110–120.

Leinweber, P., Reuter, G., & Brozio, K. (1993). Cation exchange capacity of organo-mineral particle size fractions in soils from long-term experiments. *Journal of Soil Science, 44*, 111–119.

Lelieveld, J., & Dentener, F. J. (2000). What controls tropospheric ozone? *Journal of Geophysical Research, 105*, 3531–3551.

Leng, R. A. (1991). Improving ruminant production and reducing methane emissions from ruminants by strategic supplementation. USEPA Report 400/1–91/004. Washington, DC: Office of Air and Radiation.

Levy, P. E., Wendler, R., van Oijen, M., Cannell, M. G. R., & Millaer, P. (2004). The effect of nitrogen enrichment on the carbon sink in coniferous forests: Uncertainty and sensitivity analyses of three ecosystem models. *Water, Air, and Soil Pollution, 4*, 67–74.

Lewandrowski, J., Peters, M., Jones, C., House, R., Sperow, M., Eve, M., et al. (2004). Economics of sequestering carbon in the US agricultural sector, Technical Bulletin No. (TB1909), Washington, DC: Department of Agriculture, Economic Research Service.

Li, C., Frolking, S., & Frolking, T. A. (1992). A model of nitrous oxide evolution from soil driven by rainfall events: Model structure and sensitivity. *Journal of Geophysical Research*, 97.

Li, C., Frolking, S., & Harris, N. (1994). Modelling carbon biogeochemistry in agricultural soils. *Global Biogeochemical Cycles, 8*, 237–254.

Li, C., Narayanan, V., & Harris, R. (1996). Model estimates of nitrous oxide emissions from agricultural lands in the United States. *Global Biogeochemical Cycles, 10*, 297–306.

Li, C., Aber, J., Stange, F., Butterbach-Bahl, K., & Papen, H. (2000). A process-oriented model of N_2O and NO emissions from forest soils: 1. Model development. *Journal of Geophysical Research, 105*, 4369–4384.

Li, C., Zhuang, Y., Cao, M., Crill, P., Dai, Z., Frolking, S., et al. (2001). Comparing a national inventory of N_2O emissions from arable lands in China developed with a process-based agroecosystem model to the IPCC methodology. *Nutrient Cycling in Agroecosystems, 60*, 159–175.

Li, C., Mosier, A., Wassmann, R., Cai, Z., Zheng, X., Huang, Y., et al. (2004). Modeling greenhouse gas emissions from rice-based production systems: Sensitivity and upscaling. *Global Biogeochemical Cycles, 18*, 1–19.

Li, S. X., Cun, D. G., Gao, Y. J., He, H. X., & Li, S. Q. (1993). Mineral nitrogen introduced into soil by precipitation on loess plateau. *Agricultural Research in Arid Areas, 11*, 83–92 (in Chinese).

Li, Y. H. (1972). Geochemical mass balance among lithosphere, hydrosphere, and atmosphere. *American Journal of Science, 272*, 119–137.

Liao, C. F. H., & Bartholomew, W. W. (1974). Relationship between nitrate absorption and water transpiration by corn. *Soil Science Society of America Proceedings, 38*, 472–477.

Lieth, H. (1973). Primary production: Terrestrial ecosystems. *Humus Ecology, 1*, 303–332.

Likens, G. E., Wright, R. F., & Galloway, J. N. (1979). Acid rain. *Scientific American, 241*, 39–48.

Lindau, C. W., & Bollich, P. K. (1993). Methane emissions from Louisiana Ist and ratoon crop rice. *Soil Science, 156*, 42–48.

Lindemann, W. C., Connell, G., & Urquhart, N. S. (1988). Previous sludge addition effects on nitrogen mineralization in freshly amended soil. *Soil Science Society of America Journal, 52*, 109–112.

Lipschultz, F., Zafiriou, O. C., Wofsys, S. C., McElroy, M. B., WaLois, F. W., & Watson, S. W. (1981). Production of NO and N_2O by soil nitrifying bacteria. *Nature, 294*, 641–643.

Lipson, D. A, Schmidt, S. K., & Monson, R. K. (2000). Carbon availability and temperature control the post-snowmelt decline in alpine soil microbial biomass. *Soil Biology and Biochemistry, 32*, 441–448.

Lipson, D. A., Schadt, C. W., & Schmidt, S. K. (2002). Changes in soil microbial community structure and function in an alpine dry meadow following spring snow melt. *Microbial Ecology, 43*, 307–314.

Liski, J., & Westman, C. J. (1997). Carbon storage in forest soil of Finland, 1. Effect of thermoclimate. *Biogeochemistry, 36*, 239–260.

Liu, S., Bliss, N., Sundquist, E., & Huntington, T. G. (2003). Modelling carbon dynamics in vegetation and soil under the impact of erosion and deposition. *Global Biogeochemical Cycles, 17*(2), 1074, doi:10.1029/2002GB002010.

Liu, Y. J., Laird, D. A., & Barak, P. (1997). Release and fixation of ammonium and potassium under long-term fertility management. *Soil Science Society of America Journal, 61*, 310–314.

Lloyd, D., Thomas, K. L., Benstead, J., Davies, K. L., Lloyd, S. H., Arah, J. R. M., et al. (1998). Methanogenesis and CO_2 exchange in an ombrotrophic peat bog. *Atmospheric Environment, 32*, 3229–3238.

Lobe, I., Amelung, W., Du Preez, C. C. (2001). Losses of carbon and nitrogen with prolonged arable cropping from sandy soils of the South African Highveld. *European Journal of Soil Science, 52*, 93–101.

Logan, T. J., Lal, R., & Dick, W. A. (1991). Tillage systems and soil properties in North America. *Soil and Tillage Research, 20*, 241–270.

Loranger, G., Ponge, J. F., & Lavalle, P. (2003). Humus forms in two secondary semi-evergreen tropical forests. *European Journal of Soil Science, 54*, 17–24.

Lorenz, M. G., & Wackernagel, W. (1987). Adsorption of DNA to sand and variable degradation rates of adsorbed DNA. *Applied and Environmental Microbiology, 53*, 2948–2952.

Lorimer, G. H. (1981). The carboxylation and oxygenation of ribulose 1,5-biphosphate: The primary events in photosynthesis and photorespiration. *Annual Reviews of Plant Physiology, 32*, 349–383.

Lovett, G. M., & Lindberg, E. (1986). Dry deposition of nitrate to a deciduous forest. *Biogeochemistry, 2*, 137–148.

Lu, Y., Arah, J. R. M., Wassmann, R., & Neue, H. U. (2000a). Simulation of methane production in anaerobic rice soils by a simple two-pool model. *Nutrient Cycling in Agroecosystems, 58*, 277–284.

Lu, Y., Wassmann, R., Neue, H. U., & Huang, C. (2000b). Dynamics of dissolved organic carbon and methane emissions in a flooded rice soil. *Soil Science Society of America Journal, 64*, 2011–2017.

Lugo, A. E., & Brown, S. (1993). Management of tropical soils as sinks or sources of atmospheric carbon. *Plant and Soil, 149*, 27–41.

Lundi, A., Mermut, A. R., & Anderson, D. W. (2003). Origin and rate of pedogenic carbonate accumulation in Saskatchewan soils, Canada. *Geoderma, 117*, 143–156.

Lynch, J. M. (1990). *The Rhizosphere*. Chichester, UK: Wiley Interscience.

Macdonald, J. A., Dise, N. B., Matzner, E., Armbruster, M., & Gundersen, P. (2002). Nitrogen input together with ecosystem nitrogen enrichment predict nitrate leaching from European forests. *Global Change Biology, 8*, 1028–1033.

Machefert, S. E., Dise, N. B., Goulding, K. W. T., & Whitehead, P. G. (2002). Nitrous oxide emission from a range of land uses across Europe. *Hydrology and Earth System Sciences, 63*, 325–337.

MacLaren, J. P. (1996). Plantation forestry– its role as a carbon sink: Conclusions from calculations based on New Zealand's planted forest estate. In M. J. Apps, & D. T. Price (Eds.), *Forest ecosystems, forest management and the global carbon cycle. NATO ASI Series* (Vol. I 40, pp. 255–270). Berlin: Springer.

Mäder, P., Fließbach, A., Dubois, D., Gunst, L., Fried, P., & Niggli, U. (2002). Soil fertility and biodiversity in organic farming. *Science, 296*, 1694–1697.

Magid, J., Müller, J., Jensen, L. S., & Nielsen, N. E. (1997). Modelling the measurable: Interpretation of field-scale CO_2 and N-mineralization, soil microbial biomass and light fractions as indicators of oilseed rape, maize and barley straw decomposition. In G. Cadisch, & K. E. Giller (Eds.), *Driven by nature: Plant litter quality and decomposition* (pp. 349–362). Walllingford, UK: CABI.

Mahieu, N., Powlson, D. S., & Randall, E. W. (1999). Statistical analysis of published carbon-13 CPMAS NMR spectra of soil organic matter. *Soil Science Society of America Journal, 63*, 307–319.

Mahieu, N., Olk, D. C., & Randall, E. W. (2000). Accumulation of heterocyclic nitrogen in humified organic matter: A ^{15}N-NMR study of lowland rice soils. *European Journal of Soil Science, 51*, 379–389.

Maillard, L. C. (1913). Formation de la matière humique par action de polypeptide sur sucres. *Comptes Rendus De L Academie Des Sciences, 156*, 148–149.

Makundi, W., & Okitingati, A. (1995). Carbon flows and economic evaluation of mitigation options in Tanzania's forest sector. *Biomass Bioenergy, 8*(5), 381–393.

Malcolm, R. L., & McCarthy, P. (1991). The individuality of humic substances in diverse environments. *Advances in soil organic matter research: The impact of agriculture and the environment*, 23–34.

Malhi, Y., Baldocchi, D. D., & Jarvis, P. G. (1999). The carbon balance of tropical, temperate and boreal forests. *Plant, Cell and Environment, 22*, 715–740.

Malik, K. A., & Haider, K. (1977). Decomposition of carbon-14-labelled plant materials in saline-sodic soils. In *Soil organic matter studies* (pp. 215–225). Vienna: International Atomic Energy Agency, IAEA-SM-211/23, IAEA.

Mann, L. K. (2004). Changes in soil carbon storage after cultivation. *Soil Science, 142*, 279–288.

Mao, J. D., Hu, W. G., Schmidt-Rohr, K., Davies, G., Ghabbour, E. A., & Xing, B. (2000). Quantitative characterization of humic substances by solid-state carbon-13 nuclear magnetic resonance. *Soil Science Society of America Journal, 64*, 873–884.

Margesin, R., Zimmerbauer, A., & Schinner, A. (2000). Monitoring of bioremediation by soil biological activities. *Chemosphere, 40*, 339–346.

Marion, G. M., Schlesinger, W. H., & Fonteyn, P. J. (1985). CALDEP: A regional model for soil $CaCO_3$ (caliche) deposition in south western deserts. *Soil Science, 139*, 468–481.

Marland, G., Boden, T.A., & Andres, R. J. (2006). Global, regional, and national CO_2 emissions. In *Trends: A compendium of cata on global change*. Carbon Dioxide Information Analysis Center, Oak Ridge National Laboratory. Oak Ridge, TN: US Department of Energy. Available online http://cdiac.esd.gov/trends/emis/tre_glob.htm

Marschner, B., & Kalbitz, K. (2003). Controls of bioavailability and biodegradability of dissolved organic matter in soils. *Geoderma, 113*, 211–235.

Marschner, B., & Wilczynski, A. (1991). The effect of liming on quantity and chemical composition of soil organic matter in a pine forest in Berlin, Germany. *Plant and Soil, 137*, 229–236.

Martens, R. (1985). Limitations in the application of the fumigation technique for biomass estimations in amended soils. *Soil Biology and Biochemistry, 17*, 57–63.

Martens, R. (1990). Contribution of rhizodeposits to the maintainance and growth of soil microbial biomass. *Soil Biology and Biochemistry, 22*, 141–147.

Martin, A., Mariotti, A., Balesdent, J., Lavelle, P., & Vuattoux, R. (1990). Estimate of the organic matter turnover rate in a savanna soil by the ^{13}C natural abundance. *Soil Biology and Biochemistry, 22*, 517–523.

Martin, A., Mariotti, A., Balesdent, J., Lavelle, P., & Vuattoux, R. (1990). Estimate of the organic matter turnover rate in a savanna soil by the ^{13}C natural abundance. *Soil Biology and Biochemistry, 22*, 517–523.

Martin, J. F., & Reddy, K. R. (1997). Interaction and spatial distribution of wetland nitrogen processes. *Ecological Modelling, 105*, 1–21.

Martin, J. K., & Puchridge, D. W. (1982). Carbon flow through the rizosphere of wheat crops in South Australia. In I. E. Galbally, & J. R. Freney (Eds.), *The cycling of carbon, nitrogen, sulphur and phosphorus in terrestrial and aquatic ecosystems* (pp. 77–81). Canberra: Australian Academy of Science.

Martínez-Yrízar, A. (1995). Biomass distribution and primary productivity of tropical deciduous forests. In S. H. Bullock, H. Money, & E. Medina (Eds.), *Seasonally dry tropical forests*. Cambridge: Cambridge University Press.

Masera, O., Bellon, M., & Segura, G. (1995). Forest management options for sequestering carbon in Mexico. *Biomass Bioenergy, 8*(5), 357–368.

Mather, A. S. (1990). Global forest resources. In *Historical Prespectives on Forest Resource Use* (pp. 30–57). Portland, OR: Timber.

Matson, P. (1997). NO_x emission from soils and its consequences for the atmosphere and biosphere: Critical gaps and research direction for the future. *Nutrient Cycling in Agroecosystems, 48*, 1–6.

Matson, P. A., Mcdowell, W. H., Townsend, A. R., & Vitousek, P. M. (1999). The globalization of N deposition: Ecosystem consequences in tropical environments. *Biogeochemistry, 46*, 67–83.

Matthews, H. D., Wassmann, R., & Arah, J. R. M. (2000a). Using a crop/soil simulation model and GIS techniques to assess methane emission from rice fields in Asia. I. Model development. *Nutrient Cycling in Agroecosystems, 58*, 141–159.

Matthews, H. D., Wassmann, R., Buendia, L. V., & Knox, J. W. (2000b). Using a crop/soil simulation model and GIS techniques to assess methane emissions from rice fields in Asia. II. Model validation and sensitivity analysis. *Nutrient Cycling in Agroecosystems, 58*, 161–177.

Matthews, H. D., Wassmann, R., Knox, J. W., & Buendia, L. V. (2000c). Using a crop/soil simulation model and GIS techniques to assess methane emissions from rice fields in Asia. IV. Upscaling to national levels. *Nutrient Cycling in Agroecosystems, 58*, 201–217.

Matthews, H. D., Weaver, A. J., Meissner, K. J., Gillett, N. P., & Eby, M. (2004). Natural and anthropogenic climate change: Incorporating historical land cover change, vegetation dynamics and the global carbon cycle. *Climate Dynamics, 22*, 461–479.

Matthews, R., & Wassmann, R. (2003). Modelling the impacts of climate change and methane emission reductions on rice production: A review. *European Journal of Agronomy, 19*, 573–598.

Mattson, K. G. (1995). CO_2 efflux from coniferous forest soils: Comparison of measurement methods and effect of added nitrogen. In R. Lal, J. Kimble, E. Levine, & B. A. Stewart (Eds.), *Soils and global change* (pp. 329–341). Boca Raton, FL: CRC/Lewis.

Matzner, E., & Murach, D. (1995). Soil changes induced by air pollutant deposition and their implication for forests in central Europe. *Water, Air, and Soil Pollution, 85*, 63–76.

Mayer, H. P., & Conrad, R. (1990). Factors influencing the population of methanogenic bacteria and the initiation of methane production upon flooding of paddy soil. *FEMS Microbiology Letters, 73*, 103–111.

Mba-Chibogu, B., Meyer, B., & Narain, P. (1975). Boden- und Dünger-(^{15}N-) Stickstoff in Ackerparabraunerden aus Löß: Jahresbilanz und Umverteilung auf verschiedene N-Bindungsformen in Abhängigkeit von Bewuchs, N-Düngungsform und Zusatz von Nitrifikationshemmern. *Göttinger Bodenkundliche Berichte, 34*, 1–67.

MBG net (Missouri Botanical Garden) (2002). *Wetlands*. Available online http://www.mbgnet.net/fresh/wetlands/index.htm

McBride, M. B. (1994). *Environmental soil chemistry*. New York: Oxford University Press.

McCalla, T. M., Peterson, J. R., & Lue-Hing, C. (1977). Properties of agricultural and municipal wastes. In L. F. Elliott, & F. J. Stevenson (Eds.), *Soils for management of organic wastes and waste waters* (pp. 9–43). Madison, WI: American Society of Agronomy.

McCarl, B. A., & Schneider, U. A. (2001). The cost of grenhouse gas mitigation in US agriculture and forestry. *Science, 294*, 2481–2482.

McCown, R. L., Hammer, G. L., Hargreaves, J. G. N., Holzworth, D. P., & Freebrain, D. M. (1996). APSIM: A novel software systems for model development, model testing and simulation in agricultural systems research. *Agricultural Systems, 50*, 255–271.

McCracken, K. L., McDowell, W. H., Harter, R. D., & Evans, C. V. (2002). Dissolved organic carbon retention in soils: Comaprison of solution and soil measurements. *Soil Science Society of America Journal, 66*, 563–568.

McDowell, W. H., & Likens, G. E. (1988). Origin, composition and flux of dissolved organic carbon in the Hubbard Brook Valley. *Ecology Monograph, 58*, 177–195.

McGill, W. B. (1996). Review and classification of ten soil organic matter (SOM) models. In D. S. Powlson, P. Smith, & J. U. Smith (Eds.), *Evaluation of soil organic matter models using existing long-term datasets* (pp. 111–132). Heidelberg, Germany: Springer.

McGill, W. B., & Cole, C. V. (1981). Comparative aspects of cycling of organic C, N, S, and P through soil organic matter. *Geoderma, 26*, 267–286.

McGill, W. B., Hunt, H. W., Woodmansee, R. G., & Reuss, J. O. (1981). PHOENIX, a model of the dynamics of carbon and nitrogen in grassland soils. *Ecology Bulletin (Stockholm), 33*, 49–115.

McGuire, A. D., Melillo, J. M., Kicklighter, D. W., Pan, Y., Xiao, X., Helfrich, J., et al. (1997). Equilibrium responses of global net primary production and carbon storage to doubled atmospheric carbon dioxide: Sensitivity to changes in vegetation nitrogen concentration. *Global Biochemical Cycles, 11*, 173–189.

McInnes, A., Thies, J. E., Abbott, L. K., & Howieson, J. G. (2004). Structure and diversity among rhizobial strains, populations and communities – a review. *Soil Biology and Biochemistry, 36*, 1295–1308.

McKeague, J. A., Cheshire, M. V., Andreux, F., & Berthelin, J. (1986). Organo-mineral complexes in relation to pedogenesis, In P. M. Huang, & M. Schnitzer, (Eds.), *Interactions of soil minerals with natural organics and microbes* (pp. 549–592). Madison, WI: Soil Science Society of America.

McLenaghen, R. D., Cameron, K. C., Lampkin, N. H., Daly, M. L., & Deo, B. (1996). Nitrate leaching from ploughed pasture and the effectiveness of winter catch crops in reducing leaching losses. *New Zealand Journal of Agricultural Research, 39*, 413–420.

McNabb, D. H., & Cromack, K. (1983). Dinitrogen fixation by a mature *Ceanothus velutinus* (Dougl.) stand in the western Oregon Cascades. *Canadian Journal of Micribiology, 29*, 1014–1021.

MEA (Millennium Ecosystem Assessment) (2005). *Ecosystems and human well-being: Scenarios. Findings of the Scenarios Working Group*. Washington, DC: Island.

Meesenburg, H., Meiwes, K. J., & Schultz-Sternberg, R. (1994). Entwicklung der atmogenen Stoffeinträge in niedersächsische Waldbestände. *Forst und Holz, 49*, 236–238.

Meesenburg, H., Meiwes, K. J., & Bartens, H. (1999). Veränderung der Elementvorräte im Boden von Buchen- und Fichtenökosystemen im Solling. *Freiburger Forstliche Forschung, 7*, 109–114.

Meharg, A. A. (1994). A critical review of labelling techniques used to quantify rhizosphere carbon flow. *Plant and Soil, 166*, 55–62.

Melillo, J. M. (1983). Will increases in atmospheric concentrations affect litter decay rates? In *The ecosystems center annual report* (pp. 10–11). Woods Hole, MA: Marine Biological Laboratory.

Melillo, J. M., Aber, J. D., & Muratore, J. F. (1982). Nitrogen and lignin control of hardwood leaf litter decomposition dynamics. *Ecology, 63*, 621–626.

Melillo, J. M., Aber, J. D., Linkins, A. E., Ricca, A., Fry, B., & Nadelhoffer, K. J. (1989). Carbon and nitrogen dynamics along the decay continuum: Plant litter to soil organic matter. *Plant and Soil, 115*, 189–198.

Melillo, J. M., Callaghan, T. V., Woodward, F. I., Salati, E., & Sinha, S. K. (1990). Effects on ecosystems. In J. T. Houghton, G. J. Jenkins, & J. J. Ephraums (Eds.), *Climate change: The IPCC scientific assessment* (pp. 283–310). Cambridge: Cambridge University Press.

Menault, J. C., Gignoux, J., Prado, C., & Clobert, J. (1990). Tree community dynamics in a humid savanna of the Cote dÍvoire: Modeling the effects of fire and competition with grass and neighbours. *Journal of Biogeography, 17*, 471–481.

Mengel, K., Horn, D., & Tributh, H. (1990). Availability of interlayer ammonium as related to root vicinity and mineral type. *Soil Science, 149*, 131–137.

Merr, J. L., & Roger, P. (2001). Production, oxidation, emission and consumption of methane by soils: A review. *European Journal of Soil Biology, 37*, 25–50.

Mertens, J., Vanderborght, J., Kasteel, R., Thomas, P., Merck, R., Feyen, J., et al. (2007). Dissolved organic carbon fluxes under bare soil. *Journal of Environmental Quality, 36*, 597–606.

Metherell, A. K., Cambardella, C. A., Parton, W. J., Peterson, G. A., Harding, L. A., & Cole, C. V. (1995). Simulation of soil organic matter dynamics in dryland wheat-fallow cropping systems. In R. Lal, J. Kimble, E. Levine, & B. A. Stewart (Eds.), *Soil management and greenhouse effect* (pp. 259–270). Boca Raton, FL: CRC.

Michalzik, B., & Matzner, E. (1999). Fluxes and dynamics of dissolved organic nitrogen and carbon in a spruce (Picea abies Karst.) forest ecosystem. *European Journal of Soil Science, 50*, 579–590.

Michalzik, B., Kalbitz, K., Park, J. H., Solinger, S., & Matzner, E. (2001). Fluxes and concentrations of dissolved organic carbon and nitrogen- a synthesis for temperate forests. *Biogeochemistry, 52*, 173–205.

Middleburg, J., Soetart, K., Herman, P., & Heip, C. (1996). Denitrification in marine sediments: A model study. *Global Biogeochemical Cycles, 10*, 661–673.

Mikaloff Fletcher, S. E., Bruhwiler, L. M., Miller, J. B., & Heimann, M. (2004). CH_4 sources determined from atmospheric observations of CH_4 and its $^{13}C/^{12}C$ isotopic ratios: 1. Inverse modeling of source processes. *Global Biogeochemical Cycles, 18*, GB4004, doi:10.1029/2004GB002223.

Mikan, C., Schimel, J., & Doyle, A. (2002). Temperature controls of microbial respiration above and below freezimg in Arctic tundra soils. *Soil Biology and Biochemistry, 34*, 1785–1795.

Min, H., Chen, Z. W., Wu, W. X., & Chen, M. C. (2002). Microbial aerobic oxidation of methane in paddy soil. *Nutrient Cycling in Agroecosystems, 64*, 79–85.

Minami, K. (1995). The effect of fertilizer use and other practices in methane emission from flooded rice. *Fertilizer Research, 40*, 71–85.

Mingxing, W., & Jing, L. (2002). CH_4 emission and oxidation in Chinese rie paddies. *Nutrient Cycling in Agroecosystems, 64*, 43–55.

Mishra, S., Rath, A. K., Adhya, T. K., Rao, V. R., & Sethunathan, N. (1997). Effect of continuous and alternate water regimes on methane efflux from rice under greenhouse conditions. *Biology and Fertility of Soils, 24*, 399–405.

Mitra, S., Wassmann, R., & Vlek, P. L. G. (2005). An appraisal of global wetland area and ist orgaic carbon stock. *Current Science*, 88, 25–35.

Mizota, C., &. Van Reeuwijk, L. P. (1989). Clay mineralogy and chemistry of soils formed under volcanic material in diverse climatic regions. *Soil Monograph 2*. Wageningen, The Netherlands: International Soil Reference and Information Centre.

Moeller, M. B., & Vlek, P. L. G. (1982). The chemical dynamics of ammonia volatilization from aqueous solutions. *Atmospheric Environment, 16*, 709–717.

Mohammed, I. C. H. (1979). Fixed ammonium in Lybian soils and its availability to barley seedlings. *Plant and Soil, 53*, 185–196.

Molina, J. A. E., Clapp, C. E., & Larson, W. E. (1980). Potentially mineralizable nitrogen in soil: The simple exponential model does not apply for the first 12 weeks of incubation. *Soil Science Society of America Journal, 44*, 442–444.

Molina, J. A. E., Clapp, C. E., Shaffer, M. J., Chichester, F. W., & Larson, W. E. (1983). NCSOIL, a model of nitrogen and carbon transformations in soil:description, calibration and behavior. *Soil Science Society of America Journal, 47*, 85–91.

Molina, J. A. E., Crocker, G. J., Grace, P. R., Klir, J., Körschens, M., Poulton, P. R., et al. (1997). Simulating trends in soil organic carbon in long-term experiments using the NCSOIL and NCSWAP models. *Geoderma, 81*, 91–107.

Monaghan, R., & Barraclough, D. (1995). Contributions to gross N mineralization from ^{15}N-labelled soil macroorganic matter during laboratory incubation. *Soil Biology and Biochemistry, 27*, 1623–1628.

Moncrieff, B., & Fang, C. (1999). A model for soil CO_2 production and transport 2. Application to a Florida *Pinus elliotte* plantation. *Agriculture and Forest Meteorology, 95*, 237–256.

Monger, H. C., Daughrty, L. A., Lindemann, W. C., & Liddel, C. M. (1991). Microbial precipitation of pedogenic calcite. *Geology, 19*, 997–1000.

Monger, H. C. (2002). Pedogenic carbonates: Links between biotic and abiotic $CaCO_3$. In *17th World Congress of Soil Science*, 14–21 August, 2002, Thailand. Symposium #20, Oral paper #891.

Monk, C. D., & Day, F. P. (1988). Biomass, primary production, and selected nutrient budgets for an undisturbed watershed. In W. T. Swank, & D. A. Crossley (Eds.), *Forest hydrology and ecology at coweeta* (pp. 151–159). *Ecological Studies 66*. Berlin: Springer.

Monteney, G. J., Bannink, A., & Chadwick, D. (2006). Greenhouse gas abatement strategies for animal husbandry. *Agriculture, Ecosystems and Environment, 112*, 163–170.

Moore, J. C., Walter, D. E., & Hunt, H. W. (1988). Arthropod regulation of micro- and mesobiota in belowground food webs. *Annual Review of Entomology, 33*, 419–439.

Moore, T. R., & Jackson, R. J. (1989). Dynamics of dissolved organic carbon in forested and distributed catchments, Westland, New Zealand: 2. Larry river. *Water Resources Research, 25*, 1330–1341.

Moore, T. R., & Roulet, N. T. (1993). Methane flux: Water table relations in northern wetlands. *Geophysical Research Letters, 20*, 587–690.

Moran, M. A., Pomeroy, L. R., Sheppard, L. P., Atkinson, L. P., & Hodson, R. E. (1991). Distribution of terrestrially derived dissolved organic matter in the southeastern US continental shelf. *Limnology and Oceanography, 36*, 1134–1149.

Morgan, J. A., & Parton, W. J. (1989). Characteristics of ammonia volatilization from spring wheat. *Crop Science, 29*, 726–731.

Moser, M., Prentice, C., & Frazier S. (1998). A global overview of wetland loss and degradation. *Wetlands International, Technical Session B of the 6th Meeting of the Conference of the Contracting Parties in Brisbane 1996*, Australia.

Mosier, A. R., Chapman, S. L., & Freney, J. R. (1989). Determination of dinitrogen emission and retention in floodwater and porewater of a lowland rice field fertilized with ^{15}N urea. *Fertilizer Research, 19*, 127–136.

Mosier, A. R., Kroeze, C., Nevison, C., Oenema, O., Seitzinger, S., & Cleemput, O. V. (1998a). Closing the global N_2O budget: Nitrous oxide emissions through the agricultural nitrogen cycle. *Nutrient Cycling in Agroecosystems, 52*, 225–248.

Mosier, A. R., Duxbury, J. M., Freney, J. R., Heinemeyer, O., Minami, K., & Johnson, D. E. (1998b). Mitigating agricultural emissions of methane. *Climatic Change, 40*, 39–80.

Mosier, A. R., Doran, J. W., & Freney, J. R. (2002). Managing soil denitrification. *Journal of Soil and Water Conservation, 57*, 505–513.

Mosier, A. R., Syers, J. K., & Freney, J. R. (Eds.) (2004). *Agriculture and the nitrogen cycle* (p. 296). Washington, DC: Island.

Motavalli, P. P., Palm, C. A., Parton, W. J., Elliott, E. T., & Frey, S. D. (1994). Comparison of laboratory and modelling simulation methods for estimating soil carbon pools in tropical forest soils. *Soil Biology and Biochemistry, 26*, 935–944.

Mouillot, F., Narasimha, A., Balkanski, Y., Lamarque, J. F., & Field, C. B. (2006). Global carbon emissions from biomass burning in the 20th century. *Geophysical Research Letters, 33*, L01801, doi:10.1029/2005GL024707.

Moulton, R., & Richards, K. (1990). Costs of sequestering carbon through tree planting and forest management in the United States. *General Technical Report WO-58*. Washington, DC: US Department of Agriculture.

Moura Costa, P., & Wilson, C. (2000). An equivalence factor between CO_2 avoided emissions and sequestration: Description and applications in forestry. *Mitigation and Adaptation Strategies for Global Change, 5*, 51–60.

Moyano, A., & Gallardo, J. F. (1988). Fixed ammonium determination in some clay soils. *Communications in Soil Science and Plant Analysis, 19*, 225–238.

Müeller, T. (1992). Zeitgang der mikrobiellen Biomasse in der Ackerkrume einer mitteleuropäischen Löss-Parabraunerde. *Ph.D. thesis*, University of Göttingen, Germany.

Mulholland, S. R., & Prior, C. A. (1993). AMS radiocarbon dating of phytoliths. In D. M. Pearsall, & D. R. Piperno (Eds.), *MASC research papers in science and archaeology* (pp. 21–23). Philadelphia: University of Pennsylvania.

Müller, J. (1992). Geographical distibution and seasonal variation of surface emissions and deposition velocities of atmospheric trace gases. *Journal of Geophysical Research, 97*, 3787–3804.

Müller-Wegener, U. (1988). Interactions of humic substances with biota. In *Humic substances and their role in the environment* (pp. 179–192). New York: Wiley.

Murach, D. (1983). Die Reaktionen von Fichtenfeinwurzeln auf zunehmende Bodenversauerung. *Allgemeine Forstzeitschrift, 38*, 683–686.

Murakami, T., Owa, N., & Kumazawa, K. (1987). The effects of soil conditions and nitrogen form on nitrous oxide evolution by denitrification. *Soil Science and Plant Nutrition, 33*, 35–42.

Murphy, D. V., Macdonald, A. J., Stockdale, E. A., Goulding, K. W. T., Fortune, S., Gaunt, J. L., et al. (2000). Soluble organic nitrogen in agricultural soils. *Biology and Fertility of Soils, 30*, 374–387.

Murray, B. C., Sohngen, B., & Ross, M. T. (2007). Economic consequences of consideration of permanence, leakage and additionality for soil carbon sequestration projects. *Climatic Change, 80*, 127–143.

Murty, D., Kirschbaum, M. K., McMurtie, R. E., & McGilvray, H. (2002). Does conversion of forest to agricultural land change soil carbon and nitrogen? A review of the literature. *Global Change Biology, 8*, 105–123.

Myrold, D. D. (1988). Denitrification in ryegrass and winter wheat cropping systems of western Oregon. *Soil Science Society of America Journal, 52*, 412–416.

Nagel, H. D., & Gregor, H. D. (Eds) (1999). *Ökologische Belastungsgrenzen – Critical Loads and Levels*. Berlin/Heidelberg, Germany: Springer.

Nambiar, K. K. M. (1994). *Soil fertility and crop productivity under long-term fertilizer use in India*. New Delhi: Indian Council for Agricultural Research.
Nannipieri, P., Muccini, L., & Ciardi, C. (1983). Microbial biomass and enzyme activities:production and persistence. *Soil Biology and Biochemistry, 15*, 679–685.
Nannipieri, P., Landi, L., & Badalucco, L. (1995). La capacità metabólica e la qualità del suolo. *Agronomia, 29*, 312–316.
Nannipieri, P., Sequi, P., & Fusi, P. (1996). Humus and enzyme activity. In A. Piccolo (Ed.), *Humic substances in terrestrial ecosystems* (pp. 293–328). Amsterdam: Elsevier.
Nannipieri, P., Pietramellara, G., Falchini, L., Bradford, M. A., Parekh, N. R., Pinzari F., et al. (2001). Microbial diversity in soil: Meaning, limits and adavantages of the present methods. In R. M. Rees, B. C. Ball, C. D. Campbell, & C. A. Watson (Eds.), *Sustainable management of soil organic matter* (pp. 393–412). Wallingford, UK:CABI.
Nannipieri, P., Kandeler, E., & Ruggiero, P. (2002). Enzyme activities and microbiological and biochemical processes in soil. In R. G. Burns, & R. Dick (Eds.), *Enzymes in the environment: Activity, ecology, and application* (pp. 1–34). New York: Marcel Dekker.
NASA Earth Observatory (2007). Available online http://earthobservatory.nasa.gov/
Nashölm, T., Huss-Danell, K., & Hogberg, P. (2000). Uptake of organic nitrogen in the field by four agriculturally important plant species. *Ecology, 81*, 1155–1161.
National Wetlands Working Group, Canada Committee on Ecological Land Classification (1988). *Wetlands of Canada*. Ecological land classification series, No. 24. Sustainable Development Branch, Canadian Wildlife Service, Environment Canada.
Neff, J. C., & Asner, G. (2001). Dissolved organic carbon in terrestrial ecosystems:synthesis and model. *Ecosystems, 4*, 29–48.
Neff, J. C., Holland, E. A., Dentener, F. J., Mcdowell, W. H., & Russell, K. M. (2002). The origin, composition and rates of organic nitrogen deposition: A missing piece of the nitrogen cycle. *Biogeochemistry, 57/58*, 99–136.
NEIC (National Energy Information Center) (1998). Land use issues. Available online http://www.eia.doe.gov/ oiaf/1605/gg98rpt/land.html
Neil, C., & Davidson, E. A. (2000). Soil carbon accumulation or loss following deforestation for pasture in the Brazilian Amazons. In R. Lal, J. M. Kimble, & B. A. Steward (Eds.), *Global climate change and tropical ecosystems* (pp. 197–211). Boca Raton, FL: CRC.
Nelson, P. N., & Oades, J. M. (1998). Organic matter, sodicity and soil structure. In M. E. Sumner, & R. Naidu (Eds.), *Sodic soils: Distribution, processes, management and environmental consequences* (pp. 67–91). New York: Oxford University Press.
Neue, H. U. (1992). Agronomic practices affecting methane fluxes from rice cultivation. In D. S. Ojima, & B. H. Svensson (Eds.), *Trace gas exchange in a global perspective, Ecological Bulletin*. 42, 174–182.
Newell, R., & Stavins, R. (2000). Climate change and forest sinks: Factors affecting the costs of carbon sequestration. *Journal of Environmental Economics and Management, 40*, 211–235.
Newell, S. Y. (1996). Established and potential impacts of eukaryotic mycelial decomposers in marine/terrestrial ecotones. *Journal of Experimental Marine Biology and Ecology, 200*, 187–206.
Newman, E. I. (1985). The rhizosphere-C sources and microbial populations. In A. H. Fitter (Ed.), *Ecological interactions in soil* (pp. 107–121). Oxford: Blackwell.
New York State Energy Office (1991). Analysis of carbon reduction in New York State. *Report of the New York Stata Energy Office, in consultation with NYS Department of Environmental Conservation and NYS Department of Public Service*. New York: NYS Energy Office.
Nguyen, C. (2003). Rhizodeposition of organic C by plants: Mechanisms and controls. *Agronomie, 23*, 375–396.
Nicolardot, B., Molina, J. A. E., & Allard, M. R. (1994). C and N fluxes between pools of soil organic matter: Model calibration with long-term incubation data. *Soil Biology and Biochemistry, 26*, 235–243.
Nieder, R. (1998). Bodenbearbeitung und Naehrstoffaustrag. *KTBL Arbeitsbericht, 266*, 91–116.

Nieder, R. (2000). Nährstoffanreicherung in Ackerkrumen vor dem Hintergrund des Boden-, Klima- und Gewässerschutzes. *Zeitschrift für Kulturtechnik und Landentwicklung, 41*, 49–56.

Nieder, R. (2004). Carbon and nitrogen turnover in North German forest soils. *Proceedings of the 3rd International Nitrogen Conference*, October 12–16, 2004, Nanjing, China, 152–158.

Nieder, R., & Benbi, D. K. (1996). Fixation of ammonium in arable soils: An overview. In *Transactions of the 9th Nitrogen Workshop* (pp. 47–50). Germany: Braunschweig.

Nieder, R., & Richter, J. (1989). Die Bedeutung der Umsetzung von Weizenstroh im Hinblick auf den C- und N-Haushalt von Löss-Ackerböden. *Zeitschrift für Pflanzenernährung und Bodenkunde, 152*, 415–420.

Nieder, R., & Richter, J. (2000). C and N accumulation in arable soils of West Germany and its influence on the environment – Developments 1970 to 1998. *Journal of Plant Nutrition and Soil Science, 163*, 65–72.

Nieder, R., Schollmayer, G., & Richter, J. (1989). Denitrification in the rooting zone of cropped soils with regard to methodology and climate: A review. *Biology and Fertility of Soils, 8*, 219–226.

Nieder, R., Kersebaum, K. C., Widmer, P., & Richter, J. (1993). Untersuchungen zur Stickstoff-Immobilisation in mineralisch gedüngten Ackerböden aus Löss während der Vegetationszeit von Winter-Weizen. *Zeitschrift für Pflanzenernährung und Bodenkunde, 156*, 293–300.

Nieder, R., Kersebaum, K. C., & Richter, J. (1995a). Significance of nitrate leaching and long-tern N immobilization after deepening the plough layer for the N regime of arable soils in N.W. Germany. *Plant and Soil, 173*, 167–175.

Nieder, R., Neugebauer, E., Willenbockel, A., & Richter, J. (1995b). Significance of microbial biomass and non-exchangeable ammonium with respect to the nitrogen transformations in loess soils of Niedersachsen during the growing season of winter wheat. I. Change of pool sizes. *Zeitschrift für Pflanzenernährung und Bodenkunde, 158*, 469–475.

Nieder, R., Neugebauer, E., Willenbockel, A., Kersebaum, K. C., & Richter, J. (1996). Nitrogen transformation in arable soils of north-west Germany during the cereal growing season. *Biology and Fertility of Soils, 22*, 179–183.

Nieder, R., Wachter, H., & Isermann, K. (2000). Erhöhte Stoffausträge bald auch aus Waldböden? *Allgemeine Forstzeitschrift/Der Wald, 11*, 594–599.

Nieder, R., Benbi, D. K., & Isermann, K. (2003a). Soil organic matter dynamics. In D. K. Benbi, & R. Nieder (Eds.), *Handbook of processes and modeling in the soil-plant system* (pp. 345–408). New York: Haworth.

Nieder, R., Brinkmann, S., & Böttcher J. (2003b). C and N Sequestration and depletion in pine forest floors: Impacts on drainage water quality. In W. Walther, F. Reinstor, & E. Worch (Eds.), *Proceedings of the International Conference "Diffuse Input of Chemicals into Soil and Groundwater – Assessment & Management"* (pp. 335–340), *Dres-den 27*.

Nieder, R., Köster, W., Dauck, H. P., & Brinkmann, S. (2003c). Nährstoff-Überschüsse in Deutschland von 1950 bis 2000: Quellen, Senken und Wirkungen auf die Umwelt, I. N-Überhang der Landwirtschaft. *Landnutzung und Landentwicklung, 44*, 172–178.

Nieder, R., Köster, W., & Kersebaum, K. C. (2007). Beitrag der Landwirtschaft zu diffusen N-Einträgen. *WasserWirtschaft, 1–2*, 53–57.

Nieder, R., Harden, T., Martens, M., & Benbi, D. K. (2008) (in press). Microbial biomass in arable soils of Germany during the growth period of annual crops. *Journal of Plant Nutrition and Soil Science*.

Niederbudde, E. A. (1983). Das Tonmineral-NH_4, seine Stellung zu anderen N-Bindungsformen sowie seine Bewertung für N-Immobilisierung und -Mobilisierung in Böden. *Kali-Briefe (Büntehof), 16*, 365–378.

Niederbudde, E. A., & Friedrich, R. (1984). Einfluß von Gülle und Mineraldüngung auf N- und K-spezifische Bodeneigenschaften einer Geschiebemergel-Braunerde. *Bayerisches Landwirtschaftliches Jahrbuch, 61*, 781–789.

Nielsen, K. E., Dalgaard, K., & Nornberg, P. (1987). Effects on soils of an oak invasion of a calluna heath, Denmark. I. Morphology and chemistry. *Geoderma, 41*, 79–95.

Niklaus, P. A., & Falloon, P. (2006). Estimating soil carbon sequestration under elevated CO_2 by combining carbon isotope labelling with soil carbon cycle modelling. *Global Change Biology, 12*, 1909–1921.

Nilsson, J., & Grennfelt, P., (Eds). (1988). Critical loads for sulphur and nitrogen, In: *Nord 1988:97* (p. 418) Copenhagen: Nordic Councli of Ministers.

Nilsson, L. G. (1986). Data of yield and soil analysis in the long-term soil fertility experiments. *Journal of the Royal Swedish Academy of Agriculture and Forestry, 18*, 32–70.

Nilsson, S. I. (2003). Soil acidification. In D. K. Benbi, & R. Nieder (Eds.), *Handbook of processes and modeling in the soil-plant system* (pp. 177–198). New York: Haworth.

Nilsson, S. I., Miller, H. G., & Miller, J. D. (1982). Forest growth as a possible cause of soil and water acidification: An examination of the concepts. *Oikos, 39*, 40–49.

Nissen, T. M., & Wander, M. M. (2003). Management and soil-quality effects on fertilizer-use efficiency and leaching. *Soil Science Society of America Journal, 67*, 1524–1532.

Nommik, H. (1967). Distribution of forms of nitrogen in a podsolic soil profile from Garpenberg, Central Sweden. *Journal of Soil Science, 18*, 301–308.

Nommik, H., & Vahtras, K. (1982). Retention and fixation of ammonium and ammonia in soils. In F. J. Stevenson (Ed.), *Nitrogen in agricultural soils* (pp. 123–171). *Agronomy 22.*

Norby, R. J., Gunderson, C. A., Wullschleger, S. D., O'Neill, E. G., & McCraken, M. K. (1992). Productivity and compensatory responses of yellow-poplar trees in elevated CO_2. *Nature, 357*, 322–324.

Nordhaus, W. (1991). The cost of slowing climate change: A survey. *Energy Journal, 12*(1), 37–65.

Nordhaus, W., & Boyer, J. (2000). *Warming the world: Economic models of global warming.* Cambridge, MA: MIT Press.

Nordmeyer, H., & Richter, J. (1985). Incubation experiments on nitrogen mineralization in loess and sandy soils. *Plant and Soil, 83*, 433–445.

Nordt, L. C., Wildung, L. P., & Drees, L. R. (2000). Pedogenic carbonate transformations in leaching soil systems:implications for the global C cycle. In R. Lal, J. M. Kimble, H. Eswaran, & B. A. Stewart (Eds.), *Global climate change and pedogenic carbonates* (pp. 43–64). Boca Raton, FL: Lewis.

Norman, R. J., & Gilmour, J. T. (1987). Utilization of anhydrous ammonia fixed by clay minerals and soil organic matter. *Soil Science Society of America Journal, 51*, 959–962.

Nortcliff, S. (2002). Standardisation of soil quality attributes. *Agriculture, Ecosystems and Environment, 88*, 161–168.

Northup, R. R., Dahlgren, R. A., & McColl, R. A. (1998). Polyphenols as regulators of plant-litter-soil interactions in northern California's pygmy forest: A positive feedback? *Biogeochemistry, 42*, 189–220.

Nouchi, I., Hosono, T., Aoki, K., & Minami, K. (1994). Seasonal variation in methane flux from rice paddies assiciated with methane concentration in soil water, rice biomass and temperature, and its modeling. *Plant and Soil, 161*, 195–208.

NRDC (Natural Resources Defence Council) (2004). Forest facts. Available online http://www.nrdc.org/land/ forests/fforestf.asp

Nuske, A., & Richter, J. (1981). N mineralization in loess-parabrownearths: Incubation experiments. *Plant and Soil, 59*, 237–247.

Nye, P. H., & Greenland, D. J. (1960). The soil under shifting cultivation. In *Commonwealth bureau of soils, Farnham Royal* (p. 156). *Technical Communication 51.*

Nye, P. H., & Greenwood, D. J. (1960). Soil under shifting cultivation. *Technical Communication 51* (pp. 73–126). Harpenden, UK: Commonwealth Bureau of Soils.

Nyman, J. A., & DeLaune, R. D. (1991). CO_2 emissions and soil Eh responses to different hydrological conditions in fresh, brackish, and saline marsh soils. *Limnological Oceanography, 36*, 1406–1414.

Oades, J. M. (1984). Soil organic matter and water-stable aggregates in soils. *Plant and Soil, 76*, 319–337.

Oades, J. M. (1989). An introduction to organic matter in mineral soils. In J. B. Dixon, & S. B. Weed (Eds.), *Minerals in soil environments* (pp. 89–159). Madison, WI: Soil Science Society of America.

Oades, J. M. (1993). The role of biology in the formation, stabilization and degradation of soil structure. *Geoderma, 56*, 377–400.

Oades, J. M., & Waters, A. G. (1991). Aggregate hierarchy in soils. *Australian Journal of Soil Research, 29*, 815–828.

Oades, J. M., Vassallo, A. M., Waters, A. G., & Wilson, M. A. (1987). Characterization of organic matter in particle size and density fractions from a red-brown earth by solid-state ^{13}C NMR. *Ausralian, Journal of Soil Research, 25*, 71–82.

O'Donnell, A. G. (1997). Microbial diversity and the maintainance of ecosystem services: Towards an understanding of community structure:function relationships in soils. In R. Shield (Ed.), *Proceedings of British Soil Science Golden Jubilee Meeting* (pp. 41–42). United Kingdom: University of Newcastleupon Tye.

OECD/IUCN (1996). *Guidelines for aid agencies for improved conservation and sustainable use of tropical and sub-tropical wetlands*. Paris: OECD.

Oechel, W. C., & Billings, W. D. (1992). Effects of global change on the carbon balance of arctic plants and ecosystems. In F. S. Chapin, J. L. Jefferies, J. F. Reynolds, G. R. Shaver, & J. Svoboda (Eds.), *Arctic ecosysems in a changing climate, an ecophysiological perspective*, (pp. 139–168). San Diego, CA: Academic.

Oenema, O., Wrage, N., Velthof, G. L., van Groenigen, J. W., Dolfing, J., & Kuikman, P. J. (2005). Trends in global nitrous oxide emissions from animal production systems. *Nutrient Cycling in Agroecosystems, 72*, 51–65.

Olander, L. P., & Vitousek, P. M. (2000). Regulation of soil phosphatase and chitinase activity by N and P availability. *Biogeochemistry, 49*, 175–190.

O'Leary, J. W., Glenn, E. P., & Watson, M. C. (1985). Agricultural production of halophytes irrigated with seawater. *Plant and Soil, 89*, 311–321.

Olivier, J. G. J., Bouwman, A. F., van der Hoek, K. W., & Berdowski, J. J. M. (1998). Global air emission inventories for anthropogenic sources of NO_x, NH_3 and N_2O in 1990. *Environmental Pollution, 102*, 135–148.

Olivier, J. G. J., van Aardenne, J. A., Dentener, F., Pagliari, V., Ganzeveld, L. N., & Peters, J. A. H. W. (2005). Recent trends in global greenhouse emissions: Regional trends 1970–2000 and spatial distribution of key sources in 2000. *Environmental Sciences, 2*, 81–99.

Olson, J. S., Watts, J. A., & Allison, L. J. (1983). Carbon on live vegetation of major world ecosystems. *Report ORNL-5862 Oak Ridge*. Oak Ridge, TN: National Laboratory.

Olson, M. W., Frye, R. J., & Glenn, E. P. (1996). Effect of salinity and plant species on CO_2 flux and leaching of dissolved organic carbon during decomposition of plant residue. *Plant and Soil, 179*, 183–188.

O'Neill, E. G., & Norby, R. J. (1996). Litter quality and decomposition rates of foliar litter produced under CO_2 enrichment. In *Carbon dioxide and terrestrial ecosystems* (pp. 87–103). San Diego, CA: Academic.

Osborne, G. J. (1976). The significance of intercalary ammonium in representative surface and subsoils from Southern New south Wales. *Australian Journal of Soil Research, 14*, 381–388.

OTA (1990). *Beneath the bottom line: Agricultural approaches to reduce agrichemical contamination of groundwater* (p. 337). Washington, DC: Office of Technological Assessment. US Congress.

Owensby, C. E., Coyne, P. I., Ham, J. M., Auen, L. M., & Knapp, A. K. (1993). Biomass production in a tallgrass prairy ecosystem exposed to ambient and elevated CO_2. *Ecological Applications, 3*, 644–653.

Pacholski, A. S. (2003). Calibration of a simple method for determining ammonia volatilization in the field – experiments in Henan Province, China, and modeling results. *FAL Agricultural Research, Special Issue 249*, Braunschweig, Germany, 172 pp.

Pales, J. C., & Keeling, C. D. (1965). The concentration of atmospheric carbon dioxide in Hawaii. *Journal of Geophysical Research, 70*, 6053–6076.

Pan, Y., Hom, J., Birdsey, R., & Mccullough, K. (2004). Impacts of rising nitrogen deposition on N exports from forests to surface waters in the Chesapeake Bay watershed. *Environmental Management, 33*, S120–S131.

Pankhurst, C. E. (1994). Biological indicators of soil health and sustainable productivity. In D. J. Greenland, & I. Szabolcs (Eds.), *Soil resilience and sustainable land use* (pp. 331–351). Wallingford, UK: CABI.

Pankhurst, C. E., Hawke, B. G., McDonald, H. J., Kirkby, C. A., Buckerfield, J. C., Michelsen, P., et al. (1995). Evaluation of soil biological properties as potential bioindicators of soil health. *Australian Journal of Experimental Agriculture, 35*, 1015–1028.

Parashar, D. C., Rai, J., Gupta, P. K., & Singh, N. (1991). Parameters affecting methane emission from paddy fields. *Indian Journal of Radio Space Physics, 20*, 12–17.

Parks, P., & Hardie, I. (1995). Least-cost forest carbon reserves: Cost-effective subsidies to convert marginal agricultural land to forests. *Land Economics, 71*(1), 122–136.

Parr, J. F., & Sullivan, L. A. (2005). Soil carbon sequestration in phytoliths. *Soil Biology and Biochemistry, 37*, 117–124.

Parr, J. F., Papendick, R. I., Hornick, S. B., & Meyer, R. E. (1992). Soil quality: Attributes and relationship to alternative and sustainable agriculture. *American Journal of Alternative Agriculture, 7*, 5–11.

Parton, W. J., & Rasmussen, P. E. (1994). Long-term effects of crop management in wheat-fallow: II CENTURY model simulations. *Soil Science Society of America Journal, 58*, 530–536.

Parton, W. J., Schimel, D. S., Cole, C. V., & Ojima, D. S. (1987). Analysis of factors controlling soil organic matter levels in Great Plains grasslands. *Soil Science Society of America Journal, 51*, 1173–1179.

Parton, W. J., Mosier, A. R., & Schimel, D. S. (1988). Rates and pathways of nitrous oxide production in a shortgrass steppe. *Biogeochemistry, 6*, 45–48.

Parton, W. J., Ojima, D. S., Cole, C. V., & Schimel, D. S. (1994). A general model for soil organic matter dynamics: Sensitivity to litter chemistry, texture and management. In *Quantitative modeling of soil forming processes* (pp. 147–167). Madison, WI, USA: Soil Science Society of America Special Publication 39.

Parton, W. J., Mosier, A. R., Ojima, D. S., Valentine, D. W., Schimel, D. S., Weier, K., et al. (1996). Generalized model for N_2 and N_2O production from nitrification and denitrification. *Global Biogeochemical Cycles, 10*, 401–412.

Parton, W. J., Hartman, M. D., Ojima, D. S., & Schimel, D. S. (1998). DAYCENT: Its land surface submodel: Description and testing. *Global Planetary Change, 19*, 35–48.

Parton, W. J., Holland, E. A., Del Grosso, S. J., Hartman, M. D., Martin, R. E., Mosier, A. R., et al. (2001). Generalized model for NO_x and N_2O emissions from soils. *Journal of Geophysical Research, 106*, 17403–17420.

Pastor, J., & Post, W. M. (1986). Influence of climate, soil moisture, and succession on forest carbon and nitrogen cycles. *Biogeochemistry, 2*, 3–27.

Paterson, E., Hall, J. M., Rattray, E. A. S., Griffiths, B. S., Ritz, K., & Killham, K. (1997). Effect of elevated CO_2 on rhizosphere carbon flow and soil microbial processes. *Global Change Biology, 3*, 363–377.

Patra, D. D., Brookes, P. C., Colemann, K., & Jenkimson, D. S. (1990). Seasonal changes of soil microbial biomass in an arable and a grassland soil which have been under uniform management for many years. *Soil Biology and Biochemistry, 22*, 739–742.

Patra, D. D., Chand, S., & Anwar, M. (1995). Seasonal changes in microbial biomass in soils cropped with palmarosa (*Cymbopogon martini*) and Japanese mint (*Mentha arvensis*) in subtropical India. *Biology and Fertility of Soils, 19*, 193–196.

Pattey, E., Trzcinski, M. K., & Desjardins, R. L. (2005). Quantifying the reduction of greenhouse gas emissions as a result of composting dairy and beef cattle manure. *Nutrient Cycling in Agroecosystems, 72*, 173–187.

Paul, E. A., & van Veen, J. A. (1978). The use of tracers to determine the dynymic nature of organic matter. *11th Congress of the International Soil Science Society, 3*, 61–102.

Paul, E. A., & Voroney, R. P. (1984). Field interpretation of microbial biomass activity measurements. In M. J. Klug, & C. A. Reddy (Eds.), *Current perspectives in microbial ecology* (pp. 509–514). Washington, DC: American Society of Microbiology.

Paul, E. A., Howarth, R. D., Harris, D., Follett, R., Leavitt, S. W., Kimball, B. A., et al. (1995). Establishing the pool size and fluxes in CO_2 emissions from soil organic matter. In R. Lal, J. Kimble, E. Levine, & B. A. Steward (Eds.), *Soils and global change* (pp. 297–305). Boca Raton, FL: Lewis.

Paul, E. A., Follett, R. F., Leawitt, S. W., Halvorson, A. D., Peterson, G. A., & Lyon, D. (1997a). Carbon isotope ratios of Great Plains soils in wheat-fallow systems. *Soil Science Society of America Journal, 61*, 1058–1067.

Paul, E. A., Paustian, K., Elliott, E. T., & Cole, C. V. (Eds) (1997b). *Soil organic matter in temperate ecosystems: Long-term experiments in North America*. Boca Raton, FL: CRC.

Paul, W., & Domsch, K. H. (1972). Ein mathematisches modell für den nitrifikatioprozeß in den boden. *Archives Microbiology, 87*, 77–92.

Paustian, K. (1994). Modeling soil biology and biochemical processes for sustainable agriculture research. In Z. E. Pankhurst, B. M. Doube, V. V. S. R. Gupta, & P. R. Grace (Eds.), *Soil biota management in sustainable farming systems* (pp. 182–193). Melbourne, Australia: CSIRO Information Services.

Paustian, K., & Schnürer, J. (1987). Fungal growth response to carbon and nitrogen limitation: A theoretical model. *Soil Biology and Biochemistry, 19*, 613–620.

Paustian, K., Parton, W. J., & Persson, J. (1992). Modeling soil organic matter in organic-amended and nitrogen-fertilized long-term plots. *Soil Science Society of America Journal, 56*, 476–488.

Paustian, K., Ågren, G. I., & Bosatta, E. (1997a). Modelling litter quality effects on decomposition and soil organic matter dynamics. In G. Cadish, & K. E. Giller (Eds.), *Driven by nature: Plant litter quality and decomposition* (pp. 313–335). Wallingford, UK: CABI.

Paustian, K., Andren, O., Janzen, H. H., Lal, R., Smith, P., Tian, G., et al. (1997b). Agricultural soils as a sink to mitigate CO_2 emissions. *Soil Use and Management, 13*, 230–244.

Paustian, K., Collins, B., & Paul, E. A. (1997c). Management controls on soil carbon. In E. A. Paul, K. Paustian, E. A. Elliot, & C. V. Cole (Eds.), *Soil organic matter in temperate agroecosystems: Long-term experiments in North Amreica* (pp. 15–49). Boca Raton, FL: CRC.

Paustian, K., Levine, E., Post, W. M., & Ryzhova, I. M. (1997d). The use of models to integrate information and understanding of soil C at the regional scale. *Geoderma, 79*, 227–260.

Paustian, K., Cole, C. V., Sauerbeck, D. R., & Sampson, N. (1998). CO_2 mitigation by agriculture: An overview. *Climatic Change, 40*, 135–162.

Paustian, K., Six, J., Elliott, E. T., & Hunt, H. W. (2000). Management options for reducing CO_2 emissions from agricultural soils. *Biogeochemistry, 48*, 147–163.

Paustian, K., Babcock, B. A., Hatfield, J., Lal, R., McCarl, B. A., McLaughlin, S., Mosier, A., Rice, C., Robertson, G. P., Rosenberg, N. J., Rosenzweig, C., Schlesinger, W. J., & Zilberman, D. (2004). Agricultural mitigation of greenhouse gases: Science and policy options. *CAST (Council on Agricultural Science and Technology) Report* (120 pp.), R141 2004, ISBN 1-887383-26-3.

Paustian, K., Antle, J. M., Sheehan, J., & Paul, E. A. (2006). *Agriculture's role in greenhouse gas mitigation* (76 pp.). Arlington, VA: Pew Center on Global Climate Change.

Pautsch, G. R., Kurkalova, L. A., Babcock, B. A., & Kling, C. L. (2001). The efficiency of sequestering carbon in agricultural soils. *Contemporary Economic Policy, 19*(2), 123–134.

Perakis, S. S., & Hedin, L. O. (2002). Nitrogen loss from unpolluted South American forests mainly via dissolved organic compounds. *Nature, 415*, 416–419.

Pérez, T., Romero, J., & Sanhueza, E. (2007). Effect of conversion of natural grassland to cropland on nitric oxide emissions from Venezuelan savanna soils: A four-year monitoring study. *Nutrient Cycling in Agroecosystems, 77*, 101–113.

Persson, O., Eriksson, H., & Johansson, U. (1995). An attempt to predict long-term effects of atmospheric nitrogen deposition on the yield of Norway spruce stands (*Picea abies* (L.) Karst.) in southwestern Sweden. *Plant and Soil, 168/169*, 249–254.

Persson, T. (1989). The role of animals in C and N mineralization. In M. Clarholm, & L. Bergström (Eds.), *Ecology of arable land* (pp. 185–189). Dordrecht, The Netherlands/Boston, MA/London: Kluwer.

Petersburgsky, A. V., & Smirnov, P. M. (1966). Ammonium fixation by some soils in the USSR and the availability of this ion to plants. *Plant and Soil, 25*, 119–128.

Petersen, B. J., Bernsten, J., Hansen, S., & Jensen, L. S. (2005). CN-SIM- a model for the turnover of soil organic matter. I. Long-term carbon and radiocarbon development. *Soil Biology and Biochemistry, 37*, 359–374.

Peterson, G. A., & Vetter, D. J. (1971). Soil nitrogen budget of the wheat-fallow area. *Farm Ranch Home, 12*, 23–25.

Peterson, G. A., Halvorson, A. D., Havlin, J. L., Jones, O. R., Lyon, D. J., & Tanaka, D. L. (1998). Reduced tillage and increasing cropping intensity in the Great Plains conserves soil C. *Soil and Tillage Research, 47*, 207–218.

Pettersson, B. D., & Von Wistinghausen, E. (1979). Effects of organic and inorganic fertilizers on soils and crops, results of long term experiments in Sweden. *Woods End Agricultural Institute, 1*, 44.

Piccolo, A. (2002). The supramolecular structure of humic substances: A novel understanding of humus chemistry and implications in soil science. *Advances in Agronomy, 75*, 57–134.

Piccolo, A., Nardi, S., & Concheri, G. (1996a). Macromolecular changes of humic substances induced by interaction with organic acids. *European Journal of Soil Science, 47*, 319–328.

Piccolo, A., Nardi, S., & Concheri, G. (1996b). Micelle-like conformation of humic substances as revealed by size exclusion chromatography. *Chemoshpere, 33*, 595–602.

Piccolo, A., Conte P., Cozzolino, A., & Spaccini, R. (2003). The conformational structure of humic substances. In D. K. Benbi, & R. Nieder (Eds.), *Handbook of processes and modeling in the soil-plant system* (pp. 83–120). New York: Food Products Press/The Haworth Reference Press.

Pierzynski, G. M., Sims, J. T., & Vance, G. F. (Eds) (2000). *Soils and environmental quality* (459 pp.). Boca Raton, FL: CRC.

Pietramellara, G., Dal Canto, L., Vettori, C., Gallori, E., & Nannipieri, P. (1997). Effects of air-drying and wetting cycles on the transforming ability of DNA bound on clay minerals. *Soil Biology and Biochemistry, 29*, 55–61.

Plantinga, A., Mauldin, T., & Miller, D. (1999). An econometric analysis of the costs of sequestering carbon in forests. *American Journal of Agricultural Economy, 81*, 812–824.

Platt, U., Allan, W., & Lowe, D. C. (2004). Hemispheric average Cl atom concentration from $^{13}C/^{12}C$ ratios in atmospheric methane. *Atmospheric Chemistry and Physics, 4*, 2393–2399.

Pocknee, S., & Sumner, M. E. (1997). Cation and nitrogen contents of organic matter determine its soil liming potential. *Soil Science Society of America Journal, 61*, 86–92.

Polglase, P. J., Keryn, I. P., Khanna, P. K., Nyakuengama, J. G., O'Connell, A. M., Grove, T. S., & Battaglia, M. (2000). Change in soil carbon following afforestation and reforestation. *Australian Greenhouse Office, National Carbon Accounting System Technical Report No. 20* (177 pp.). Available online http://www.greenhouse.gov.au/ncas/reports/pubs/tr20final.pdf

Ponge, J. F. (2003). Humus forms in terrestrial ecosystems: A framework to biodiversity. *Soil Biology and Biochemistry, 35*, 935–945.

Ponnamperuma, F. N. (1972). The chemistry of submerged soils. *Advances in Agronomy, 24*, 29–96.

Post, A. M., Emanuel, W. R., Zinke, P. J., & Strangenbberger, A. G. (1982). Soil carbon pools and world life zones. *Nature, 298*, 156–159.

Post, W. M., & Kwon, K. C. (2000). Soil carbon sequestration and land-use change: Processes and potential. *Global Change Biology, 6*, 317–327.

Post, W. M., & Mann, L. K. (1990). Changes in soil organic carbon and nitrogen as a result of cultivation. In A. F. Bouwman (Ed.), *Soils and the greenhouse effect* (pp. 401–407). Chichester/Sussex/England: Wiley.

Post, W. M., Emanuel, W. R., & King, A. W. (1992). Soil organic matter dynamics and the glo-bal carbon cycle. In N. H Batjes, & E. M. Bridges (Eds.), *WISE Report 3* (pp. 107–119). Wageningen, The Netherlands: International Soil Reference and Information Centre.

Poth, M., & Focht, D. D. (1985). ^{15}N kinetic analysis of N_2O production by *Nitrosomonas europaea*: An examination of nitrifier denitrification. *Applied Environmental Microbiology, 49*, 1134–1141.
Potter, C. S. (1997). An ecosystem simulation model for methane production and emission from wetlands. *Global Biogeochemical Cycles, 11*, 495–506.
Potter, C. S., Randerson, J. T., Field, C. B., Matson, P. A., Vitousek, P. M., Mooney, H. A., et al. (1993). Terrestrial ecosystem production: A process model based on global satellite and surface data. *Global Biogeochemical Cycles, 7*, 811–841.
Potter, C. S., Matson, P. A., Vitousek, P. M., & Davidson, E. A. (1996). Process modeling of controls on nitrogen trace gas emissions from soils worldwide. *Journal of Geophysical Research, 101*, 1361–1377.
Potter, C. S., Riley, R. H., & Klooster, S. A. (1997). Simulation modeling of nitrogen trace gas emissions along an age gradient of tropical forest soils. *Ecological Modelling, 97*, 179–196.
Powlson, D. S. (1996). Why evaluate soil organic matter models? In D. S. Powlson, P. Smith, & J. U. Smith (Eds.), *Evaluation of soil organic matter models using existing long-term datasets* (pp. 3–11). Heidelberg, Germany: Springer.
Powlson, D. S., Poulton, P. R., Addiscott, T. M., & McCann, D. S. (1989). Leaching of nitrate from soils receiving organic or inorganic fertilizers continuously for 135 years. In J. A. A. Hansen, & K. Henriksen (Eds.), *Nitrogen in organic wastes applied to soils* (pp. 334–345). London: Academic.
Powlson, D. S., Smith, P., & Smith, J. U. (Eds.) (1996). *Evaluation of soil organic matter models using existing long-term datasets*. Heidelberg, Germany: Springer.
Powlson, D. S., Smith, P., Coleman, K., Smith, J. U., Glendining, M. J., Körschens, M., et al. (1998). A European network of long-term sites for studies on soil organic matter. *Soil and Tillage Research, 47*, 263–274.
Pratt, P. F. (1984). Nitrogen use and nitrate leaching in irrigated agriculture. In R. D. Hauck (Ed.), *Nitrogen in crop production* (pp. 319–333). Madison, WI: ASA, CSSA, SSSA.
Praveen-Kumar, Tripathi, K. P., & Aggarwal, R. K. (2002). Influence of crops, crop residues and manure on amino acid and amino sugar fractions of organic nitrogen in soil. *Biology and Fertility of Soils, 35*, 210–213.
Pressenda, L. C. R., Aravena, R., Mefi, A. J., Telles, E. C. C., Boulet, R., Vanencia, E. P. E., et al. (1996). The use of carbon isotopes (C-13, C-14) in soil to evaluate vegetation changes during the Holocene in central Brazil. *Radiocarbon, 38*, 191–201.
Preston, C. M. (1996). Application of NMR to soil organic matter analysis: History and prospects. *Soil Science, 161*, 144–166.
Preston, C. M., & Schnitzer, M. (1984). Effects of chemical modification and extractants on the ^{13}C NMR spectra of humic materials. *Soil Science Society of America Journal, 48*, 305–311.
Prihar, S. S., Gajri, P. R., Benbi, D. K., & Arora, V. K. (2000). *Intensive cropping: Efficient use of water, nutrients and tillage* (p. 264). New York: Food Products Press.
Proctor, J. (1983). Mineral nutrients in tropical forests. *Progress in Physical Geography, 7*, 422–431.
Prud'homme, M. (2005). Global nitrogen fertilizer supply and demand outlook. *Science in China (Life Sciences), 48*, 818–826.
Puget, P., Angers, D. A., & Chenu, C. (1999). Nature of carbohydrates associated with water-stable aggregates of two cultivated soils. *Soil Biology and Biochemistry, 31*, 55–63.
Puhlmann, M., Kuka, K., & Franko, U. (2006). Comparison of methods for the estimation of inert carbon suitable for initialisation of the CANDY model. *Nutrient Cycling in Agroecosystems, 74*, 295–304.
Pumpanen, J., Ilvesniemi, H., & Hari, P. (2003). A process-based model for predicting soil carbon dioxide efflux and concentration. *Soil Science Society of America Journal, 67*, 402–413.
Qafoku, N. P., Sumner, M. E., & Radcliff, D. E. (2000). Anion transport in columns of variable charge subsoils: nitrate and chloride. *Journal of Environmental Quality, 29*, 484–493.
Qi-Xiao, W., Li-Li, C., & Xiao-Hua, Z. (1995). Fixed ammonium contents and NH_4^+ fixation capacities of some cultivated soils in China. *Pedosphere, 5*, 315–323.
Qualls, R. G. (2000). Comparison of the behaviour of soluble organic and inorganic nutrients in forest soils. *Forest Ecology Management, 138*, 29–50.

Qualls, R. G., & Bridgham, S. D. (2005). Mineralization rate of ^{14}C-labelled dissolved organic matter from leaf litter in soils of a weathering chronosequence. *Soil Biology and Biochemistry, 37*, 905–916.

Qualls, R. G., & Haines, B. L. (1992). Biodegradability of dissolved organic matter in forest throughfall, soil solution, and stream water. *Soil Science Society of America Journal, 56*, 578–586.

Qualls, R. G., Haines, B. L., & Swank, W. T. (1991). Fluxes of dissolved organic nutrients and humic substances in a deciduous forest. *Ecology, 72*, 254–266.

Quiroga, A. R., Buschiazzo, D. E., & Peinemann, N. (1996). Soil organic matter particle size fractions in soils of the semiarid Argentinian pampas. *Soil Science, 161*, 104–108.

Rabenhorst, M. C. (1995). Carbon storage in tidal marsh soils. In R. Lal, J. Kimble, E. Levine, & B. A. Steward (Eds.), *Soils and the global change* (pp. 93–103). Boca Raton, FL: Lewis.

Rabenhorst, M. C., & Wilding, L. P. (1986). Pedogenesis on the Edwards Plateau, Texas. III. A new model for the formation of petrocalcic horizons. *Soil Science Society of America Journal, 50*, 693–699.

Raber, B., Kögel-Knabner, I., Stein, C., & Klem, D. (1998). Partitioning of polycyclic aromatic hydrocarbons to dissolved organic matter from different soils. *Chemosphere, 36*, 79–97.

Rachhpal-Singh, & Nye, P. H. (1986). A model of ammonia volatilization from applied urea. I. Development of the model. *Journal of Soil Science, 37*, 9–20.

Raffaelle, J. B., McGregor K. C., Foster, G. R., & Cullum, R. F. (1997). Effect of narrow grass strips on conservation reserve land converted to cropland. Available online http://cat.inist.fr/?aModele=afficheN&cpsidt=2163164

Raich, J. W., & Potter, C. S. (1995). Global patterns of carbon dioxide emissions from soils. *Global Biogeochemical Cycles, 9*, 23–36.

Raich, J. W., & Schlesinger, W. H. (1992). The global carbon dioxide flux in soil respiration and its relationship to climate. *Tellus, 44B*, 81–99.

Raich, J. W., Potter, C. S., & Bhagawati, D. (2002). Interannual variability in global soil respiration. *Global Change Biology, 8*, 800–812.

Raich, T. M., Rastetter, E. B., Melillo, J. M., Kicklighter, D. W., Steudler, P. A., Peterson, B. J., et al. (1991). Potential net primary productivity in South America: An application of a global model. *Ecological Applications, 1*, 399–429.

Rasmussen, P. E., Douglas, C. L., Colins, H. P., & Albrecht, S. L. (1998). Long-term cropping system effects on mineralizable nitrogen in soil. *Soil Biology and Biochemistry, 13*, 1829–1837.

Ratering, S., & Conrad, R. (1998). Effects of short-term drainage and aeration on the production of methane in submerged rice soil. *Global Change Biology, 4*, 397–407.

Ravindranath, N., & Somashekhar, B. (1995). Potential and economics of forestry options for carbon sequestration in India. *Biomass Bioenergy, 8*(5), 323–336.

Reddy, K. R., & Patrick, W. H. J. (1980). Losses of applied $^{15}NH_4$-N, urea-^{15}N and organic ^{15}N in flooded soil. *Soil Science, 130*, 326–333.

Reddy, K. R., Feijtel, T. C., & Patrick, W. H. (1986). Effect of soil redoxcoditions on microbial oxidation of organic matter. In Y. Chen, & Y. Avnimelech (Eds.), *The role of organic matter in modern agriculture* (pp. 117–155). The Netherlands: Martinus Nijhoff.

Reddy, K. R. & D'Angelo, E. M. (1994). Soil Processes regulating water quality in wetlands. In W. Mitsch (Ed.), Global Wetlands-Old World and Nerv (pp. 309–324). New York: Elsevier.

Reddy, K. R., D'Angelo, E. M., & Harris, W. G. (2000). Biogeochemistry of wetlands. In M. E. Sumner (Ed.), *Handbook of soil science* (pp. G-89–G-119). Boca Raton, FL: CRC.

Redfield, A. C. (1972). Development of a New England salt marsh. *Ecological Monograms, 42*, 201–237.

Redfield, A. C., & Rubin, M. (1962). The age of salt marsh peat and its relation to recent changes in sea level at Barnstable, MA. *Proceedings of the National Academy of Sciences of the United States of America, 48*, 1728–1735.

Reichardt, W., Meepetch, S., Briones, A. M., Ruaysoongnern, S., & Naklang, K. (1998). Microbial biomass, its assessment, functions, and management in drought-prone, nutrient-poor rainfed lowland soils. In J. K. Ladha, L. Wade, A. Dobermann, W. Reichardt, G. Kirk, & C. Piggin

(Eds.), *Rainfed lowland rice: Advances in nutrient management research* (pp. 141–159). Manila: International Rice Research Institute (IRRI).
Reicosky, D. C., Kemper, W. D., Langdale, G. W., Douglas, C. L., & Rasmussen, P. E. (1995). Soil organic matter changes resulting from tillage and biomass production. *Journal of Soil Water Conservation, 50*, 253–261.
Reicosky, D. C., Dugas, W. A., & Torbert, H. A. (1997). Tillage-induced soil carbon dioxide loss from different cropping systems. *Soil and Tillage Research, 41*, 105–118.
Reilly, J. M., & Fuglie, K. O. (1998). Future yield growth in field crops: What evidence exists? *Soil and Tillage Research, 47*, 275–290.
Reilly, J. M., & Schimmelpfennig, D. (1999). Agricultural impact assessment, vulnerability, and the scope for adaptation. *Climatic Change, 43*, 745–788.
Remde, A., & Conrad, R. (1991). Role of nitrification and denitrification for NO metabolism in soil. *Biogeochemistry, 12*, 189–205.
Renault, P., & Stengel, P. (1994). Modeling oxygen diffusion in aggregated soils: I. Anaerobiosis inside the aggregates. *Soil Science Society of America Journal, 58*, 1017–1023.
Renger, M., & Strebel, O. (1976). Nitratantlieferung an die Pflanzenwurzel als Funktion der Tiefe und der Zeit. *Landwirtshaftliche Forschung SH, 33*(II), 13–19.
Renwick, W. H., Smith, S. V., Sleezer, R. O., & Buddemeier, R. W. (2004). Comment on managing soil carbon (II). *Science, 305*, 1567c.
Restrepo, C. & Naranjo, L. G. (1987). Recuente histórico de la disminución de humedales y la desaparacón de aves acuáticas en la Valle Cauca, Colombia. In Alvarez, H., Kattan, G. & Murcia, C. (Eds.). Memorias III Congreso de Ornitología Neotropical. Cali: ICBP-USFWS-SVO. Pp. 43–45.
Ribaudo, M., Colaccico, D., Barbarika, A., & Young, C. (1989). The economic efficiency of voluntary soil conservation programs. *Journal of Soil and Water Conservation, 44*, 40.
Richards, K. (1997). The time value of carbon in bottom-up studies. *Critical Reviews in Environmental Science and Technology, 27*, 279–292.
Richards, K., & Anderson, K. (2001). The leaky sink: Persistent obstacles to a forest carbon sequestration program based on individual projects. *Climate Policy, 1*, 41–54.
Richards, K., & Stokes, C. (2004). A review of forest carbon sequestration cost studies: A dozen years of research. *Climatic Change, 63*, 1–48.
Richards, K., Moulton, R., & Birdsey, R. (1993). Costs of creating carbon sinks in the US *Energy Conservation and Management, 34*(9–11), 905–912.
Richter, J., & Benbi, D. K. (1996). Modeling of nitrogen transformations and translocations. *Plant and Soil, 181*, 109–121.
Richter, J., & Roelcke, M. (2000). The N-cycle as determined by intensive agriculture – examples from central Europe and China. *Nutrient Cycling in Agroecosystems, 57*, 33–46.
Richter, J., Nuske, A., Habenicht, W., & Bauer, J. (1982). Optimized N-mineralization parameters of loess soils from incubation experiments. *Plant and Soil, 68*, 379–388.
Ridgwell, A., Marshall, S. J., & Gregson, K. (1999). Consumption of atmospheric methane by soils: A process-based model. *Global Biogeochemical Cycles, 13*, 59–70.
Rijtema, P. E., & Kroes, J. G. (1991). Some results of nitrogen simulations with model ANIMO. *Fertilizer Research, 27*, 189–198.
Riley, W. J., & Matson, P. A. (2000). NLOSS: A mechanistic model of denitrified N_2O and N_2 evolution from soil. *Soil Science, 165*, 237–249.
Robertson, G. P. (2004). Abatement of nitrous oxide, methane and other non-CO_2 greenhouse gases: The need for a systems approach. In C. B. Field, & M. R. Raupach (Eds.), *The global carbon cycle. Integrating humans, climate and the natural world* (pp. 493–506). Washington, DC: SCOPE 62, Island.
Robie, R. A., Hemingway, B. S., & Fischer, J. R. (1978). Thermodynamic properties of minerals and related substances at 298.15 K and 1 bar (10^5 Pascal) pressure and at high temperatures *US Geological Bulletin*, 1452.
Rochette, P., & Janzen, H. H. (2005). Towards a revised coefficient for estimating N_2O emissions from legumes. *Nutrient Cycling in Agroecosystems*, 73.

Rodrigues, G. (1954). Fixed ammonium in tropical soils. *Journal of Soil Science, 5*, 264–274.

Roelcke, M., Han, Y., Li, S. X., & Richter, J. (1996). Laboratory measurements and simulations of ammonia volatilization from urea applied to calcareous Chinese loess soils. *Plant and Soil, 181*, 123–129.

Roelcke, M., Li, S. X., Tian, X. H., Gao, Y., & Richter, J. (2002). In situ comparisons of ammonia volatilization from N fertilizers in Chinese loess soils. *Nutrient Cycling in Agroecosystems, 62*, 73–88.

Roger, P. A. (1996). *Biology and management of the floodwater ecosystem in ricefields*. Manila: International Rice Research Institute, ORSTOM.

Röhricht, C., Müller, A., & Schmidt, A. (2006). Prüfung stndortangepasster Anbausysteme unter Berücksichtigung des Klimawandels. In J. Mathews, St. Marahrens, & S. Schmidt (Eds.), *UBA-workshop: Klimaänderungen – Herausforderungen für den Bodenschutz, September 28–29, 2005* (pp. 117–121). Umweltbundesamt Dessau.

Romanya, J., Cortina, J., Falloon, P., Coleman, K., & Smith, P. (2000). Modelling changes in soil organic matter after planting fast-growing *Pinus radiata* on mediterranean agricultural soils. *European Journal of Soil Science, 51*, 627–641.

Römkens, P. F. A. M., Van der Plicht, J., Hassink, J. (1985). Soil organic matter dynamics after the conversion of arable land to pasture. *Biology and Fertility of Soils, 28*, 277–284.

Roose, E., & Barthes, B. (2001). Organic matter management for soil conservation and productivity restoration in Africa: A contribution from Francophone research. *Earth and Environmental Science, 61*, 159–170.

Roper, J. (2003). *Tropical forests and climate change*. Available online http://www.rcfa-cfan.org/english/issues.13.html

Rosegrant, M., Paisner, M. S., & Meijer, S. (2001). *Long-term prospects for agriculture and the resource base. The world bank rural development family*. Rural development strategy background paper 1. Washington, DC: The World Bank.

Rosenberg, N. J. (1981). The increasing CO_2 concentration in the atmosphere and ist implications on agricultural productivity, I. Effects on photosynthesis, transpiration and water use efficiency. *Climatic Change, 3*, 265–279.

Rosenzweig, C., & Hillel, D. (1995). Potential impacts of climate change on agriculture and food supply. In *Consequences* (Vol. 1, No. 2, Summer 1995). Washington, DC: US Global Change Research Information Office.

Ross, D. J. (1989). Estimation of soil microbial C by a fumigation-extraction procedure: Influence of soil moisture content. *Soil Biology and Biochemistry, 21*, 767–772.

Rouhier, H., Billes, G., El Kohen, A., Mousseau, M., & Bottner, P. (1994). Effect of elevated CO_2 on carbon and nitrogen distribution within a tree (*Castanea sativa* Mill.)-soil system. *Plant and Soil, 162*, 281–292.

Roy, R. N., Misra, R. V., & Montanez, A. (2002). Decreasing reliance on mineral nitrogen – yet more food. *Ambio, 31*, 177–183.

RSPB. (1993). *Wet grasslands – what future?* Sandy, UK: RSPB.

Ruimy, A., Jarvis, P. G., Baldocchi, D. D., & Saugier, B. (1995). CO_2 fluxes over plant canopies and solar radiation: A review. *Advanced Ecology Research, 26*, 1–68.

Rumpel, C., Kögel-Knabner, I., & Hüttl, R. F. (1999). Organic matter composition and degree of humification in lignite-rich mine soils under a chronosequence of pine. *Plant and Soil, 213*, 161–168.

Ryden, J. C. (1981). Nitrous oxide exchange between a grassland soil and the atmosphere. *Nature, 292*, 235–237.

Sabine, C. L., Heimann, M., Artaxo, P., Bakker, D. C. E., Chen, C. T. A., Field, C. B., et al. (2004). Current status and past trends of the global carbon cycle. In C. B. Field, & M. R. Raupach (Eds.), *The global carbon cycle* (pp. 17–44). Washington, DC: Island.

Sadeghi, A. M., McInnes, K. J., Kissel, D. E., Cabrera, M. L., Koelliker, J. M., & Kanemasu, E. T. (1988). Mechanistic model for predicting ammonia volatilization from urea. In B. R. Bock, & D. E. Kissel (Eds.), *Bulletin Y-206, Ammonia volatilization from urea fertilizers* (pp. 67–92). Muscle Shoals, AL: National Fertilizer Development Center, Tennessee Valley Authority.

Saffigna, P. G., Powlson, D. S., Brookes, P. C. & Thomas, G. A. (1989). Influence of sorghum residues and tillage on soil organic matter and soil organic matter and soil microbial biomass in an Australian Vertisol. Soil Biology and Biochemistry 21, 759–765.

Sage, R. F., & Monson, R. K. (1999). *C_4 plant Biolog*. San Diego, CA: Academic.

Saggar, S., Tate, K. R., Feltham, C. W., Childs, C. W., & Parshotam, A. (1994). Carbon turnover in a range of allophonic soils amended with ^{14}C-labelled glucose. *Soil Biology and Biochemistry, 26*, 1263–1271.

Saggar, S., Hedley, C., & Mackay, A. D. (1997). Partitioning and translocation of photosyntetically fixed ^{14}C in grazed hill pastures. *Biology and Fertility of Soils, 25*, 152–158.

Saggar, S., Andrew, R. M., Tate, K., Hedley, C. B., Rodda, N. J., & Townsend, A. R. (2004a). Modelling nitrous oxide emissions from dairy-grazed pastures. *Nutrient Cycling in Agroecosystems, 68*, 243–255.

Saggar, S., Hedley, C. B., Giltrap, D. L., Tate, K., Lambie, S., & Li, C. (2004b), Nitrous oxide emissions from grazed pastures. In *SuperSoil 2004: 3rd Australian New Zealand Soils Conference* [published on CDROM]. Australia: University of Sydney. Website www.regional.org.au/au/assi/

Sahrawat, K. L. (1995). Fixed ammonium and carbon-nitrogen ratios of some semi-arid tropical Indian soils. *Geoderma, 68*, 219–224.

Salinas-Garcia, J. R., Hons, F. N:, & Matocha, J. E. (1997). Long-term effects of tillage and fertilization on soil organic matter dynamics. *Soil Science Society of America Journal, 61*, 152–159.

Sallih, Z., & Pansu, M. (1993). Modeling of soil carbon forms after organic amendments under controlled conditions. *Soil Biology and Biochemistry, 25*, 1755–1762.

Sanders, J., Southgate, D., & Lee, J. (1995). *The economics of soil degradation: Technological change and policy alternatives* (Technical Monograph Number 22). Washington, DC: USAID Soil Management Support Service.

Sanhueza, E. (1997). Impact of human activity on NO soil fluxes. *Nutrient Cycling in Agroecosystems, 48*, 61–68.

Sanhueza, E., Cardenas, L., Donoso, L., & Santana, M. (1994). Effect of plowing on CO_2, CO, CH_4, N_2O and NO fluxes from tropical Savannah soils. *Journal of Geophysical Research, 99*, 16429–16434.

Sardinha, M., Mueller, T., Schmisky, H., & Jörgensen, R. G. (2003). Microbial performance in a temperate floodplain soil along a salinity gradient. *Applied Soil Ecology, 23*, 237–244.

Sass, R. L., & Fisher, F. M. J. (1994). CH_4 emission from paddy fields in the United States gulf coast area. In K. Minami, A. Mosier, & R. L. Sass (Eds.), *CH_4 and N_2O: Global emissions and controls from rice fields and other agricultural and industrial sources* (pp. 65–77). Tsukuba, Japan: NIAES Series 2.

Sass, R. L., Fisher, F. M., Wang, Y., Turner, F. T., & Jund, M. F. (1992). Methane emission from rice fields: The effect of floodwater management. *Global Biogeochemical Cycles, 6*, 249–262.

Sass, R. L., Fisher, F. M., & Huang, Y. (2000). A process-based model for methane emission from irrigated rice fields: Experimental basis and assumptions. *Nutrient Cycling in Agroecosystems, 58*, 249–258.

Sass, R. L., Andrews, J. A., Ding, A., & Fisher, F. M. J. (2002). Spatial and temporal variability in methane emissions from rice paddies: Implications for assessing regional methane budgets. *Nutrient Cycling in Agroecosystems, 64*, 3–7.

Satoo, T. (1983). Temperate broad-leaved evergreen forests of Japan. In J. D. Ovington (Ed.), *Temperate broadleaved evergreen forests* (pp. 169–189). Amsterdam: Elsevier.

Sauerbeck, D. (1993). CO_2-emissions from agriculture: Sources and mitigation potentials. *Water, Air, and Soil Pollution, 70*, 381–388.

Sauerbeck, D., & Gonzales, M. A. (1977). Field decomposition of carbon-14-labelled plant residues in various soils of the Federal Republic of Germany and Costa Rica. In *Soil organic matter studies, Proceedings of the FAO/IAEA Symposium* (Vol. I, pp. 159–170). Vienna: IAEA.

Sauerbeck, D. R. (2001). CO_2 emissions and C sequestration by agriculture-perspectives and limitations. *Nutrient Cycling in Agroecosystems, 60*, 253–266.

Saviozzi, A., Levi-Minzi, R., Cardelli, R., & Riffaldi, R. (2001). A comparison of soil quality in adjacent cultivated, forest and native grassland soils. *Plant and Soil, 233*, 251–259.

Scanlon, D., & Moore, T. (2000). Carbon dioxide production from peatland soil profiles: The influence of temperature, oxic/anoxic conditions and substrate. *Soil Science, 165*, 153–160.

Schachtschabel, P. (1961). Bestimmung des fixierten Ammoniums im Boden. *Zeitschrift für Pflanzenernährung, Düngung und Bodenkunde, 93*, 125–136.

Schadt, C. W., Martin, A. P., Lipson, D. A., & Schmidt, S. K. (2003). Seasonal dynamics of previously unknown fungal lineages in tundra soils. *Science, 301*, 1359–1361.

Schäfer, M. (1991). Fauna of the European temperate deciduous forest. In E. Röhrig, & B. Ulrich (Eds.), *Ecosystems of the world, VII. Temperate deciduous forests* (pp. 503–525). Amsterdam: Elsevier.

Schafer, W. M., & Nielsen, G. A. (1979). Soil development and plant succession on 1- to 50-year old strip mine spoils in southeastern Montana. In M. K. Wali (Ed.), *Ecology and coal resource development* (pp. 541–649). New York: Pergamon.

Schäffer, A. (1993). Pesticide effects on enzyme activities in the soil ecosystem. In J. M. Bollag, & G. Stotzky (Eds.), *Soil biochemistry* (pp. 273–340). New York: Marcel Dekker.

Scharpenseel, H. W., Schomaker, M., & Ayoub, A. (1990). *Soils on a warmer earth*. Amsterdam: Elsevier.

Scharpenseel, H. W., Neue, H. U., & Singer, A. S. (1992). Biotransformations in different climatic belts; source-sink relationships. In J. Kubat (Ed.), *Humus, its structure and role in agriculture and environment*. Amsterdam: Elsevier Science.

Scharpenseel, H.W., Mtimet, A., & Freytag, J. (2000). Soil inorganic carbon and global change. In R. Lal, J. M. Kimble, H. Eswaran, & B. A. Stewart (Eds.), *Global climate change and pedogenic carbonates* (pp. 27–42). Boca Raton, FL: Lewis.

Scheffer, F., & Meyer, B. (1970). Zielsetzungen und Analyse eines Tiefumbruch-Versuches auf einer grundwasserfernen Löß-Parabraunerde in Süd-iedersachsen. *Göttinger Bodenkundliche Berichte, 16*, 1–169.

Scheffer, F., & Schachtschabel, P. (1984). *Lehrbuch der Bodenkunde*. Stuttgart, Germany: Ferdinand Enke Verlag.

Scherer, H. W. (1993). Dynamics and availability of the non-exchangeable NH_4-N- a review. *European Journal of Agronomy, 2*, 149–160.

Scherer, H. W., & Mengel, K. (1979). Der Gehalt an fixiertem Ammoniumstickstoff auf einigen repräsentativen hessischen Standorten. *Landwirtschaftliche Forschung, 32*, 416–424.

Schimel, D. S. (1995). Terrestrial ecosystems and the carbon cycle. *Global Change Biology, 1*, 77–91.

Schimel, D. S., & Porter, C. S. (1995). Process modelling and spatial extrapolation. In P. A. Matson, & R. C. Harriss (Eds.), *Biogenic trace gases: Measuring emissions from soil and water* (pp. 358–383). Oxford: Blackwell.

Schimel, D. S., Enting, I. G., Heimann, T. M. L., Wigley, T., Raynaud, D., Alves, D., et al. (1995). CO_2 and the carbon cycle. In J. T. Houghton, L. G. Meira Filho, J. Bruce, H. Lee, B. A. Callander, E. Haites, et al. (Eds.), *Climate change 1994* (pp. 35–71). Cambridge: Cambridge University Press.

Schjoerring, J. K., & Mattsson, M. (2001). Quantification of ammonia exchange between agricultural cropland and the atmosphere: Measurements over two complete growth cycles of oilseed, rape, wheat, barley and pea. *Plant and Soil, 228*, 105–115.

Schjoerring, J. K., Husted, S., & Mattsson, M. (1998). Physiological parameters controlling plant-atmosphere ammonia exchange. *Atmospheric Environment, 32*, 491–498.

Schjønning, P., Elmholt, S., Munkholm, L. J., & Debosz, K. (2002). Soil quality aspects of humid sandy loams as influenced by organic and conventional long-term management. *Agriculture, Ecosystems and Environment, 88*, 195–214.

Schlegel, H. J., &Havlin, J. L. (1997). Green fallow for the central Great Plains. *Agronomy Journal, 89*, 762–767.

Schlesinger, H. W. (1982). Carbon storage in the caliche of the arid world: A case study from Arizona. *Soil Science, 133*, 247–255.

Schlesinger, H. W. (1984). Soil organic matter: A source of atmospheric CO_2. In G. M. Woodwell (Ed.), *The role of terrestrial vegetation in the global carbon cycle* (pp. 111–127). New York: Wiley.

Schlesinger, H. W. (1997). *Biogeochemistry, an analysis of global change* (2nd ed.). San Diego, CA: Academic.

Schlesinger, W. H. (1982). Carbon storage in the calche of arid soils: A case study from Arizona. *Soil Science, 133*, 247–255.

Schlesinger, W. H. (1986). Changes in soil carbon storage and associated properties with disturbance and recovery. In J. R. Trabalka, & D. E. Reichle (Eds.), *The changing carbon cycle: A global analysis* (pp. 194–220). New York: Springer.

Schlesinger, W. H. (1999). Carbon sequestration in soils. *Science, 284*, 2095.

Schlichting, E., Blume, H. P., & Stahr, K. (1995). *Bodenkundliches Praktikum* (2nd ed.) (295 p.). Berlin: Blackwell.

Schmidt, M. W. I., Knicker, H., Hatcher, P. G., & Kögel-Knabner, I. (1997). Improvement of ^{13}C and ^{15}N CPMAS NMR spectra of bulk soils, particle size fractions and organic material by treatment with 10% hydrofluoric acid. *European Journal of Soil Science, 48*, 319–328.

Schnitzer, M. (1978). Humic substances: Chemistry and reactions. In M. Schnitzer, & S. U. Khan (Eds.), *Soil organic matter* (pp. 1–64). Amsterdam: Elsevier.

Schnitzer, M. (1986). Binding of humic substances by soil mineral colloids. In P. M. Huang, & M. Schnitzer (Eds.), *Interactions of soil minerals with natural organics and microbes* (pp. 77–101). Madison, WI: Soil Science Society of America.

Schnitzer, M. (2000). A lifetime perspective on the chemistry of soil organic matter. *Advances in Agronomy, 68*, 1–58.

Schnitzer, M., & Khan, S. U. (1972). *Humic substances in the environment*. New York: Marcel Dekker.

Scholten, T. (1990). Vergleich der Humusformen in zwei Buchenbeständen unter dem Einfluss hoher atmosphärischer Stoffeinträge. *Mitteilungen der Deutschen Bodenkundlichen Gesellschaft, 62*, 175–182.

Schulten, H. R., & Leinweber, P. (2000). New insights into organic-mineral particles: Composition, properties and models of molecular structure. *Biology and Fertility of Soils, 30*, 399–432.

Schulten, H. R., & Schnitzer, M. (1993). A state of the art structural concept for humic substances. *Naturwissenschaften, 80*, 29–30.

Schulten, H. R., & Schnitzer, M. (1997). Chemical model structures fo soil organic matter and soils. *Soil Science, 162*, 115–130.

Schulten, H. R., & Schnitzer, M. (1998). The chemistry of soil organic nitrogen: A review. *Biology and Fertility of Soils, 26*, 1–15.

Schulten, H. R., Plage, B., & Schnitzer, M. (1991). A chemical structure for humic substances. *Naturwissenschaften, 78*, 311–312.

Schulten, H. R., Sorge-Lewin, C., & Schnitzer, M. (1997). Structure of "unknown" soil nitrogen investigated by analytical pyrolysis. *Biology and Fertility of Soils, 24*, 249–254.

Schultz, J. (2000). *Handbuch der Ökozonen*. Stuttgart, Germany: Verlag Eugen Ulmer.

Schulze, E. D., Mooney, H. A., Sala, O. E., Jobbagy, E., Buchmann, N., Bauer, G., et al. (1996). Rooting depth, water availability, and vegetation cover along an aridity gradient in Patagonia. *Oecologia, 87*, 72–79.

Schütz, H., Seiler, W., & Renneneberg, W. (1990). Soil and land use related sources and sinks of methane (CH_4) in the context of global methane budget. In A. F. Bouwman (Ed.), *Soils and the greenhouse effect* (pp. 269–289). Chichester/Sussex, England: Wiley.

Schütz, M., Holzapfel-Pschorn, A., Conrad, R., Rennenberg, H., & Seiler, W. (1989). A three year continuous record on the influence of daytime, season and fertilizer treatment on methane emission rates from an Italian rice paddy field. *Journal of Geophysical Research, 94*, 16405–16416.

Schütz, M., Seiler, W., & Conrad, R. (1990). Influence of soil temperature on methane emission from rice paddy field. *Biogeochemistry, 11*, 77–95.

Scott, D. A. (Ed.). (1993). *A directory of wetlands in Oceania*. Slimbridge, UK: IWRB and Kuala Lumpur, Malaysia: AWB.

Sedjo, R. A. (1999). *Potential for carbon forest plantations in marginal timber forests: The case of Patagonia, Argentina* (Discussion Paper 99-27). Washington, DC: Resources for the Future.

Sedjo, R. A. (2001). *Forest carbon sequestration: Some issues for forest investments* (RFF Discussion Paper, 01-34). Washington, DC: Resources for the Future.

Sedjo, R. A., & Solomon, A. (1989, July). Greenhouse warming: Abatement and adaptation. In P. Crosson, J. Darmstadter, W. Easterlingand, & N. Rosenberg (Eds.), *RFF Proceedings* (pp. 110–119).

Seifert, J. (1980). The effect of temperature on nitrification intensity in soil. *Folia Microbiology, 25*, 144.

Seifritz, W. (1990). CO_2 disposal by means of silicates. *Nature, 345*, 486.

Seiler, W., & Conrad, R. (1987). Contribution of tropical ecosystems to the global budgets of trace gases especially CH_4, H_2, CO and N_2O. In R. E. Dickinson (Ed.), *Geophysiology of Amazonia vegetation and climate interactions* (pp. 133–160). New York: Wiley.

Seitzinger, S. P., & Kroeze, C. (1998). Global distribution in nitrous oxide production and N inputs in freshwater and coastal marine ecosystems. *Global Biogeochemical Cycles, 12*, 93–113.

Senwo, Z. N., & Tabatabai, M. A. (1998). Amino acid composition of soil organic matter. *Biology and Fertility of Soils, 26*, 235–242.

Serraj, R., Sinclair, T. R., & Purcell, L. C. (1999). Symbiotic N_2 fixation response to drought. *Journal of Experimental Botany, 50*, 143–155.

Sexstone, A. J., Revsbech, N. P., Parkin, T. B., & Tiedje, J. M. (1985). Direct measurement of oxygen profiles and denitrification rates in soil aggregates. *Soil Science Society of America Journal, 49*, 645–651.

Sextone, A. J., & Mains, C. N. (1990). Production of methane and ethylene in organic horizons of spruce forest soils. *Biology and Fertility of Soils, 22*, 135–139.

Seyfried, M. S., & Rao, P. S. C. (1988). Kinetics of nitrogen mineralization in Costa Rican soils: Model evaluation and pretreatment effects. *Plant and Soil, 106*, 159–169.

Shaffer, M.J., Rojas, K., deCoursey, D. G., & Hebson, C. S. (1992). *NTRM, a soil-crop simulation model for nitrogen, tillage, and crop-residue management* (USDA-ARS, Conservation Research Report 34-1). Springfield, VA: National Technical Information Service.

Shao, K. S., & Li, Z. (1997). Effect of rice cultivars and fertilizer management on methane emissin in a rice paddy in Beijing. *Nutrient Cycling in Agroecosystems, 49*, 139–146.

Sharpley, A. N., & Smith, S. J. (1995). Nitrogen and phosphorus in soils receiving manure. *Soil Science, 159*, 253–258.

Shaxson, T. F. (1998). Concepts and indicators for assessment of sustainable land use. *Advances in Geological Ecology, 31*, 11–19.

Shen Shan-Min, Brookes, P. C., & Jenkinson, D. S. (1987). Soil respiration and the measurement of microbial biomass C by the fumigation technique in fresh and air-dried soil. *Soil Biology and Biochemistry, 19*, 153–158.

Shen, S. M., Pruden, G., & Jenkinson, D. S. (1984). Mineralization and immobilization of nitrogen in fumigated soils and the measurement of microbial biomass nitrogen. *Soil Biology and Biochemistry, 16*, 437–444.

Shepherd, M. A. (1999). The effectiveness of cover crops during eight years of a U.K. sandland rotation. *Soil Use and Management, 15*, 41–48.

Shevchenko, S. M., & Bailey, G. W. (1996). Life after death: Lignin-humic relationships reexamined. *Critical Reviews in Environmental Science and Technology, 26*, 95–153.

Shibahara, F., & Inubishi, K. (1995). Measurement of microbial biomass C and N in paddy soils by the fumigation-extraction method. *Soil Science and Plant Nutrition, 41*, 681–689.

Shibahara, F., & Inubishi, K. (1997). Effect of organic matter application on microbial biomass and available nutrients in various types of paddy soils. *Soil Science and Plant Nutrition, 43*, 191–203.

Shields, J. A., & Paul, E. A. (1973). Decomposition of ^{14}C-labelled plant material under field conditions. *Canadian Journal of Soil Science, 53*, 297–306.

Siebert, S. (2005). *Global-scale modeling of nitrogen balances at the soil surface* (Frankfurt Hydrological Paper 2, p. 27). Frankfurt, Germany: University of Frankfurt/Main. Available online http:/www.geo.uni-frankfurt.de/ipg/ag/dl/publikationen

Siemens, J., & Kaupenjohann, M. (2002). Contribution of dissolved organic nitrogen to N leaching from four German agricultural soils. *Journal of Plant Nutrition and Soil Science, 165*, 675–681.

Sigren, L. K., Lewis, S. T., Fisher, F. M., & Sass, R. L. (1997). Effects of field drainage on soil parameters related to methane production and emission from rice paddies. *Global Biogeochemical Cycles, 11*, 151–162.

Silva, R. G., Cameron, K. C., Di, H. J., & Hendry, T. (1999). A lysimeter study of the impact of cow urine, dairy shed effluent and nitrogen fertilizer on drainage water quality. *Australian Journal of Soil Research, 37*, 357–369.

Silver, W. L., Ostertag, R., & Lugo, A. E. (2000). The potential for carbon sequestration through reforestation of abandoned tropical agricultural and pasture lands. *Restoration Ecology, 8*, 394–407.

Šimek, M., & Cooper, J. E. (2002). The influence of soil pH on denitrification: Progress towards the understanding of this interaction over the last 50 years. *European Journal of Soil Science, 53*, 345–354.

Simpson, A. J., Salloum, M. J., Kingery, W. L., & Hatcher, P. G. (2002). Improvements in the two-dimensional nuclear magnetic resonance spectroscopy of humic substances. *Journal of Environmental Quality, 31*, 388–392.

Sims, P. L., & Coupland (1979). Natural temperate grasslands: Producers. In R. T. Coupland (Ed.), *Grassland ecosystems of the world: Analysis of grasslands and their uses* (pp. 49–72). Cambridge: Cambridge University Press.

Šimunek, J., & Suarez, D. L. (1993). Modelling of carbon dioxide transport and production in soil 1. Model development. *Water Resources Research, 29*, 487–497.

Sinsabaugh, R. L., Antibus, R. K., Linkins, A. E., McClaugherty, C. A., Rayburn, L., Repert, D., et al. (1993). Wood decomposition: Nitrogen and phosphorus dynamics in relation to extracellular enzyme activity. *Ecology, 74*, 1586–1593.

Sinsabaugh, R. L., Moorhead, D. L., & Likins, A. E. (1994). The enzymatic basis of plant litter decomposition: Emergence of an ecological process. *Applied Soil Ecology, 1*, 97–111.

Six, J., Feller, C., Denef, K., Ogle, S. M., de Moraes Sa, J. C., & Albrecht, A. (2001). *Soil organic matter, biota and aggregation in temperate and tropical soils – Effects of no-tillage*. Available online http://www.edpsciences.org/10.1051/agro:2002043

Six, J., Conant, R. T., Paul, E. A., & Paustian, K. (2002). Stabilization mechanisms of soil organic matter: Implications for C-saturation of soils. *Plant and Soil, 241*, 155–176.

Skiba, U., & Ball, B. (2002). The effect of soil texture and soil drainage on emissions of nitric oxide and nitrous oxide. *Soil Use and Management, 18*, 56–60.

Skiba, U., Smith, K. A., & Fowler, D. (1993). Nitrification and denitrification as sources of nitric oxide and nitrous oxide in a sandy loam soil. *Soil Biology and Biochemistry, 25*, 1527–1536.

Skiba, U., Fowler, D., & Smith, K. A. (1994). Emission of NO and N_2O from soils. In J. van Ham, L. J. M. H. Janssen, & R. J. Swart (Eds.), *Non-CO_2 greenhouse gases* (pp. 153–158). Dordrecht, The Netherlands: Kluwer.

Skiba, U., Fowler, D., & Smith, K. A. (1997). Nitric oxide emissions from agricultural soils in temperate and tropical climates: Sources, controls and mitigation options. *Nutrient Cycling in Agroecosystems, 48*, 139–153.

Skjemstad, J. O., Le Feuvre, R. P., & Prebble, R. E. (1990). Turnover of soil organic matter under pasture as determined by ^{13}C natural abundance. *Australian Journal of Soil Research, 28*, 267–276.

Smil, V. (1997, July). Global population and nitrogen cycle. *Scientific American*, 76–81.

Smil, V. (1999). Nitrogen in crop production: An account of global flows. *Global Biogeochemical Cycles, 13*, 647–662.

Smil, V. (2001). *Enriching the Earth: Fritz Haber, Carl Bosch, and the transformation of world food* (339 pp.). Cambridge, MA: MIT Press.

Smith, C. J., DeLaune, R. D., & Patrick, W. H. (1983). Carbon dioxide emission and carbon accumulation in coastal wetlands. *Coastal Shelf Sciences, 17*, 21–29.

Smith, J., Smith, P., Wattenbach, M., Gottschalk, P., Romanenkov, V. A., Shevtsova, L. K., et al. (2007). Projected changes in the organic carbon stocks of cropland mineral soils of European Russia and Ukraine, 1990–2070. *Global Change Biology, 13*, 342–356.

Smith, J. U., Bradbury, N. J., & Addiscott, T. M. (1995). SUNDIAL: Simulation of nitrogen dynamics in arable land. A user friendly PC-based version of the Rothamsted nitrogen turn over model. *Agronomy Journal, 88*, 38–43.

Smith, J. U., Smith, P., Monaghan, R., & MacDonald, A. J. (2002). When is a measured soil organic matter fraction equivalent to a model pool? *European Journal of Soil Science, 53*, 405–416.

Smith, J. U., Smith, P., Wattenbach, M., et al. (2005). Projected changes in mineral soil carbon of European croplands and grasslands, 1990–2080. *Global Change Biology, 11*, 2141–2152.

Smith, K. A. (1990). Anaerobic zones and denitrification in soil: Modelling and measurement. In N. P. Revsbech, & J. Sorensen (Eds.), *Denitrification in soil and sediment* (pp. 229–244). New York: Plenum.

Smith, K. A., Dobbie, K. E., Ball, B., Bakken, L. R., Sitaula, B. K., Hansen, S., et al. (2000a). Oxidation of atmospheric methane in Northern European soils, comparison with other ecosystems, and uncertainties in the global terrestrial sink. *Global Change Biology, 6*, 791–803.

Smith, O. L. (1979). An analytical model of the decomposition of soil organic matter. *Soil Biology and Biochemistry, 11*, 585–606.

Smith, P., Powlson, D. S., Glending, M. J., & Smith, J. (1997a). Potential for carbon sequestration in European soils: Preliminary estimates for five scenarios using results from long-term experiments. *Global Change Biology, 3*, 67–79.

Smith, P., Powlson, D. S., Smith, J. U., & Elliott, E. T. (Eds.). (1997b). Evaluation and comparison of soil organic matter models using datasets from seven long-term experiments. *Geoderma (Special Issue), 81*, 1–225.

Smith, P., Powlson, D. S., Smith, J. U., & Falloon, P. D. (1997c). SOMNET- a global network and database of soil organic matter models and long-term experimental datasets. *Globe, 38*, 4–5.

Smith, P., Smith, J. U., Powlson, D. S., McGill, W. B., et al. (1997d). A comparison of the performance of nine soil organic matter models using datasets from seven long-term experiments. In P. Smith, D. S. Powlson, J. U. Smith, & E. T. Elliott (Eds.), *Evaluation and comparison of soil organic matter models using datasets from seven long-term experiments, Geoderma* (Vol. 81, pp. 153–225). Amsterdam: Elsevier.

Smith, P., Powlson, D. S., Smith, J., Falloon, P., & Coleman, K. (2000b). Meeting Europe's climate change commitments: Quantitative estimates of the potential for carbon mitigation by agriculture. *Global Change Biology, 6*, 525–539.

Smith, P., Smith, J., Wattenbach, M., Meyer, J., Lindner, M., Zaehle, S., et al. (2006). Projected changes in mineral soil carbon of European forests, 1990–2100. *Canadian Journal of Soil Science, 86*, 159–169.

Smith, S. J., Power, J. F., & Kemper, W. D. (1994). Fixed ammonium and nitrogen availability indexes. *Soil Science, 158*, 132–140.

Smith, W. N., Desjardins, R. L., Grant, B., Li, C., Lemke, R., Rochette, P., et al. (2002). Testing the DNDC model using N_2O emissions at two experimental sites in Canada. *Canadian Journal of Soil Science, 82*, 365–374.

Sobecki, T. M., & Wilding, L. P. (1983). Formation of calcic and argillic horizons in selected soils of the Texas coast Prairie. *Soil Science Society of America Journal, 47*, 707–715.

Söderlund, R., & Svensson, B. H. (1976). The global nitrogen cycle. In B. H. Svensson, & R. Söderlund (Eds.), *Nitrogen, phosphorus and sulfur – global cycles, Ecological Bulletin No. 22* (pp. 23–73). Stockholm: SCOPE, Swedish Natural Science Research Council.

Sohi, S. P., Mahieu, N., Arah, J. R. M., Powlson, D. S., Madari, B., & Gaunt, J. L. (2001). A procedure for isolating soil organic matter fractions suitable for modeling. *Soil Science Society of America Journal, 65*, 1121–1128.

Soil Science Society of America. (1995, June). Statement on soil quality. *Agronomy News*.

Soil Survey Staff. (1995). *Keys to soil taxonomy* (305 pp.). Washington, DC: USDA Soil Conservation Service.

Solinger, S., Kalbitz, K., & Matzner, E. (2001). Controls on the dynamics of dissolved organic carbon and nitrogen in a Central European deciduous forest. *Biogeochemistry, 55*, 327–349.

Sombroek, W. G. (1990). Soils on a warmer earth: Tropical and subtropical regions. In H. W. Scharpenseel, M. Schomaker, & A. Ajoub (Eds.), *Soils on a warmer earth* (pp. 157–174). Amsterdam: Elsevier.

Sombroek, W. G., Nachtergaele, F. O., & Hebel, A. (1993). Amounts, dynamics and sequestering of carbon in tropical and subtropical soils. *Ambio, 22*, 417–426.

Sonder, K. (2004). Soil erosion in Andean cropping systems: The impact of rainfall erosivity. *Ph.D thesis*, University of Hohenheim.

Sorensen, L. H. (1972). Stabilization of newly formed amino acid metabolites in soil by clay minerals. *Soil Science, 114*, 5–11.

Sörensen, L. H. (1983). Size and persistence of microbial biomass formed during the humification of glucose, hemicellulose, and straw in soils containing different amounts of clay. *Plant and Soil, 75*, 121–130.

Southworth, J., Randolph, J. C., Habeck, M., Doering, O. C., Pfeiffer, R. A., Rao, D. G., et al. (2000). Consequences of future climate change and changing climate variability on maize yields in the midwestern United States. *Agriculture, Ecosystems and Environment, 82*, 139–158.

Sowden, F. J. (1959). Investigations on the amounts of hexosamines found in various soils and methods for their determination. *Soil Science, 88*, 138–143.

Sowden, F. J., Chen, Y., & Schnitzer, M. (1977). The nitrogen distribution in soils formed under widely differing climatic conditions. *Geochimica et Cosmochimica Acta, 41*, 1524–1526.

Sowden, F. J., Maclean, A. A., & Ross, G. J. (1978). Native clay fixed ammonium content and the fixation of added ammonium of some soils of Eastern Canada. *Canadian Journal of Soil Science, 58*, 27–38.

Spahni, R., Chappellaz, J., Stocker, T. F., Loulergue, L., Hausammann, G., Kawamura, K., et al. (2005). Atmospheric methane and nitrous oxide of the late Pleistocene from Antarctic ice cores. *Science, 310*, 1317–1321.

Sparks, D. L., Blevins, R. L., Bailey, H. H., & Barhisel, R. I. (1979). Relationship of ammonium nitrogen distribution to mineralogy in a Hapludalf soil. *Soil Science Society of America Journal, 43*, 786–789.

Sparling, G. P. (1997). Soil microbial biomass, activity, and nutrient cycling as indicators of soil health. In C. E. Pankhurst, B. M. Doube, & V. V. S. R. Gupta (Eds.), *Biological indicators of soil health* (pp. 97–119). Wallingford, UK: CABI.

Sparling, G. P., & West, A. W. (1989). Importance of soil water content when estimating soil microbial C, N and P by the fumigation-extraction methods. *Soil Biology and Biochemistry, 21*, 245–253.

Sparling, G. P., West, A. W., & Whale, K. N. (1985). Interference of plant roots in the estimation of soil microbial ATP, C, N and P. *Soil Biology and Biochemistry, 17*, 275–278.

Sparling, G. P., Speir, T. W., & Whale, K. N. (1986). Changes of microbial biomass C, ATP content, soil phospho-monoesterase and phospho-diesterase activity following air-drying of soils. *Soil Biology and Biochemistry, 18*, 363–370.

Sparling, G. P., Shepherd, T. G., & Kettles, H. A. (1992). Changes in soil organic C, microbial C and aggregate stability under continuous maize and cereal cropping, and after restoration to pasture in soils from the Manawatu region, New Zealand. *Soil and Tillage Research, 24*, 225–241.

Sperow, M., Eve, M., & Paustian, K. (2003). Potential soil C sequestration on US agricultural soils. *Climatic Change, 57*, 319–339.

Spycher, G., Sollins, P., & Rose, S. (1983). Carbon and nitrogen in the light fraction of a forest soil: Vertical distribution and seasonal fluctuations. *Soil Science, 135*, 79–87.

SSSA (Soil Science Society of America). (1997). *Glossary of soil science terms* (138 pp.). Madison, WI: American Society of Agronomy.

Stahl, P. D., & Parkin, T. B. (1996). Relationship of soil ergosterol concentration and fungal biomass. *Soil Biology and Biochemistry, 28*, 847–855.

Stallard, R. F. (1998). Terrestrial sedimentation and the carbon cycle: Coupling weathering and erosion to carbon burial. *Global Biogeochemical Cycles, 12*, 231.

Stanford, G., & Smith, S. J. (1972). Nitrogen mineralization potentials of soils. *Soil Science Society of America Proceedings, 36*, 465–472.

Stange, F., Butterbach-Bahl, K., Papen, H., Zechmeister-Boltenstern, S., Li, C., & Aber, J. (2000). A process-oriented model of N_2O and NO emissions from forest soils, 2. Sensitivity analysis and validation. *Journal of Geophysical Research, 105*, 4385–4398.

Starr, M., Westman, C. J., & Ala-Reini, J. (1996). The acid buffer capacity of some Finnish forest soils: Results of acid addition laboratory experiments. *Water, Air and Soil Pollution, 89*, 147–157.

Stavins, R. (1999). The costs of carbon sequestration: A revealed-preference approach. *American Economic Review, 89*, 994–1009.
Stavins, R. N., & Richards, K. (2005). *The cost of US forest-based carbon sequestration*. Arlington, VA: Pew Center on Global Climate Change.
Steelink, C. (1985). Implication of elemental characteristics of humic substances. In G. R. Aiken, D. M. McKnight, R. L. Wershaw, & P. MacCarthy (Eds.), *Humic substances in soil, sediment, and water* (pp. 457–476). New York: Wiley-Interscience.
Stefanic, G. (1994). Biological definition, quantifying method and agricultural interpretation of soil fertility. *Romanian Agriculture Research, 2*, 107–116.
Steffens, D., & Sparks, D. L. (1997). Kinetics of nonexchangeable ammonium release from soils. *Soil Science Society of America Journal, 61*, 455–462.
Stehfest, E. (2006). Modelling of global crop production and resulting N_2O emissions. *Ph.D. thesis* (150 pp.), University of Kassel.
Stehfest, E., & Bouwman, L. (2006). N_2O and NO emission from agricultural fields and soils under natural vegetation: Summarizing available measurement data and modeling of global annual emissions. *Nutrient Cycling in Agroecosystems, 74*, 207–228.
Steudler, P. A., Bowden, R. D., Melillo, J. M., & Aber, J. D. (1989). Influence of nitrogen fertilization on methane uptake in temperate forest soils. *Nature, 341*, 314–316.
Stevenson, F. J. (1982). *Nitrogen in agricultural soils*. Madison, WI: Soil Science Society of America.
Stevenson, F. J. (1986). *Cycles of soil carbon, nitrogen, phosphorus, sulfur, micronutrients* (380 pp.). New York: Wiley.
Stevenson, F. J. (1994). *Humus chemistry. genesis, composition, reactions* (2nd ed.). New York: Wiley.
Steward, J. W. B., & Cole, C. V. (1989). Influence of elemental interactions and pedogenic processes in organic matter dynamics. *Plant and Soil, 115*, 199–209.
Stotzky, G. (1997). Soil as an environment for microbial life. In J. D. van Elsas, J. T. Trevors, & E. M. H. Wellington (Eds.), *Modern soil microbiology* (pp. 1–372). New York: Marcel Dekker.
Strebel, O., Grimme, H., Renger, M., & Fleige, H. (1980). A field study with nitrogen-15 of soil and fertilizer nitrate uptake and of water withdrawal by spring wheat. *Soil Science, 130*, 205–210.
Suhaldoc, M., Schroll, R., Gattinger, A., Schloter, M., Munch, J. C., & Lestan, D. (2004). Effects of modified Pb-, Zn-, and Cd- availability on the microbial communities and on the degradation of isoproturon in a heavy metal contaminated soil. *Soil Biology and Biochemistry, 36*, 1943–1954.
Sulce, S., Palma-Lopez, D., Jaquin, F., Vong, P. C., & Guiraud, G. (1996). Study of immobilization and remobilization of nitrogen fertilizer in cultivated soils by hydrolytic fractionation. *European Journal of Soil Science, 47*, 249–255.
Sutton, M. A., Schjørring, J. K., & Wyers, G. P. (1995). Plant-atmosphere exchange of ammonia. *Philosophical Transactions of Royal Society of London A, 351*, 261–278.
Swaine, M. D. (1992). Characteristics of dry forest in West Africa and the influence of fire. *Journal of Vegetation Science, 3*, 365–374.
Swift, M. J., Heal. O. W., & Anderson, J. M. (1979). *Decomposition in terrestrial ecosystems*. Berkeley, CA: University of California Press.
Swift, R. S. (2001). Sequestration of carbon by soil. *Soil Science, 166*, 858–871.
Swinnen, J., van Veen, A., & Merckx, R. (1995). Carbon fluxes in the rhizosphere of winter wheat and spring barley with conventional vs. integrated farming. *Soil Biology and Biochemistry, 27*, 811–820.
Szabolcs, I. (1989). *Salt affected soils* (74 pp.). Boca Raton, FL: CRC.
Szabolcs, I. (1994). The concept of soil resilience. In D. J. Greenland, & I. Szabolcs (Eds.), *Soil resilience and sustainable land use* (pp. 33–39). Madison, WI: Soil Science Society of America.
Takahashi, T., Sutherland, S. C., Sweeny, C., Poisson, A., Metzl, N., Tilbrook, B., et al. (2002). Global sea-air CO_2 flux based on climatological surface ocean pCO_2, and seasonal biological and temperature effects. *Deep Sea Research, 49*, 1601–1623.
Takao, S. (1965). Organic acid production by basidiomycetes. I. Screening of acid-producing strains. *Applied Microbiology, 13*, 732–737.

Tan, K. H. (1986). Degradation of soil minerals by organic acids. In P. M. Huang, & M. Schnitzer (Eds.), *Interactions of soil minerals with natural organics and microbes*. Madison, WI: Soil Science Society of America.

Tan, K. H. (1994). *Environmental soil science*. New York: Marcel Dekker.

Tasman Institute (1994). *A framework for trading carbon credits from New Zealand's forests*. Report C6. Melbourne, Australia: Tasman Economic Research PTY.

Tate, K. R., Scott, N. A., Ross, D. J., Parshotam, A., & Claydon, J. J. (2000). Plant effects on soil carbon storage and turnover in a montane beech *(Nothofagus)* forest and adjacent tussock grassland in New Zealand. *Australian Journal of Soil Research, 38*, 685–698.

Taylor, R. D., Howard, G. W., & Begg, G. W. (1995). Developing wetland inventories in Southern Africa: A review. *Vegetatio, 118*, 57–79.

Templer, P. H., Groffman, P. M., Flecker, A. S., & Power, A. G. (2005). Land use change and soil nutrient transformations in the Los Haitises region of the Dominican Republic. *Soil Biology and Biochemistry, 37*, 215–225.

Tenny, F. G., & Waksman, S. A. (1929). Composition of natural organic materials and their decomposition in the soil: IV. The nature and rapidity of decomposition on the various organic complexes in different plant materials under aerobic conditions. *Soil Science, 28*, 55–84.

Terry, R. E. (1986). Nitrogen transformations in Histosols. In Y. Chen, & Y. Avnimelech (Eds.), *The role of organic matter in modern agriculture* (pp. 55–69). Martinus Nijhoff.

Theng, B. K. G., Tate, K. R., Sollins, P., Moris, N., Nadkarni, N., & Tate, R. L. (1989). Constituents of organic matter in temperate and tropical soils, In D. C Coleman, J. M. Oades, & G. Uehara (Eds.), *Dynamics of soil organic matter in tropical ecosystems* (pp. 5–32). Honolulu, HI: University of Hawaii Press.

Thiele-Bruhn, S., & Beck, I. C. (2005). Effects of sulphonamide and tetracycline antibiotics on soil microbial activity and microbial biomass. *Chemosphere, 59*, 457–465.

Thomas, G. W., & Phillips, R. E. (1979). Consequences of water movement in macropores. *Journal of Environmental Quality, 8*, 149–152.

Thuille, A., Buchmann, N., & Schulze, E. D. (2000). Carbon stocks and soil respiration rates during deforestation, grassland use and subsequent Norway spruce afforestation in the Southern Alps, Italy. *Tree Physiology, 20*, 849–857.

Tiedje, J. M. (1988). Ecology of denitrification and dissimilatory nitrate reduction to ammonium, In A. J. B. Zehnder (Ed.), *Biology of anaerobic microorganisms* (pp. 179–245). New York: Wiley.

Tiessen, H., Steward, J. B. W., & Moir, J. O. (1983). Changes in organic and inorganic phosphorus composition of two grassland soils and their particle size fraction during 60–70 years of cultivation. *Journal of Soil Science, 34*, 815–823.

Tiessen, H., Steward, J. W. B., & Anderson, D. W. (1994). Determinants of resilience in soil nutrient dynamics. In D. J. Greenland, & I. Szabolcs (Eds.), *Soil resilience and sustainable land use* (pp. 157–170). Wallingford, UK: CABI.

Tiessen, H., Feller, C., Sampaio, E. V. S. B., & Garin, P. (1998). Carbon sequestration and turnover in semiarid savannas and dry forests. *Climatic Change, 40*, 105–117.

Tieszen, L. L., Reed, B. C., Bliss, N. B., Wylie, B. K., & Dejong, D. D. (1997). NDVI, C_3 and C_4 production, and distributions in Great Plains grassland land cover classes. *Ecological Applications, 7*, 59–78.

Tilman, D. (1996). Biodiversity: Population vs ecosystem stability. *Ecology, 77*, 97–107.

Tisdall, J. M., & Oades, J. M. (1982). Organic matter and water-stable aggregates in soils. *Journal of Soil Science, 33*, 141–163.

Tolonen, K. (1979). Peat as a renewable resource. Long-term accumulation rates in North European mires. In *Proceedings of the International Symposium on Classification of Peat and Peatland*. Helsinki: International Peat Society.

Tolonen, K., & Turunen, J. (1996). Accumulation rates of carbon in mires in Finland and implications for climate change. *Holocene, 6*, 171–178.

Torbert, H. A., Prior, S. A., & van de Rogers, H. H. (1995). Elevated atmospheric carbon dioxide effects on cotton plant residue decomposition. *Soil Science Society of America Journal, 59*, 1321–1328.

Townsend, A. R., Braswell, B. H., Holland, E. A., & Penner, J. E. (1996). Spatial and temporal patterns in terrestrial carbon storage due to deposition of fossil fuel nitrogen. *Ecological Applications, 6*, 806–814.

Trasar-Cepeda, C., Leiros, M., Seoane, S., & Gil-Sotres, F. (2000). Limitations of soil enzymes as indicators of soil pollution. *Soil Biology and Biochemistry, 32*, 1867–1875.

Trebs, I., Lara, L., Zeri, L. L. M., Gatti, L. V., Artaxo, P., Dlugi, R., et al. (2006). Dry and wet deposition of inorganic nitrogen compounds to a tropical pasture site (Rondônia, Brazil). *Atmospheric Chemistry and Physics, 6*, 447–469.

Treter, U. (1993). Die borealen Waldländer, *Das Geographische Seminar* (210 pp.). Braunschweig, Germany: Westermann.

Trolldenier, G. (1995). Methanogenesis during rice grwoth as related to the water regime between crop seasons. *Biology and Fertility of Soils, 19*, 84–86.

Tsirkunov, V. V., Nikrangrov, A. M., Laznik, M. M., & Zhu, D. W. (1992). Analysis of long-term and seasonal river quality changes in Latvia. *Water Research, 26*, 1203–1216.

Tsuruta, H., Kanda, K., & Hirose, T. (1997). Nitrous-oxide emission from rice paddy field in Japan. *Nutrient Cycling in Agroecosystems, 49*, 51–58.

Turner, R. E., & Rabalais, N. N. (1991). Changes in Mississippi River water quality this century. *BioScience, 41*, 140–147.

Turunen, J., Pitkänen, A., Tahvanainen, T., & Tolonen, K. (2000, August 6–12). Carbon accumulation in West Siberian Peatlands. In *Sustaining our peatlands: Proceedings of the 11th International Peat Congress* (153–154). Québec City, Canada, Shippagan, Jyväskylä: Canadian Society of Peat and Peatlands/International Peat Society.

Ulrich, B. (1980). Die Wälder in Mitteleuropa:Messergebnisse ihrer Umweltbelastung, Theorie ihrer Gefährdung, Prognose ihrer Entwicklung. *Allgemeine Forstzeitschrift, 35*, 1198–1202.

Ulrich, B. (1987). Stability, elasticity, and resilience of terrestrial ecosystems with respect to matter balance. *Ecological Studies, 61*, 11–49.

Umweltbundesamt (1999). Annual Report 1998 (336 pp.), Berlin, Germany.

UNEP (1992). *World atlas of desertification*. London: Arnold.

Unger, P. W. (1991). Organic matter, nutrient, and pH distribution in no- and conventional-tillage semiarid soils. *Agronomy Journal, 83*, 186–189.

Uriyo, A. P., & Kessaba, A. (1975). Amounts and distribution of organic phosphorus in some soil profiles in Tanzania. *Geoderma, 13*, 201–210.

Urquiaga, S., Cruz, K. H. S., & Boddey R. M. (1992). Contribution of nitrogen fixation to sugar cane: Nitrogen-15 and nitrogen-balance estimates. *Soil Science Society of America Proceedings, 56*, 105–114.

Usback, S., & James, R. (1993). *A directory of important wetlands in Australia*. Canberra: Australian Nature Conservation Agency.

US EPA (2007, April). Inventory of US greenhouse gas emissions and sinks: 1990–2005, Washington, DC.

USGCRP (2006). Land use/land cover change. Available online http://www.usgcrp.gov/usgcrp/ProgramElements/land.htm

Vågen, T. G., Lal, R., & Singh, B. R. (2004). Soil carbon sequestration in sub-Saharan Africa: a review. *Land Degradation and Development, 16*, 53–71.

Vallini, G., Pera, A., Agnolucci, M., & Valdrighti, M. M. (1997). Humic acids stimulate growth and activity of in vitro tested axenic cultures of soil autotrophic nitrifying bacteria. *Biology and Fertility of Soils, 24*, 243–248.

Van Aardenne, J. A., Dentener, F. J., Olivier, J. G. J., Goldewijk, K., Klein, C. G. M., & Lelieveld, J. (2001). A 1°×1° resolution dataset of historical anthropogenic trace gas emissions for the period 1890–1990. *Global Biogeochemical Cycles, 15*, 909–928.

Van Breemen, N., & Feijtel, T. C. J. (1990). Soil processes and properties involved in the production of greenhouse gases with special reference to soil taxonomic systems. In A. F. Bouwman, (Ed.), *Soils and the greenhouse effect* (pp. 195–223). Chichester, UK: Wiley.

Van Breemen, N., Jenkins, A., Wright, R. F., Beerling, D. J., Arp, W. J., Berendse, F., et al. (1998). Impacts of elevated carbon dioxide and temperature on a boreal forest ecosystem (CLIMEX project). *Ecosystems, 1*, 345–351.

Vance, G. F., & David, M. B. (1989). Effect of acid treatment on the leachate chemistry of a New England spodosol: Importance of the B horizon on dissolved organic carbon retention. *Soil Science Society of America Journal, 53*, 1242–1247.

Van Cleemput, O., Uytterhaegen, L., & Baert, L.(1987). Chemo-denitrification of groundwater. In E. Welte, & I. Szabolcs (Eds.), *5th International CIEC Symposium:Protection of water quality from harmful emissions with special regard to nitrate and heavy metals* (pp. 301–311). Hungary: Balatonfüred. Van Cleve, K., Oliver, L., Schlentner, R., Viereck, L. A., & Dyrness, C. T. (1983). Productivity and nutrient cycling in taiga forest ecosystems. *Canadian Journal of Forest Research, 13*, 747–766.

Van Dam, D., & van Breemen, N. (1995). NICCE- a model for cycling of nitrogen and carbon isotopes in coniferous forest ecosystems. *Ecological Modelling, 79*, 255–275.

Van den Bos, R.M. (2000). *Human influence on carbon fluxes in coastal peatlands*. The Netherlands: Urije Univrsiteit Amsterdam.

Van den Bygaart, A. J., Yang, X. M., Kay, B. D., & Aspinall, D. (2002). Variability in carbon sequestration potential in no-till soil landscapes of southern Ontario. *Soil and Tillage Research, 65*, 231–241.

Van der Salm, C., van der Gon, H., Wieggers, R., Bleeker, A., & van der Toorn, A. (2006). The effect of afforestation on water recharge and nitrogen leaching in the Netherlands. *Forest Ecology and Management, 221*, 170–182.

Van der Weerden, T. J., Sherlock, R. R., Williams, P. H., & Cameron, K. C. (1999). Nitrous oxide emission and methane oxidation by soil following cultivation of two different leguminous pastures. *Biology and Fertility of Soils, 30*, 52–60.

Van der Werf, G. R., Randerson, J. T., Collatz, G. J., Giglio, L., Kasibhatla, S. P., Arellano, A. F. J., et al. (2004). Continental-scale partitioning of fire emissions during 1997 to 2001 El Nino/La Nina period. *Science, 303*, 73–76.

Van Dijk, H. (1966). *The use of isotopes in soil organic matter studies* (129 pp.). New York: Pergamon.

Van Dijk, H. (1971). Cation binding of humic acids. *Geoderma, 5*, 53.

Van Dijk, H. (1982). Survey of Dutch soil organic matter research with regard to humification and degradation rates in arable land. In D. Boels, D. B. Davies, & A. E. Johnston (Eds.), *Land use seminar on soil degradation, Wageningen, October 1980* (pp. 133–143). Rotterdam: Balkeme.

Van Drecht, G., Bouwman, A. F., Knoop, J. M., Beusen, A. H. W., & Meinardi, C. R. (2003). Global modeling of the fate of nitrogen from point and nonpoint sources in soils, groundwater, and surface water. *Global Biogeochemical Cycles, 17*(4), doi:10.1029/2003GB002060.

Van Drecht, G., Bouwman, A. F., Boyer, E. W., Green, P., & Siebert, S. (2005). A comparison of global spatial distributions of nitrogen inputs for nonpoint sources and effects on river nitrogen export. *Global Biogeochemical Cycles, 19*, doi:10.1029/2005GB002454.

Van Gestel, M., Mercks, R., & Vlassak, K. (1996). Distribution of 14C-labelled biomass and microbial products in microaggregates of a silty loam soil. *Soil Biology and Biochemistry, 28*, 1113–1115.

Van Ginkel, J. A., Gorrisen, A., & Van Veen, J. A. (1997). Carbon and nitrogen allocation in *Lolium perenne* in response to elevated atmospheric CO_2 with emphasis on soil carbon dynamics. *Plant and Soil, 188*, 299–308.

Van Keulen, H. (2001). (Tropical) soil organic matter modelling: Problems and prospects. *Nutrient Cycling in Agroecosystems, 61*, 33–39.

Van Kooten, G., Arthur, L., & Wilson, W. (1992). Potential to sequester carbon in Canadian forests: Some economic considerations. *Canadian Public Policy, 18*(2), 127–138.

Van Kooten, G., Stennes, B., Kremar-Nozic, E., & van Gorkom, R. (2000). Economics of afforestation for carbon sequestration in western Canada. *The Forestry Chronicle, 76*(1), 165–172.

Van Veen, J. A., & Frissel, M. J. (1981). Simulation model of the behavior of N in soil. In M. J. Frissel, & J. A. van Veen (Eds.), *Simulation of nitrogen behavior of soil plant systems* (pp. 126–144). Wageningen, The Netherlands: Pudoc.

Van Veen, J. A., & Kuikman, P. J. (1990). Soil structural aspects of decomposition of organic matter by microorganisms. *Biogeochemistry, 11*, 213–233.

Van Veen, J. A., & Paul, E. A. (1981). Organic carbon dynamics in grassland soils. I. Background information and computer simulation. *Canadian Journal of Soil Science, 61*, 185–201.

Van Veen, J. A., Ladd, J. N., & Frissel, M. J. (1984). Modelling C and N turnover through the microbial biomass in soil. *Plant and Soil, 76*, 257–274.

Varanka, M. W., Zablocki, Z. M., &. Hinsley, T. D. (1976). The effect of digested sludge on soil biological activity. *Journal of Water Pollution Control Federation, 48*, 1728–1740.

Vareschi, V. (1980). *Vegetationsökologie der Tropen*. Stuttgart, Germany: Ulmer.

Vaughan, D., & Malcolm, R. E. (1985). *Soil organic matter and biological activity*. Boston, MA: Junk Publisher.

Veldkamp, E., & Keller, M. (1997). Fertilizer induced nitric oxide emissions from agricultural soils. *Nutrient Cycling in Agroecosystems, 48*, 69–77.

Velthof, G. L., Brader, A. B., & Oenema, O. (1996). Seasonal variations in nitrous oxide losses from managed grasslands in the Netherlands. *Plant and Soil, 181*, 263–274.

Verberne, E. L. J., Hassink, J., de Willigen, P., Groot, J. J. R., & van Veen, J. A. (1990). Modelling organic matter dynamics in different soils. *Netherlands Journal of Agricultural Science, 38*, 221–238.

Vereecken, H., Maes, J., Feyen, J., & Darus, P. (1989). Estimating the soil moisture retention characteristic from texture, bulk density, and carbon content. *Soil Science, 148*, 389–403.

Vereecken, H., Vanclooster, M., Swerts, M., & Diels, J. (1991). Simulating water and nitrogen behavior in soils cropped with winter wheat. *Fertilizer Research, 27*, 233–243.

Vigerust, E. (1983). Physical properties in sewage sludge and sludge treated soils, In G. Catroux, P. L'Hermite, & E. Suess (Eds.), *The influence of sewage sludge application on physical and biological properties of soils* (pp. 107–119). Dordrecht, The Netherlands/Holland: D. Reidel Publication Company.

Vigil, M. F., & Nielsen, D. C. (1998). Winter wheat yield depression from legume green fallow. *Agronomy Journal, 90*, 727–734.

Visser, S. A., & Parkinson, D. (1992). Soil biological criteria as indicators of soil quality: Soil microorganisms. *American Journal of Alternative Agriculture, 7*, 33–37.

Visser, S. A. (1995). Physiological action of humic substances on microbial cells. *Soil Biology and Biochemistry, 17*, 457–462.

Vleeshouwers, L. M., & Verhagen, A. (2002). Carbon emission and sequestration by agricultural land use:a model study for Europe. *Global Change Biology, 8*, 519–530.

Vlek, P. L. G., & Craswell, C. T. (1979). Effect of nitrogen source and management on ammonia volatilization losses from flooded rice soil systems. *Soil Science Society of America Journal, 43*, 352–358.

Vlek, P. L. G., & Stumpe, J. W. (1978). Effects of solution chemistry and environmental conditions on ammonia volatilization losses from aqueous systems. *Soil Science Society of America Journal, 42*, 416–421.

Vogt, K. A., Vogt, D. J., Palmiotto, P. A., Boon, P., O'Hara, J., & Asbjornsen, H. (1996). Review of root dynamics in forest ecosystems grouped by climate, climatic forest type and species. *Plant and Soil, 187*, 159–219.

Volkhoff, B., & Cerri, C. C. (1988). L'humus des sols du Brésil-Nature et relations avec l'environnent. *Cahier ORSTOM, Série Pédologique, 24*, 83–95.

von Luetzow, M. (1993). Jahreszeitliche Fluktuation der mikrobiellen Biomasse und ihres Stickstoffgehaltes in konventionell und biologisch-dynamisch bewirtschafteten Parabraunerden der Friedberger Wetterau, *Ph.D. thesis*, University of Giessen, Germany.

von Lützow, M., & Ottow, J. C. G. (1994). Einfluss von konventioneller und biologisch-dynamischer auf die mikrobielle Biomasse und deren Stickstoff-Dynamik in Parabraunerden der Friedberger Wetterau. *Zeitschrift für Pflanzenernährung und Bodenkunde, 157*, 359–367.

von Lützow, M., Kögel-Knabner, I., Ekschmitt, K., Flessa, H., Guggenberger, G., Matzner, E., et al. (2007). SOM fractionation methods: Relevance to functional pools and to stabilization mechanisms. *Soil Biology and Biochemistry, 39*, 2183–2207.

von Zezschwitz, E. (1976). Ansprachemerkmale der terrestrischen Waldhumusformen des norddeutschen Mittelgebirgsraumes. *Geologisches Jahrbuch*, F *3*, 53–105.

von Zezschwitz, E. (1980). Analytische Kennwerte typischer Humusformen westfälischer Bergwälder. *Zeitschrift für Pflanzenernährung und Bodenkunde, 143*, 692–700.

von Zezschwitz, E. (1985). Immissionsbedingte Änderungen analytischer Kennwerte nordwestdeutscher Mittelgebirgsböden. *Geologisches Jahrbuch*, F *20*, 3–41.

Voos, G., & Groffman, P. M. (1997). Relationships between microbial biomass and dissipation of 2,4-D and dicamba in soil. *Biology and Fertility of Soils, 24*, 106–110.

Wada, K., & Aomine, S. (1975). Soil development during the quaternary. *Soil Science, 116*, 170–177.

Waksman, S. A. (1936). *Humus*. Baltimore: Williams & Wilkins.

Wali, M. K. (1999). Ecological succession and the rehabilitation of disturbed terrestrial ecosystems. *Plant and Soil, 213*, 195–220

Wallander, H., Massicotte, H. B., & Nylund, J. E. (1997). Seasonal variation in protein, ergosterol and chitin in five morphotypes of Pinus sylvestris L. ectomycorrhizae in a mature Swedish forest. *Soil Biology and Biochemistry, 29*, 45–53.

Walter, B., & Heimann, M. (2000). A process-based, climate-sensitive model to derive methane emissions from natural wetlands: Application to five wetland soils, sensitivity to model parameters, and climate. *Global Biogeochemical Cycles, 14*, 745–765.

Walter, H. (1964). *Die Vegetation der Erde in öko-physiologischer Betrachtung*. Jena, Germany: Gustav Fischer.

Walter, H. (1979). *Vegetation und Klimazonen, Die Ökologische Gliederung der Geo-Biosphäre*. Stuttgart, Germany: Ulmer.

Walther, W. (1989). The nitrate leaching out of soils and their significance for groundwater, results of long-term tests. In J. A. Hansen, & K. Henriksen (Eds.), *Nitrogen in organic wastes applied to soils* (pp. 346–356). London: Academic.

Wander, M. (2004). Soil organic matter fractions and their relevance to soil function. In F. Magdoff, & R. Weil (Eds.), *Soil organic matter in sustainable agriculture* (pp. 67–102). Boca Raton, FL: CRC.

Wander, M. M., & Traina, S. J. (1996). Organic matter fractions from organically and conventionally managed soils. I. Carbon and nitrogen distribution. *Soil Science Society of America Journal, 60*, 1081–1087.

Wander, M. M., & Yang, X. (2000). Influence of tillage on the dynamics of loose- and occluded-particulate and humified organic matter fractions. *Soil Biology and Biochemistry, 32*, 1151–1160.

Wang, J. S., Logan, J. A., McElroy, M. B., Duncan, B. N., Megretskaia, I. A., & Yantosaca, R. (2004). A 3-D model analysis of the slowdown and interannual variability in the methane growth rate from 1988 to 1997. *Global Biogeochemical Cycles, 18*, GB3011, doi:10.1029/3003GB002180.

Wang, P. G., Ji, G. L., & Yu, T. R. (1987). Electrostatic adsorption of anions. In T. R. Yu (Ed.), *Chemistry of variable charge soils* (pp. 113–139). New York: Oxford University Press.

Wang, S., Grant, R. F., Verseghy, D. L., & Black, T. A. (2002). Modeling carbon dynamics of boreal forest ecosystems using the Canadian land surface scheme. *Climatic Change, 55*, 451–477.

Wang, Z. P., Delaune, R. D., Masscheleyn, P. B., & Patric, W. H. J. (1993). Soil redox and pH effects on methane production in a flooded rice soil. *Soil Science Society of America Journal, 57*, 382–385.

Wangwacharakul, V., & Bowonwiwat, R. (1995). Economic evaluation of CO_2 response options in the forestry sector: The case Thailand. *Biomass Bioenergy, 8*, 293–308.

Wardle, D. A. (1992). A comparative assessment of factors which influence microbial biomass carbon and nitrogen levels in soil. *Biological Reviews, 67*, 321–358.

Wardle, D. A. (1998). Controls of temporal variability of the soil microbial biomass: A global-scale synthesis. *Soil Biology and Biochemistry, 30*, 1627–1637.

Wardle, D. A., & Nicholson, K. S. (1996). Synergistic effects of grassland plant species on soil microbial biomass and and activity: Implications for ecosystem-level effects of enriched plant diversity. *Functional Ecology, 10*, 410–416.

Warkentin, B. P. (1995). The changing concept of soil quality. *Journal of Soil and Water Conservation, 50*, 226–228.

Waschkies, C., & Huettl, R. F. (1999). Microbial degradation of geogenic organic C and N in mine spoils. *Plant and Soil, 213*, 221–230.

Wassmann, R., & Aulakh, M. S. (2000). The role of rice plants in regulating mechanisms of methane emissions. *Biology and Fertility of Soils, 31*, 20–29.

Wassmann, R., Neue, H. U., Lantin, R. S., Aduna, J. B., Alberto, M. C. R., Andales, M. J., et al. (1994). Temporal patterns of methane emissions from wetland ricefields treated by different modes of N-application. *Journal of Geophysical Research-Atmosphere, 99*, 16457–16462.

Wassmann, R., Neue, H. U., Lantin, R., Makarim, K., Chareonsilp, N., Buendia, L. V., et al. (2000). Characterization of methane emissions from rice fields in Asia. II. Differences among irrigated, rainfed, and deepwater rice. *Nutrient Cycling in Agroecosystems, 58*, 13–22.

Watanabe, I., Hashimoto, T., & Shimoyama, A. (1997). Mathane oxidizing activities and methanotrophic populations associated with wetland rice soils. *Biology and Fertility of Soils, 24*, 261–265.

Watson, C. A., Ross, J. M., Bagnaresi, U., Minotta, G. F., Roffi, F., Atkinson, D., et al. (2000). Environment-induced modifications to root longevity in Lolium perenne and Trifolium repens. *Annals of Botany, 85*, 397–401.

Watson, R. T., Zinyovera, M. C., Moss, R. H., & Dokken, D. J., (Eds). (1995). *Climate change 1995, impacts, adaptations and mitigation of climate change*. Intergovernmental Panel of Climate Change (IPCC). Cambridge: Cambridge University Press.

Watts, D. G., & Hanks, R. J. (1978). A soil-water-nitrogen model for irrigated corn on sandy soils. *Soil Science Society of America Journal, 42*, 492–499.

Weber, D. F., and Gainey, P. L. (1962). Relative sensitivity of nitrifying organisms to hydrogen ions in soils and in solutions. *Soil Science, 94*, 138–145.

Weier, K. L., Doran, J. W., Power, J. F., & Walters, D. T. (1993). Denitrification and the dinitrogen/nitrous oxide ratio as affected by soil water, available carbon, and nitrate. *Soil Science Society of America Journal, 57*, 66–72.

Weigand, W., Auerswald, K., & Beck, T. (1995). Microbial biomass in agricultural topsoils after 6 years of bare fallow. *Biology and Fertility of Soils, 19*, 129–134.

Welte, E., & Timmermann, F. (1976). Fertilité du sol e bilan de l'azote dans l'essai permanent de fumure "Ewiger Roggenbau"(culture continue de seigle) á Halle/Saale. *Annales Agronomiques, 27*, 721–742.

Wershaw, R. L. (1989). Size and shape of humic substances by scattering techniques. In M. H. B. Hayes, P. McCarthy, R. L. Malcolm, & R. S. Swift (Eds.), *Humic substances II. In search of structure* (pp. 545–559). New York: Wiley.

West, N. E., Stark, J. M., Johnson, D. W., Abrams, M. M., Wright, J. R., Heggem, D., et al. (1994). Effects of climatic change on the edaphic features in arid and semiarid lands of western North America. *Arid Soil Research and Rehabilitation, 8*, 307–351.

West, T. O., & McBride, A. C. (2005). The contribution of agricultural lime to carbon dioxide emissions in the United States:dissolution, transport, and net emissions. *Agriculture, Ecosystems and Environment, 108*, 145–154.

West, T. O., & Post, W. M. (2002). Soil organic carbon sequestration rates by tillage and crop rotation: A global data analysis. *Soil Science Society of America Journal, 66*, 1930–1946.

Westermann, P. (1993). Wetland and swamp microbiology. In T. E. Ford (Ed.), *Aquatic microbio logy* (pp. 215–238). Oxford: Blackwell.

Wetzel, G. (1984). Detrital dissolved and particulate organic carbon functions in aquatic ecosystems. *Bulletin of Marine Sciences, 35*, 503–509.

Whalen, J. K., Bottomley, P. J., & Myrold, D. D. (2000). Carbon and nitrogen mineralization from light- and heavy-fraction additions to soil. *Soil Biology and Biochemistry, 32*, 1345–1352.

Wheeler, T. R., Craufurd, P. Q., Ellis, R. H., Porter, J. R., & Vara Prasad, P. V. (2000). Temperature variability and the yield of annual crops. *Agriculture, Ecosystems and Environment, 82*, 159–167.

Whipps, J. M. (1990). Carbon economy. In J. M. Lynch (Ed.), *The Rhizosphere* (pp. 59–97). Chichester, UK: Wiley.

White, D. S., & Howes, B. L. (1994). Translocation, remineralization and turnover of nitrogen in the roots and rhizomes of Spartinia alterniflora. *American Journal of Botany, 81*, 1225–1234.

White, J. R., & Reddy, K. R. (1999). Influence of nitrate and phosphorus loading on denitrifying enzyme activity in Everglades wetland soils. *Soil Science Society of America Journal, 63*, 1945–1954.

White, J. R., & Reddy, K. R. (2000). The effects of phosphorus loading on organic nitrogen mineralization of soils and detritus along a nutrient gradient in the northern Everglades, Florida. *Soil Science Society of America Journal, 64*, 1525–1534.

White, R., Murray S., & Rohweder, M. (2000). Pilot analysis of global ecosystems: Grassland ecosystems. Available online http://www.wri.org/biodiv/pubs_description.cfm?PubID = 3057

White, R. E. (1985). A model for nitrate leaching in undisturbed structured clay soil during unsteady flow. *Journal of Hydrology, 79*, 37–51.

White, R. E. (1997). *Principles and practice of soil science:The soil as a natural resource* (348 pp.). Oxford: Blackwell.

White, R. P., & Vanasselt, W. (2000). Grassland in pieces: Modification and conversion take a toll. Available online http://earthtrends.wri.org/features/view_feature.php?fid=8&theme=9

Whitehead, D. C. (1995). *Grassland nitrogen* (397 pp.). Guildford, UK: CABI/Biddles.

Whiting, G. J., & Chanton, J. P. (1993). Primary production control of methane emission from wetlands. *Nature, 364*, 794–795.

Whyte, R. O. (1976). *Land and land appraisal*. The Hague, The Netherlands: Junk Publishers.

Wickland, K., Striegel, R., Neff, J., & Sachs, T. (2006). Effects of permafrost melting on CO_2 and CH_4 exchange of a poorly drained black spruce lowland. *Journal of Geophysical Research, 111*, G02011, doi:10.1029/2025JG000099.

Widmer, P. (1993). Zeitliche und räumliche Variabilität der mikrobiellen Biomasse in niedersächsischen Böden. *Ph.D. thesis*, University of Hannover, Germany.

Wiechmann, H. (1978). *Stoffverlagerung in Podsolen*. Stuttgart, Germany: Ulmer.

Williams, E. J., Hutchinson, G. L., & Fehsenfeld, F. C. (1992). NO_x and N_2O emissions from soil. *Global Biogeochemical Cycles, 6*, 351–358.

Williams, J. R., & Gray, T. R. G. (1974). Decomposition of litter on the soil surface. In C. H. Dickinson, & G. J. Pugh (Eds.), *Biology of plant litter decomposition*. New York: Academic.

Williams, J.R., & Renard, K. G. (1985). Assessment of soil erosion and crop productivity with process models (EPIC). In R. F. Follett, & B. A. Stewart (Eds.), *Soil erosion and crop productivity* (pp. 67–103). Madison, WI: ASA, CSSA, SSSA.

Williams, J. R., Pulman, J. W., & Dyke, P. T. (1985). Assessing the effect of soil erosion on productivity with EPIC. In *Proceedings National Symposium on Erosion and Soil Productivity*. New Orleans, LA/St Joseph, MI: American Society of Agricultural Engineers.

Wilson, M. A. (1987). *NMR techniques and applications in geochemistry and soil chemistry* (340 pp.). Oxford: Pergamon.

Windt, A., & Wollenweber, D. (2005). Grunddüngung zu Zuckerrüben. *Zuckerrübe, 54*, 206–208.

Winter, K., & Beese, F. (1995). The spatial distribution of soil microbial biomass in a permanent row crop. *Biology and Fertility of Soils, 19*, 322–326.

Wlotzka, F. (1961). *Untersuchungen zur Geochemie des Stickstoffs. Ph.D thesis*, Germany: University of Göttingen.

Woledge, J., & Parsons, A. J. (1986). Temperate grasslands. In N. R. Baker, & S. P. Long (Eds.), *Photosynthesis in contrasting environments* (pp. 173–197). Amsterdam: Elsevier.

Wolters, V. (1988). Effects of *Mesenchytraeus glandulosus* (Oligochaeta, Enchytraeidae) on decomposition processes. *Pedobiologia, 32*, 387–398.

Wolters, V. (1991). Soil invertebrates – effects on nutrient turnover and soil structure– a review. *Zeitschrift für Pflanzenernährung und Bodenkunde, 154*, 389–402.

Wong-Chong, G. M., & Loehr, R. C. (1975). The kinetics of microbial nitrification. *Water Resources, 9*, 1099–1106.

Woodruff, C. M. (1949). Estimating the nitrogen delivery of soil from the soil organic matter determination as reflected by Sanbourn field. *Soil Science Society of America Proceedings, 14*, 208–212.

Woodwell, G. M. (1984). The carbon dioxide problem. In G. M. Woodwell (Ed.), *The role of terrestrial vegetation in the global carbon cycle: Measurement by remote sensing* (pp. 3–17). New York: Wiley.

WRI (World Resources Institute, Earth Trends). (2007). Net flux of carbon to the atmosphere from land-use change, 2000. Available online http://earthtrends.wri.org/maps_spatial/maps_detail_static.php?map_select=485&theme=3

Wu, J., & Brookes, P. C. (2005). The proportional mineralization of microbial biomass and organic matter caused by air-drying and rewetting of a grassland soil. *Soil Biology and Biochemistry, 37*, 507–515.

Wuebbles, D. J., & Hayhoe, K. (2002). Atmospheric methane and global change. *Earth Science Review, 57*, 177–210.

Xu, C., Shaffer, M. J., & Al-Kaisi, M. (1998). Simulating the impact of management practices on nitrous oxide emissions. *Soil Science Society of America Journal, 62*, 736–742.

Xu, D. (1995). The potential for reducing atmospheric carbon by large-scale afforestation in China and related cost/benefit analysis. *Biomass Bioenergy, 8*, 337–344.

Xu, J., & Juma, N. G. (1994). Relations of shoot C, root C and root length with root-released C of two barley cultivars and the decomposition of root-released C in soil. *Canadian Journal of Soil Science, 74*, 17–22.

Xu, Z., Zheng, X., Wang, Y., Han, S., Huang, Y., Zhu, J. et al. (2004). Effects of elevated CO_2 and N fertilization on CH_4 emission from paddy fields. *Global Biogeochemical Cycles, 18*, GB3009, doi:10.129/2004GB002233.

Yaalon, D. H., & Feigin, A. (1970). Non-exchangeable ammonium ions in some clays and shales of Israel. *Israelian Journal of Chemistry, 8*, 425–434.

Yagi, K. (2002). Methane emission in rice, mitigation options. In *Encyclopedia of soil science* (pp. 814–818). New York: Marcel Dekker.

Yagi, K., & Minami, K. (1990). Effects of organic matter application on methane emission from some Japanese paddy fields. *Soil Science and Plant Nutrition, 36*, 599–610.

Yagi, K., Kumagai, K., Tsuruta, H., & Minami, K. (1994). Emission, production and oxidation of methane in a Japanese rice paddy field. In R. Lal et al. (Eds.), *Soil management and greenhouse effect* (pp. 231–244). Boca Raton, FL: CRC/Lewis.

Yagi, K., Tsuruta, H., Kanda, K., & Minami, K. (1996). Effect of water management on methane emission from a Japanese paddy field: Automated methane monitoring. *Global Biogeochemical Cycles, 10*, 255–267.

Yakovchenko, V. I., Sikora, L. J., & Rauffmann, D. D. (1996). A biologically based indicator of soil quality. *Biology and Fertility of Soils, 21*, 245–251.

Yamane, I., & Sato, K. (1964). Decomposition of glucose and gas formation in flooded soil. *Soil Science and Plant Nutrition, 10*, 127–133.

Yan, X., Akimoto, H., & Ohara, T. (2003a). Estimation of nitrous oxide, nitric oxide and ammonia emissions from croplands in East, Southeast and South Asia. *Global Change Biology, 9*, 1080–1096.

Yan, X., Cai, Z., Ohara, T., & Akimoto, H. (2003b). Methane emission from rice fields in mainland China: Amount and seasonal and spatial distribution. *Journal of Geophysical Research, 108*(D16), 4505, doi:10.1029/2002JD003182.

Yan, X., Ohara, T., & Akimoto, H. (2003c). Development of region-specific emission factors and estimation of methane emission from rice fields in the East, Southeast and South Asian countries. *Global Change Biology, 9*, 237–254.

Yan, X., Yagi, K., Akiyama, H., & Akimoto, H. (2005). Statistical analysis of the major variables controlling methane emission from rice fields. *Global Change Biology, 11*, 1131–1141.

Yang, H. S., & Janssen, B. H. (2000). A mono-component model of carbon mineralization with a dynamic rate constant. *European Journal of Soil Science, 51*, 517–529.

Yano, Y., McDowell, W. H., & Kinner, N. E. (1998). Quantification of biodegradable dissolved organic carbon in soil solution with flow-through bioreactors. *Soil Science Society of America Journal, 62*, 1556–1564.

Yao, H., & Conrad, R. (2001). Thermodynamics of propionate degradation in anoxic paddy soil from different rice-growing regions. *Soil Biology and Biochemistry, 33*, 359–364.

Yienger, J. J., & Levy, H. H. (1995). Empirical model of the global soil-biogenic NOx emissions. *Journal of Geophysical Research, 100*, 11447–11464.

Youngberg, C. T., & Wollum, A. G. (1976). Nitrogen accretion in developing Ceanothus velutinus stands. *Soil Sience Society of America Journal, 40*, 109–112.

Yu, Z., Zhang, Q., Kraus, T. E. C., Dahlgren, R. A., Anastasio, C., & Zasoski, R. J. (2002). Contribution of amino compounds to dissolved organic nitrogen in forest soils. *Biogeochemistry, 61*, 173–198.

Zak, D. R., Pregitzer, K. S., Curtis, P. S., Teeri, J. A., Fogel, R., & Randlett, D. L. (1993). Elevated atmospheric CO_2 and feedback between carbon and nitrogen cycles. *Plant and Soil, 151*, 105–117.

Zamolodchikov, D., & Karelin, D. V. (2001). An empirical model of carbon fluxes in Russian tundra. *Global Change Biology, 7*, 147–161.

Zavarzin, G. A. (2002). Microbial geochemical calcium cycle. *Microbiology, 71*, 5–22.

Zavitovski, J., & Newton, M. (1968). Ecological importance of snowbrush Caenothus velutinus in the Oregon Cascades. *Ecology, 49*, 1134–1145.

Zbell, B., & Höhn, A. (1995). Einflüsse von Landnutzungsänderungen auf das CO_2-Minderungspotential von Agrarlandschaften des norddeutschen Tieflands, In W. Merbach, & R. Bork (Eds.), *Kohlenstoff und Stickstoffumsatz im System Pflanze-Boden* (pp.135–140). ZALF Bericht Nr. 23, Selbstverlag des Zentrums für Agrarlandschafts- und Landnutzungsforschung (ZALF) e.V. Müncheberg.

Zech, W., Senesi, N., Guggenberger, G., Kaiser, K., Lehmann, J., Miano, T. M., et al. (1997). Factors controlling humification and mineralization of soil organic matter in the tropics. *Geoderma, 79*, 117–161.

Zedler, J. B., & Kercher, S. (2005). Wetland resources: Status, trends, ecosystem services, and restorability. *Annual Review of Environment and Resources, 30*, 39–74.

Zeikus, J. G. (1981). Lignin metabolism and the carbon cycle. In M. Alexander (Ed.), *Advances in microbial ecology* (Vol. 5, pp. 211–243). New York: Plenum.

Zhang, W. L., Stuetzel, H., & Kolbe, H. (1998). Strategies to reduce ground water pollution from nitrogen fertilization in intensive cropping area of China. In N. El Bassam, R. K. Behl, & B. Prochnow (Eds.), *Sustainable agriculture for food, energy and industry: Strategies towards achievement* (Vol. 1, pp. 354–358). London: James & James (Science Publishers).

Zhang, Y., & Scherer, H. (2000). Mechanisms of fixation and release of ammonium in paddy soils after flooding. *Biology and Fertility of Soils, 31*, 517–521.

Zhang, Y., Li, C., Trettin, C. C., Li, H., & Sun, G. (2002). An integrated model of soil, hydrology, and vegetation for carbon dynamics in wetland ecosystems. *Global Biogeochemical Cycles, 16*, 1061, doi:10.1029/2001GB001838.

Zheng, X., Wang, M. X., Wang, Y., Shen, R., & Li, J. (2000). Mitigation options on greenhouse gas (CH_4, N_2O, NO) emissions from cropland. *Advances in Atmospheric Science, 17*, 83–92.

Zhu, Z. L., & Wen, Q. X. (1992). *Nitrogen in soils of China*, Nanjing, China (in Chinese): Jiangsu Science and Technology Publishing House.

Ziechmann, W., & Müller-Wegener, U. (1990). *Bodenchemie*. Mannheim, Germany: Wissenschaftsverlag.

Ziechmann, W. (1996). *Huminstoffe und ihre Wirkung*, spektrum, Heidelberg.

Ziegler, F. (1989). Changes of lipid content and lipid composition in forest humus layers derived from Norway spruce. *Soil Biology and Biochemistry, 21*, 237–243.

Ziegler, F., & Zech, W. (1989). Distribution pattern of total lipids and lipid fractions in forest humus. *Zeitschrift für Pflanzenernährung und Bodenkunde, 152*, 287–290.

Zielke, R. C., & Christensen, D. R. (1986). Organic carbon and nitrogen changes in soil under selected cropping systems. *Soil Science Society of America Journal, 50*, 363–367.

Zinke, P. J., Stangenberger, A. G., Post, W. M., Emmanuel, W. R., & Olson, J. S. (1984). *Worldwide organic soil carbon and nitrogen data*. ORNL/TM-8857. Oak Ridge, TN: Oak Ridge National Laboratory.

Zsolnay, A. (1996). Dissolved humus in soil waters. In A. Piccolo (Ed.), *Humic substances in terrestrial ecosystems* (pp. 171–223). Amsterdam: Elsevier.

Index

D

Q

R